CAMBRIDGE MONOGRAPHS
ON MATHEMATICAL PHYSICS

General Editors: W.H.McCrea, D.W.Sciama, J.C.Polkinghorne

AN INTRODUCTION TO REGGE THEORY AND HIGH ENERGY PHYSICS

This title, first published in 1977, has been reissued as an Open Access publication on Cambridge Core.

AN INTRODUCTION TO REGGE THEORY & HIGH ENERGY PHYSICS

P. D. B. COLLINS

Physics Department, University of Durham

CAMBRIDGE UNIVERSITY PRESS

CAMBRIDGE

LONDON · NEW YORK · MELBOURNE

CAMBRIDGE UNIVERSITY PRESS

CAMBRIDGE
UNIVERSITY PRESS

Shaftesbury Road, Cambridge CB2 8EA, United Kingdom

One Liberty Plaza, 20th Floor, New York, NY 10006, USA

477 Williamstown Road, Port Melbourne, VIC 3207, Australia

314–321, 3rd Floor, Plot 3, Splendor Forum, Jasola District Centre, New Delhi – 110025, India

103 Penang Road, #05–06/07, Visioncrest Commercial, Singapore 238467

Cambridge University Press is part of Cambridge University Press & Assessment, a department of the University of Cambridge.

We share the University's mission to contribute to society through the pursuit of education, learning and research at the highest international levels of excellence.

www.cambridge.org
Information on this title: www.cambridge.org/9781009403283

DOI: 10.1017/9781009403269

First published 1977
Reissued as OA 2023
First paperback edition 2024

A catalogue record for this publication is available from the British Library

ISBN 978-1-009-40329-0 Hardback
ISBN 978-1-009-40328-3 Paperback

Contents

Preface

In 1959 Regge showed that, when discussing solutions of the Schroedinger equation for non-relativistic potential scattering, it is useful to regard the angular momentum, l, as a complex variable. He proved that for a wide class of potentials the only singularities of the scattering amplitude in the complex l plane were poles, now called 'Regge poles'. If these poles occur for positive integer values of l they correspond to bound states or resonances, and they are also important for determining certain technical aspects of the dispersion properties of the amplitudes. But it soon became clear that his methods might also be applicable in high energy elementary particle physics, and it is in fact here that the theory of the complex angular momentum plane, usually called 'Regge theory' for short, is now most fruitfully employed.

Apart from the leptons (electron, muon and neutrinos) and the photon, all the very large number of elementary particles which have been found, baryons and mesons, enjoy the strong interaction (i.e. the nuclear force which *inter alia* binds nucleons into nuclei) as well as the less forceful electromagnetic, weak and gravitational interactions. Such particles are called 'hadrons', from the Greek ἁδρός meaning large. Some are stable, but most are highly unstable and decay rapidly into other hadrons and leptons. They can be classified according to their various quantum numbers such as baryon number, charge, strangeness etc., but for a given set of quantum numbers sequences of particles have been found which differ only in their spin. For example resonances similar to the rho-meson (which is an unstable particle and decays into pi-mesons, viz $\rho \to \pi\pi$) occur with spins $\sigma = \hbar, 2\hbar, 3\hbar, \ldots$, the mass increasing with the spin.

If one were to try and 'explain' such resonances as being like bound states produced by a potential $V(r)$ acting between the pions (fig. i (a)), the radial Schroedinger equation would contain an effective potential

$$V_{\text{eff}}(r) = V(r) + \frac{l(l+1)}{r^2},$$

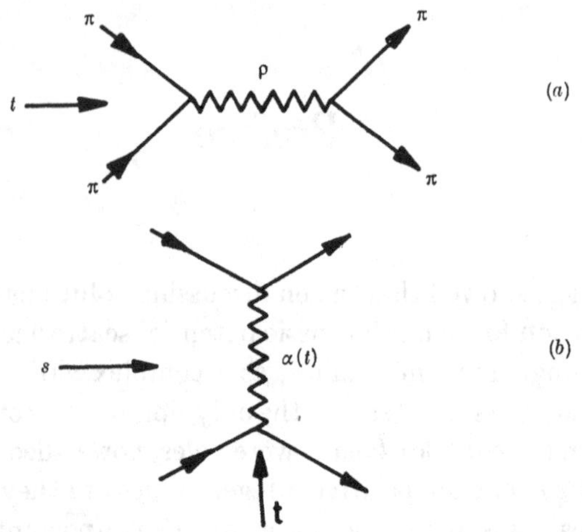

FIG. i (*a*) The binding of two pi-mesons to produce an unstable ρ resonance which subsequently decays into two pi-mesons again. (*b*) The exchange of a trajectory $\alpha(t)$ which gives the high energy behaviour of the scattering amplitude.

which provides less strong binding as the orbital angular momentum of the pions, l, is increased, because of the centrifugal barrier term, $l(l+1)r^{-2}$. So the potential is less effective for high l, which explains why high-spin resonances have higher masses. In fact one could solve the equation for arbitrary complex values of l, and the eigenvalues would vary continuously along a trajectory in the l plane connecting the various physical solutions which occur for $l = n\hbar$ (n integer). Of course such a non-relativistic model is quite hopeless for high energy physics, but the basic idea, that sequences of composite particles of mass m_i and spin σ_i ($i = 1, 2, 3, \ldots$) will lie on a given Regge trajectory $l = \alpha(t)$, where t is the square of the centre-of-mass energy, such that, for all i, $\alpha(m_i^2) = \sigma_i$, successfully inter-relates many sets of resonances. Indeed it is now widely believed that all the hadrons are composite particles lying on such trajectories, and are not really 'elementary' at all.

Also it is well established that the strong-interaction forces are due to the exchange of particles. This is a generalization of Yukawa's hypothesis that the long-range part of the inter-nucleon force is due to the exchange of pi-mesons. But rather than consider the exchange of individual particles it is more useful to consider the exchange of a complete trajectory of particles. Regge theory predicts that the high energy behaviour of a scattering amplitude $A(s,t)$ will be

$$A(s,t) \sim s^{\alpha(t)}$$

(where now s is the square of the centre-of-mass energy, and $-t$ is the square of the momentum transferred (fig. i(b)). This is found to hold in a great variety of processes.

So Regge theory is concerned with the particle spectrum, the forces between particles, and the high energy behaviour of scattering amplitudes; in fact with almost all aspects of strong interactions. Hence an understanding of Regge theory has become essential for those who wish to work on high energy physics, and the aim of this book is to provide an introduction to the subject.

In the first chapter we discuss the kinematics of scattering processes, introduce scattering amplitudes, and review their analytic structure as functions of the energy and momentum transfer. In chapter 2 we define partial-wave amplitudes for a given l, and show how and why it is useful to make an analytic continuation in l. We explain why Regge poles in l, which lie on Regge trajectories, correspond to particles. In chapter 3 we examine the occurrence of Regge poles in potential scattering, in field theories, and in other models of strong interactions. Then in chapter 4 we introduce the somewhat more complicated formalism needed to discuss spin problems, before presenting in chapter 5 evidence for the Regge classification of particles on trajectories. Chapter 6 is devoted to a discussion of Regge pole predictions for the high energy behaviour of scattering amplitudes, while in chapter 7 we explore the hypothesis that there exists a 'duality' between resonance poles and Regge-trajectory exchanges. Chapter 8 is concerned with the more complicated effects of Regge cuts, singularities in the angular-momentum plane associated with the simultaneous exchange of two or more Regge trajectories. Then in chapter 9 we look at Regge-theory predictions for the behaviour of many-particle scattering processes, and in chapter 10, those for 'inclusive' reactions in which only a few of the final-state particles are actually detected. This is a field which has provided abundant evidence for the success of Regge theory in recent years. In chapter 11 we examine various models for the behaviour of high energy cross-sections, and the self-consistency of strong interactions under the hypothesis that Regge exchanges provide the binding forces between particles which in their turn generate Regge trajectories: the so-called 'bootstrap' mechanism. The final chapter is devoted to a rather brief discussion of the implications of Regge theory for electromagnetic and weak interactions. There are also mathematical appendices on Legendre functions and rotation functions.

The book is intended mainly for those who are just starting to concern themselves with elementary-particle physics, and as far as possible only a good background of undergraduate physics is assumed; that is, quantum theory and especially scattering theory to the level of, say, Schiff's *Quantum mechanics* (1968), special relativity, and the basic concepts of elementary-particle physics such as resonance scattering and isotopic spin (as in, for example, Bransden, Evans and Major (1973)). Also a knowledge of complex-variable theory and the special functions of mathematical physics is required. But in places some of the ideas of quantum field theory (mainly Feynman diagrams) are employed with only the briefest introduction, and the beginner will either have to accept what is said or consult the reference texts. Similarly a more detailed treatment of the Lie groups SU(2) and SU(3) than we have space for here is desirable. But it is hoped that those who read this book in conjunction with some of the references will not experience too many difficulties. They are strongly advised to skip the most difficult parts at a first reading, and refer back when necessary. (To assist this I have marked with a * sections which might be omitted.) It is also hoped that more experienced research workers may find here a useful compendium of the basic ideas and results of Regge theory.

When writing a book on a subject which is developing so fast it is always hard to guess which aspects will stand the test of time, and which will be found wanting. In the early 1960s it seemed to some people that the whole of Regge theory might fall into the latter category, but now many features seem securely established, and I have tried to concentrate on these, with only occasional excursions to glimpse what is happening near the rapidly moving frontier. The greatest consolidation has been possible with those aspects of the theory which directly pertain to experiment, and so I have included a good deal of 'Regge phenomenology', especially in chapters 5–10, but I have tried not to overlook completely the various hints which Regge theory provides as to the long-sought fundamental theory of strong interactions.

I have not attempted to give complete references to the voluminous literature on the subject. Indeed, except for a few of the historically most important papers, I have not referred much to the original literature on the early developments, but such references can readily be found in the various books and review articles which are mentioned. With more recent material I have attempted to give a wider selection of useful references, but only to illustrate the text and certainly not to apportion credit for particular discoveries. I can only apologize to those whose work has been overlooked or inadequately represented.

This book owes much to an earlier work on Regge poles which Professor E. J. Squires and I wrote some years ago (Collins and Squires 1968) and to a review article (Collins 1971), as well as various lecture courses I have given at Durham and elsewhere. But, while I have not changed the presentation just for the sake of it, I have tried to think afresh as to the best way of introducing the subject, stealing ideas from the many excellent review articles and lecture notes which are now available. Also I have tried to simplify as much as possible.

In conclusion I would like to express my indebtedness and gratitude to many people; to Professor G. F. Chew who first introduced me to this subject; to Professor E. J. Squires from whom I have learned many of its intricacies; to my colleagues in Durham who provide a stimulating environment for the study of elementary-particle theory, and much else; to Professor J. C. Polkinghorne, F.R.S. who induced me to write this book; to Professor E. J. Squires, Professor J. C. Polkinghorne, Dr A. D. Martin and Dr W. J. Zakrzewski for many useful comments on the text; to Mr T. D. B. Wilkie and Mr A. D. M. Wright for much help with correcting and improving it; to Margaret and Andrew for providing the rest frame which made it possible; and to Mrs Diana Philpot who has coped wonderfully with a very difficult typescript.

Physics Department, University of Durham P. D. B. COLLINS
August, 1975

1
The scattering matrix

1.1 Introduction

In a typical scattering experiment, performed at an accelerator laboratory, a particle from the accelerated beam strikes another particle in the target material (usually a proton) and the result may be the production of several different types of particles, travelling in various directions, as in fig. 1.1. Thus, before the interaction, we have an initial state $|i\rangle$ composed of two free particles (beam and target), and when the interaction is over, a final state $|f\rangle$ consisting often of many particles. A complete quantum-mechanical theory of the scattering process, if it existed, would allow us to deduce the probability of achieving any particular final state from the given initial state.

We define the scattering operator, S, such that its matrix elements between the initial and final states $\langle f|S|i\rangle$, give us the probability P_{fi} that $|f\rangle$ will be the final state resulting from $|i\rangle$, i.e.

$$P_{fi} = |\langle f|S|i\rangle|^2 = \langle i|S^\dagger|f\rangle\langle f|S|i\rangle \qquad (1.1.1)$$

where $S^\dagger$ is the Hermitian adjoint of S. A knowledge of the full scattering matrix (or S-matrix for short) containing the matrix elements connecting any conceivable initial state to any conceivable final state would clearly constitute a complete description of all possible particle interactions, which is, of course, our ultimate goal.

Unfortunately, there is as yet no fundamental theory for the strong interactions of elementary particles, so it is not possible to present the subject deductively, but we shall try in this chapter to explain briefly the assumptions on which we will be relying for our subsequent development of Regge theory, i.e. the general principles such as analyticity and crossing, which, though not rigorously verified, have stood the test of time, and will form the basis for our discussion. We shall try to make them plausible by showing how they are incorporated both in non-relativistic potential scattering and quantum field theories, which therefore provide useful sources of intuition.

In a field theory like quantum electrodynamics, these S-matrix ele-

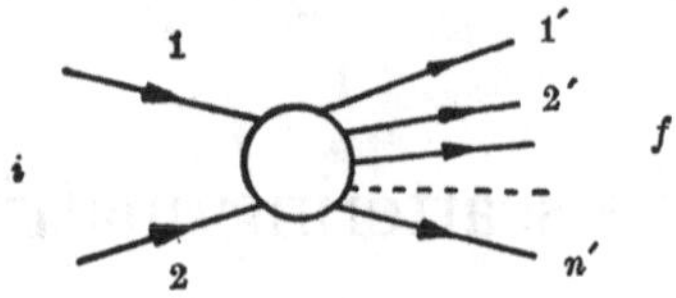

FIG. 1.1 A scattering process with two particles in the initial state and n in the final state.

ments can be deduced, at least in principle, from the basic Lagrangian describing the interactions of the fundamental particles. But for strong interactions there are many problems with this sort of approach, such as the failure of re-normalization methods and the lack of convergence of the perturbation series. However, the S-matrix elements themselves are always evaluated between the so-called asymptotic states at times $t = \pm\infty$; or, more accurately, the initial state a long time before the interaction commences, and the final state a long time afterwards (i.e. long compared with the duration of the interaction, typically $\approx 10^{-22}$ s). What goes on during the interaction is clearly not directly observable. It is thus certainly very useful, and some (see for example Chew (1962)) would claim more in accord with the philosophy of quantum mechanics, to try to develop a theory for the S-matrix directly. Others still feel that one should start from the interactions of quantized fields, and that our goal should be to obtain for strong interactions something akin to quantum electrodynamics (see for example Bjorken and Drell (1965) for a review of this subject). We are still so far from a complete theory that such disputes seem premature. Here we shall adopt mainly an S-matrix viewpoint, chiefly because in working with S-matrix elements one is concerned with (almost) directly measurable quantities, and so the S-matrix provides an excellent vantage point from which to survey the confrontation of theoretical speculation with experimental fact.

In the following sections we introduce the basic ideas of S-matrix theory, the unitarity equations and the analyticity properties of scattering amplitudes. We show how these analyticity assumptions allow one to write dispersion relations for the scattering amplitudes, and discuss the ambiguities which such dispersion relations frequently possess because they involve divergent integrals. We also briefly consider Feynman perturbation field theory and Yukawa potential-scattering models, and show how they incorporate many of these features. This will set the stage for the introduction of Regge theory in the next chapter.

We shall employ the usual units for particle physics, in which the velocity of light, c, and Planck's constant, $\hbar$, are both set equal to unity. Energies, momenta and masses are all expressed in electron volts, or more conveniently in $\text{GeV} \equiv 10^9\,\text{eV}$. This unit can be converted into a time or length using

$$\hbar = 6.58 \times 10^{-25}\,\text{GeV s}$$

$$\hbar c = 1.97 \times 10^{-16}\,\text{GeV m}$$

A convenient alternative unit of length is the fermi

$$1\,\text{fm} \equiv 10^{-15}\,\text{m} = \frac{10}{1.97\,\text{GeV}} \approx 5\,\text{GeV}^{-1}$$

Cross-sections are usually measured in millibarns; $1\,\text{mb} = 10^{-31}\,\text{m}^2$ which may be converted into GeV units using

$$\text{GeV}^{-2} = 0.389\,\text{mb}.$$

1.2 The S-matrix

S-matrix theory starts from the following basic assumptions.

Postulate (i)

Free particle states, containing any number of particles, satisfy the superposition principle of quantum mechanics, so that if $|\psi_\alpha\rangle$ and $|\psi_\beta\rangle$ are physical states so is $|\psi_\gamma\rangle \equiv a|\psi_\alpha\rangle + b|\psi_\beta\rangle$ where a and b are arbitrary complex numbers. (There are in fact superselection rules such as charge and baryon-number conservation which violate this rule but they will not trouble us here; see Martin and Spearman (1970).)

Postulate (ii)

Strong interaction forces are of short range. We know from nuclear physics that the strong interaction is not felt at distances greater than a few times 10^{-15} m (a few pion Compton wavelengths). This means that we can regard the particles as free (i.e. non-interacting) except when they are very close together, and so the asymptotic states, before and after an experiment is performed, consist of just free particles. (We regard a bound state such as the deuteron as a single particle.) Clearly this is only justified if we neglect long-range forces such as electromagnetism and gravitation. In fact, they cannot be incorporated into the S-matrix framework without considerable difficulty and we shall mainly ignore these weaker interactions and suppose ourselves to be dealing with an idealized world where they have been 'switched off'.

To define completely a single free-particle state we must first specify all its internal quantum numbers, i.e. its charge Q, baryon number B, isospin I, strangeness S, parity P (and for a non-strange meson the G-parity G, and charge conjugation C_n), and its spin σ (where the eigenvalue of $\boldsymbol{\sigma}^2$ is $[\sigma(\sigma+1)]$). (The classification of particles in terms of these quantum numbers is discussed in chapter 5.) We denote these quantum numbers collectively by the 'particle type' T. We must also specify the component of its spin along a chosen quantization axis, say, σ_3, and its mass m, energy E, and momentum $\boldsymbol{p}$, in some chosen Lorentz frame.

Postulate (iii)

The scattering process, and hence the S-matrix, is invariant under Lorentz transformations. It is thus convenient to regard $E \equiv p_0$ as the time component of a relativistic four-vector whose space components are p_1, p_2 and p_3, i.e.

$$p_\mu \equiv (p_0, \boldsymbol{p}), \quad \mu = 0, 1, 2, 3 \tag{1.2.1}$$

Since we are always concerned with free particles for which the total energy is given by

$$E^2 = \boldsymbol{p}^2c^2 + m^2c^4 \tag{1.2.2}$$

where m is the particle's rest mass, and as we work in units where $c \equiv 1$, the four-momentum satisfies the 'mass-shell' constraint

$$\sum_\mu p_\mu p^\mu \equiv p^2 = p_0^2 - \boldsymbol{p}^2 = E^2 - \boldsymbol{p}^2 = m^2 \tag{1.2.3}$$

so only three of its four components are independent once the mass is given.

In this book we shall adopt the commonly used convention that the spin quantization axis will be the direction of motion of the particle in the chosen frame of reference. The component of the spin along this axis is called the helicity, λ, and is defined by

$$\lambda \equiv \frac{\boldsymbol{\sigma}\cdot\boldsymbol{p}}{|\boldsymbol{p}|} \tag{1.2.4}$$

Clearly λ can take any of the $2\sigma+1$ possible values, $\sigma, \sigma-1, \ldots, -\sigma$.

Thus a single-particle state is denoted by

$$|T, \lambda, p_\mu\rangle \equiv |P\rangle \tag{1.2.5}$$

and such states are irreducible representations of the Lorentz group (for proof see for example Martin and Spearman (1970)).

Obviously states corresponding to different momenta, different intrinsic quantum numbers, or different helicities must be orthogonal to each other, so their scalar products take the form

$$\langle P'|P\rangle \equiv \langle T',\lambda',p'_\mu|T,\lambda,p_\mu\rangle = N\,\delta^3(\mathbf{p}'-\mathbf{p})\,\delta_{T'T}\,\delta_{\lambda'\lambda} \quad (1.2.6)$$

where $\delta^3(\mathbf{p}'-\mathbf{p})$ is a short-hand notation for

$$\delta(p'_1-p_1)\,\delta(p'_2-p_2)\,\delta(p'_3-p_3),$$

and N is a normalization factor.

We want to normalize our state vectors in a Lorentz invariant manner. The normalization of the state will tell us the number of particles in a given phase-space volume element $\mathrm{d}^3\mathbf{p}$ about the vector $\mathbf{p}$, but this is clearly not a Lorentz invariant quantity because the size of such a volume element $\mathrm{d}^3\mathbf{p}$ is not invariant. However, the volume element $\mathrm{d}^4p\delta(p^2-m^2)$ is manifestly invariant, while the δ-function ensures that the mass-shell constraint (1.2.3) is obeyed. In fact, it can be re-expressed as

$$\mathrm{d}^4p\delta(p^2-m^2) = \frac{\mathrm{d}^3\mathbf{p}}{2p_0}\,\theta(p_0) \quad (1.2.7)$$

because, with the usual rules for manipulating the Dirac δ-function, i.e.

$$\delta(ax) = 1/a\,\delta(x)$$

we find

$$\delta(p^2-m^2) \equiv \delta(p_0^2-\mathbf{p}^2-m^2) = \frac{1}{2p_0}\,\delta[p_0-\sqrt{(\mathbf{p}^2+m^2)}]$$
$$-\frac{1}{2p_0}\,\delta[p_0+\sqrt{(\mathbf{p}^2+m^2)}] \quad (1.2.8)$$

and we shall always restrict our integrations to positive p_0 only. Hence it is convenient to choose N in (1.2.6) such that

$$\langle P'|P\rangle = (2\pi)^3\,2p_0\delta^3(\mathbf{p}'-\mathbf{p})\,\delta_{T'T}\,\delta_{\lambda'\lambda} \quad (1.2.9)$$

The factor $(2\pi)^3$ is purely a matter of convention, but the presence of p_0 ensures, through (1.2.7), that our normalization remains invariant under Lorentz transformations.

A state consisting of n free particles may be written as a direct product of single particle states

$$|T_1,\lambda_1,p_1;\,T_2,\lambda_2,p_2;\,\ldots;\,T_n,\lambda_n,p_n\rangle$$
$$\equiv |P_1\ldots P_n\rangle = |P_1\rangle\otimes|P_2\rangle\otimes\ldots\otimes|P_n\rangle \quad (1.2.10)$$

and has the normalization, from (1.2.9),

$$\langle P'_{1'} \ldots P'_{n'} | P_1 \ldots P_n \rangle = \prod_{i=1}^{n} (2\pi)^3 \, 2p_{0i} \, \delta^3(\mathbf{p}'_i - \mathbf{p}_i) \, \delta_{T'_i T_i} \delta_{\lambda'_i \lambda_i} \delta_{n'n} \quad (1.2.11)$$

Postulate (iv)

The scattering matrix is unitary. This follows if the free particle states $|m\rangle$, $m = 1, 2, \ldots$ constitute a complete orthonormal set of basis states satisfying the completeness relation

$$\sum_m |m\rangle \langle m| = 1 \quad (1.2.12)$$

since starting from any given state $|i\rangle$ the probability that there will be *some* final state must be unity. So from (1.1.1)

$$\sum_m P_{mi} = \sum_m |\langle m| S |i\rangle|^2 = \sum_m \langle i| S^\dagger |m\rangle \langle m| S |i\rangle$$
$$= \langle i| S^\dagger S |i\rangle = 1 \quad (1.2.13)$$

and as this must be true for any state $|i\rangle$ we have

$$S^\dagger S = 1 = S S^\dagger \quad (1.2.14)$$

so S is a unitary matrix.

For our many-particle states with normalization (1.2.11) the completeness relation (1.2.12) reads

$$\sum_{m=1}^{\infty} \prod_{i=1}^{m} \sum_{\lambda_i} \sum_{T_i} (2\pi)^{-3} \int \frac{d^3 p_i}{2p_{0i}} |P_1 \ldots P_m\rangle \langle P_1 \ldots P_m| = 1 \quad (1.2.15)$$

since the summation must run over all possible numbers, types and helicities of particles, as well as over all their possible momenta. So in terms of these states the unitarity relation (1.2.13) becomes

$$\sum_{m=1}^{\infty} \prod_{i=1}^{m} \sum_{\lambda_i} \sum_{T_i} (2\pi)^{-3} \int \frac{d^3 q_i}{2q_{0i}} \langle P'_1 \ldots P'_{n'} | S | Q_1 \ldots Q_m \rangle$$
$$\times \langle Q_1 \ldots Q_m | S^\dagger | P_1 \ldots P_n \rangle = \langle P'_1 \ldots P'_{n'} | P_1 \ldots P_n \rangle \quad (1.2.16)$$

where $Q_i \equiv \{T_i, \lambda_i, q_{\mu i}\}$ is used to label the intermediate-state particles with four-momenta q_{μ_i}. Note that in these equations we have treated the particles as non-identical as we shall continue to do below. For identical particles one must sum over the $n!$ ways of pairing the momenta in (1.2.11), and correspondingly $(n!)^{-1}$ appears in the completeness relation (1.2.15), and hence in (1.2.16).

This unitarity equation (1.2.16) is of fundamental importance in determining the nature of the S-matrix. However, it is also rather complicated, and it becomes much easier to understand, and to utilize, if we represent it diagrammatically in terms of 'bubble diagrams'. (A more complete account of this subject will be found in Eden *et al.* (1966).)

1.3 Bubble diagrams and scattering amplitudes

The summation over different types of particles and their different helicities in (1.2.16) adds unnecessarily to the notational complexity of the equation. For the rest of this chapter we shall only be concerned with the momentum-space properties of the S-matrix, so we shall cease to refer to T and λ, and write all our equations as though there existed only a single type of particle of zero spin. Thus an n-particle state will be written as just $|p_1 \dots p_n\rangle$. Each integration over a momentum should therefore be regarded as implying also a summation over all the different types of particles which can contribute, given the restrictions required by quantum number conservation, and over all the $2\sigma_i + 1$ possible helicities available to a particle of spin σ_i.

We denote each S-matrix element representing a scattering process by a 'bubble' with lines corresponding to the incoming and outgoing particles, viz.

$$\langle p'_1 \dots p'_{n'}| S |p_1 \dots p_n\rangle \equiv$$ 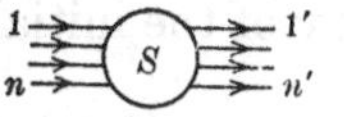

(1.3.1)

and $$\langle p'_1 \dots p'_{n'}| S^\dagger |p_1 \dots p_n\rangle \equiv$$ 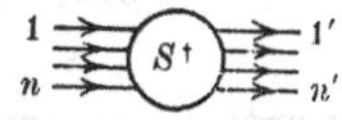

(1.3.2)

The intermediate states appearing in a unitarity equation such as (1.2.16) are denoted by

$$\prod_{i=1}^{m} \int (2\pi)^{-3} \frac{d^3 q_i}{2q_{0i}} \equiv$$ 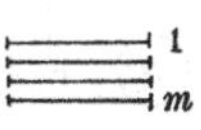

(1.3.3)

the bars on the ends indicating that such lines must be attached to bubbles. The overlap between states (1.2.11) is written

$$\langle p'_1 \dots p'_{n'}|p_1 \dots p_n\rangle =$$ 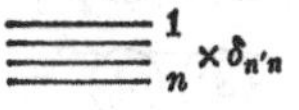

(1.3.4)

Because of Lorentz invariance (postulate (iii)) we know that energy and momentum are conserved in a scattering process, and hence an S-matrix element such as (1.3.1) vanishes unless

$$\sum_{i=1}^{n} p_{\mu_i} = \sum_{i=1}^{n'} p'_{\mu_i}, \quad \mu = 0, 1, 2, 3 \tag{1.3.5}$$

This implies that for example in (1.2.16) only intermediate states with $\left(\sum_{i=1}^{m} m_i\right)^2 \leqslant \left(\sum_{i=1}^{n} p_i\right)^2$ contribute to the sum. The equality occurs at the threshold energy for the process $|p_1 \dots p_n\rangle \rightarrow |q_1 \dots q_m\rangle$.

Thus suppose we have, as will always be the case in practice, a two-particle initial state, and suppose that for simplicity we take all the hadrons to have the same mass, m. (This would mean of course that they were all stable as they would have no state of lower mass into which to decay.) Then for $(2m)^2 \leqslant (p_1+p_2)^2 \leqslant (3m)^2$, i.e. above the two-particle threshold but below that for three particles, only a two-particle intermediate state, and only a two-particle final state, can occur in the unitarity equation (1.2.16) which becomes

$$\int \prod_{i=1}^{2} (2\pi)^{-3} \frac{\mathrm{d}^3 q_i}{2q_{0i}} \langle p_1' p_2' | S | q_1 q_2 \rangle \langle q_1 q_2 | S^\dagger | p_1 p_2 \rangle = \langle p_1' p_2' | p_1 p_2 \rangle \tag{1.3.6}$$

and with the above rules it may be rewritten as

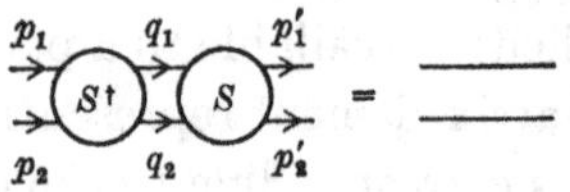

(1.3.7)

But if the energy of the initial state is increased, so

$$(3m)^2 \leqslant (\sum_i p_i)^2 \leqslant (4m)^2,$$

two- or three-particle states are possible for the initial state (in principle) and for the intermediate and final states (in practice), so (1.2.16) gives us the set of unitarity equations.

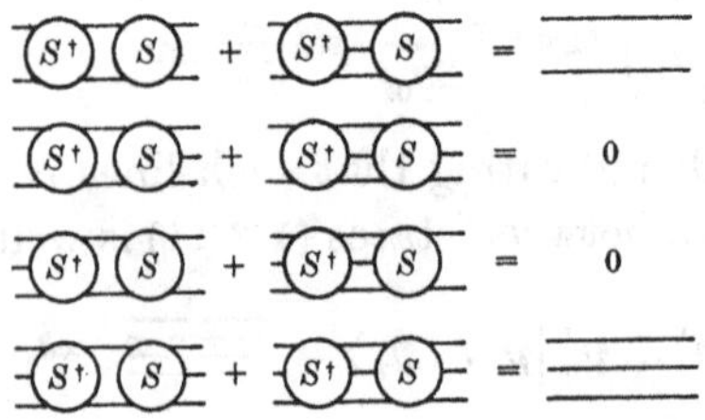

(1.3.8)

The generalization to higher energies where even more particles can occur should be obvious.

The finite range of the strong interaction force (postulate (ii)) permits a further development of these equations. For example, the S-matrix element with two particles in both the initial and final states can be decomposed as follows:

(1.3.9)

$$= \langle p_1', p_2' | p_1, p_2 \rangle + \langle p_1', p_2' | S_c | p_1, p_2 \rangle$$

Here the first term applies if the two particles never get close enough to interact, while the second, the so-called 'connected part', represents the interaction of the two particles. (The + sign is used for the connected part of S for reasons which will become apparent below.) These are quite distinct because in the first term each particle has the same energy and momentum in the final state as it had in the initial state, while with the second term only the total energy and total momentum of the two particles need be conserved. Putting in the conservation δ-functions of (1.2.11) and four-momentum conservation for 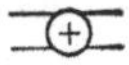 explicitly, (1.3.9) gives

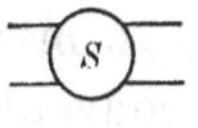

$$= (2\pi)^6 4p_{01}p_{02}\,\delta^3(\mathbf{p}_1' - \mathbf{p}_1)\,\delta^3(\mathbf{p}_2' - \mathbf{p}_2) + i(2\pi)^4\,\delta^4(p_1 + p_2 - p_1' - p_2')\,\langle p_1' p_2' | A | p_1 p_2 \rangle \quad (1.3.10)$$

The factor $i(2\pi)^4$ is included to give a conventional normalization to the A-matrix or 'scattering amplitude' representing $\oplus$.

On the other hand the $2 \to 3$ S-matrix element is only possible if the two particles actually scatter, so

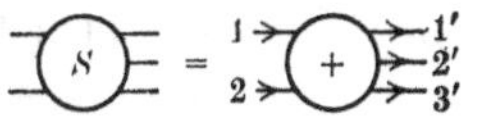

$$= i(2\pi)^4\,\delta^4(p_1 + p_2 - p_1' - p_2' - p_3') \times \langle p_1' p_2' p_3' | A^+ | p_1 p_2 \rangle \quad (1.3.11)$$

If there are more external lines there may be more disconnected parts, thus

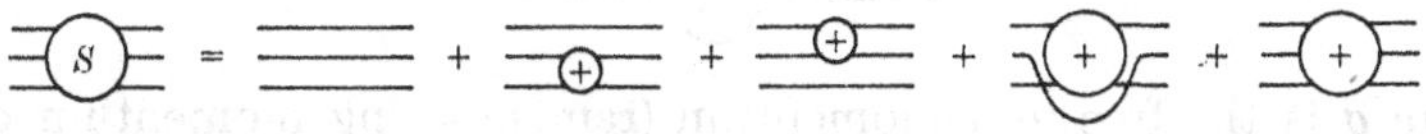

(1.3.12)

For $S^\dagger$ we write correspondingly

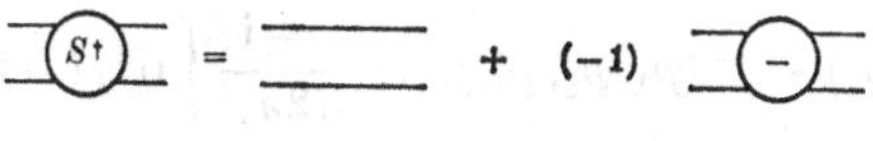

(1.3.13)

where

$$= i(2\pi)^4\,\delta^4(p_1 + p_2 - p_1' - p_2')\,\langle p_1' p_2' | A^- | p_1 p_2 \rangle \quad (1.3.14)$$

the minus signs again being conventional.

This disconnectedness property allows a considerable further simplification of the unitarity equations. Thus, on substituting

(1.3.9) and (1.3.13), (1.3.7) becomes

(1.3.15)

which, on multiplying out and cancelling identical terms, gives the two-particle unitarity equation

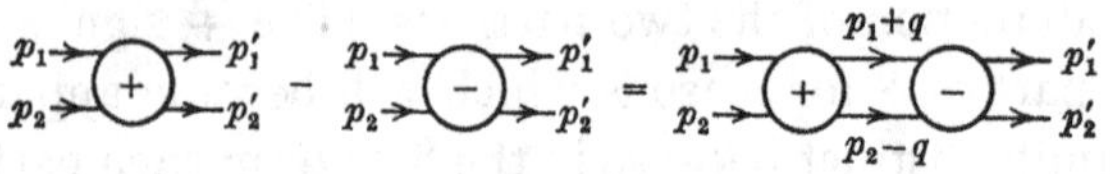

(1.3.16)

Similarly above the three-particle threshold the first equation of (1.3.8) gives

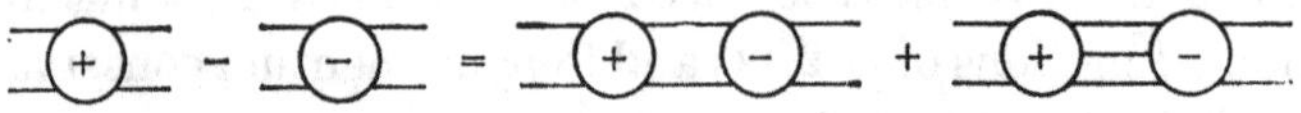

(1.3.17)

In such equations the δ-functions of overall energy and momentum conservation are of course the same for each term, and so may be cancelled, along with various factors of i, 2π etc. (our conventions have been designed to assist this) and we end up with the following simpler set of rules for the diagrams:

For each connected bubble 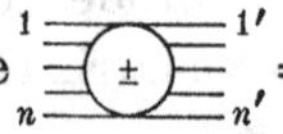 $= (-1)A^{\pm}(p_1 \dots p_n; p_1' \dots p_{n'}')$ (1.3.18)

For each internal line 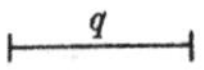 $= -2\pi \mathrm{i}\,\delta(q^2-m^2)$ (1.3.19)

For each closed loop 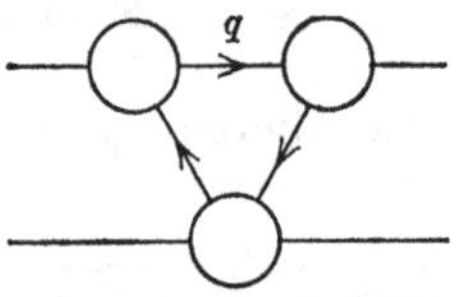 $= \dfrac{\mathrm{i}}{(2\pi)^4}\displaystyle\int \mathrm{d}^4 q$ (1.3.20)

where q is the free four-momentum (remembering momentum conservation at each vertex – see for example (1.3.16)). Thus for example (1.3.16) becomes

$$A^+(p_1, p_2, p_1', p_2') - A^-(p_1, p_2, p_1', p_2') = \frac{-\mathrm{i}}{(2\pi)^4}\int \mathrm{d}^4 q\,(-2\pi\mathrm{i})^2$$
$$\times\,\delta((p_1+q)^2-m^2)\,\delta((p_2-q)^2-m^2)\,A^+(p_1, p_2, p_1+q, p_2-q)$$
$$\times\,A^-(p_1+q, p_2-q, p_1', p_2') \quad (1.3.21)$$

These unitarity equations greatly restrict the form of the scattering amplitude, as we shall see.

1.4 The analyticity properties of scattering amplitudes

We have so far written the scattering amplitudes, $A^{\pm}(p_1 \ldots p_n; p'_1 \ldots p'_{n'})$ as arbitrary functions of the four-momenta of the particles involved. However, Lorentz invariance implies that A must be a Lorentz scalar, and hence may be written as a function of Lorentz scalars only. As long as we are neglecting spin this means that A is a function only of scalar products of the momenta.

Thus for the four-line process $1+2 \to 3+4$ the amplitude $A(p_1, p_2; p_3, p_4)$ will be a function of Lorentz scalars such as $(p_1+p_2)^2$, $(p_1+p_3)^2$, $(p_1+p_2+p_3)^2$ etc. (Remember $p_i^2 = m_i^2$, $i = 1, \ldots, 4$, are not variables.) However, not all these are independent quantities, since, for example $(p_1+p_2)^2 = (p_3+p_4)^2$ by four-momentum conservation. In general for an n-line process there are $4n$ variables (the components of the n four-vectors), but n mass-shell constraints of the form $p_i^2 = m_i^2$, 4 constraints for overall energy and momentum conservation, and 6 constraints for rotational invariance in the four-dimensional Minkowski space, leaving us with $3n-10$ independent variables. Thus, if we regard a single particle propagator as a 'scattering process' $1 \to 2$, 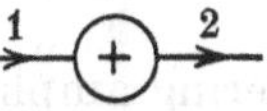, we have $n = 2$ so there are -4 degrees of freedom, i.e. the 4 constraints $p_{1\mu} = p_{2\mu}$, $\mu = 0, 1, 2, 3$. For the more realistic process $1+2 \to 3+4$, $n = 4$, and so there are two independent variables, while $1+2 \to 3+4+5$ depends on 5 variables, and so on. We denote these variables by the Lorentz invariants

$$s_{ijk\ldots} \equiv (\pm p_i \pm p_j \pm p_k \ldots)^2.$$

But what sort of function of these invariants is A? This brings us to the next postulate of S-matrix theory.

Postulate (v): Maximal analyticity of the first kind

The scattering amplitudes are the real boundary values of analytic functions of the invariants $s_{ijk}\ldots$ regarded as complex variables, with only such singularities as are demanded by the unitarity equations.

Thus although obviously only real values of the $s_{ijk}\ldots$ make physical sense we are going to treat them as complex variables, and suppose that the amplitudes are analytic functions of the s_{ijk}, so that we can obtain the physical scattering amplitude by taking the limit $s \to$ real.

A simple understanding of why the amplitudes may plausibly be expected to have such analyticity properties can be obtained from the following argument. Consider the scattering of a wave packet travelling initially along the z axis with velocity v,

$$\psi_{\text{in}}(z,t) = \frac{1}{(2\pi)^{\frac{1}{2}}}\int_{-\infty}^{\infty} \mathrm{d}\omega\phi(\omega)\,\mathrm{e}^{-\mathrm{i}\omega(t-z/v)} \tag{1.4.1}$$

where ω is the energy ($\hbar \equiv 1$), and, taking the Fourier inverse,

$$\phi(\omega) = \frac{1}{(2\pi)^{\frac{1}{2}}}\int_{-\infty}^{\infty} \mathrm{d}t\,\psi(0,t)\,\mathrm{e}^{\mathrm{i}\omega t} \tag{1.4.2}$$

To make physical sense this integral must converge for real ω, but it defines $\phi(\omega)$ for all complex values of ω. If the wave packet does not reach $z = 0$ until $t = 0$ then $\psi(0,t) = 0$ for $t < 0$ so

$$\phi(\omega) = \frac{1}{(2\pi)^{\frac{1}{2}}}\int_{0}^{\infty} \mathrm{d}t\,\psi(0,t)\,\mathrm{e}^{\mathrm{i}\omega t} \tag{1.4.3}$$

This means that $\phi(\omega)$ is an analytic function of ω regular in the upper-half plane (i.e. for $\operatorname{Im}\{\omega\} > 0$) since in this region the integral (1.4.3) must certainly converge (because it exists for real ω, and we get even better convergence from $\mathrm{e}^{-(\operatorname{Im}\{\omega\})t}$ for $\operatorname{Im}\{\omega\} > 0$). Similarly for the scattered wave we have

$$\psi_{\text{out}}(\boldsymbol{r},t) = \frac{1}{(2\pi)^{\frac{1}{2}}}\frac{1}{r}\int_{-\infty}^{\infty} \mathrm{d}\omega\, A(\omega)\,\phi(\omega)\,\mathrm{e}^{-\mathrm{i}\omega(t-r/v)} \tag{1.4.4}$$

where, by definition, $A(\omega)$ is the scattering amplitude for scattering at a given energy (see for example Schiff (1968)). If the scattering process is causal the scattered wave cannot have reached a distance r from the scattering centre until time $t = r/v$ has elapsed so

$$\psi_{\text{out}}(\boldsymbol{r},t) = 0 \quad \text{for} \quad t < r/v,$$

which from the Fourier inverse of (1.4.4), with repetition of the argument (1.4.1) to (1.4.3), implies that $A(\omega)$ is also an analytic function of ω in the upper-half plane.

The difficulty with an argument such as this is of course that it assumes that it makes sense to talk about the precise distribution of the wave packet in time despite the fact that we are also assuming that the energy is known with precision, so it is not obvious how far this concept of microscopic causality makes sense. Clearly, no quantum-mechanical measurement could establish what the time distribution of a wave packet is, even in principle. However, we shall see below that we only seem to require micro-causality in the classical limit.

Attempts have been made to deduce the analyticity properties (and singularities) of scattering amplitudes from axiomatic field theory (see for example Goldberger and Watson (1964)), and axiomatic S-matrix theory (see Eden *et al.* 1966), but there are many difficulties in discovering how to continue round the various singularities. Only for physical-region singularities is the situation reasonably clear (Bloxam, Olive and Polkinghorne 1969). If the scattering amplitude can be written as a perturbation series (a sum of Feynman diagrams) the analyticity properties of the individual terms in the series can be found (at least for the lower orders), but of course we are concerned with strong interactions where such a perturbation series is not expected to converge. However, since S-matrix theory and perturbation theory seem to possess similar singularity structures it is often useful to employ Feynman-diagram models (see section 1.12). Here we shall simply assume that the singularity structure which can be deduced heuristically from the S-matrix postulates is in fact correct.

1.5 The singularity structure

The most important type of singularity which can be identified in the unitarity equations is a simple pole which corresponds to the exchange of a physical particle. The occurrence of such poles can be deduced from the $3 \to 3$ unitarity equations (1.3.8), for example, in which we find the term

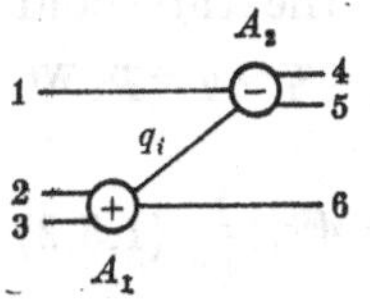

$$\sim A_1^+(-2\pi i\,\delta(q_i^2-m_i^2))\,A_2^-, \quad q_i = p_2+p_3-p_6$$
$$= p_4+p_5-p_1 \quad (1.5.1)$$

The δ-function occurs because of course it is only precisely when $(p_2+p_3-p_6)^2 = m_i^2$ that particle i can be exchanged between the bubbles. Now since

$$\frac{1}{q_i^2-m_i^2 \pm i\epsilon} = P\,\frac{1}{q_i^2-m_i^2} \pm \pi i\,\delta(q_i^2-m_i^2)$$

(where P = principal part), the amplitudes must contain pole contributions of the form

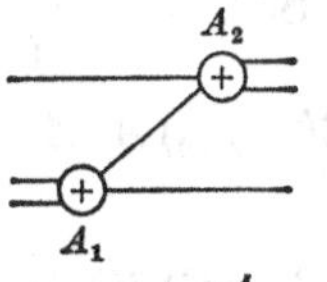

$$= A_1^+\,\frac{1}{q_i^2-m_i^2+i\epsilon}\,A_2^+,$$

and

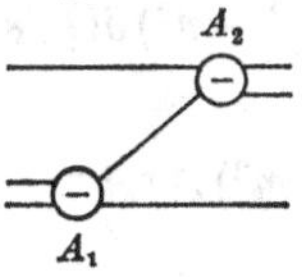

$$= A_1^-\,\frac{1}{q_i^2-m_i^2-i\epsilon}\,A_2^- \quad (1.5.2)$$

so that 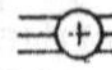 −  contains the δ-function of (1.5.1) in the limit $\epsilon \to 0$. This result is not unexpected because in perturbation theory the Feynman propagator for a spinless particle takes the form of a pole $(q_i^2 - m_i^2 + i\epsilon)^{-1}$ (see section 1.12 below). Also, we are familiar in nuclear physics with unstable particles (or resonances) which give rise to amplitudes of the Breit–Wigner form $\sim (q_i^2 - m_i^2 + im_i \Gamma_i)^{-1}$ where Γ_i is the width of the resonance, giving a complex pole at $q_i^2 = m_i^2 - im_i \Gamma_i$.

The additional feature which we can observe in (1.5.2) is that the residue of the pole at $q_i^2 = m_i^2$ can be 'factorized' into the amplitudes for the two separate scatterings involving particle i, viz. $1 + i \to 4 + 5$ and $2 + 3 \to i + 6$. It is sometimes said that this factorization is a consequence of unitarity, but really it stems from the disconnectedness postulate (ii) since (1.5.2) can represent successive scattering processes which are completely independent of each other and occurring at two well separated places ($\gg$ 1 fm).

We thus find that the exchange of a particle gives a pole in q^2 in the S-matrix; and vice versa the presence of a pole in q^2 indicates the presence of a particle, stable if it occurs for real q^2, unstable if it occurs for complex q^2, as in the Breit–Wigner formula.

The next-simplest singularity is due to the exchange of two particles, as in (1.3.21). This gives rise to a branch point at the threshold $(p_1 + p_2)^2 = (2m)^2$. Transforming the integration variable $q \to q - p_1$ we get

$$A^+ - A^- = \frac{i}{(2\pi)^2} \int d^4 q \, \delta(q^2 - m^2) \, \delta((p_1 + p_2 - q)^2 - m^2) \, A^+ A^- \quad (1.5.3)$$

In the centre-of-mass system $p_1 = (p_{01}, \mathbf{p})$ and $p_2 = (p_{02}, -\mathbf{p})$, so

$$(p_1 + p_2) = (p_{01} + p_{02}, \mathbf{0}) \equiv (\sqrt{s}, \mathbf{0}) \quad (1.5.4)$$

where we have defined $\sqrt{s}$ to be the total energy in the centre-of-mass system. Putting $q = (q_0, \mathbf{q})$, the argument of the second δ-function in (1.5.3) becomes

$$(p_1 + p_2 - q)^2 - m^2 = s - 2(\sqrt{s}) \, q_0 + q^2 - m^2 = s - 2(\sqrt{s}) \, q_0 \quad (1.5.5)$$

since the first δ-function gives $q^2 = m^2$. So

$$\begin{aligned} A^+ A^- &= \frac{i}{(2\pi)^2} \int d^4 q \, \delta(q^2 - m^2) \, \delta(s - 2(\sqrt{s}) \, q_0) \, A^+ A^- \\ &= \frac{i}{(2\pi)^2 \, 2\sqrt{s}} \int dq_0 \, d^3\mathbf{q} \, \delta(q_0^2 - |\mathbf{q}|^2 - m^2) \, \delta(\tfrac{1}{2}\sqrt{s} - q_0) \, A^+ A^- \\ &= \frac{i}{(2\pi)^2 \, 2\sqrt{s}} \int d^3\mathbf{q} \, \delta(\tfrac{1}{4}s - |\mathbf{q}|^2 - m^2) \, A^+ A^- \end{aligned} \quad (1.5.6)$$

Putting $d^3\boldsymbol{q} = \frac{1}{2}\int |\boldsymbol{q}|\, d|\boldsymbol{q}|^2\, d\Omega$, where $d\Omega$ is the element of solid angle associated with the direction of $\boldsymbol{q}$, this gives

$$A^+ - A^- = i\frac{\sqrt{(\frac{1}{4}s - m^2)}}{(4\pi)^2\sqrt{s}}\int d\Omega\, A^+ A^- \qquad (1.5.7)$$

Below the threshold the unitarity equation can be extended to read

$$= 0 \quad \text{or} \quad A^+ - A^- = 0 \qquad (1.5.8)$$

so A^+ and A^- can be regarded as the same function $A(s \pm i\epsilon, \ldots)$ analytically continued above or below the two-particle threshold at $s \equiv (p_1 + p_2)^2 = 4m^2$ where there is a branch point, the discontinuity across the square-root branch cut being given by (1.5.7) (see fig. 1.2). The physical amplitude is of course to be evaluated with s real, but we have a choice of approaching the real axis from above or below. We choose (by convention) the $+i\epsilon$ prescription for A^+ to the effect that

$$\text{Physical}\, A^+(s, \ldots) = \lim_{\epsilon \to 0} A^+(s + i\epsilon, \ldots) \qquad (1.5.9)$$

and draw the branch cut along the real s-axis as shown in fig. 1.2. The sheet of the s plane exhibited in fig. 1.2 is called the 'physical sheet'.

Since A is real below threshold it is clear from the Schwarz reflection principle (Titchmarsh 1939) that $A(s^*, \ldots) = A^*(s, \ldots)$, and that A^- is just the complex conjugate of A^+, and so

$$\text{Physical}\, A^-(s, \ldots) = \lim_{\epsilon \to 0} A(s - i\epsilon, \ldots) \qquad (1.5.10)$$

An amplitude satisfying this reflection relation is said to be 'Hermitian analytic', or 'real analytic'.

These results may be generalized to give us the discontinuity across the branch cut associated with an arbitrary number of particles, 1 up to n, in the intermediate state (fig. 1.3) which according to Cutkosky (1960, 1961) is

$$\text{Disc}\{A\} = \int \prod_{l=1}^{n-1} \frac{i d^4 k_l}{(2\pi)^4} \prod_{i=1}^{n} [-2\pi i \delta(q_i^2 - m_i^2)]\, A_1^+ A_2^- \qquad (1.5.11)$$

where the integration is over the $n-1$ independent loops l which are formed by the n intermediate lines. Since

$$\frac{1}{q_i^2 - m_i^2 \pm i\epsilon} = P\frac{1}{q_i^2 - m_i^2} \pm \pi i \delta(q_i^2 - m_i^2) \qquad (1.5.12)$$

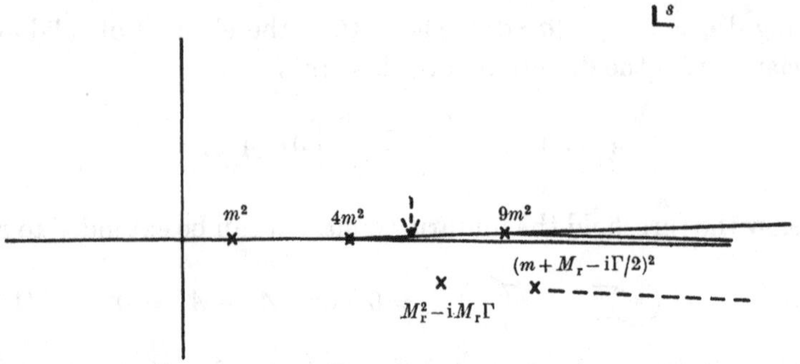

FIG. 1.2 Singularities of the scattering amplitude in the complex s plane, showing the pole at $s = m^2$, the threshold branch points at $s = 4m^2, 9m^2, \ldots$, a resonance pole at $s = M_r^2 - iM_r\Gamma$ on the unphysical sheet reached through the branch cut, and the $m + M_r$ threshold branch cut. The physical value for A^+ is obtained by approaching the real axis from above, as shown by the arrow.

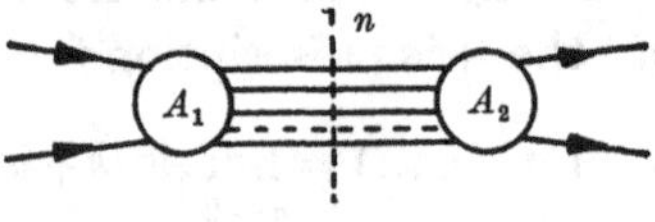

FIG. 1.3 The discontinuity across an n-particle intermediate state.

(where $P \equiv$ principal part) it proves possible to rewrite (1.5.11) as

$$\text{Disc}\{A\} = \text{Disc} \int \prod_{l=1}^{n-1} \frac{\text{id}^4 k_l}{(2\pi)^4} \prod_{i=1}^{n} \frac{1}{(q_i^2 - m_i^2)} A_1^+ A_2^- \tag{1.5.13}$$

This is in fact the same as the discontinuity obtained using Feynman propagators for the intermediate-state particles (see section 1.12 below).

The singularities of integrals like (1.5.13) have been investigated in detail (see Eden *et al.* 1966) and their positions are given by the Landau rules (Landau (1959); see section 1.12 below):

(i) $q_i^2 = m_i^2$ for all $i = 1, \ldots, n$; (1.5.14)

(ii) $\sum_{\text{loop } l} \alpha_i q_i = 0$ for some constants α_i, the summation going right round each closed loop, and $\alpha_i \neq 0$ for any i in the loop.

It is thus possible to identify all the singularities of an amplitude by drawing all the (infinite number of) different intermediate states composed of all the various particles in the theory which can take us from the initial state to the final state. We shall consider some further examples below. The positions and discontinuities across the cuts are all calculable (in principle) from these Landau and Cutkosky rules once we know the particle poles.

These singularities include the poles on the real axis due to the stable particles, and branch points also on the real axis due to the various stable-particle thresholds. We have also noted that an unstable particle or resonance gives rise to a pole below the real axis at $q_i^2 = m_i^2 - im_i\Gamma_i$ where Γ_i is its decay width. Since the real part of the resonance mass must obviously be greater than the threshold energy of the channel into which the particle can decay, this pole will not be on the physical sheet, but on the sheet reached by going down through the threshold branch point. Branch cuts involving such particles will also be off the physical sheet (see fig. 1.2).

We have mentioned that these singularities are supposed to stem from causality. Coleman and Norton (1965) have shown that in the physical region the Landau equations (1.5.14) correspond to the kinematic conditions for the event represented by the given diagram to occur classically. That is to say, if we regard each internal propagator as representing a pointlike particle having momentum q_i, then the vertices where the particle is emitted and absorbed can be regarded as having a space–time separation

$$\Delta_i = q_i\alpha_i$$

where α_i is the proper time elapsing between emission and absorption. If $\alpha_i = 0$ these two points are coincident. For it to be possible for a particle to pass round a closed loop we clearly need $\sum_{\text{loop}} \Delta_i = 0$ which is just (1.5.14) (ii). And (1.5.14) (i) is just the mass-shell condition for the four-momentum. Hence a physical region singularity occurs only when the relevant Feynman diagram can represent a real physical process for pointlike, classical relativistic particles. Micro-causality thus seems to be needed in S-matrix theory only in the correspondence-principle limit when quantum mechanics approaches classical mechanics.

1.6 Crossing

A very important result of the above analyticity property is a relation it implies between otherwise quite separate scattering processes. This relation is known as 'crossing'.

If we consider the amplitude for $1+2 \rightarrow 3+4+5$ it is intuitively rather obvious that it will have the same set of singularities as the amplitude for $1+2+\bar{5} \rightarrow 3+4$, where $\bar{5}$ is the anti-particle of 5, since all we have to do is reverse the direction of the line corresponding to particle 5, i.e. we cross it over, viz.

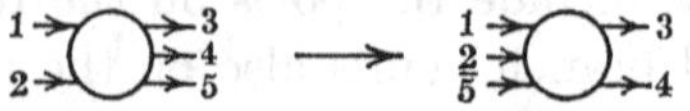

The intermediate states in these two bubbles will be exactly similar.

It is clear that $\bar{5}$ has to be the anti-particle of 5 because it must have the opposite sign for all the additive quantum numbers if both processes are to be possible. Of course these two processes occur for different regions of the variables since the first requires (*inter alia*) $\sqrt{s_{12}} \geqslant \sqrt{s_{34}} + m_5$ while the second needs $\sqrt{s_{34}} \geqslant \sqrt{s_{12}} + m_{\bar{5}}$. However, since the two amplitudes have the same singularities it should, in principle, be possible to obtain one from the other by analytic continuation.

Furthermore, if we rotate all the legs

we get back to the same region of the variables, and so the amplitudes for $1+2 \to 3+4+5$ and $\bar{3}+\bar{4}+\bar{5} \to \bar{1}+\bar{2}$ should be identical. This is an example of *TCP* invariance since it requires that the *S*-matrix be unchanged by the combined operations of time reversal *T*, charge conjugation *C*, and parity inversion *P* (which is obviously what we need to get the anti-particles going backwards in space and time).

Unfortunately, it is not possible to prove the above results as we cannot be sure that analytic continuation from the physical region of one process will necessarily take us onto the physical sheet of the other process. We have to assume that the continuations can be made without leaving the physical sheet of the s variables. However, such results do hold in perturbation theory, and seem very plausible also in particle physics.

1.7 The $2 \to 2$ amplitude

As an example, which will be of considerable use to us later, we consider in some detail the kinematics and singularities of the scattering process $1+2 \to 3+4$ (fig. 1.4 (*a*)). The channels are named after their respective energy invariants, to be introduced below.

By crossing and the *TCP* theorem all the six processes

$$\left.\begin{array}{lll} 1+2 \to 3+4 & \bar{3}+\bar{4} \to \bar{1}+\bar{2} & (s\text{-channel}) \\ 1+\bar{3} \to \bar{2}+4 & 2+\bar{4} \to \bar{1}+3 & (t\text{-channel}) \\ 1+\bar{4} \to \bar{2}+3 & 2+\bar{3} \to \bar{1}+4 & (u\text{-channel}) \end{array}\right\} \quad (1.7.1)$$

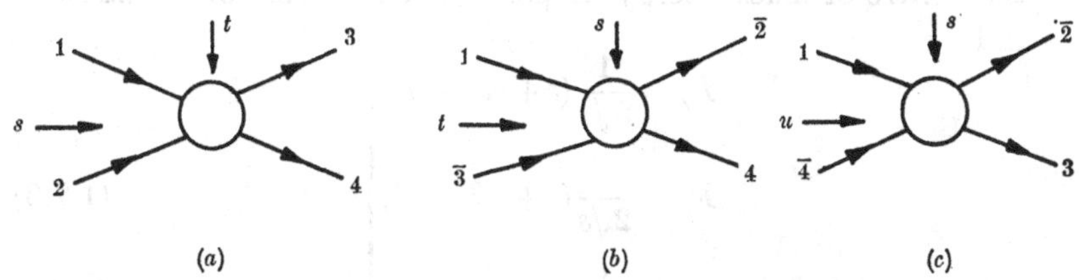

FIG. 1.4 The scattering processes in the s, t and u channels of (1.7.1).

will share the same scattering amplitude, but the pairs of channels labelled s, t and u will occupy different regions of the variables.

In the centre-of-mass system for particles 1 and 2 we write their four-momenta as

$$p_1 = (E_1, \mathbf{q}_{s12}), \quad p_2 = (E_2, -\mathbf{q}_{s12}) \tag{1.7.2}$$

$\mathbf{q}_{s12}$ being the three-momentum, equal but opposite for the two particles. Similarly for the final state

$$p_3 = (E_3, \mathbf{q}_{s34}), \quad p_4 = (E_4, -\mathbf{q}_{s34}) \tag{1.7.3}$$

Since the initial and final states involve only free particles the mass-shell constraints must be satisfied:

$$\left.\begin{aligned} p_1^2 &= E_1^2 - q_{s12}^2 = m_1^2 \\ p_2^2 &= E_2^2 - q_{s12}^2 = m_2^2 \\ p_3^2 &= E_3^2 - q_{s34}^2 = m_3^2 \\ p_4^2 &= E_4^2 - q_{s34}^2 = m_4^2 \end{aligned}\right\} \tag{1.7.4}$$

We define the invariant

$$\left.\begin{aligned} s &= (p_1 + p_2)^2 = (p_3 + p_4)^2 \\ &= (E_1 + E_2)^2 = (E_3 + E_4)^2 \end{aligned}\right\} \tag{1.7.5}$$

which is the square of the total centre-of-mass energy for the s-channel processes. Now combining (1.7.5) and (1.7.4)

$$s = p_1^2 + p_2^2 + 2p_1 \cdot p_2 = m_1^2 + m_2^2 + 2p_1 \cdot p_2 \tag{1.7.6}$$

where the dot denotes a four-vector product. Similarly

$$p_1 \cdot (p_1 + p_2) = m_1^2 + p_1 \cdot p_2 = E_1 \sqrt{s} \tag{1.7.7}$$

using (1.7.2) and (1.7.5). Then combining (1.7.6) and (1.7.7) we get

$$E_1 = \frac{1}{2\sqrt{s}} (s + m_1^2 - m_2^2) \tag{1.7.8}$$

for the centre-of-mass energy of particle 1 in terms of s. Likewise we find

$$\left.\begin{aligned} E_2 &= \frac{1}{2\sqrt{s}}(s+m_2^2-m_1^2) \\ E_3 &= \frac{1}{2\sqrt{s}}(s+m_3^2-m_4^2) \\ E_4 &= \frac{1}{2\sqrt{s}}(s+m_4^2-m_3^2) \end{aligned}\right\} \tag{1.7.9}$$

Then from (1.7.8) and (1.7.4) we get

$$q_{s12}^2 = E_1^2 - m_1^2 = \frac{1}{4s}[s-(m_1+m_2)^2][s-(m_1-m_2)^2] \tag{1.7.10}$$

It is convenient to introduce the 'triangle function'

$$\lambda(x,y,z) \equiv x^2+y^2+z^2-2xy-2yz-2xz \tag{1.7.11}$$

so that
$$q_{s12}^2 = \frac{1}{4s}\lambda(s,m_1^2,m_2^2) \tag{1.7.12}$$

and similarly we find $\quad q_{s34}^2 = \dfrac{1}{4s}\lambda(s,m_3^2,m_4^2)$

We next introduce the invariant

$$t = (p_1-p_3)^2 = (p_4-p_2)^2 \tag{1.7.13}$$

This is evidently the square of the total centre-of-mass energy in the t channel, remembering that we have to change the sign of p_3 and p_2 on crossing. For this process we have

$$E_1 = \frac{1}{2\sqrt{t}}(t+m_1^2-m_3^2) \tag{1.7.14}$$

$$q_{t13}^2 = \frac{1}{4t}\lambda(t,m_1^2,m_3^2) \quad \text{etc.} \tag{1.7.15}$$

and the threshold occurs at $t = (m_1+m_3)^2$. However, as far as the s channel is concerned t represents the momentum transferred in the scattering process, i.e. the difference between the momenta of particles 1 and 3. So from (1.7.13), using (1.7.2) and (1.7.3)

$$\begin{aligned} t &= m_1^2+m_3^2-2p_1\cdot p_3 \\ &= m_1^2+m_3^2-2E_1E_3+2\mathbf{q}_{s12}\cdot\mathbf{q}_{s34} \\ &= m_1^2+m_3^2-2E_1E_3+2q_{s12}q_{s34}\cos\theta_s \end{aligned} \tag{1.7.16}$$

where θ_s is the scattering angle between the directions of motion of particles 1 and 3 in the s-channel centre-of-mass system (fig. 1.4(*a*)). And on substituting (1.7.8) and (1.7.9) we get

$$z_s \equiv \cos\theta_s = \frac{s^2+s(2t-\Sigma)+(m_1^2-m_2^2)(m_3^2-m_4^2)}{4sq_{s12}q_{s34}}$$

$$= \frac{s^2+s(2t-\Sigma)+(m_1^2-m_2^2)(m_3^2-m_4^2)}{\lambda^{\frac{1}{2}}(s,m_1^2,m_2^2)\,\lambda^{\frac{1}{2}}(s,m_3^2,m_4^2)} \tag{1.7.17}$$

from (1.7.12), (1.7.13), where we have defined

$$\Sigma = m_1^2+m_2^2+m_3^2+m_4^2 \tag{1.7.18}$$

Similarly, as far as the t-channel is concerned s represents the momentum transfer and we find

$$z_t \equiv \cos\theta_t = \frac{t^2+(2s-\Sigma)+(m_1^2-m_3^2)(m_2^2-m_4^2)}{4tq_{t13}q_{t24}}$$

$$= \frac{t^2+t(2s-\Sigma)+(m_1^2-m_3^2)(m_2^2-m_4^2)}{\lambda^{\frac{1}{2}}(t,m_1^2,m_3^2)\,\lambda^{\frac{1}{2}}(t,m_2^2,m_4^2)} \tag{1.7.19}$$

Finally, for the u-channel process the centre-of-mass energy squared is

$$u \equiv (p_1-p_4)^2 = (p_3-p_2)^2 = m_1^2+m_4^2-2p_1\cdot p_4 \tag{1.7.20}$$

and we can write down similar expressions for the energies, momenta and scattering angle of the particles in this channel.

However, we know from section 1.4 that the four-line amplitude depends only on two independent invariants, so there must be a relation between s, t and u. In fact, combining (1.7.6), (1.7.16) and (1.7.20) we find

$$s+t+u = m_1^2+m_2^2+m_3^2+m_4^2+2m_1^2+2p_1\cdot(p_2-p_3-p_4)$$

but momentum conservation requires $p_1+p_2 = p_3+p_4$, and using (1.7.4), (1.7.18) we get

$$s+t+u = \Sigma \tag{1.7.21}$$

We shall usually work with s and t as the independent variables.

These formulae greatly simplify for equal-mass scattering $m_1 = m_2 = m_3 = m_4$ since

$$\lambda^{\frac{1}{2}}(s,m^2,m^2) = [s(s-4m^2)]^{\frac{1}{2}}$$

giving

$$\left.\begin{aligned} q_{s12}^2 = q_{s34}^2 = \frac{s-4m^2}{4}; \quad z_s = 1+\frac{2t}{s-4m^2} = -1-\frac{2u}{s-4m^2} \\ q_{t13}^2 = q_{t24}^2 = \frac{t-4m^2}{4}; \quad z_t = 1+\frac{2s}{t-4m^2} = -1-\frac{2u}{t-4m^2} \end{aligned}\right\} \tag{1.7.22}$$

The physical region for the s channel is given by

$$s \geqslant \max\{(m_1+m_2)^2, (m_3+m_4)^2\}$$

(i.e. the threshold for the process) and $-1 \leqslant \cos\theta_s \leqslant 1$. This boundary is conveniently expressed by the function

$$\phi(s,t) \equiv 4sq_{s12}^2 q_{s34}^2 \sin^2\theta_s = 0 \tag{1.7.23}$$

which using (1.7.12), (1.7.13), (1.7.17) and a little algebra gives

$$\begin{aligned}\phi(s,t) = stu - s(m_1^2-m_3^2)(m_2^2-m_4^2) - t(m_1^2-m_2^2)(m_3^2-m_4^2)\\ -(m_1^2m_4^2-m_3^2m_2^2)(m_1^2+m_4^2-m_3^2-m_2^2) = 0\end{aligned} \tag{1.7.24}$$

or

$$\begin{vmatrix} 0 & 1 & 1 & 1 & 1 \\ 1 & 0 & m_2^2 & t & m_1^2 \\ 1 & m_2^2 & 0 & m_3^2 & s \\ 1 & t & m_3^2 & 0 & m_4^2 \\ 1 & m_1^2 & s & m_4^2 & 0 \end{vmatrix} = 0 \tag{1.7.25}$$

Despite the unsymmetrical appearance of equation (1.7.24), we also find

$$\phi(s,t) = 4tq_{t13}^2 q_{t24}^2 \sin^2\theta_t = 4uq_{u14}^2 q_{u23}^2 \sin^2\theta_u \tag{1.7.26}$$

and so $\phi(s,t) = 0$ gives the boundaries of the physical regions for the s, t and u channels. For equal-mass scattering (1.7.24) reduces to $stu = 0$, so the boundaries are just the lines $s = 0$, $t = 0$ and $u = 0$. For unequal masses the boundary curves become asymptotic to these lines. Some examples are shown in fig. 1.5 where s, t and u are plotted subject to the constraint (1.7.21).

The various singularities may also be plotted on the Mandelstam diagram. Thus, if all the masses are equal we may expect bound state poles at $s = m^2$, $t = m^2$ and $u = m^2$, the two-particle branch point at s, t or $u = 4m^2$, and further thresholds at $9m^2$, $16m^2$ etc. due to 3, 4 and more particle intermediate states. For the more realistic $\pi N \rightarrow \pi N$ scattering we show in fig. 1.5(b) the nucleon pole and various resonances (ignoring isospin complications).

Because of the crossing property the nearby singularities in the t and u channels may be expected to control the behaviour of the s-channel scattering amplitude near the forward and backward directions ($z_s = \pm 1$ respectively). Thus in πN scattering there is a forward peak at $t = 0$ due to the $\pi\pi$ threshold branch cut, and in particular due to the dominant resonances, ρ, f etc., which occur in the $\pi\pi$ channel, and a backward peak for $u \approx 0$ due to the exchange of N, Δ and other

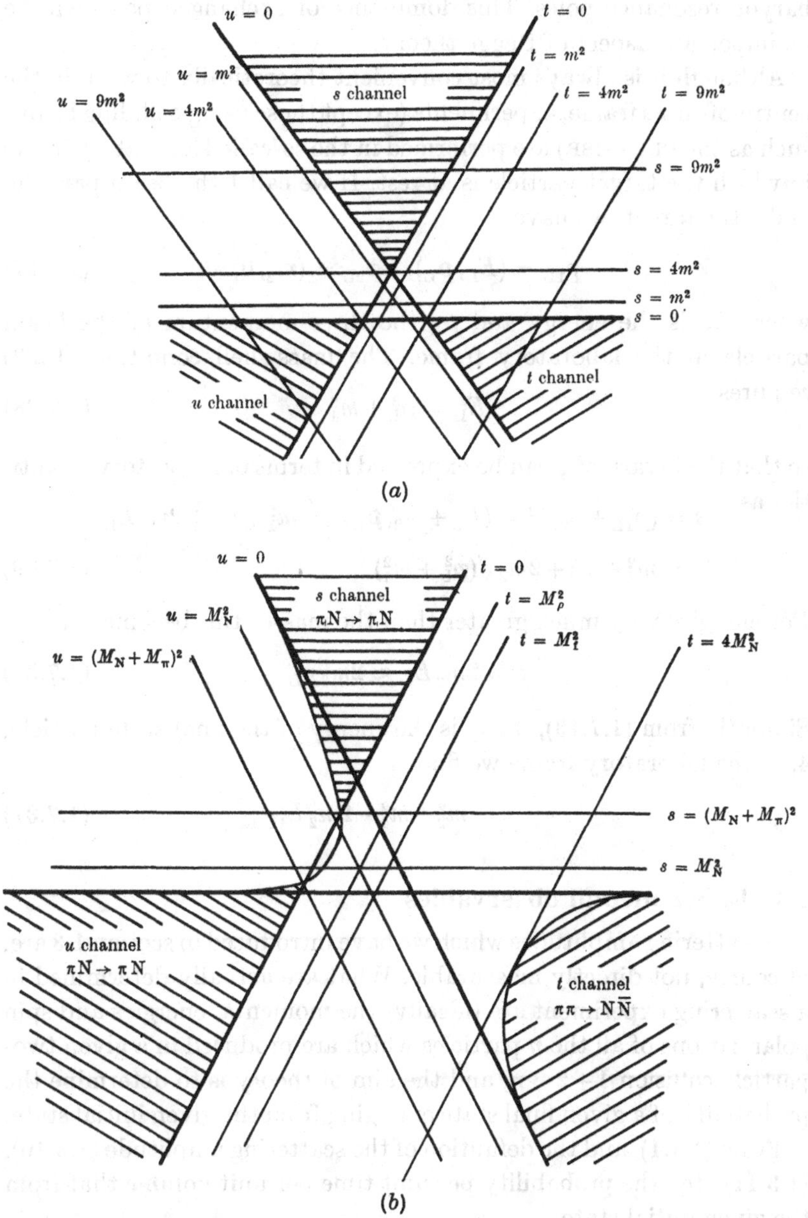

Fig. 1.5 (*a*) The Mandelstam *s–t–u* plot for equal mass-scattering, showing the positions of the pole at m^2, and the branch points at $4m^2$, $9m^2$, ... in each channel. The three physical regions are shown shaded. (*b*) The Mandelstam plot for πN scattering (ignoring isospin), showing the physical regions and some of the nearest singularities, the nucleon poles in the *s* and *u* channels, and the ρ and f poles in the *t* channel (not to scale).

baryon resonance poles. This dominance of exchanged poles will be an important aspect of Regge theory.

Although it is always most convenient theoretically to work in the centre-of-mass frame, experiments (except those using colliding beams such as the CERN–ISR) are performed in the so-called laboratory frame in which the target particle is at rest. If we call 1 the beam particle, and 2 the target, we have

$$p_{1L} = (E_L, \mathbf{p}_L), \quad p_{2L} = (m_2, 0) \tag{1.7.27}$$

where E_L is the energy and $\mathbf{p}_L$ the three-momentum of the beam particle in the laboratory frame. The mass-shell condition (1.2.3) requires

$$E_L^2 = \mathbf{p}_L^2 + m_1^2 \tag{1.7.28}$$

so that the invariant s can be expressed in terms of laboratory quantities as

$$\begin{aligned} s &= (p_{1L}+p_{2L})^2 = (E_L+m_2, \mathbf{p}_L)^2 = m_1^2+m_2^2+2m_2E_L \\ &= m_1^2+m_2^2+2m_2\sqrt{(p_L^2+m_1^2)} \end{aligned} \tag{1.7.29}$$

For energies very much greater than the masses this becomes

$$s \approx 2m_2E_L \approx 2m_2p_L \tag{1.7.30}$$

Similarly from (1.7.13), if E_{4L} is the energy of the final-state particle, 4, in the laboratory frame we find

$$t = m_2^2+m_4^2-2m_2E_{4L} \tag{1.7.31}$$

1.8 Experimental observables

The scattering amplitudes which we have introduced in section 1.3 are, of course, not directly measurable. What are actually determined in a scattering experiment are (ideally) the momenta, energies and spin polarizations of all the n particles which are produced in a given two-particle collision $1+2\to n$, and the aim of theory is to determine the probability of a given final state emerging from the given initial state.

From (1.1.1), and the definition of the scattering amplitude (1.3.10), (1.3.11) etc., the probability per unit time per unit volume that from the given initial state

$$|i\rangle = |P_1, P_2\rangle$$

we shall get the final state $|f_n\rangle = |P_1' \dots P_n'\rangle$ is the transition rate

$$R_{fi} = (2\pi)^4\delta^4(p_f-p_i)|\langle f_n|A|\mathrm{i}\rangle|^2 \tag{1.8.1}$$

The scattering cross-section, $\sigma_{12\to n}$, for this process is defined as the total transition rate per unit incident flux. The flux of incident particles, F, i.e. the number incident per unit area per unit time, is just given by the relative velocity of the two particles, $|\mathbf{v}_1-\mathbf{v}_2|$, divided by the invariant normalization volume V, i.e. the volume of phase space occupied by the two single particles, which from (1.2.11) is just

$$V = (2p_{01}2p_{02})^{-1}$$

So in the centre-of-mass system we have

$$F = 4E_1E_2\,|\mathbf{v}_1-\mathbf{v}_2| \tag{1.8.2}$$

The centre-of-mass velocities are, from (1.7.2),

$$\mathbf{v}_1 = \frac{q_{s12}}{E_1}, \quad \mathbf{v}_2 = -\frac{q_{s12}}{E_2} \tag{1.8.3}$$

so
$$F = 4E_1E_2\left(\frac{q_{s12}}{E_1}+\frac{q_{s12}}{E_2}\right) = 4(E_1+E_2)\,q_{s12} = 4(\sqrt{s})\,q_{s12} \tag{1.8.4}$$

which is, of course, invariant. To obtain the total transition rate we have to sum over all the possible final states $|f_n\rangle$ which contain the n particles, so

$$\begin{aligned}\sigma_{12\to n} &= \sum_f \frac{R_{fi}}{F} = \frac{1}{4q_{s12}\sqrt{s}}\sum_f (2\pi)^4\,\delta(p_f-p_i)\,|\langle f_n|\,A\,|\mathrm{i}\rangle|^2 \\ &= \frac{1}{4q_{s12}\sqrt{s}}\int \prod_{i=1}^{n}\frac{\mathrm{d}^4p'_i}{(2\pi)^3}\,\delta(p_i'^2-m_i^2)\,(2\pi)^4\,\delta^4\left(\sum_{i=1}^{n}p'_i-p_1-p_2\right) \\ &\qquad \times \sum_{\text{spins}} |\langle P'_i \ldots P'_n|\,A\,|P_1P_2\rangle|^2 \end{aligned} \tag{1.8.5}$$

where we have integrated over all possible momenta of the n final-state particles remembering the normalization (1.2.11), and (1.2.7). For the time being we shall continue to deal only with spinless particles, and drop the $\sum_{\text{spin}}$ and replace the P'_i by p'_i. The factor

$$\begin{aligned}\mathrm{d}\Phi_n &\equiv \prod_{i=1}^{n}\left(\frac{\mathrm{d}^4p'_i}{(2\pi)^3}\,\delta(p_i'^2-m_i^2)\right)(2\pi)^4\,\delta^4\left(\sum_{i=1}^{n}p'_i-p_1-p_2\right) \\ &= \prod_{i=1}^{n}\left(\frac{\mathrm{d}^3\mathbf{p}'_i}{2p_{i0}(2\pi)^3}\right)(2\pi)^4\,\delta^4\left(\sum_{i=1}^{n}p'_i-p_1-p_2\right)\end{aligned} \tag{1.8.6}$$

represents the volume of phase space available to the n final-state particles, and the integral in (1.8.5) is over this volume.

The total scattering cross-section for particles 1 and 2 is obtained by summing (1.8.5) over all possible final states containing different numbers of particles, viz.

$$\sigma_{12}^{\text{tot}} = \sum_{n=2}^{\infty} \sigma_{12\to n} \tag{1.8.7}$$

If there are only two particles, 3 and 4, in the final state, with centre-of-mass four-momenta given by (1.7.3) we have, from (1.8.5),

$$\sigma_{12\to 34} = \frac{1}{4q_{s12}(\sqrt{s})(2\pi)^2}\int \frac{\mathrm{d}^3\mathbf{p}_3\,\mathrm{d}^3\mathbf{p}_4}{2E_3\,.\,2E_4}|\langle p_3p_4|A|p_1p_2\rangle|^2\,\delta^4(p_3+p_4-p_1-p_2) \tag{1.8.8}$$

Since the three-momenta of the particles are equal and opposite in (1.7.3) we can use the δ-function in (1.8.8) to perform one of the integrations, leaving

$$\sigma_{12\to 34} = \frac{1}{4q_{s12}(\sqrt{s})(2\pi)^2}\int \frac{\mathrm{d}^3\mathbf{q}_{s34}}{2E_3\,.\,2E_4}\,\delta(E_3+E_4-\sqrt{s})\,|\langle p_3p_4|A|p_1p_2\rangle|^2 \tag{1.8.9}$$

We can express the momentum volume element in polar coordinates $\mathrm{d}^3\mathbf{q}_{s34} = q_{s34}^2\,\mathrm{d}q_{s34}\,\mathrm{d}\Omega$, where $d\Omega = \sin\theta_s\,\mathrm{d}\theta_s\,\mathrm{d}\phi$ is the element of solid angle associated with the direction of particle 3, say, the polar angles being defined with respect to the beam direction, the z axis. Then defining

$$E = E_3+E_4 = (m_3^2+q_{s34}^2)^{\frac{1}{2}}+(m_4^2+q_{s34}^2)^{\frac{1}{2}} \tag{1.8.10}$$

gives

$$\mathrm{d}E = \left(\frac{q_{s34}}{E_3}+\frac{q_{s34}}{E_4}\right)\mathrm{d}q_{s34} = \frac{q_{s34}E}{E_3E_4}\,\mathrm{d}q_{s34} \tag{1.8.11}$$

and so

$$\int \frac{q_{s34}^2\,\mathrm{d}q_{s34}}{E_3E_4}\,\delta(E-\sqrt{s}) = \int \frac{q_{s34}\,\mathrm{d}E}{E}\,\delta(E-\sqrt{s}) = \frac{q_{s34}}{\sqrt{s}} \tag{1.8.12}$$

and we end up with

$$\sigma_{12\to 34} = \frac{q_{s34}}{64\pi^2 s q_{s12}}\int |\langle p_3p_4|A|p_1p_2\rangle|^2\,\mathrm{d}\Omega \tag{1.8.13}$$

It is therefore useful to introduce the 'differential cross-section'

$$\frac{\mathrm{d}\sigma}{\mathrm{d}\Omega} = \frac{q_{s34}}{64\pi^2 s q_{s12}}\,|\langle p_3p_4|\,A\,|p_1p_2\rangle|^2 \tag{1.8.14}$$

which gives the probability of particle 3 being scattered into $\mathrm{d}\Omega$, per unit incident flux.

As we are at the moment only considering spinless particles the scattering probability will always be independent of the azimuthal angle ϕ, as there is nothing to select any particular direction perpendicular to the beam, and from (1.7.16) at fixed s

$$\mathrm{d}\Omega = \mathrm{d}(\cos\theta_s)\,\mathrm{d}\phi = \frac{\mathrm{d}t}{2q_{s12}q_{s34}}\mathrm{d}\phi \tag{1.8.15}$$

so, since $\int \mathrm{d}\phi = 2\pi$, we can more conveniently take as the differential cross-section

$$\frac{\mathrm{d}\sigma}{\mathrm{d}t} = \frac{1}{64\pi q_{s12}^2 s}|A(s,t)|^2 \tag{1.8.16}$$

In general we can obtain the partial (or differential) cross-sections with respect to any invariant simply by inserting a δ-function into (1.8.5). Thus defining $t' \equiv (p_1 - p_i')^2$ we have

$$\frac{\mathrm{d}\sigma}{\mathrm{d}t'} = \frac{1}{4q_{s12}\sqrt{s}}\int \prod_{i=1}^{n} \frac{\mathrm{d}^4 p_i'}{(2\pi)^3}\delta(p_i'^2 - m_i^2)\,(2\pi)^4\,\delta^4(\Sigma p_i' - p_1 - p_2)$$
$$\times\,\delta(t' - (p_1 - p_i')^2) \sum_{\text{spins}} |\langle P_i' \ldots P_n'|\, A\, |P_1 P_2\rangle|^2 \tag{1.8.17}$$

and clearly this can be repeated to give the partial cross-section with respect to any number of independent invariants.

1.9 The optical theorem

The total cross-section (1.8.7) satisfies a remarkable unitarity relation called the 'optical theorem' of which we shall make frequent use below.

The unitarity equation (1.2.14) reads, for a particular initial and final state,

$$(SS^\dagger)_{fi} = \sum_n S_{fn} S_{ni}^\dagger = \delta_{fi} \tag{1.9.1}$$

For elastic scattering $1+2 \rightarrow 1+2$ we have from (1.3.10)

$$S_{fi} = \delta_{fi} + \mathrm{i}(2\pi)^4\,\delta^4(p_f - p_i)\,\langle f|\, A\, |i\rangle \tag{1.9.2}$$

which with (1.3.13) gives us, from (1.9.1),

$$\mathrm{i}(\langle f|\, A^+ \,|i\rangle - \langle f|\, A^- \,|i\rangle) = -(2\pi)^4 \sum_n \delta^4(p_n - p_i)\,\langle f|\, A^+ \,|n\rangle \langle n|\, A^- \,|i\rangle \tag{1.9.3}$$

and if the initial and final states are identical we get (remembering (1.5.10))

$$2\,\mathrm{Im}\,\{\langle i|\, A\, |i\rangle\} = (2\pi)^4 \sum_n \delta^4(p_n - p_i)\,|\langle n|\, A^+ \,|i\rangle|^2 \tag{1.9.4}$$

But the right-hand side is the same as (1.8.7) with (1.8.5) apart from the flux factor so we obtain the relation

$$\sigma_{12}^{\text{tot}} = \frac{1}{2q_{s12}\sqrt{s}}\,\mathrm{Im}\,\{\langle i|\, A\, |i\rangle\} \tag{1.9.5}$$

Since the final state must be identical with the initial state, $\langle i|\, A\, |i\rangle$ is the forward elastic scattering amplitude $(1+2 \to 1+2)$ with the directions of motion of the particles unchanged, i.e. $\theta_s = 0$, which means (from (1.7.16) with $m_3 = m_1$, $m_2 = m_4$) that $t = 0$, so

$$\sigma_{12}^{\text{tot}} = \frac{1}{2q_{s12}\sqrt{s}} \operatorname{Im}\{A^{\text{el}}(s, t=0)\} \tag{1.9.6}$$

This optical theorem is well known in non-relativistic potential scattering (see for example Schiff (1968)) where it tells us that because of the conservation of probability the magnitude of the wave function in the 'shadow' behind the target at $(\theta_s = 0)$ must be reduced relative to the incoming wave by an amount equal to the total scattering in all directions. Equation (1.9.5) is just this same conservation requirement extended to the relativistic situation where particle creation can also occur. Note that it is only the elastic amplitude for $1+2 \to 1+2$ which appears on the right-hand side, but the total cross-section for $1+2 \to$ anything is on the left-hand side.

We can understand how this relation occurs diagrammatically from fig. 1.6, where the last step follows from (1.5.11) since we are taking the discontinuity of 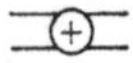 across the n-particle cut and summing over all possible intermediate states (compatible with four-momentum conservation). The real analyticity of A implies that

$$\operatorname{Disc}\{A\} = \operatorname{Im}\{A\}.$$

This optical theorem is one of the most useful constraints which unitarity imposes on a scattering amplitude. We shall also consider some generalizations in chapter 10.

1.10 Single-variable dispersion relations

According to our discussion in section 1.5 the only singularities which appear on the physical sheet are believed to be the poles corresponding to stable particles, and the threshold branch points. Thus, if we consider equal-mass scattering, and if we hold t fixed at some small, real, negative value (see fig. 1.5) in the s plane we find the singularities shown in fig. 1.7. On the right-hand side, for $\operatorname{Re}\{s\} > 0$, we have the s-channel bound-state pole and the various s-channel thresholds. On the left, for $\operatorname{Re}\{s\} < 0$, we meet the u-channel pole and the u-channel thresholds. The spacing between the two clearly depends on the relation (1.7.21)

$$s = 4m^2 - t - u \tag{1.10.1}$$

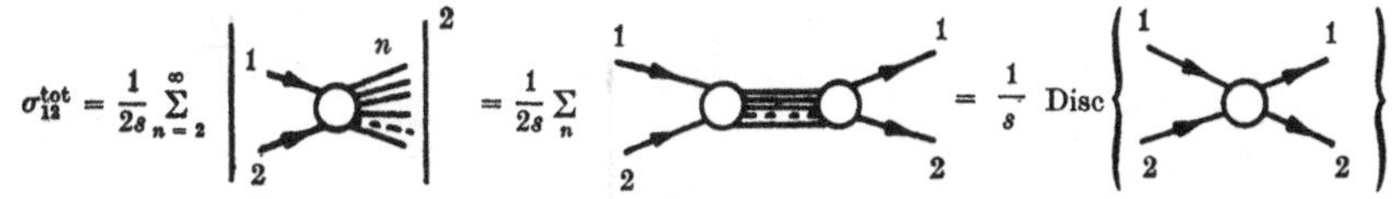

FIG. 1.6 The optical theorem. The factor $(2s)^{-1}$ is the large-s expression for the flux (1.8.4).

(and if we had taken t sufficiently negative these singularities would overlap).

We have drawn the branch cuts for the s-channel thresholds along the real axis towards $\text{Re}\{s\} \to +\infty$ (but slightly displaced for greater visibility), and the u singularities towards $\text{Re}(s) \to -\infty$. Thus the sheet we are looking at in fig. 1.7 (a) is the physical sheet on which the s-channel physical amplitude is obtained by approaching the real s axis from above, $\lim_{\epsilon \to 0} s + \text{i}\epsilon$, and similarly the u-channel amplitude is obtained from $\lim_{\epsilon \to 0} u + \text{i}\epsilon$, which corresponds to approaching the real s axis from below because of the relation (1.10.1).

We define the discontinuity functions

$$\left.\begin{aligned} D_s(s,t) &\equiv \frac{1}{2\text{i}}(A(s_+,t,u) - A(s_-,t,u)) \\ D_u(u,t) &\equiv \frac{1}{2\text{i}}(A(s,t,u_+) - A(s,t,u_-)) \end{aligned}\right\} \tag{1.10.2}$$

where $s_\pm \equiv s \pm \text{i}\epsilon$, and the discontinuity is taken across all the cuts. We have suppressed the third dependent variable in D_s and D_u. Because of the real analyticity of A (see section 1.5) we have

$$A(s^*,t,u) = A^*(s,t,u)$$

and so
$$D_s(s,t) = \text{Im}\{A(s,t,u)\} \tag{1.10.3}$$

along the s branch cuts and

$$D_u = \text{Im}\{A(s,t,u)\}$$

along the u branch cuts.

The idea of dispersion relations is simply to express the scattering amplitude in terms of the Cauchy integral formula

$$F(z) = \frac{1}{2\pi\text{i}}\oint \frac{\text{d}z'}{z'-z} F(z')$$

(see Titchmarsh 1939), so that

$$A(s,t,u) = \frac{1}{2\pi\text{i}}\oint \frac{\text{d}s'}{s'-s} A(s',t,u') \tag{1.10.4}$$

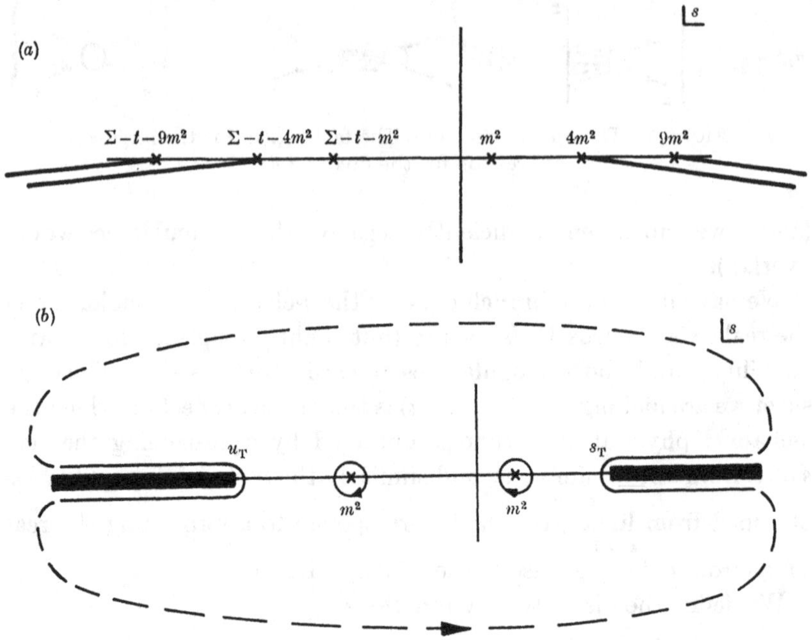

FIG. 1.7 (*a*) The physical-sheet singularities in s for fixed t ($\Sigma = 4m^2$). (*b*) The integration contour in the complex s plane, expanded to infinity but enclosing the cuts and poles on the real axis.

where the integral is evaluated over any closed anti-clockwise contour in the complex s plane enclosing the point s such that $A(s,t,u)$ is regular (holomorphic) inside and on that contour (fig. 1.7(*b*)). We then expand the contour so that it encircles the poles and encloses the branch cuts, as shown, giving

$$A(s,t,u) = \frac{g_s(t)}{m^2-s} + \frac{g_u(t)}{m^2-u} + \frac{1}{2\pi i}\int_C \frac{ds'}{s'-s} A(s',t,u) \qquad (1.10.5)$$

(Remember s' and u' are related by $s'+t+u' = 4m^2$.) Then if

$$|A(s,t,u)| \underset{s\to\infty}{\to} |s|^{-\epsilon}, \quad \epsilon > 0 \qquad (1.10.6)$$

the contribution from the circle at infinity will vanish, and we end up with

$$A(s,t,u) = \frac{g_s(t)}{m^2-s} + \frac{g_u(t)}{m^2-u} + \frac{1}{\pi}\int_{s_T}^{\infty} \frac{D_s(s',t)}{s'-s} ds' + \frac{1}{\pi}\int_{u_T}^{\infty} \frac{D_u(u',t)}{u'-u} du' \qquad (1.10.7)$$

where s_T and u_T are the s- and u-channel thresholds, respectively.

Such dispersion relations were originally derived for the scattering of light by free electrons by Kramers (1927) and Kronig (1926), and provide the crucial test of the analyticity assumptions which we introduced in section 1.5. They agree with experiment within the accuracy of the available experimental data (see for example Eden (1971)). Theoretically, they are of great importance because we have found that once we are given the particle poles all the other singularities of the scattering amplitudes and their discontinuities can be found from the unitarity equations (at least in principle). So the unitarity equations give us $\mathrm{Im}\{A\}$, but not $\mathrm{Re}\{A\}$. But, once we know all the discontinuities of an amplitude, by using dispersion relations we can determine the real part of the amplitude too, and so unitarity plus analyticity determines the amplitudes completely, given the particle poles.

However, the convergence requirement (1.10.6) is frequently not satisfied, in which case we have to resort to subtractions. Thus if we have (neglecting the other terms in (1.10.7) for simplicity)

$$A(s,t,u) = \frac{1}{\pi}\int_{s_{\mathrm{T}}}^{\infty} \frac{D_s(s',t)}{s'-s}\,\mathrm{d}s' \tag{1.10.8}$$

but the integrand diverges as $s' \to \infty$, we write instead a dispersion relation for $A(s,t,u)[(s-s_1)(s-s_2)\ldots(s-s_n)]^{-1}$ including sufficient terms in the bracket to ensure convergence (assuming a finite number will suffice). So

$$A(s,t,u)\prod_{i=1}^{n}(s-s_i)^{-1} = \sum_{j=1}^{n}\frac{A(s_j,t,u_j)}{(s-s_j)}\prod_{\substack{i=1\\ i\neq j}}^{n}(s_j-s_i)^{-1}$$

$$+\frac{1}{\pi}\int_{s_{\mathrm{T}}}^{\infty}\frac{D_s(s',t)}{(s'-s_1)\ldots(s'-s_n)(s'-s)} \tag{1.10.9}$$

since we pick up an extra contribution from each of the poles at $s = s_1, s_2, \ldots, s_n$. Hence

$$A(s,t,u) = F_{n-1}(s,t) + \frac{1}{\pi}\prod_{i=1}^{n}(s-s_i)\int_{s_{\mathrm{T}}}^{\infty}\frac{D_s(s',t)}{(s'-s_1)\ldots(s'-s_n)(s'-s)}\,\mathrm{d}s' \tag{1.10.10}$$

where $F_{n-1}(s,t)$ is an arbitrary polynomial in s of degree $n-1$, but now the integral converges if $D_s(s,t) \underset{s\to\infty}{\to} s^{n-\epsilon}, \epsilon > 0$. Thus the divergence problem is solved at the expense of introducing an arbitrary polynomial which is not determined (at least directly) by the unitarity equations. One of the main purposes of Regge theory is to close this gap by determining the subtractions.

A particularly useful form of these dispersion relations is for forward elastic scattering, such as $\pi N \to \pi N$, at $t = 0$ where $u = \Sigma - s$. From the optical theorem (1.9.5)

$$\left.\begin{aligned} D_s(s,0) &= \mathrm{Im}\{A^{\mathrm{el}}(s,0)\} = 2q_{s12}(\sqrt{s})\,\sigma_{12}^{\mathrm{tot}}(s) \\ D_u(u,0) &= \mathrm{Im}\{A^{\mathrm{el}}(u,0)\} = 2q_{u14}(\sqrt{u})\,\sigma_{14}^{\mathrm{tot}}(u) \end{aligned}\right\} \tag{1.10.11}$$

and these cross-sections will be identical if particles 2 and $\bar{2}\,(=4)$ are the same. It can be shown (section 2.4) that $\sigma_{12}^{\mathrm{tot}}(s) \underset{s\to\infty}{\to}$ constant (modulo possible $\log s$ factors) so only two subtractions are needed in (1.10.7). So making the subtractions at $s = 0$ we get (neglecting any pole contributions) for real s above the s-channel threshold

$$\begin{aligned} \mathrm{Re}\{A^{\mathrm{el}}(s,0)\} = a_0 + a_1 s + \frac{s^2}{\pi} P \int_{s_{\mathrm{T}}}^{\infty} \mathrm{d}s'\,(\sqrt{s'})\,q'_{s12}\,\sigma_{12}^{\mathrm{tot}}(s') \\ \times\left(\frac{1}{s'^2(s'-s)} + \frac{1}{(s'-\Sigma)^2\,(s'+s-\Sigma)}\right) \end{aligned} \tag{1.10.12}$$

(where $P \equiv$ principal value – see (1.5.2)). Thus a knowledge of the total cross-section (with guesses as to its behaviour for very large s where it has not been measured) allows us to find $\mathrm{Re}\{A(s,0)\}$ in terms of just two unknowns, the subtraction constants a_0 and a_1. Since $\mathrm{Re}\{A(s,0)\}$ can also be determined directly by Coulomb interference experiments (see for example Eden (1967)) the validity of these forward dispersion relations can be tested.

1.11 The Mandelstam representation

The single-variable dispersion relations were obtained by keeping one invariant fixed (t fixed in (1.10.4)) and representing the amplitude as a contour integral round the singularities in the other invariant (s). But $D_s(s,t)$ will have singularities in t, corresponding to the t-channel thresholds etc. Thus in fig. 1.8 (a) we display these t-channel exchanges in the s-channel unitarity equation. It will also have u-channel threshold branch points, but of course u is not an independent variable, through (1.7.21), and so at fixed positive s these will appear at negative t values (see fig. 1.5).

One expects these singularities to lie on the real t axis, and so one can write a dispersion relation for $D_s(s,t)$ similar to that for $A(s,t,u)$

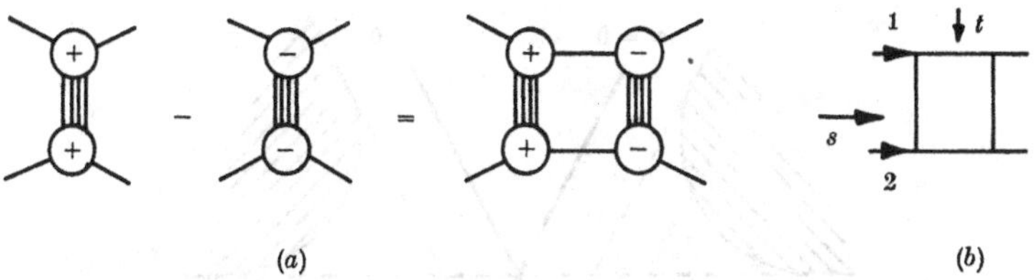

FIG. 1.8 (a) The contribution of t-channel intermediate states to the s-channel two-body unitary equation. (b) The 'box' diagram, the simplest diagram contributing to $\rho_{st}(s,t)$.

itself. We define the discontinuity of $D_s(s,t)$ across the t thresholds as

$$\rho_{st}(s,t) = \frac{1}{2\mathrm{i}}(D_s(s,t_+) - D_s(s,t_-)), \quad t > b_1(s) > 0 \qquad (1.11.1)$$

and across the u thresholds as

$$\rho_{su}(s,u) = \frac{1}{2\mathrm{i}}(D_s(s,u_+) - D_s(s,u_-)), \quad u > b_2(s) > 0 \qquad (1.11.2)$$

The boundary functions $b_{1,2}(s)$ are given by the position of the singularity of the lowest order diagram which contributes to ρ, usually the box diagram fig. 1.8 (b). We shall find in the next section that

$$b_1(s) = b_2(s) = 4m^2 + \frac{4m^4}{s-4m^2} \qquad (1.11.3)$$

for equal-mass kinematics, giving the boundaries shown in fig. 1.9. Hence we can write a dispersion relation at fixed s,

$$D_s(s,t) = \frac{1}{\pi}\int_{b_1(s)}^{\infty} \frac{\rho_{st}(s,t'')}{t''-t}\,\mathrm{d}t'' + \frac{1}{\pi}\int_{b_2(s)}^{\infty} \frac{\rho_{su}(s,u'')}{u''-u}\,\mathrm{d}u'' \qquad (1.11.4)$$

Similarly the u-discontinuity has branch cuts corresponding to the s- and t-thresholds, so we can write

$$D_u(u,t) = \frac{1}{\pi}\int_{b_1(u)}^{\infty} \frac{\rho_{tu}(u,t'')}{t''-t}\,\mathrm{d}t'' + \frac{1}{\pi}\int_{b_2(u)}^{\infty} \frac{\rho_{su}(s'',u)}{s''-s}\,\mathrm{d}s'' \qquad (1.11.5)$$

If these expressions are substituted into (1.10.7) (neglecting the pole terms for simplicity) we end up with

$$A(s,t,u) = \frac{1}{\pi^2}\iint^{\infty} \frac{\rho_{st}(s',t'')}{(s'-s)(t''-t)}\,\mathrm{d}s'\,\mathrm{d}t'' + \frac{1}{\pi^2}\iint^{\infty} \frac{\rho_{su}(s',u'')}{(s'-s)(u''-u')}\,\mathrm{d}s'\,\mathrm{d}u''$$
$$+ \frac{1}{\pi^2}\iint^{\infty} \frac{\rho_{tu}(u',t'')}{(u'-u)(t''-t)}\,\mathrm{d}u'\,\mathrm{d}t'' + \frac{1}{\pi^2}\iint \frac{\rho_{su}(s'',u')}{(u'-u)(s''-s')}\,\mathrm{d}u'\,\mathrm{d}s'' \qquad (1.11.6)$$

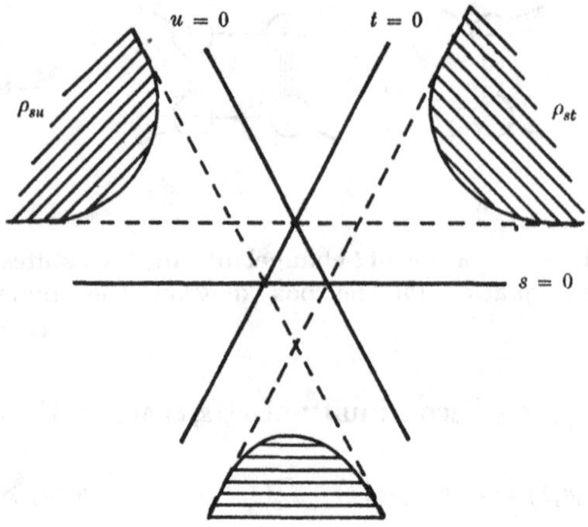

FIG. 1.9 The Mandelstam plot for equal-mass scattering (cf. fig. 1.5(*a*)), showing the double spectral functions (shaded areas). The boundary of ρ_{st} is given by (1.11.3).

It must be remembered that this relation, like (1.10.7), is written at fixed t, so that in the second and fourth terms we have to make use of the relations

$$s+t+u = s'+t+u' = \Sigma \tag{1.11.7}$$

in introducing primes into the variables which come from the denominators in (1.11.4) and (1.11.5). The primed variables are, of course, dummy variables of integration, so we are free to interchange primes in the fourth term, and then add it to the second term giving

$$\iint^{\infty} \rho_{su}(s', u'') \left(\frac{1}{(s'-s)(u''-u')} + \frac{1}{(u''-u)(s'-s'')}\right) ds'\, du'' \tag{1.11.8}$$

which can be rewritten, using (1.11.7), as

$$\iint^{\infty} \frac{\rho_{su}(s', u'')}{(s'-s)(u''-u)} ds'\, du''$$

so (1.11.6) becomes

$$A(s,t,u) = \frac{1}{\pi^2}\iint^{\infty} \frac{\rho_{st}(s', t'')}{(s'-s)(t''-t)} ds'\, dt'' + \frac{1}{\pi^2}\iint^{\infty} \frac{\rho_{su}(s', u'')}{(s'-s)(u''-u)} ds'\, du''$$
$$+ \frac{1}{\pi^2}\iint^{\infty} \frac{\rho_{tu}(u', t'')}{(u'-u)(t''-t)} du'\, dt'' \tag{1.11.9}$$

The functions ρ_{st}, ρ_{su}, ρ_{tu} are called 'double spectral functions', and (1.11.9) is a double dispersion relation. This representation of the scattering amplitude in terms of its double spectral functions is called the 'Mandelstam representation' (Mandelstam 1958, 1959). We do not know enough about the singularities of the scattering amplitude to be sure that such dispersion relations are valid. In particular we do not know that all the physical sheet singularities lie on the real axis. Indeed, it has been found that with diagrams where the masses of the intermediate states are smaller than those of the external states, anomalous thresholds appear at complex positions on the physical sheet, and the integration contour would have to make an excursion into the complex plane to include them. (A discussion of this problem may be found in Eden *et al.* (1966).) But it seems likely that (1.11.9) will at least be a good approximation for most practical purposes.

We chose to derive (1.11.9) from the fixed-t dispersion relation (1.10.7). However, the final result is symmetrical in the three variables s, t and u, and could equally well have been obtained starting from fixed-s or fixed-u dispersion relations. This is because the double spectral function is, from (1.11.1) and (1.10.2),

$$\rho_{st}(s,t) = \frac{1}{2\mathrm{i}}\left[\frac{1}{2\mathrm{i}}(A(s_+,t_+)-A(s_-,t_+))-\frac{1}{2\mathrm{i}}(A(s_+,t_-)-A(s_-,t_-))\right]$$

$$= -\tfrac{1}{4}(A(s_+,t_+)+A(s_-,t_-)-A(s_-,t_+)-A(s_+,t_-)) \qquad (1.11.10)$$

which can be taken to be

$$\frac{1}{2\mathrm{i}}(D_t(s_+,t)-D_t(s_-,t)) \quad \text{or} \quad \frac{1}{2\mathrm{i}}(D_s(s,t_+)-D_s(s,t_-)) \qquad (1.11.11)$$

There are two complications about the use of (1.11.9). There is the rather trivial point that we have omitted bound-state poles which may occur in any of the three channels, s, t or u. These should simply be added as necessary, as in (1.10.7). The more serious problem concerns the possible divergence of the integrand, as s', t'' etc. tend to infinity. Like (1.10.7), (1.11.9) is only defined up to the various subtractions which may be needed to make the integrals converge. We may thus be forced to introduce apparently arbitrary subtractions into the Mandelstam representation. However we shall find in the next chapter that the hypothesis of analytic continuation in angular momentum enables us to determine these subtractions too.

1.12 The singularities of Feynman integrals

We have remarked in section 1.5 that the unitarity equations imply that scattering amplitudes have similar singularities to the Feynman diagrams of perturbation quantum field theory. This is not surprising because such field theories give Lorentz invariant scattering amplitudes with the same sort of connectedness properties, and they also satisfy unitarity at least perturbatively. Of course, we do not expect such a perturbation approach to be valid for strong interactions where, since the couplings are not small, the perturbation series will not converge, and where we cannot apply the usual re-normalization techniques. However, one can hope to gain some insight into the form of strong interaction amplitudes from field-theoretical analogies.

The spin properties of the particles will not be very important for our purposes so we shall only consider spinless scalar mesons of mass m interacting through a Lagrangian $\mathscr{L}_{\text{int}} = g\phi^3$. The Feynman rules for such particles are very simple (see Bjorken and Drell 1965)). For a given diagram we include a factor $\mathrm{i}[(2\pi)^4(q^2-m^2+\mathrm{i}\epsilon)]^{-1}$ for each internal line of momentum q, a factor g for each vertex, a factor $(2\pi)^4\delta^4(q_1+q_2-q_3)$ for momentum conservation at each vertex $1+2\to 3$, and we integrate over the four-momenta of each internal line. The δ-functions mean that only closed loops have free momenta, however, and one δ-function of over-all energy-momentum conservation can be factored out in the definition of the scattering amplitude, as in (1.3.10).

Hence the contribution to the amplitude of the single particle exchange Born diagram fig. 1.10 (a) is just

$$\frac{g^2}{q^2-m^2+\mathrm{i}\epsilon}, \quad q^2 = (p_1+p_2)^2 \tag{1.12.1}$$

while that of the box diagram. fig. 1.10 (b) is

$$-\mathrm{i}\frac{g^4}{(2\pi)^4}\int \mathrm{d}^4k\{[(k+p_1)^2-m^2+\mathrm{i}\epsilon]\,[(k-p_2)-m^2+\mathrm{i}\epsilon]$$
$$\times[(k+p_1-p_3)^2-m^2+\mathrm{i}\epsilon]\,[k^2-m^2+\mathrm{i}\epsilon]\}^{-1} \tag{1.12.2}$$

And an arbitrary diagram gives (neglecting the normalization factors)

$$A \propto \int \frac{\mathrm{d}^4k_1 \ldots \mathrm{d}^4k_l}{\prod\limits_{i=1}^{n}(q_i^2-m_i^2+\mathrm{i}\epsilon)} \tag{1.12.3}$$

where the k_l are the independent loop momenta, and the q's are

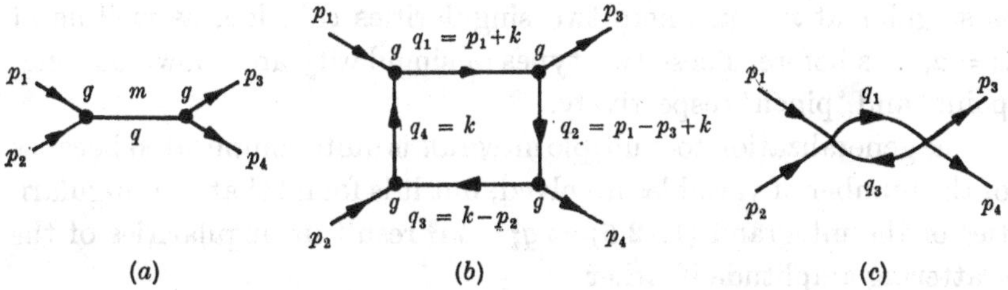

FIG. 1.10 (*a*) The Feynman diagram for single particle exchange in the *s*-channel. (*b*) The box diagram. (*c*) The contracted box diagram when the lines q_2 and q_4 are short-circuited by setting $\alpha_2 = \alpha_4 = 0$.

constrained by the δ-functions at each vertex. Using the Feynman relation

$$\frac{1}{u_1 u_2 \dots u_n} = (n-1)! \int_0^1 \mathrm{d}\alpha_1 \dots \mathrm{d}\alpha_n \frac{\delta\left(1-\sum\limits_{i=1}^{n}\alpha_i\right)}{\left[\sum\limits_{i=1}^{n}\alpha_i u_i\right]^n} \qquad (1.12.4)$$

we can rewrite (1.12.3) as

$$A \propto \int_0^1 \mathrm{d}\alpha_1 \dots \mathrm{d}\alpha_n \int \mathrm{d}^4k_1 \dots \mathrm{d}^4k_l \frac{\delta(1-\Sigma\alpha_i)}{[\Sigma\alpha_i(q_i^2-m_i^2)+\mathrm{i}\epsilon]^n} \qquad (1.12.5)$$

The singularities of such integrals are studied in detail in Eden *et al.* (1966). If a function $F(x)$ is represented by an integral such as

$$F(x) = \int_a^b f(x,z)\,\mathrm{d}z \qquad (1.12.6)$$

it will not necessarily have a singularity just because $f(x,z)$ does, since the contour of integration can be displaced in the complex z plane to avoid the singularity, and by Cauchy's theorem all such continuations are equivalent. Singularities arise for two reasons. (i) The singularity in $f(x,z)$ occurs at an end point of integration, a or b, so the contour cannot be deformed to avoid it. Thus

$$F(x) = \int_a^b \frac{1}{z-x}\,\mathrm{d}z = \log\left(\frac{b-x}{a-x}\right) \qquad (1.12.7)$$

is singular at $x = a$ or b. (ii) Two or more singularities of f approach the contour from different sides (or a singularity moves off to infinity), thus pinching the contour so that it cannot avoid them. Thus

$$F(x) = \int_a^b \frac{\mathrm{d}z}{(z-x)(z-x_0)} = \frac{1}{(x-x_0)} \log\left[\left(\frac{b-x_0}{a-x_0}\right)\left(\frac{b-x}{a-x}\right)\right] \qquad (1.12.8)$$

is singular at $x = x_0$ where two singularities coincide, as well as at $x = a, b$ as before. These two types of singularity are known as 'end-point' and 'pinch' respectively.

The generalization to multiple integrals is quite complicated because of the number of variables involved, but it is found that the singularities of the integrand (1.12.5) at $q_i^2 = m_i^2$ result in singularities of the scattering amplitude if either

$$q_i^2 = m_i^2 \quad \text{or} \quad \alpha_i = 0, \quad \text{for all} \quad i = 1, \ldots, n,$$

and
$$\frac{\partial}{\partial k_j} \sum_{i=1}^{n} \alpha_i (q_i^2 - m_i^2) = 0 \quad \text{for} \quad j = 1, \ldots, l$$

But since (see for example (1.12.2)) each q is linear in the k's the latter condition is equivalent to $\sum_i \alpha_i q_i = 0$ for each loop j. These are the Landau equations (1.5.14).

Thus for the box diagram fig. 1.10(*b*) we have either $q_i^2 = m_i^2$ or $\alpha_i = 0$ for $i = 1, \ldots, 4$ and

$$\alpha_1 q_1 + \alpha_2 q_2 + \alpha_3 q_3 + \alpha_4 q_4 = 0 \tag{1.12.9}$$

To take any $\alpha_i = 0$ is equivalent to removing that line from consideration, so for example if $\alpha_2, \alpha_4 = 0$ we have fig. 1.10(*c*). This requires $q_i^2 = q_3^2 = m^2$ and $\alpha_1 q_1 + \alpha_2 q_3 = 0$ so $q_1 = -q_3$ and the singularity is at $s = (q_1 - q_3)^2 = 4q_1^2 = 4m^2$, i.e. at the threshold. If none of the α's vanish (1.12.9) must hold. Multiplying (1.12.9) successively by each of the q_i ($i = 1, \ldots, 4$) gives us four linear equations for the α's, and a solution with $\alpha_i \neq 0$ is possible only if the determinant of the coefficients vanishes, i.e.

$$\det(q_i \cdot q_j) = 0, \quad i, j = 1, \ldots, 4$$

Since $s = (q_1 - q_3)^2$ and $t = (q_2 - q_4)^2$ we find the singularity is at

$$(s - 4m^2)(t - 4m^2) = 4m^4 \tag{1.12.10}$$

This is the boundary of the Mandelstam double spectral function (1.11.3), because it gives us the curve where the discontinuity across the s-threshold cut has a discontinuity in t due to the t-threshold. Note that as $s \to \infty$ this boundary moves to the threshold at $t = 4m^2$. More complex singularities, involving larger numbers of particles in the intermediate states, will occur at larger values of the invariants. We shall not pursue the subject further here, and readers seeking a more detailed discussion should consult Eden *et al.* (1966). We shall want to make use of some of these results below.

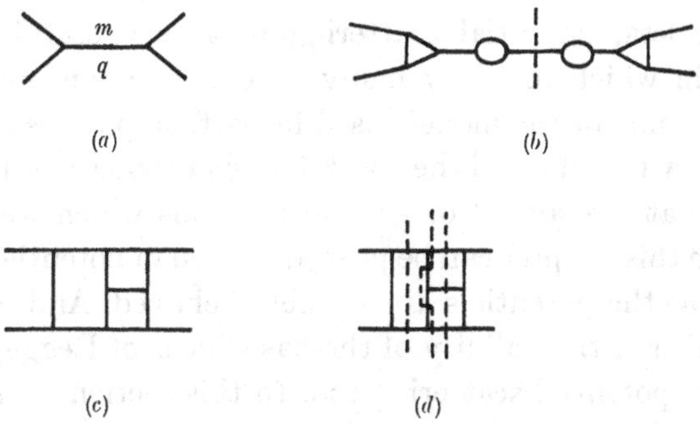

FIG. 1.11 (*a*) The unitarity diagram for single particle exchange giving a pole discontinuity of the form $\delta(q^2-m^2)$. (*b*) One of the (infinite) set of Feynman diagrams which, when cut across the single-particle propagator as shown by the dashed line contributes to the discontinuity in (*a*). (*c*) A Feynman diagram. (*d*) Three different ways of cutting (*c*) showing that it contributes to the two-, three- and four-particle unitarity diagrams.

It should be noted that the correspondence between Feynman diagrams and unitarity diagrams is always many-to-one. Thus the single particle exchange unitarity diagram fig. 1.11 (*a*) corresponds to the discontinuity of the sum of the infinite sequence of Feynman diagrams like fig. 1.11 (*b*) which give the re-normalization of the vertices, and of the mass of the exchanged particle. And a more complicated Feynman diagram like fig. 1.11 (*c*) will contribute to several different unitary diagrams because the discontinuity across this diagram can be taken in different ways as in fig. 1.11 (*d*). This must be borne in mind when interpreting Feynman-diagram models for strong interaction processes.

1.13 Potential scattering

It is rather obvious that non-relativistic potential-scattering theory can have at most limited relevance to particle physics. This is not just a matter of the failure to incorporate relativistic kinematics, but because the very idea of a potential which is a function of the spatial co-ordinates is very difficult to generalize to the relativistic situation. In fact the occurrence of a local causal interaction through a potential field always implies, because of Lorentz invariance, radiation of the field quanta too. And in particle physics, except at very low energies, it is always likely that inelastic processes involving the production of new particles will occur, which clearly cannot readily be incorporated into the framework of potential scattering.

None the less, potential scattering is a very useful theoretical laboratory in which to study many aspects of quantum scattering theory, and some of the models used in particle physics are founded on analogies with potential theory. For our purposes it is particularly important that the sort of dispersion relations which we have been discussing in this chapter can be proved to hold in potential scattering provided that the potentials are suitably behaved. And in chapter 3 we shall find that the validity of the basic ideas of Regge theory can be proved in potential scattering too. In this section we shall try to bring out the similarities between the singularity structure of Yukawa potential-scattering amplitudes and those of the strong-interaction S-matrix.

The Schroedinger equation for two particles interacting via a local potential $V(r)$, in the centre-of-mass system, is (Schiff 1968)

$$\left[\frac{\hbar^2}{2M}\nabla^2+\frac{\hbar^2}{2M}k^2-V(r)\right]\psi(\boldsymbol{r})=0 \tag{1.13.1}$$

where k is the wave number (energy $E=\hbar^2k^2/2M$), and M is the reduced mass. It is convenient to introduce

$$U(r)=V(r)\frac{2M}{\hbar^2} \tag{1.13.2}$$

so that (1.13.1) becomes

$$(\nabla^2+k^2-U(r))\,\psi(\boldsymbol{r})=0 \tag{1.13.3}$$

The initial state is represented by a plane wave, wave vector $\boldsymbol{k}$, along the z axis (fig. 1.12)

$$\psi(\boldsymbol{r})=\mathrm{e}^{\mathrm{i}\boldsymbol{k}\cdot\boldsymbol{r}}=\mathrm{e}^{\mathrm{i}kz} \tag{1.13.4}$$

and we seek a solution to this equation subject to the boundary condition that as $r\to\infty$

$$\psi(\boldsymbol{r})\to\mathrm{e}^{\mathrm{i}kz}+A(\boldsymbol{k},\boldsymbol{k}')\frac{\mathrm{e}^{\mathrm{i}\boldsymbol{k}'\cdot\boldsymbol{r}}}{r} \tag{1.13.5}$$

where the second term is the outgoing scattered wave, with wave vector $\boldsymbol{k}'$ in the direction of unit vector $\hat{\boldsymbol{r}}$, and $A(\boldsymbol{k},\boldsymbol{k}')$ is the scattering amplitude. For elastic scattering $|\boldsymbol{k}|=|\boldsymbol{k}'|=k$.

The solution to (1.13.3) with the boundary condition (1.13.5) is given by the Lippman–Schwinger equation

$$\psi(\boldsymbol{r})=\mathrm{e}^{\mathrm{i}kz}+\int G_0(\boldsymbol{r},\boldsymbol{r}')\,U(r')\,\psi(\boldsymbol{r}')\,\mathrm{d}\boldsymbol{r}' \tag{1.13.6}$$

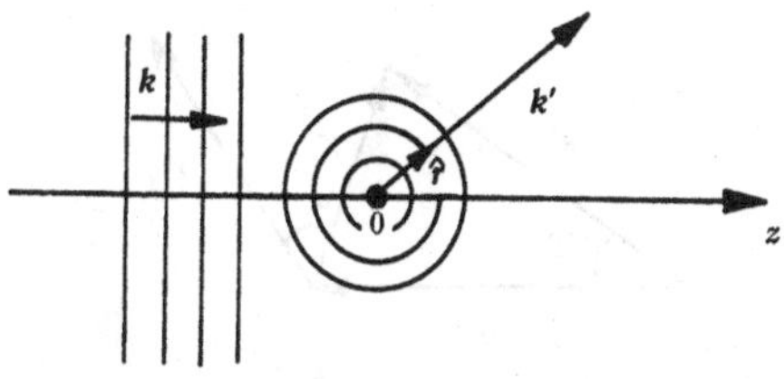

FIG. 1.12 Incident plane wave, wave vector $\boldsymbol{k}$ along z axis, scattered by a potential centred at $z = 0$ into the direction $\hat{\boldsymbol{r}}$, with wave vector $\boldsymbol{k}'$

where the Green's function is

$$G_0(\boldsymbol{r}, \boldsymbol{r}') \equiv -\frac{1}{4\pi}\frac{e^{ik|\boldsymbol{r}-\boldsymbol{r}'|}}{|\boldsymbol{r}-\boldsymbol{r}'|} \qquad (1.13.7)$$

That (1.13.6) is a solution of (1.13.3) may be checked by direct substitution, remembering that

$$\nabla^2\left(\frac{1}{|\boldsymbol{r}-\boldsymbol{r}'|}\right) = -4\pi\,\delta(\boldsymbol{r}-\boldsymbol{r}') \qquad (1.13.8)$$

And provided $r V(r) \underset{r\to\infty}{\rightarrow} 0$ we find, since $|\boldsymbol{r}-\boldsymbol{r}'| \approx r - \boldsymbol{r}'\cdot\hat{\boldsymbol{r}}$,

$$\psi(\boldsymbol{r}) \rightarrow e^{ikr} - \frac{e^{ikr}}{4\pi r}\int e^{-i\boldsymbol{k}'\cdot\boldsymbol{r}'}U(r')\,\psi(\boldsymbol{r}')\,d\boldsymbol{r}' \qquad (1.13.9)$$

which by comparison with (1.13.5) gives

$$A(\boldsymbol{k}, \boldsymbol{k}') = -\frac{1}{4\pi}\int e^{i\boldsymbol{k}'\cdot\boldsymbol{r}'}U(r')\,\psi(\boldsymbol{r}')\,d\boldsymbol{r}' \qquad (1.13.10)$$

The Born approximation, appropriate at high energies, is obtained by approximating $\psi(\boldsymbol{r}')$ in (1.3.10) by the incoming plane wave (1.13.4), assuming the scattering to be small, giving

$$A^B(\boldsymbol{k}, \boldsymbol{k}') = -\frac{1}{4\pi}\int e^{i(\boldsymbol{k}-\boldsymbol{k}')\cdot\boldsymbol{r}'}U(r')\,d\boldsymbol{r}' \qquad (1.13.11)$$

It is convenient to introduce (like our previous notation) $s = k^2$ for the total energy (in units where $\hbar^2 = 2M = 1$), and

$$t = -K^2 = -(\boldsymbol{k}-\boldsymbol{k}')^2 = -2k^2(1-\cos\theta)$$

where $\boldsymbol{K}$ is the momentum transfer vector. Then

$$A^B(\boldsymbol{k}, \boldsymbol{k}') \equiv A^B(s, t) = -\frac{1}{4\pi}\int e^{i\boldsymbol{K}\cdot\boldsymbol{r}'}U(r')\,d\boldsymbol{r}' \qquad (1.13.12)$$

Then putting $$\int d\boldsymbol{r}' = \int_0^\infty r'^2\,dr'\int_0^\pi \sin\alpha\,d\alpha\int_0^{2\pi} d\beta \qquad (1.13.13)$$

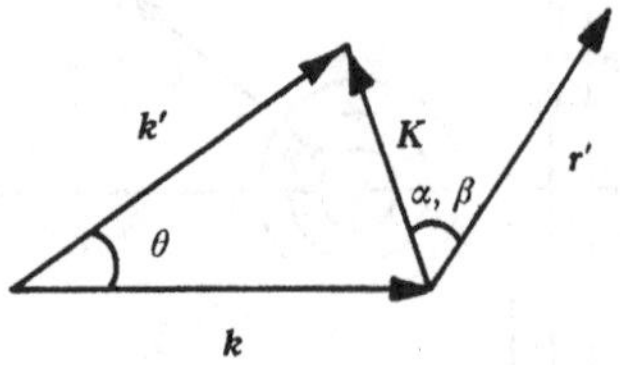

FIG. 1.13 The wave vectors $|\boldsymbol{k}| = |\boldsymbol{k}'|$ so $|\boldsymbol{K}| = 2|\boldsymbol{k}|\sin\frac{1}{2}\theta$. The angles α, β are the polar angles of $\boldsymbol{r}'$ with respect to the $\boldsymbol{K}$ axis.

and $\boldsymbol{K}.\boldsymbol{r}' = Kr'\cos\alpha$, where α, β are polar angles about the $\boldsymbol{K}$ axis (fig. 1.13), the angular integration is readily performed, since $U = U(r')$ only, giving

$$A^{\mathrm{B}}(s,t) = -\frac{1}{K}\int_0^\infty \sin(Kr')\,U(r')\,r'\,\mathrm{d}r' \qquad (1.13.14)$$

The simplest form of potential which has the short-range character appropriate to strong interactions is the Yukawa potential

$$U(r) = g^2\frac{\mathrm{e}^{-\mu r}}{r} \qquad (1.13.15)$$

where g^2 is the coupling strength and μ^{-1} is the range, for which we find

$$A^{\mathrm{B}}(s,t) = \frac{g^2}{\mu^2+K^2} = \frac{g^2}{\mu^2-t} \qquad (1.13.16)$$

So the Born approximation to the Yukawa scattering amplitude is just a pole at $t = \mu^2$ whose residue is given by the coupling strength. Of course if we have more complicated potentials the analyticity properties will not be so simple, but a large class of potentials can be represented by a superposition of Yukawa's

$$U(r) = \frac{1}{r}\int_m^\infty \rho(\mu)\,\mathrm{e}^{-\mu r}\,\mathrm{d}\mu \qquad (1.13.17)$$

where ρ is a weight function, giving

$$A^{\mathrm{B}}(s,t) = \int_m^\infty \mathrm{d}\mu\,\frac{\rho(\mu)}{\mu^2-t} \qquad (1.13.18)$$

which is obviously holomorphic in s, and cut in t for $t = m^2 \to \infty$.

To proceed further we note that since

$$(\nabla^2+k^2)\,\mathrm{e}^{\mathrm{i}\boldsymbol{k}\cdot\boldsymbol{r}} = 0 \qquad (1.13.19)$$

(1.13.3) can be written

$$(\nabla^2+k^2)\,\psi = (\nabla^2+k^2)\,\mathrm{e}^{\mathrm{i}\boldsymbol{k}\cdot\boldsymbol{r}} + U\psi \qquad (1.13.20)$$

and so formally
$$\psi = e^{i\boldsymbol{k}\cdot\boldsymbol{r}} + \frac{1}{\nabla^2 + k^2} U\psi \tag{1.13.21}$$

which by successive re-substitution becomes

$$\psi = e^{i\boldsymbol{k}\cdot\boldsymbol{r}} + \frac{1}{\nabla^2 + k^2} U e^{\boldsymbol{k}\cdot\boldsymbol{r}} + \frac{1}{\nabla^2 + k^2} U \frac{1}{\nabla^2 + k^2} U e^{i\boldsymbol{k}\cdot\boldsymbol{r}} + \dots \tag{1.13.22}$$

and so in (1.13.10) we get

$$A(\boldsymbol{k}, \boldsymbol{k}') = -\frac{1}{4\pi} \int e^{-i\boldsymbol{k}'\cdot\boldsymbol{r}'} U \left(e^{i\boldsymbol{k}\cdot\boldsymbol{r}'} + \frac{1}{\nabla^2 + k^2} U e^{i\boldsymbol{k}\cdot\boldsymbol{r}'} + \dots \right) d\boldsymbol{r}' \tag{1.13.23}$$

The first term is just the Born approximation (1.13.11) which we can denote by
$$A^{\mathrm{B}}(\boldsymbol{k}, \boldsymbol{k}') = \langle \boldsymbol{k}' | U | \boldsymbol{k} \rangle \tag{1.13.24}$$

where the states $|\boldsymbol{k}\rangle$ are momentum eigenstates such that

$$\nabla^2 |\boldsymbol{k}\rangle = -k^2 |\boldsymbol{k}\rangle.$$

Then using the completeness relation to write

$$\frac{1}{\nabla^2 + k^2} = \frac{1}{(2\pi)^3} \int |\boldsymbol{p}\rangle \frac{d^3\boldsymbol{p}}{k^2 - p^2} \langle \boldsymbol{p}| \tag{1.13.25}$$

the Born series (1.13.23) becomes

$$A(\boldsymbol{k}, \boldsymbol{k}') = \langle \boldsymbol{k}' | U | \boldsymbol{k} \rangle + \frac{1}{(2\pi)^3} \int \langle \boldsymbol{k}' | U | \boldsymbol{p} \rangle \frac{d^3\boldsymbol{p}}{k^2 - p^2} \{ \langle \boldsymbol{p} | U | \boldsymbol{k} \rangle + \dots \} \tag{1.13.26}$$

Since the term in brackets { } is just the Born expansion of $A(\boldsymbol{k}, \boldsymbol{p})$ we can rewrite (1.13.26) as the Lippman–Schwinger equation for the scattering amplitude

$$A(\boldsymbol{k}, \boldsymbol{k}') = A^{\mathrm{B}}(\boldsymbol{k}, \boldsymbol{k}') + \frac{1}{(2\pi)^3} \int A(\boldsymbol{k}, \boldsymbol{p}) \frac{d^3\boldsymbol{p}}{k^2 - p^2} A^{\mathrm{B}}(\boldsymbol{p}, \boldsymbol{k}') \tag{1.13.27}$$

which is represented diagrammatically in fig. 1.14.

For our Yukawa potential, using (1.13.16) for (1.13.24), (1.13.26) gives

$$A(\boldsymbol{k}, \boldsymbol{k}') = \frac{g^2}{\mu^2 + (k' - k)^2} + \frac{g^4}{(2\pi)^3} \int \frac{d^3\boldsymbol{p}}{[(k' - p^2 + \mu^2][k^2 - p^2][(p - k)^2 + \mu^2]} + \dots \tag{1.13.28}$$

a power series in the coupling constant which is reminiscent of the

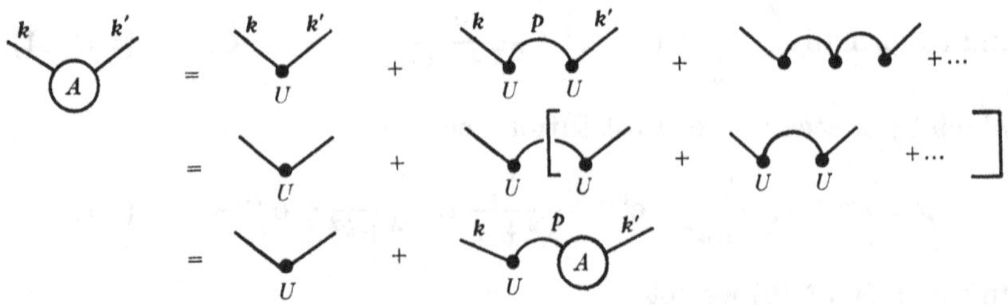

FIG. 1.14 Diagrammatic representation of the Lippman–Schwinger equation as a Born series in which the potential acts an arbitrary number of times.

Feynman rules for the diagrams in fig. 1.10, but of course in three dimensions. The second term has a cut in $k^2 = s$ for $k^2 > 0$ where the denominator $(p^2 - k^2)^{-1}$ vanishes. The first term has a pole at $t = \mu^2$; the second has a cut beginning at $t = 4\mu^2$, and in fact has a Mandelstam double spectral function boundary at

$$t = b(s) = 4\mu^2 + \frac{\mu^4}{s} \tag{1.13.29}$$

Thus Yukawa potential scattering, or simple generalizations like (1.13.18), have a singularity structure very similar to that of ϕ^3 quantum field theory. The principal differences are of course the absence of u-channel singularities (which would correspond to a Majorana type of exchange potential), the absence of inelastic thresholds in s, and the fact that the elastic threshold branch point is at $s = 0$ because we are using the non-relativistic kinematics $s = E = k^2$, rather than the relativistic $s = E^2 = k^2 + m^2$.

1.14 The eikonal expansion*

A useful approximation method, which we shall make use of in chapter 8, is the so-called 'eikonal' expansion of the scattering amplitude. It can readily be derived in potential scattering where it is appropriate for energies much greater than the interaction potential, i.e. $E \gg V$, or $k^2 \gg U$ in (1.13.3) (see Glauber 1959, Jochain and Quigg 1974).

In this situation we expect that there will be very little scattering in the backward direction, and so we can write the solution of (1.13.3) as

$$\psi(\boldsymbol{r}) = e^{i\boldsymbol{k}\cdot\boldsymbol{r}}\phi(\boldsymbol{r}) \tag{1.14.1}$$

* This section may be omitted at first reading.

where $\phi(\boldsymbol{r})$ represents the modulation of the incoming wave caused by the potential. When (1.14.1) is substituted in (1.13.6) the equation for $\phi(\boldsymbol{r})$ becomes

$$\phi(\boldsymbol{r}) = 1 - \frac{1}{4\pi}\int e^{ik|\boldsymbol{r}-\boldsymbol{r}'| - i\boldsymbol{k}\cdot(\boldsymbol{r}-\boldsymbol{r}')} U(\boldsymbol{r}')\,\phi(\boldsymbol{r}')\,(|\boldsymbol{r}-\boldsymbol{r}'|)^{-1}\mathrm{d}\boldsymbol{r}'$$

$$= 1 - \frac{1}{4\pi}\int e^{ikr''(1-\cos\theta'')} U(\boldsymbol{r}-\boldsymbol{r}'')\,\phi(\boldsymbol{r}-\boldsymbol{r}'')\,r''\,\mathrm{d}r''\,\mathrm{d}(\cos\theta'')\,\mathrm{d}\phi'' \tag{1.14.2}$$

where in the last step we have introduced the vector $\boldsymbol{r}'' \equiv \boldsymbol{r} - \boldsymbol{r}'$, and θ'', ϕ'' are the polar angles of $\boldsymbol{r}''$ with respect to the direction of $\boldsymbol{r}$.

At high energies we can assume that the range over which $U\phi$ varies appreciably is much greater than the wavelength of the beam, λ, so we can perform the $\cos\theta''$ integration by parts, and neglect the second term, giving

$$\phi \approx 1 - \frac{1}{4\pi}\int\left(\frac{e^{ikr''(1-\cos\theta'')}}{-ikr''}\,U(\boldsymbol{r}-\boldsymbol{r}'')\,\phi(\boldsymbol{r}-\boldsymbol{r}'')\right)_{\cos\theta''=-1}^{\cos\theta''=1} r''\,\mathrm{d}r''\,\mathrm{d}\phi'' \tag{1.14.3}$$

However, the term with $\cos\theta'' = -1$ is very rapidly oscillating, and hence makes a very small contribution when we perform the integration over r'', and neglecting it we get a contribution only when $\boldsymbol{r}''$ is parallel to $\boldsymbol{k}$, i.e. along the z axis, and so (since $\int \mathrm{d}\phi'' = 2\pi$) (1.14.3) becomes

$$\phi \approx 1 - \frac{i}{2k}\int_{-\infty}^{z} U(x,y,z'')\,\phi(x,y,z'')\,\mathrm{d}z'' \tag{1.14.4}$$

for which the solution is

$$\phi(x,y,z) = \exp\left(-\frac{i}{2k}\int_{-\infty}^{z} U(x,y,z'')\,\mathrm{d}z''\right) \tag{1.14.5}$$

So if we resolve $\boldsymbol{r}$ into (see fig. 1.15)

$$\boldsymbol{r} = \boldsymbol{b} + \hat{\boldsymbol{k}}z$$

where $\boldsymbol{b}$ is a two-dimensional vector perpendicular to the unit vector $\hat{\boldsymbol{k}}$, we have

$$\psi(\boldsymbol{r}) = \exp\left[i\boldsymbol{k}\,.\,\boldsymbol{r} - \frac{i}{2k}\int_{-\infty}^{z} U(\boldsymbol{b}+\hat{\boldsymbol{k}}z'')\,\mathrm{d}z''\right] \tag{1.14.6}$$

which in (1.13.10) gives

$$A(\boldsymbol{k},\boldsymbol{k}') = -\frac{1}{4\pi}\int e^{-i\boldsymbol{k}'\cdot\boldsymbol{r}'}\,U(\boldsymbol{b}'+\hat{\boldsymbol{k}}z')$$

$$\times\exp\left(i\boldsymbol{k}\,.\,\boldsymbol{r}' - \frac{i}{2k}\int_{-\infty}^{z} U(\boldsymbol{b}'+\hat{\boldsymbol{k}}z'')\,\mathrm{d}z''\right)\mathrm{d}z'\,\mathrm{d}^2\boldsymbol{b}' \tag{1.14.7}$$

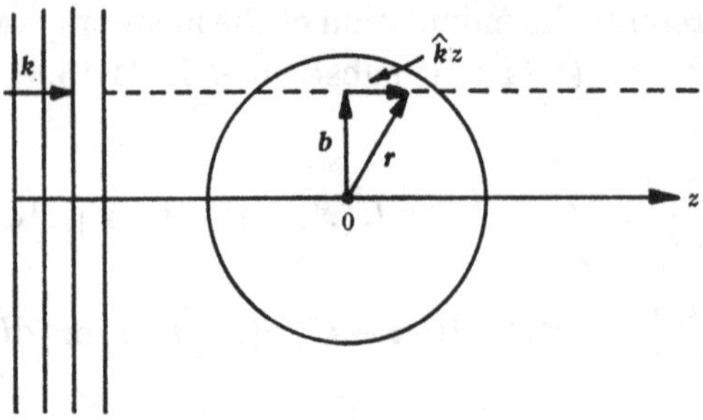

FIG. 1.15 Plane wave incident on a potential. $\boldsymbol{b}$ is the two-dimensional impact-parameter vector, perpendicular to z.

For small-angle scattering $(\boldsymbol{k}-\boldsymbol{k}').\hat{\boldsymbol{k}} \approx 0$, and in this approximation the z' integration is over an exact differential. That is because

$$\frac{\partial}{\partial z'}\left(\exp\left[-\int^{z'} U\mathrm{d}z''\right]\right) = -\left(\exp\left[-\int^{z'} U\mathrm{d}z''\right]\right) U\,\mathrm{d}z'$$

And so we obtain

$$A(\boldsymbol{k},\boldsymbol{k}') = \frac{\mathrm{i}k}{2\pi}\int \mathrm{e}^{\mathrm{i}\boldsymbol{k}\cdot\boldsymbol{b}'}(1-\mathrm{e}^{\chi(\boldsymbol{b}')})\,\mathrm{d}^2\boldsymbol{b}' \tag{1.14.8}$$

where we have introduced the 'eikonal function' defined by

$$\chi(\boldsymbol{b}) \equiv -\frac{1}{2k}\int_{-\infty}^{\infty} U(\boldsymbol{b}+\hat{\boldsymbol{k}}z'')\,\mathrm{d}z'' \tag{1.14.9}$$

For spherically symmetric potentials we can perform the angular integration in (1.14.8), since

$$\mathrm{d}^2\boldsymbol{b}' = b'\,\mathrm{d}b'\,\mathrm{d}\phi$$

$$\boldsymbol{K}.\boldsymbol{b}' = (2k\sin\tfrac{1}{2}\theta)\,b'\cos\phi = (\sqrt{t})\,b'\cos\phi$$

and (Magnus and Oberhettinger (1949) p. 26)

$$\frac{1}{2\pi}\int_0^{2\pi} \mathrm{e}^{\mathrm{i}x\cos\phi}\,\mathrm{d}\phi = J_0(x) \tag{1.14.10}$$

where J_0 is the zeroth order Bessel function, and obtain

$$A(\boldsymbol{k},\boldsymbol{k}') = -\mathrm{i}k\int_0^{\infty} J_0(b'\sqrt{-t})\,(\mathrm{e}^{\mathrm{i}\chi(b)}-1)\,b'\,\mathrm{d}b' \tag{1.14.11}$$

If the exponent is expanded powers of χ we get the eikonal series

$$A(\boldsymbol{k},\boldsymbol{k}') = -\mathrm{i}k\sum_n\int_0^{\infty} J_0(b'\sqrt{-t})\,\frac{(\mathrm{i}\chi)^n}{n!}\,b'\,\mathrm{d}b' \tag{1.14.12}$$

The eikonal function (1.14.9) can be expressed as the two-dimensional Fourier transform of the Born approximation (1.13.12) i.e.

$$\chi(\boldsymbol{b}) = \frac{1}{2\pi k}\int \mathrm{d}^2\boldsymbol{k}\, \mathrm{e}^{-i\boldsymbol{k}\cdot\boldsymbol{b}} A^{\mathrm{B}}(\boldsymbol{k}, \boldsymbol{k}')$$

$$= \frac{1}{2k}\int_{-\infty}^{0} J_0(b\sqrt{-t})\, A^{\mathrm{B}}(s,t)\, \mathrm{d}t \qquad (1.14.13)$$

and inverting (1.14.13) using (Magnus and Oberhettinger (1949) p. 35)

$$\int_0^\infty J_0(xy)\, J_0(x'y)\, \mathrm{d}y = \delta(x-x') \qquad (1.14.14)$$

we find

$$A^{\mathrm{B}}(s,t) = k\int_0^\infty \chi(b)\, J_0(b\sqrt{-t})\, b\, \mathrm{d}b \qquad (1.14.15)$$

which is just the first term in the series (1.14.12)

Thus the first term in the eikonal series is identical to the first term in the Born series (1.13.26) at high energies. The relationship between the higher order terms of the two series is more complicated (see Jochain and Quigg 1974) because for real potentials the eikonal series contains alternating real and imaginary terms, while in general all the terms of the Born series (except the first) are complex. But in the large k, fixed K, limit the two series agree. Thus the eikonal series can be regarded as an approximation to the sum of ladder diagrams (fig. 1.14) when each successive scattering is restricted to small angles only. We shall find that this is a very useful approximation in later work (see section 8.4).

2
The complex angular-momentum plane

2.1 Introduction

The new idea which Regge (1959, 1960) introduced into scattering theory was the importance of analytically continuing scattering amplitudes in the complex angular-momentum plane.

At first sight this seems rather a pointless procedure because in quantum mechanics the angular momentum of a system is restricted to integer multiples of $\hbar$ (or half-integer multiples if the particles have intrinsic spin). However, this quantization results mainly from the 'kinematics' of the process, from the invariance of the system under spatial rotations, and has little to do with the forces which determine the nature of the interaction. Thus in solving non-relativistic potential scattering problems one frequently begins by separating the Schroedinger equation into its angular and radial parts, so that one can concentrate on the radial equation (see section 3.3 below)

$$\frac{d^2\phi_l(r)}{dr^2}+\left(k^2-\frac{l(l+1)}{r^2}-U(r)\right)\phi_l(r)=0 \tag{2.1.1}$$

which contains the potential, and hence the dynamics of the interaction. The angular-momentum quantum number, l, appears simply as a parameter of this equation.

Normally, one would solve (2.1.1) only for the physically meaningful integer l values ($\geqslant 0$), but there is nothing to prevent us from considering unphysical, non-integer or indeed non-real values of l. We shall see why this is of some utility in potential scattering in the next chapter, but the basic ideas are much more general than potential scattering, and are in fact more useful in elementary-particle physics.

We begin this chapter by defining partial-wave amplitudes, and discuss some of their properties, and we then consider their continuation to complex values of angular momentum. We show that the singularities which occur in the angular-momentum plane are related to the asymptotic behaviour of the scattering amplitude, and so determine the subtractions needed in dispersion relations. It is found that moving poles in the angular-momentum plane give rise to composite particles (or resonances), so that the asymptotic behaviour of a scattering amplitude is determined by the particles which can be exchanged. This is one of the main tests of the applicability of Regge's ideas to particle physics, and provides the main topic for the rest of the book. It has also led to the introduction of the 'bootstrap hypothesis', that all strongly interacting particles may arise as a consequence of just analyticity and unitarity requirements.

2.2 Partial-wave amplitudes

In this chapter we shall only be concerned with $2 \to 2$ scattering, and will restrict ourselves to spinless particles, so that the total angular momentum of the initial state is just the relative orbital angular momentum of the two particles. Since angular momentum is a conserved quantity the orbital angular momentum of the final state must be the same as that of the initial state, so it is frequently convenient to consider the scattering amplitude for each individual angular-momentum state separately, i.e. the so-called 'partial-wave' amplitudes. However, the initial state will not in general be an eigenstate of angular momentum, but a sum over many possible angular-momentum eigenstates, and hence the total scattering amplitude will be a sum over all these partial-wave amplitudes.

For spinless particles the angular dependence of the wave function describing a state of orbital angular momentum l in the s channel is given by the Legendre function of the first kind $P_l(z_s)$ (see (A.3)). We work in the centre-of-mass system in which $z_s \equiv \cos\theta_s$ is given by (1.7.17), so at fixed s the scattering angle is just given by t (or u from (1.7.21)), so $t = t(z_s, s)$.

The centre-of-mass partial-wave scattering amplitude of angular momentum l in the s channel is defined from the total scattering amplitude by

$$A_l(s) = \frac{1}{16\pi}\frac{1}{2}\int_{-1}^{1} dz_s P_l(z_s)\, A(s, t(z_s, s)), \quad l = 0, 1, 2, \ldots \qquad (2.2.1)$$

The factor $(16\pi)^{-1}$ is purely a matter of convention and is included in order to simplify the unitarity equation (2.2.7) below. We can use the orthogonality relation (A.20) to invert (2.2.1) giving

$$A(s,t) = 16\pi \sum_{l=0}^{\infty} (2l+1)\, A_l(s)\, P_l(z_s) \qquad (2.2.2)$$

which is called the 'partial-wave series' for $A(s,t)$.

A great advantage of (2.2.2) is that at low values of s we expect only a few partial waves to contribute to the series because classically a particle with angular momentum $l > q_s R$ (where q_s is its momentum and R is the range of the force) would miss the target and so not be scattered. Thus, very approximately, with strong interactions of range about 1 fm, only S waves should be needed for $q_s \lesssim 200\,\mathrm{MeV}/c$, S, P waves for $q_s \lesssim 400\,\mathrm{MeV}/c$, and so on.

Another advantage is that each partial wave satisfies its own unitarity equation independent of the others. This can be deduced by substituting the partial-wave series (2.2.2) into the two-particle unitarity relation (1.5.7) to obtain

$$16\pi \sum_l (2l+1)\,(A_l^{if}(s_+) - A_l^{if}(s_-))\,P_l(z_s) = \frac{iq_{sn}}{16\pi^2\sqrt{s}}(16\pi)^2$$
$$\times \int_0^{2\pi} d\phi \int_{-1}^{1} dz' \sum_{l'} (2l'+1)\,A_{l'}^{in}(s_+)\,P_{l'}(z') \sum_{l''} (2l''+1) A_{l''}^{nf}(s_-)\,P_{l''}(z'') \tag{2.2.3}$$

where $z' \equiv \cos\theta_{in}$ is the cosine of the angle between the direction of motion of the particles in the initial state i and intermediate state n, and $z'' \equiv \cos\theta_{nf}$ is the corresponding angle between the intermediate and final states, and of course $z_s = \cos\theta_{if}$ (see fig. 2.1). The addition theorem of cosines gives

$$\cos\theta_{in} = \cos\theta_{if}\cos\theta_{fn} + \sin\theta_{if}\sin\theta_{fn}\cos\phi \tag{2.2.4}$$

where ϕ is the angle between the scattering planes of the processes $i \to n$ amd $n \to f$. The addition theorem for Legendre functions (Erdelyi *et al.* (1958) p. 168) is

$$P_l(z'') = P_l(z_s)\,P_l(z') + 2\sum_{m=1}^{l} (-1)^m \frac{\Gamma(l-m+1)}{\Gamma(l+m+1)} P_l^m(z_s)\,P_l^m(z')\cos m\phi \tag{2.2.5}$$

where $P_l^m(z)$ is the associated Legendre function of the first kind. The orthogonality relation (A.20) (using Erdelyi *et al.*, p. 171) gives

$$\int_0^{2\pi} d\phi \int_{-1}^{1} dz'\,P_{l'}(z')\,P_{l''}(z'') = \delta_{l'l''}\frac{4\pi}{2l'+1}P_{l'}(z_s) \tag{2.2.6}$$

so (2.2.3) becomes

$$A_l^{if}(s_+) - A_l^{if}(s_-) = \frac{4iq_{sn}}{\sqrt{s}}\,A_l^{in}(s_+)\,A_l^{nf}(s_-) \tag{2.2.7}$$

Thus only the given angular-momentum state l is involved in the unitarity relation. The absence of factors 16π is due to their inclusion in (2.2.1).

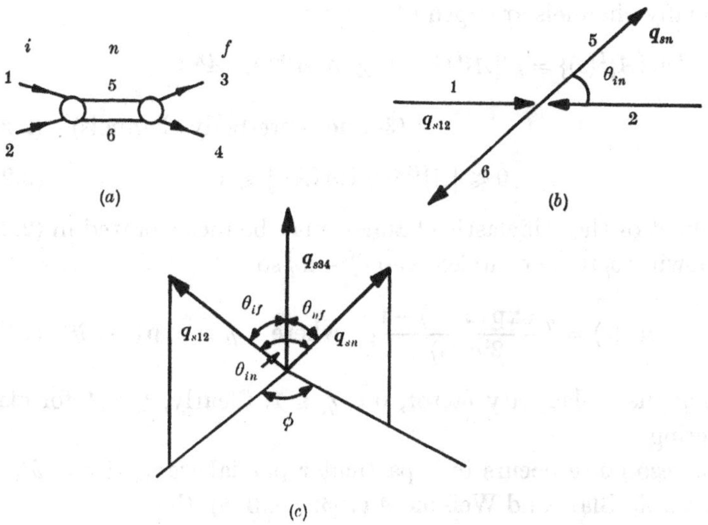

FIG. 2.1 (a) The two-body intermediate state $|n\rangle = 5+6$ in $1+2 \to 3+4$. (b) The centre-of-mass scattering angle θ_{in} in $1+2 \to 5+6$. (c) The scattering angles θ_{in}, θ_{nf} and θ_{if}. The angle ϕ is the azimuthal angle about the direction of $\mathbf{q}_{s34}$ between the plane containing $\mathbf{q}_{s12}$ and $\mathbf{q}_{s34}$ and the plane containing $\mathbf{q}_{s34}$ and $\mathbf{q}_{sn}$.

For elastic scattering, where the initial, intermediate and final states contain the same particles, (2.2.7) becomes, because of (1.10.3),

$$\operatorname{Im}\{A_l^{ii}(s)\} = \frac{2q_{si}}{\sqrt{s}}|A_l^{ii}(s)|^2 = \rho^i(s)|A_l^{ii}(s)|^2 \tag{2.2.8}$$

where

$$\rho^i(s) \equiv \frac{2q_{si}}{\sqrt{s}} \tag{2.2.9}$$

is the partial-wave phase-space factor for state i. Note that since $\rho^i(s) \leqslant 1$ for all s, (2.2.8) implies that $0 \leqslant \operatorname{Im}\{A_l^{ii}\} \leqslant 1$.

The relation (2.2.8) may be ensured by writing

$$A_l^{ii}(s) = \frac{e^{2i\delta_l(s)}-1}{2i\rho^i(s)} = \frac{e^{i\delta_l(s)}\sin\delta_l(s)}{\rho^i(s)} = \frac{1}{\rho^i(s)}\frac{1}{\cot\delta_l(s)-i} \tag{2.2.10}$$

which defines the (real) 'phase shift' $\delta_l(s)$. Below the inelastic threshold the scattering amplitude is completely specified by this function. By analysing the angular distribution of $d\sigma/dt$ it is possible to determine these phase shifts directly from the experimental data, at least for the lower partial waves at small s. However, real phase shift analysis has to cope with the problems of spin (see chapter 4) and inelasticity, and is rather more difficult.

If many channels are open (2.2.7) gives

$$\operatorname{Im}\{A_l^{ii}(s)\} = \rho^i|A_l^{ii}(s)|^2 + \sum_{n\neq i} \rho^n A^{in}(s_+)\,A^{ni}(s_-)$$
$$+\,(\text{3- and more-body channels}) \quad (2.2.11)$$

so
$$0 \leqslant |A_l^{ii}|^2 \leqslant \operatorname{Im}\{A^{ii}\} \leqslant 1 \quad (2.2.12)$$

The effect of these inelastic channels may be incorporated in (2.2.10) by allowing δ_l to be complex, $\delta_l \to \delta_l^{\mathrm{R}} + \mathrm{i}\delta_l^{\mathrm{I}}$ so

$$A_l^{ii}(s) = \frac{\eta_l \exp(2\mathrm{i}\delta_l^{\mathrm{R}}) - 1}{2\mathrm{i}\rho^i(s)}, \quad \text{where} \quad \eta_l \equiv \exp(-2\delta_l^{\mathrm{I}}) \quad (2.2.13)$$

η_l being the inelasticity factor, $0 \leqslant \eta_l \leqslant 1$. Clearly, $\eta_l = 1$ for elastic scattering.

If a resonance occurs in a particular partial wave at $s = M_{\mathrm{r}}^2$ (see for example Blatt and Weisskopf (1952) p. 398), then

$$\delta_l^{\mathrm{R}}(s) \xrightarrow[s\to M_{\mathrm{r}}^2]{} (2n+1)\frac{\pi}{2} \quad (n = \text{integer})$$

so if we put say $\quad \tan\delta_l(s) \approx \dfrac{M_r\Gamma}{M_{\mathrm{r}}^2 - s}, \quad s \approx M_{\mathrm{r}}^2$

in (2.2.10) we find

$$A_l^{ii}(s) \simeq \frac{1}{\rho^i(s)}\frac{M_r\Gamma}{M_{\mathrm{r}}^2 - s - \mathrm{i}M_r\Gamma} \simeq \frac{1}{\rho^i(s)}\frac{\Gamma/2}{M_{\mathrm{r}} - E - \mathrm{i}\,\Gamma/2}, \quad \text{where} \quad E \equiv \sqrt{s} \quad (2.2.14)$$

which is the elastic Breit–Wigner resonance formula of nuclear physics, and corresponds to a resonance of mass M_{r} and width Γ. In potential scattering the condition $\delta_l \to (2n+1)\pi/2$ is very similar to the condition for the formation of a bound state except that a resonance occurs for positive energy and so can decay (see for example Schiff (1968) p. 128). We can thus regard resonances as unstable composite particles similar to bound states. If there is inelasticity the resonance may decay into one of several channels f, the decay amplitude being

$$A_l^{if}(s) = \frac{1}{\rho_{if}}\frac{M_r(\Gamma_i\Gamma_f)^{\frac{1}{2}}}{M_{\mathrm{r}}^2 - s - \mathrm{i}M_r\Gamma}, \quad \rho_{if} = \left(\frac{2q_i q_f}{s}\right)^{\frac{1}{2}} \quad (2.2.15)$$

where Γ_f is the partial width for decay into channel f, and $\Gamma \equiv \sum_f \Gamma_f$ is the total decay width. Note the factorization of the residue of the pole. Many such resonances have been discovered in partial-wave analyses (see for example Pilkuhn (1967)).

Since $P_l(z = 1) = 1$ for all l, the optical theorem (1.9.5) with (2.2.2) reads

$$\sigma_{12}^{\text{tot}}(s) = \frac{8\pi}{q_{s12}\sqrt{s}} \sum_l (2l+1)\,\text{Im}\,\{A_l^{ii}(s)\} \tag{2.2.16}$$

while from (2.2.2) substituted in (1.8.13), after performing the angular integration using (A.20), we have for $1+2 \to 1+2$

$$\sigma_{12}^{\text{el}}(s) = \frac{16\pi}{s} \sum_l (2l+1)|A_l^{ii}(s)|^2 \tag{2.2.17}$$

Then from (2.2.8) we see that below the inelastic threshold $\sigma_{12}^{\text{tot}} = \sigma_{12}^{\text{el}}$ as of course it must.

We can obviously make an exactly similar partial-wave decomposition in the t channel, defining

$$A_l(t) = \frac{1}{16\pi}\frac{1}{2}\int_{-1}^{1} \mathrm{d}z_t\, P_l(z_t)\, A(s(z_t, t), t), \quad l = 0, 1, 2, \ldots \tag{2.2.18}$$

with inverse $$A(s,t) = 16\pi \sum_{l=0}^{\infty} (2l+1)\, A_l(t)\, P_l(z_t) \tag{2.2.19}$$

In the next section we shall be concerned with the relation between (2.2.19) and scattering in the crossed s channel.

2.3 The Froissart–Gribov projection

Equation (2.2.19) provides a representation of the scattering amplitude which is satisfactory throughout the t-channel physical region. Since $A_l(t)$ contains the t-channel thresholds and resonance poles the amplitude obtained from (2.2.19) has all the t singularities. But its s dependence is completely contained in the Legendre polynomials which are entire functions of z_t, and hence of s at fixed t. It is therefore evident that this representation must break down if we continue it beyond the t-channel physical region ($-1 \leqslant z_t \leqslant 1$) to the nearest singularity in s (or u) at $s = s_0$ say, where the series will diverge. For example the pole

$$(m^2 - s)^{-1} = m^{-2}\left(1 + \frac{s}{m^2} + \left(\frac{s}{m^2}\right)^2 + \ldots\right)$$

can be represented as a polynomial in s which diverges at $s = m^2$.

In fig. 2.2 we have plotted the nearest s- and u-channel poles and branch points in terms of the variable z_t. They always occur outside the physical region of the t channel, but it is clear from fig. 1.5 that the use of (2.2.19) is restricted to only a small region of the Mandelstam

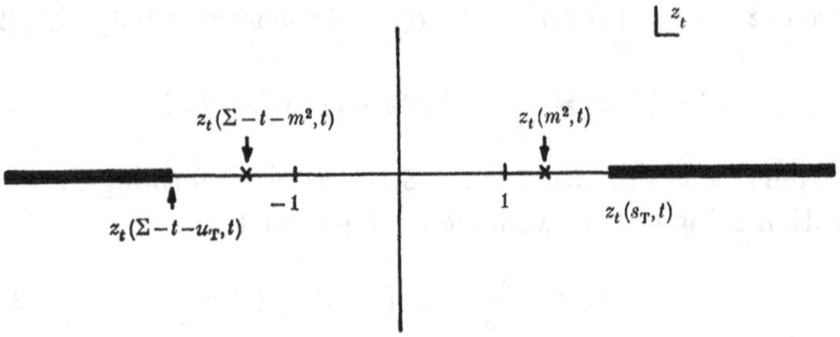

FIG. 2.2 The singularities in z_t at fixed t ($> t_T$). Outside the physical region ($-1 \leqslant z_t \leqslant 1$) these are the s-channel poles and threshold branch points for $z_t > 1$, and the u-channel singularities for $z_T < -1$, cf. fig. 1.5.

plot beyond the physical region. This greatly impedes the use of the crossing relation. For example, if the low-t region is dominated by a resonance pole of spin σ it may be a good approximation to put

$$A(s,t) \approx 16\pi(2\sigma+1)\frac{M_r \Gamma}{M_r^2 - t - \mathrm{i}M_r\Gamma} P_\sigma(z_t) \tag{2.3.1}$$

(cf. (2.4.14) with $\rho^i(s) \to 1$). However, though this may be satisfactory in the t-channel physical region, we cannot make use of it in the s-channel region ($t \leqslant 0$) because we know that the series (2.2.19), to which (2.3.1) is an approximation, will have diverged before we can reach the s channel (see fig. 1.5).

To obtain an expression for the partial-wave amplitudes which incorporates the s and u singularities, and hence is valid over the whole Mandelstam plane, we make use of the dispersion relation (1.10.7). Since from (1.7.19), (1.7.21)

$$\left.\begin{aligned} s'-s &= 2q_{t13}q_{t24}(z_t'-z_t) \\ u'-u &= -2q_{t13}q_{t24}(z_t'-z_t) \end{aligned}\right\} \tag{2.3.2}$$

we can rewrite (1.10.7) as

$$\begin{aligned} A(s,t) = {} & \frac{g_s(t)}{2q_{t13}q_{t24}(z_t(m^2,t)-z_t(s,t))} \\ & -\frac{g_u(t)}{2q_{t13}q_{t24}(z_t(\Sigma-t-m^2,t)-z_t(\Sigma-t-s,t))} \\ & +\frac{1}{\pi}\int_{z_t(s_T,t)}^{\infty}\frac{D_s(s',t)}{z_t'-z_t}\mathrm{d}z' + \frac{1}{\pi}\int_{z_t(s_T,t)}^{\infty}\frac{D_u(u',t)}{z_t'-z_t}\mathrm{d}z' \end{aligned} \tag{2.3.3}$$

but subtractions may be needed in the integrals. If (2.3.3) is substituted in (2.2.18) we can perform the z_t integration using Neumann's relation (A.14) provided the order of the two integrations can be interchanged, and we find

$$A_l(t) = \frac{1}{16\pi}\frac{g_s(t)}{2q_{t13}q_{t24}} Q_l(z_t(m^2,t)) + \frac{1}{16\pi}\frac{g_u(t)}{2q_{t13}q_{t24}} Q_l(z_t(\Sigma - t - m^2, t))$$

$$+\frac{1}{16\pi^2}\int_{z_t(s_T,t)}^{\infty} D_s(s',t)\, Q_l(z_t')\,\mathrm{d}z_t'$$

$$+\frac{1}{16\pi^2}\int_{z_t(u_T,t)}^{\infty} D_u(u',t)\, Q_l(z_t')\,\mathrm{d}z_t', \quad l = 0,1,2,\ldots \tag{2.3.4}$$

This is called the Froissart–Gribov projection (Froissart 1961, Gribov 1961), and is completely equivalent to (2.2.18) provided the dispersion relation is valid. Note, however, that (2.3.4) and (2.2.18) involve completely different regions of z_t and hence s. Since (2.2.18) requires integration only over a finite region the partial-wave amplitudes can always be so defined, at least in the t-channel physical region, but (2.3.4) involves an infinite integration and can be used only if the integral converges (so that the order of the integrations can be inverted). From (A.27) $Q_l(z) \underset{z\to\infty}{\sim} z^{-l-1}$, so if D_s (or D_u) $\sim z^N$, (2.3.4) is defined only for $l > N$. To find the lower partial waves we also need to know the subtraction functions like (1.10.10).

2.4 The Froissart bound

Froissart (1961) showed that, for amplitudes which satisfy the Mandelstam representation, s-channel unitarity limits the asymptotic behaviour of the scattering amplitude in the s-channel physical region, $t \leqslant 0$, and hence limits the number of subtractions which may be needed. This bound may be obtained as follows.

Since

$$Q_l(z) \underset{l\to\infty}{\sim} l^{-\frac{1}{2}}\,\mathrm{e}^{-(l+\frac{1}{2})\zeta(z)}, \quad \zeta(z) \equiv \log[z + \sqrt{(z^2-1)}] \tag{2.4.1}$$

(see (A.31)) the Froissart–Gribov projection (2.3.4) for s-channel partial waves gives

$$A_l(s) \underset{l,s\to\infty}{\longrightarrow} f(s)\,\mathrm{e}^{-l\zeta(z_0)} \tag{2.4.2}$$

where z_0 is the lowest t-singularity of $A(s,t)$ (threshold or bound-state pole) and $f(s)$ is some function of s. This means that all the partial waves with

$$l \gg l_{\mathrm{M}} \equiv \zeta^{-1}(z_0) \tag{2.4.3}$$

will be very small. Indeed one may define the range of the force R (see section 2.2) that such

$$Rq_s \equiv l_{\mathrm{M}} \tag{2.4.4}$$

and particles passing the target at impact parameters $b > R$ effectively miss the target and are not scattered much. Thus for nucleon–nucleon scattering, since the pion pole is the nearest t-singularity we have (cf. (1.7.22) with $t = m_\pi^2$)

$$z_0 = 1 + \frac{m_\pi^2}{2q_s^2}, \quad R = \frac{1}{q_s \zeta(z_0)} \xrightarrow[s\to\infty]{} \frac{1}{m_\pi} = \frac{\hbar}{m_\pi c} \tag{2.4.5}$$

in our units, so the range of the force, and hence the effective size of the nucleon is 1 pion Compton wavelength, as is expected from the uncertainty principle.

Hence from (2.4.2)

$$A_l(s) \xrightarrow[l,s\to\infty]{} f(s)\exp(-l/Rq_s) \to \exp\left(-\frac{2l}{R\sqrt{s}} + \log f(s)\right) \tag{2.4.6}$$

since $q_s \to \frac{1}{2}\sqrt{s}$, and so for large s we can expect that there will only be appreciable scattering in partial waves such that

$$l < (\sqrt{s})\, R \log(f(s)) \to c(\sqrt{s}) \log s \tag{2.4.7}$$

where c is some constant. Thus the partial-wave series (2.2.2) may be truncated as

$$A(s,t) \approx 16\pi \sum_{l=0}^{c(\sqrt{s})\log s} (2l+1)\, A_l(s)\, P_l(z_s) \tag{2.4.8}$$

Then using the bound (2.2.12) and $|P_l(z)| \leqslant 1$ for $-1 \leqslant z \leqslant 1$ we have

$$|A(s,t)| \leqslant 16\pi \sum_{l=0}^{c(\sqrt{s})\log s} (2l+1) \leqslant \text{const.}\, s \log^2 s, \quad s\to\infty, \quad t \leqslant 0 \tag{2.4.9}$$

on summing the arithmetic progression. With the optical theorem (1.9.5) this gives

$$\sigma^{\text{tot}}(s) \underset{s\to\infty}{\leqslant} \text{const.} \log^2 s \tag{2.4.10}$$

which is the Froissart bound. It has since been proved more rigorously from field theory by Martin (1963, 1965).

For us (2.4.9) has the very important consequence that, for fixed $t \leqslant 0$, $D_s(s,t)$, $D_u(u,t) \leqslant$ const. $s \log^2 s$, $s \to \infty$ so that $N \leqslant 1$, and the Froissart–Gribov projection (2.3.4) is defined for all $l > 1$.

Equation (2.4.6) also allows us to determine more precisely the region within which the partial-wave series (2.2.2) will converge. The asymptotic behaviour of $P_l(z)$ is given by (A.29), which with (2.4.6) shows that (2.2.2) will converge if

$$|\text{Im}\{\theta\}| \leqslant \zeta(z_0) = \cosh^{-1}(z_0) \tag{2.4.11}$$

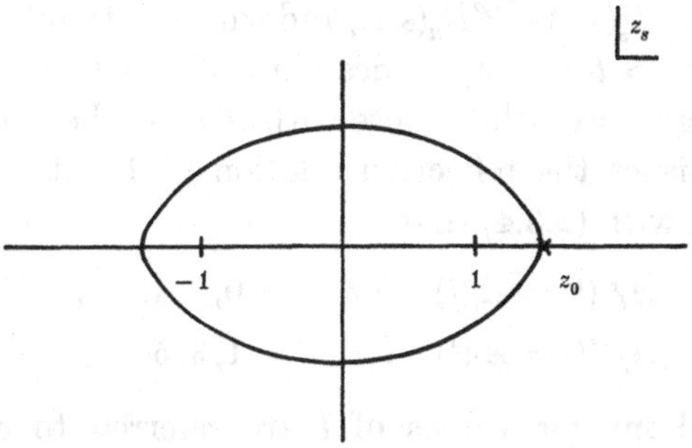

FIG. 2.3 The Lehman–Martin ellipse; the boundary of convergence of the s-channel partial-wave series in the complex z_s plane, caused by the nearest singularity at $z_s = z_0$.

which defines an ellipse in the complex z_s plane with foci at $z_s = \pm 1$ and semi-major axis z_0 (see fig. 2.3). This is often referred to as the small Lehmann–Martin ellipse (Lehmann 1958, Martin 1966).

2.5 Signature

In (2.3.4) $A_l(t)$ is defined in terms of integrals over the right-hand (s-channel) and left-hand (u-channel) cuts in z_t (fig. 2.2). The asymptotic behaviour of these contributions as $l \to \infty$ is readily obtained from (2.4.2). On the right-hand cut z_t is always > 1 so $\zeta(z)$ is always real and positive, for $t > t_{\mathrm{T}}$, so

$$A_l^{\mathrm{RH}} \xrightarrow[l \to \infty]{} f(t)\, \mathrm{e}^{-l\zeta(z_0)}, \quad z_0 \equiv z_t(s_0, t) \tag{2.5.1}$$

However, along the left-hand cut $z_t < -1$ so

$$\zeta(z) = \zeta(|z|) + \mathrm{i}\pi \quad \text{and} \quad A_l^{\mathrm{LH}} \xrightarrow[l \to \infty]{} f(t)\, \mathrm{e}^{-l\zeta(|z_0|)}\, \mathrm{e}^{-\mathrm{i}\pi l} \tag{2.5.2}$$

which is unbounded as $l \to \mathrm{i}\infty$. In section 2.7 we shall want to express the scattering amplitude as a contour integral in the complex l plane, but we should be hindered by such a divergent behaviour.

Instead, therefore, we define partial-wave amplitudes of definite signature $\mathscr{S} = \pm 1$ by (neglecting the pole terms for simplicity)

$$\begin{aligned} A_l^{\mathscr{S}}(t) &= \frac{1}{16\pi^2} \int_{z_{\mathrm{T}}}^{\infty} D_s(s', t)\, Q_l(z_t')\, \mathrm{d}z_t' + \mathscr{S} \frac{1}{16\pi^2} \int_{z_{\mathrm{T}}}^{\infty} D_u(s', t)\, Q_l(z_t')\, \mathrm{d}z_t' \\ &= \frac{1}{16\pi^2} \int_{z_{\mathrm{T}}}^{\infty} (D_s(s', t) + \mathscr{S} D_u(s', t))\, Q_l(z_t')\, \mathrm{d}z_t' \\ &\equiv \frac{1}{16\pi^2} \int_{z_{\mathrm{T}}}^{\infty} D_s^{\mathscr{S}}(s', t)\, Q_l(z_t')\, \mathrm{d}z_t' \end{aligned} \tag{2.5.3}$$

where $D_s^{\mathscr{S}}(s,t) \equiv D_s(s,t) + \mathscr{S} D_u(s,t)$, and where both integrals run over positive z_t (for $t > t_T$). Amplitudes with $\mathscr{S} = +1$ are referred to as having even signature, while those with $\mathscr{S} = -1$ have odd signature. Since $Q_l(z)$ satisfies the reflection relation (A.17) it should be clear by comparison with (2.3.4) that

$$\left.\begin{aligned} A_l^+(t) &= A_l(t) \quad \text{for} \quad l = 0,2,4,\ldots \\ A_l^-(t) &= A_l(t) \quad \text{for} \quad l = 1,3,5,\ldots \end{aligned}\right\} \tag{2.5.4}$$

These physical integer values of l are referred to as the 'right-signature points' of $A_l^{\mathscr{S}}(t)$ (i.e. even l for even signature, and vice versa) and conversely the unphysical integer values (i.e. odd l for even signature, and vice versa) are called 'wrong-signature points'. With the definition (2.5.3)

$$A_l^{\mathscr{S}}(t) \xrightarrow[l\to\infty]{} f(t)\,\mathrm{e}^{-l\zeta(z_0)}, \quad \text{for} \quad \mathscr{S} = \pm 1 \tag{2.5.5}$$

and so converges as $l \to \infty$.

We can sum the partial-wave series to give amplitudes of definite signature

$$A^{\mathscr{S}}(s,t) \equiv 16\pi \sum_{l=0}^{\infty} (2l+1)\, A_l^{\mathscr{S}}(t)\, P_l(z_t) \tag{2.5.6}$$

so the even part of $A^+(s,t)$ in z_t = even part of $A(s,t)$, and the odd part of $A^-(s,t)$ = odd part of $A(s,t)$. These amplitudes satisfy the dispersion relation (again omitting poles)

$$A^{\mathscr{S}}(s,t) = \frac{1}{\pi}\int_{s_T}^{\infty} \frac{D_s(s',t)}{s'-s}\,\mathrm{d}s' + \mathscr{S}\frac{1}{\pi}\int_{u_T}^{\infty} \frac{D_u(u',t)}{u'-s}\,\mathrm{d}u' \tag{2.5.7}$$

$$= \frac{1}{\pi}\int_{s_T}^{\infty} \frac{D_s^{\mathscr{S}}(s',t)}{s'-s}\,\mathrm{d}s' \tag{2.5.8}$$

where s has replaced u in the denominator of the second term because of the replacement $z_t \to -z_t$ in the corresponding term of (2.5.3). The Mandelstam representation for such an amplitude is from (1.11.4), (1.11.5) in (2.5.7) (with some changes of variables)

$$A^{\mathscr{S}}(s,t) = \frac{1}{\pi^2}\iint^{\infty} \frac{\rho_{st}(s,t'') + \mathscr{S}\rho_{tu}(s',t'')}{(s'-s)(t''-t)}\,\mathrm{d}s'\,\mathrm{d}t''$$

$$+ \frac{1}{\pi^2}\iint^{\infty} \frac{\rho_{su}(s',t'') + \mathscr{S}\rho_{su}(u'',s')}{(s'-s)(u''-u')}\,\mathrm{d}s'\,\mathrm{d}u'' \tag{2.5.9}$$

The lack of symmetry in s, t and u stems from the fact that we have taken definite signature in the t channel. These definite-signature amplitudes are of course unphysical because of the change of the sign of z_t involved in the definition (2.5.3). But from (2.5.6) with (2.5.4) and (A.11) it is possible to obtain the physical amplitude from them by

$$A(s,t) = \tfrac{1}{2}(A^+(z_t,t)+A^+(-z_t,t)+A^-(z_t,t)-A^-(-z_t,t)) \quad (2.5.10)$$

For analytic continuation in l we shall always use $A^{\mathscr{S}}(s,t)$ rather than $A(s,t)$.

Since with equal-mass kinematics z_t is given by (1.7.22), it has a pole at $t = t_{\mathrm{T}} \equiv 4m^2$. So, from (2.5.1), for $t < t_{\mathrm{T}}$ it is

$$\hat{A}_l^{\mathscr{S}}(t) \equiv \mathrm{e}^{\mathrm{i}\pi l} A_l^{\mathscr{S}}(t) \quad (2.5.11)$$

which has the good asymptotic l behaviour, rather than $A_l^{\mathscr{S}}(t)$ itself. But we shall find in the next section that the threshold behaviour is $A_l^{\mathscr{S}}(t) \sim (q_t^2)^l \sim (t-4m^2)^l$ so the required factor (2.5.11) is included automatically.

2.6 Singularities of partial-wave amplitudes and dispersion relations*

In the t-channel physical region we can obtain the signatured partial-wave amplitudes either from (2.2.18) and (2.5.6), i.e.

$$A_l^{\mathscr{S}}(t) = \frac{1}{32\pi}\int_{-1}^{1} A^{\mathscr{S}}(s,t)\, P_l(z_t)\, \mathrm{d}z_t, \quad l = 0,1,2,\ldots \quad (2.6.1)$$

or equivalently from (2.5.3) and (2.5.8), i.e.

$$A_l^{\mathscr{S}}(t) = \frac{1}{16\pi^2}\int_{z_{\mathrm{T}}}^{\infty} D_s^{\mathscr{S}}(s,t)\, Q_l(z_t)\, \mathrm{d}z_t, \quad l = 0,1,2,\ldots \quad (2.6.2)$$

Since $2D_s^{\mathscr{S}}(s,t)$ is the discontinuity of $A^{\mathscr{S}}(s,t)$ across the cuts in z_t, while from (A.15) the discontinuity of $Q_l(z)$ is $-\pi P_l(z)$, we can combine (2.6.1) and (2.6.2) in

$$A_l^{\mathscr{S}}(t) = \frac{1}{32\pi^2}\int_{C_1 \text{ or } C_2} A^{\mathscr{S}}(s,t)\, Q_l(z_t)\, \mathrm{d}z_t \quad (2.6.3)$$

where the contours encircle the cuts of either $Q_l(z_t)$ or $A^{\mathscr{S}}(s,t)$ as shown in fig. 2.4.

Since the integration in (2.6.1) is over a finite s region, at fixed t, it is clear that $A_l^{\mathscr{S}}(t)$ will have all the t-channel threshold branch points of $A^{\mathscr{S}}(s,t)$ which also occur at fixed t. In (2.6.2) these branch points

* This section may be omitted at first reading.

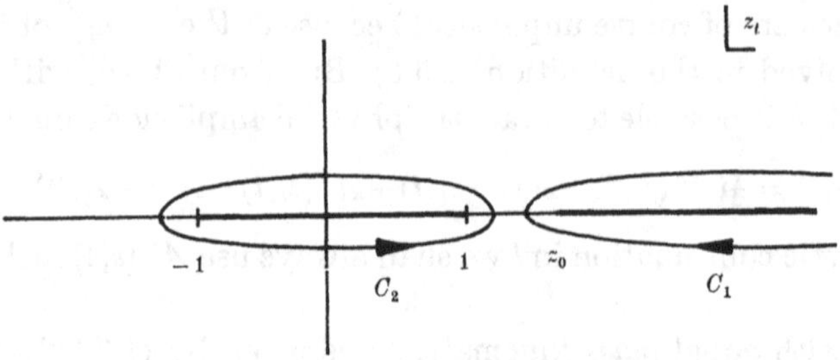

FIG. 2.4 Integration contours in the complex z_t plane used in (2.6.3).

appear in $D_s^{\mathscr{S}}(s,t)$. They are of course generated by the unitarity equations as discussed in chapter 1.

However, the partial-wave projection may introduce further threshold singularities. These arise from the vanishing of the three-momenta which appear in the expression for z_t, (1.7.19). Thus at the threshold for the initial state $t \to (m_1+m_3)^2$, $\lambda(t, m_1^2, m_3^2) \to 0$, $q_{t13} \to 0$, so $z_t \to \infty$. In view of the asymptotic behaviour of Legendre functions (A.27), $Q_l(z_t) \sim (z_t)^{-l-1}$, this means

$$Q_l(z_t)\,\mathrm{d}z_t \sim [t-(m_1+m_3)^2]^{l/2} \tag{2.6.4}$$

and so from (2.6.2) $$A_l^{\mathscr{S}}(t) \sim [t-(m_1+m_3)^2]^{l/2} \tag{2.6.5}$$

Also q_{t13} vanishes at the so-called 'pseudo-threshold' $t \to (m_1-m_3)^2$ and $q_{t14} \to 0$ at $t \to (m_2 \pm m_4)^2$, so if we introduce the notation

$$T_{ij}^{\pm}(t) \equiv [t-(m_i \pm m_j)^2]^{\frac{1}{2}} \tag{2.6.6}$$

we find $$A_l^{\mathscr{S}}(t) \sim (T_{13}^{+}(t)\,T_{13}^{-}(t)\,T_{24}^{+}(t)\,T_{24}^{-}(t))^l \tag{2.6.7}$$

If the initial- and final-state thresholds coincide, i.e. $m_1+m_2 = m_3+m_4$, there is simply a kinematical zero of order l at the threshold, but otherwise there are square-root branch points for odd values of l. What is worse, if we want to continue to non-integer values of l, (2.6.7) implies that there will always be kinematical branch points. So if we wish to write dispersion relations for the partial-wave amplitudes, integrating over just the dynamical singularities as we did for the full amplitude in (1.10.7), we must first remove these kinematical singularities by defining the 'reduced' partial-wave amplitudes

$$B_{l}^{\mathscr{S}}(t) \equiv A_l^{\mathscr{S}}(t)\,(q_{t13}\,q_{t24})^{-l} \tag{2.6.8}$$

whose threshold singularities in t are just the dynamical threshold branch points. Clearly $B_l^{\mathscr{S}}(t)$ is Hermitian analytic if $A^{\mathscr{S}}(s,t)$ is (see section 1.5).

The positive t, or right-hand cut discontinuity of this amplitude may be obtained from (2.5.9) in (2.6.2) with (2.6.8), viz.

$$\operatorname{Im}\{B_l^{\mathscr{S}}(t)\}_{\mathrm{RH}} = \frac{1}{16\pi^2}\int_{z_0}^{\infty}(\rho_{st}(s',t)+\mathscr{S}\rho_{tu}(s',t))\,Q_l(z_t')\,\mathrm{d}z_t'(q_{t13}q_{t24})^{-l} \tag{2.6.9}$$

In addition to these thresholds $A^{\mathscr{S}}(s,t)$ may also have fixed-t singularities due to bound-state poles below threshold. Thus a t-channel bound state of mass M and spin σ contributes

$$A^{\mathscr{S}}(s,t) = \frac{(2\sigma+1)\,g_t^2\,(q_{t13}q_{t24})^{\sigma}}{M^2-t}P_\sigma(z_t) \tag{2.6.10}$$

where g_t^2 is the coupling strength (the factor $(2\sigma+1)$ is purely conventional) and we have included the threshold factor $(q_{t13}q_{t24})^{\sigma}$ explicitly (so that g_t may be constant). In (2.2.18) with (A.20) and (2.6.8) this gives

$$B_l^{\mathscr{S}}(t) = \frac{1}{16\pi}\frac{g_t^2}{M^2-t}\delta_{l\sigma} \tag{2.6.11}$$

a contribution to the $l=\sigma$ partial wave only. These right-hand singularities are exhibited in fig. 2.5 where we have drawn the threshold cuts along the positive t axis.

However, there are further singularities which occur at negative values of t due to the s-channel singularities of $A^{\mathscr{S}}(s,t)$. (Remember $A^{\mathscr{S}}(s,t)$ has no u singularities as these have been folded over into the s channel by (2.5.3).) Thus suppose there is a bound-state pole in the s channel of spin σ and mass M,

$$A^{\mathscr{S}}(s,t) = \frac{(2\sigma+1)\,g_s^2(q_{s12}q_{s34})^{\sigma}P_\sigma(z_s)}{M^2-s} \tag{2.6.12}$$

$$\equiv \frac{G_s(s)}{M^2-s}P_\sigma(z_s)$$

so

$$D^{\mathscr{S}}(s,t) = \pi G_s(s)\,P_\sigma(z_s(s,t))\,\delta(s-M^2) \tag{2.6.13}$$

which substituted in (2.5.3) gives, through (2.6.8),

$$B_l^{\mathscr{S}}(t) = G_s(M^2)\,P_\sigma(z_s(M^2,t))\,Q_l(z_t(M^2,t))\,(q_{t13}q_{t24})^{-l} \tag{2.6.14}$$

Now $Q_l(z)$ has branch points in z at $z=\pm 1$ (for integer l) and so (2.6.14) has singularities at

$$z_t(M^2,t) = \pm 1$$

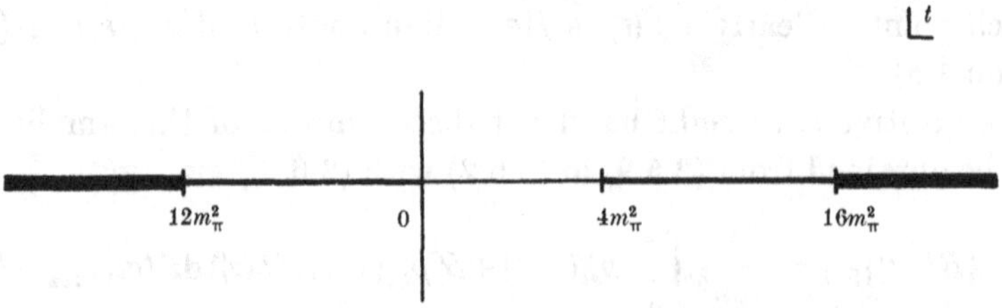

FIG. 2.5 Singularities of the t-channel partial-wave amplitudes for $\pi\pi$ scattering, showing the thresholds at $t = (2m_\pi)^2, (4m_\pi)^2, \ldots$ and the left-hand cuts at $t = 4m_\pi^2 - s_{Ti}$ where s_{Ti} are the s-channel thresholds at $s = (2m_\pi)^2, (4m_\pi)^2, \ldots$. (Note that G-parity forbids odd numbers of pions, see section 5.1.)

which from (1.7.19) requires

$$\frac{t^2 + t(2M^2 - \Sigma) + (m_1^2 - m_3^2)(m_2^2 - m_4^2)}{\lambda^{\frac{1}{2}}(t, m_1^2, m_3^2)\,\lambda^{\frac{1}{2}}(t, m_2^2, m_4^2)} = \pm 1 \tag{2.6.15}$$

For example if all the external particles have equal masses (e.g. for $\pi\pi \to \pi\pi$, $m_1 = m_2 = m_3 = m_4 = m_\pi$), this reduces to

$$1 + \frac{2M^2}{t - 4m_\pi^2} = \pm 1 \tag{2.6.16}$$

so there are branch points at $t = \infty$ and at $t = 4m_\pi^2 - M^2$, and conventionally the branch cut is drawn along the negative t axis as in fig. 2.5. Note that the s-channel pole of spin σ contributes to all the partial waves of the t channel through (2.6.14).

The singularity arises through a pinch of the singularity of $A^{\mathscr{S}}(s,t)$ with the branch points of $Q_l(z)$ in (2.5.3). All the other s-singularities, the threshold branch points etc., will give similar pinches, and hence similar left-hand branch points, at positions determined simply by replacing M^2 in (2.6.16) by the (real) threshold value of s.

For unequal-mass kinematics the mapping of the s singularities into t is much more complicated. There are four solutions to (2.6.15), two being independent of M^2, i.e. $t = 0$ and ∞. Thus for πN scattering the N exchange pole generates branch points at $t = 0, \infty, (M_N - m_\pi^2/M_N)^2$ and $M_N^2 + 2m_\pi^2$. (Note that if $m_\pi \to M_N$ these two cuts join up, giving a single cut at $t = 3M_N^2$ in agreement with (2.6.16).) (For further details see for example Martin and Spearman (1970) p. 376 *et seq.*)

Since the imaginary part of Q_l is given by (A.15) for integer l, we find from (2.5.3) that

$$\operatorname{Im}\{A_l^{\mathscr{S}}(t)\}_{\mathrm{LH}} = \frac{1}{32\pi}\int_{-1}^{z_0} P_l(z_t')\, D_s^{\mathscr{S}}(s', t)\, \mathrm{d}z_t' \tag{2.6.17}$$

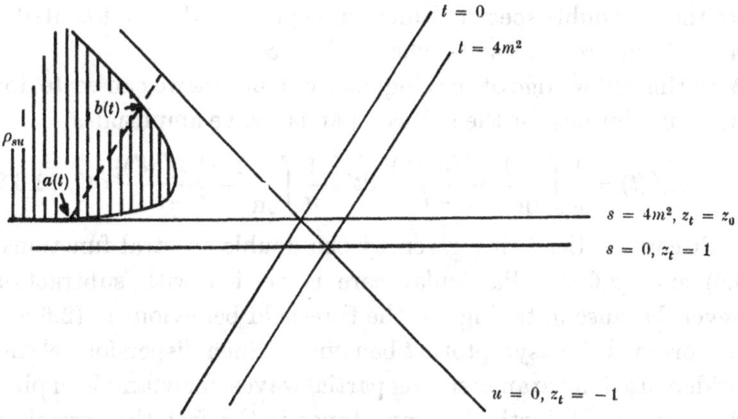

FIG. 2.6 Singularities in the Mandelstam plot involved in the partial-wave projection of a definite-signature t-channel amplitude. The left-hand cut for negative (fixed) t involves integration over the s-singularities between $z_t = z_0$ (the nearest s-singularity) and $z_t = -1$. For sufficiently negative t this includes integration over the double spectral function between the boundary points $a(t)$ and $b(t)$ as well, the dashed line being the fixed-t integration contour.

(z_0 being the lowest s-singularity – see fig. 2.6) gives the discontinuity of $A_l^{\mathscr{S}}(t)$ along its left-hand cut. For non-integer l we need to use (A.16), but we are more interested in the singularities of $B_l^{\mathscr{S}}(t)$, and, for $s > 0$, $t \pm i\epsilon$ corresponds to $z \pm i\epsilon$ (from (1.7.19)), so the branch point of $Q_l(z)$ at $z = -1$ is cancelled by that of the kinematical factor (2.6.8), i.e. $Q_l(z_t)\,(q_{t13}q_{t24})^{-l}$ has no cut for $z_t < -1$. There is a contribution from the cut of $Q_l(z_t)$ for $-1 < z_t < 1$, and another from the discontinuity of $D_s(s,t)$ in the negative t region, obtained from (2.5.9), so

$$\operatorname{Im}\{B_l^{\mathscr{S}}(t)\}_{\mathrm{LH}} = \frac{1}{32\pi}\int_{-1}^{z_0} P_l(-z_t')\, D_s^{\mathscr{S}}(s',t)\,\mathrm{d}z_t'(-q_{t13}q_{t24})^{-l}$$
$$+\frac{1}{16\pi^2}\int_{a(t)}^{b(t)} Q_l(z_t')\,(\rho_{su}(s',u')+\mathscr{S}\rho_{su}(u',s'))\,\mathrm{d}z_t'\,(q_{t13}q_{t24})^{-l} \quad (2.6.18)$$

where the regions of integration are shown in fig. 2.6. Since interchanging s and u is equivalent to changing the sign of z_t, with (A.17) (2.6.18) becomes

$$\operatorname{Im}\{B^{\mathscr{S}}(t)\}_{\mathrm{LH}} = \frac{1}{32\pi}\int_{-1}^{z_0} P_l(-z_t')\, D_s^{\mathscr{S}}(s',t)\,\mathrm{d}z_t'(-q_{t13}q_{t24})^{-l}$$
$$+\frac{1}{16\pi^2}\int_{a(t)}^{b(t)} Q_l(z_t')\,\rho_{su}(s',u')\,(1-\mathscr{S}\,\mathrm{e}^{-\mathrm{i}\pi l})\,\mathrm{d}z_t'(q_{t13}q_{t24})^{-l} \quad (2.6.19)$$

This last term, which is due to the fact that an exchange force (and hence the ρ_{su} double spectral function) is present, does not contribute at right-signature values of l where $e^{-i\pi l} = \mathcal{S}$.

With this knowledge of the singularity structure we can write down dispersion relations for the reduced partial-wave amplitudes

$$B_l^{\mathcal{S}}(t) = \frac{1}{\pi}\int_{\text{RH}} \frac{\text{Im}\{B_l^{\mathcal{S}}(t')\}}{t'-t}\,\text{d}t' + \frac{1}{\pi}\int_{\text{LH}} \frac{\text{Im}\{B_l^{\mathcal{S}}(t')\}}{t'-t}\,\text{d}t' \quad (2.6.20)$$

both discontinuities being given by the double spectral functions in (2.6.9) and (2.6.19). Particular care is needed with subtractions, however, because in taking out the threshold behaviour in (2.6.8) we have worsened the asymptotic t behaviour. Such dispersion relations are widely used in parameterizing partial waves, for example in phase-shift analyses. Of particular importance is the fact that crossing is readily incorporated because the crossed channel singularities appear in the left-hand cut. Also the right-hand cut discontinuity is given by the unitarity equation. From (2.2.7) (interchanging s and t) with (2.6.8) we find

$$B_l^{\mathcal{S}if}(t_+) - B_l^{\mathcal{S}if}(t_-) = 2\text{i}\sum_n \rho_l^n(t)\, B_l^{\mathcal{S}in}(t_+)\, B_l^{\mathcal{S}nf}(t_-)$$

$$+\,3\text{- and more-body intermediate states} \quad (2.6.21)$$

where
$$\rho_l^n(t) \equiv (q_{t13}\,q_{t24})^l \frac{q_{tn}}{\sqrt{t}} \quad (2.6.22)$$

and in the elastic region (cf. (2.2.8))

$$\text{Im}\{B_l^{\mathcal{S}ii}(t)\} = \frac{2(q_{t13})^{2l+1}}{\sqrt{t}}\,|B_l^{\mathcal{S}ii}(t)|^2 \quad (2.6.23)$$

This form of the unitarity equation will be useful for analytic continuation in l.

2.7 Analytic continuation in angular momentum

The Froissart–Gribov projection, (2.6.2), may be used to define $A_l^{\mathcal{S}}(t)$ for all values of l, not necessarily integer or even real, as we have been assuming so far. In fact, it can be used for all l such that $\text{Re}\{l\} > N(t)$, where D_s (or D_u) $\sim z^{N(t)}$, and where $N(t) \leqslant 1$ for $t \leqslant 0$ from (2.4.9). The main advantage of using (2.6.2) rather than (2.2.18) for $l \neq$ integer is that Q_l has a better behaviour than P_l as $l \to \infty$ (compare (A.28) and (A.31)).

The only singularities of $Q_l(z)$ are poles at $l = -1, -2, \ldots$ (see (A.32)), so (2.6.2) defines a function of l which is holomorphic (free of singularities) for $\text{Re}\{l\} > \max(N(t), -1)$.

It is not immediately apparent that there is much merit to this extended definition of the partial-wave amplitudes because of course it is only positive integer values of l that have physical significance, and there is clearly an infinite number of different ways of interpolating between the integers. However $A_l^{\mathscr{S}}(t)$ defined by (2.6.2) vanishes as $|l| \to \infty$ (see (2.5.5)) and a theorem due to Carlson (proved in Titchmarsh (1939) p. 186) tells us that (2.6.2) must be the unique continuation with this property.

More precisely Carlson's theorem states that: if $f(l)$ is regular, and of the form $O(e^{k|l|})$, where $k < \pi$, for $\mathrm{Re}\{l\} \geqslant n$, and $f(l) = 0$ for an infinite sequence of integers, $l = n, n+1, n+2, \ldots$, then $f(l) = 0$ identically. Thus if we were to write

$$A_l^{\mathscr{S}}(t) = A_l^{FG}(t) + f(l,t)$$

where $A_l^{FG}(t)$ is obtained from the Froissart–Gribov projection, and $f(l,t) = 0$ for integer l, the theorem tells us that either $A_l^{\mathscr{S}}(t) \nrightarrow 0$ as $|l| \to \infty$ or $f(l,t)$ vanishes everywhere. Perhaps the simplest example is

$$A_l^{\mathscr{S}}(t) = A_l^{FG}(t) + F(t) \sin \pi l$$

Remembering that $\sin \pi l = (e^{i\pi l} - e^{-i\pi l})(2i)^{-1}$ it is clear that $|A_l^{\mathscr{S}}(t)| \to \infty$ as $l \to i\infty$, due to the added term.

Hence (2.6.2) defines $A_l^{\mathscr{S}}(t)$ uniquely as a holomorphic function of l with convergent behaviour as $|l| \to \infty$, for all $\mathrm{Re}\{l\} > N(t)$. However, we are prevented from continuing below $\mathrm{Re}\{l\} = N(t)$ by the divergent behaviour of $D_s(s,t)$ as $s \to \infty$.

To proceed further we must make the additional, and crucial, assumption that the scattering amplitude $A_l(t)$ is an analytic function of l throughout the complex angular-momentum plane, with only isolated singularities. It will then be just these isolated singularities which cause the divergence problems, and we can easily continue past them.

For example suppose that $D_l^{\mathscr{S}}(s,t)$ has a leading asymptotic power behaviour

$$D_s^{\mathscr{S}}(s,t) \sim s^{\alpha(t)} + \text{lower order terms} \tag{2.7.1}$$

so $N(t) = \alpha(t)$. Then, since from (A.27) $Q_l(z) \sim z^{-l-1}$, and from (1.7.19) $z_t \xrightarrow[s\to\infty]{} s/q_{t13}q_{t24}$, the large-$s$ region of (2.6.2) ($s > s_1$ say) gives

$$A_l^{\mathscr{S}}(t) \underset{l>\alpha(t)}{\sim} \int_{s_1}^{\infty} s^{\alpha(t)} s^{-l-1} \, ds = -\frac{e^{(\alpha(t)-l)\log s_1}}{\alpha(t)-l} \tag{2.7.2}$$

Hence $A_l(t)$ has a pole at $l = \alpha(t)$. This is, by hypothesis, the rightmost singularity in the complex l plane, and it is this singularity which is preventing continuation to the left of $\mathrm{Re}\{l\} = \alpha(t)$. However, once we have isolated this pole we can continue round it to the left, until we reach the singularity due to the next term in the asymptotic expansion of $D_s^{\mathscr{S}}(s,t)$.

There may be logarithmic terms like

$$D_s^{\mathscr{S}}(s,t) \sim s^{\alpha(t)}(\log s)^{\beta(t)} \tag{2.7.3}$$

giving

$$A_l^{\mathscr{S}}(t) \underset{l>\alpha(t)}{\sim} \int_{s_1}^{\infty} s^{\alpha(t)}(\log s)^{\beta(t)} s^{-l-1}\,\mathrm{d}s = \frac{1}{(\alpha(t)-l)^{1+\beta(t)}} + \ldots, \quad \beta(t) \neq -1$$

$$= \log(\alpha(t)-l), \quad \beta(t) = -1 \tag{2.7.4}$$

so $A_l^{\mathscr{S}}(t)$ has a branch point at $l = \alpha(t)$, or a multiple pole if β is a positive integer. We shall discuss the physical significance of these poles and branch points below.

The assumption that $A_l^{\mathscr{S}}(t)$ has only isolated singularities in l, and so can be analytically continued throughout the complex angular-momentum plane, is sometimes called the postulate of 'maximal analyticity of the second kind', to distinguish it from postulate (v) of section 1.4 concerning analyticity in s and t. It is the basic assumption upon which the applicability of Regge theory to particle physics rests. It is certainly not proven, but, as we shall see in the next chapter, it is true of various plausible models for strong interactions, and, much more important, it seems to be in accord with experiment.

If it is true, then the partial-wave series (2.5.6) can be rewritten as a contour integral in the l plane (a method used by Sommerfeld (1949), following a technique of Watson (1918)), viz.

$$A^{\mathscr{S}}(s,t) = -\frac{16\pi}{2\mathrm{i}} \int_{C_1} (2l+1) A_l^{\mathscr{S}}(t) \frac{P_l(-z_t)}{\sin \pi l}\,\mathrm{d}l \tag{2.7.5}$$

The contour C_1 is shown in fig. 2.7. It embraces the positive integers and zero, but avoids any singularities of $A_l(t)$. The residues of the poles of the integrand at the integers $l = n$, where $\sin \pi l \to (-l)^n (l-n)\pi$, are

$$\frac{2\pi\mathrm{i}(2n+1) A_n^{\mathscr{S}}(t) P_n(-z_t)}{(-1)^n \pi} = 2\mathrm{i}(2n+1) A_n^{\mathscr{S}}(t) P_n(z_t) \tag{2.7.6}$$

using (A.11), so Cauchy's theorem gives, from (2.7.5)

$$A_l^{\mathscr{S}}(s,t) = 16\pi \sum_l (2l+1) A_l^{\mathscr{S}}(t) P_l(z_t) \tag{2.7.7}$$

Hence (2.7.5) is equivalent to (2.7.7) provided $A_l(t)$ has the required analyticity in l.

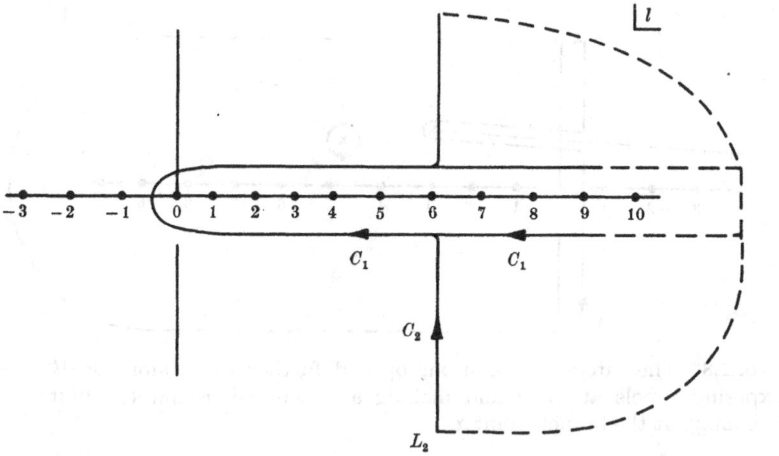

FIG. 2.7 The integration contour C_1 in the complex l plane enclosing the positive integers. This is then opened up along the line $\mathrm{Re}\{l\} = L_2$ to give the contour C_2 with a semi-circle at infinity.

Since we have found that $A_l^{\mathscr{S}}(t)$ has no singularities in $\mathrm{Re}\{l\} > N(t)$ we can displace the contour from C_1 to C_2, shown in fig. 2.7, without encountering any singularities of the integrand, provided the vertical line has $\mathrm{Re}\{l\} \equiv L_2 > N(t)$. The contribution of the semi-circle at infinity will vanish because of (2.5.5) and (A.30). Also these equations show that the region of convergence of (2.7.5) in z is much larger than the small Lehmann ellipse (2.4.11) within which (2.7.7) is valid. This region is independent of $\mathrm{Im}\{\theta\}$, and in fact, because of (2.5.11), should include the whole z plane. The s singularities of $A^{\mathscr{S}}(s,t)$ which prevent the convergence of (2.7.7) are present in $P_l(-z_t)$, $z_t > 1$, for non-integer l through (A.13).

If we displace L_2 to the left we shall encounter the l-plane singularities like (2.7.2), (2.7.4) which are responsible for the divergence of (2.6.2). Let us suppose for simplicity that we encounter just one pole at $l = \alpha(t)$ of the form $A_l(t) \sim \beta(t)(l-\alpha(t))^{-1}$, and one branch point at $l = \alpha_c(t)$ in $\mathrm{Re}\{l\} > -\frac{1}{2}$, as shown in fig. 2.8. Then we obtain

$$A^{\mathscr{S}}(s,t) = -\frac{16\pi}{2\mathrm{i}}\int_{C_2}(2l+1)\,A_l^{\mathscr{S}}(t)\,\frac{P_l(-z_t)}{\sin \pi l}\,\mathrm{d}l$$
$$-16\pi^2(2\alpha(t)+1)\,\beta(t)\,\frac{P_{\alpha(t)}(-z_t)}{\sin \pi\alpha(t)}$$
$$-\frac{16\pi}{2\mathrm{i}}\int^{\alpha_c(t)}(2l+1)\,\Delta_l^{\mathscr{S}}(t)\,\frac{P_l(-z_t)}{\sin \pi l}\,\mathrm{d}l \qquad (2.7.8)$$

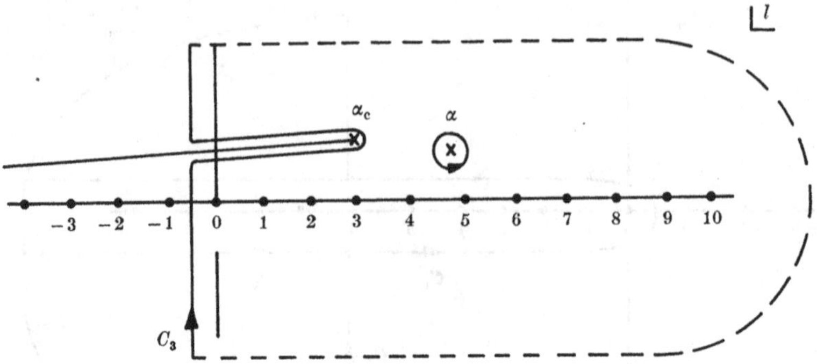

FIG. 2.8 The integration contour opened further to C_3 along $\mathrm{Re}\{l\} = -\frac{1}{2}$, exposing a pole at $l = \alpha$ and making an excursion round the branch cut beginning at the branch point α_c.

where the last term is the integration round the branch point shown in fig. 2.8, $\Delta_l^{\mathscr{S}}(t)$ being the cut discontinuity. Equation (2.7.8) is known as the Sommerfeld–Watson representation.

Because of the asymptotic z behaviour of $P_l(z)$ (see (A.25), (A.26)) it is evident that the first term, called the 'background integral', $\sim s^{-\frac{1}{2}}$ as $s \to \infty$ and so vanishes. Similarly, the pole term $\sim s^{\alpha(t)}$ like (2.7.1), while the asymptotic behaviour of the cut depends on the form of its discontinuity at the branch point $l \to \alpha_c(t)$. If $\Delta_l^{\mathscr{S}}(t)$ behaves like $(l-\alpha_c(t))^{1+\beta(t)}$ then the asymptotic form is $\sim s^{\alpha_c(t)}(\log s)^{\beta(t)}$; see (2.7.4).

In potential scattering, for well behaved potentials, there are only poles, no cuts, as Regge showed in his original papers on the subject (see chapter 3). In particle physics, we expect that there will be cuts as well, but we shall postpone detailed discussion of them until chapter 8, and for the time being concentrate on the poles.

2.8 Regge poles

The second term in (2.7.8) is called a 'Regge pole', i.e. a pole in the complex l plane. Its contribution to the scattering amplitude is

$$A^{\mathscr{S}\mathrm{R}}(s,t) = -16\pi^2(2\alpha(t)+1)\,\beta(t)\,\frac{P_{\alpha(t)}(-z_t)}{\sin\pi\alpha(t)} \qquad (2.8.1)$$

Because of (A.13) the s discontinuity takes the form

$$D_s^{\mathrm{R}}(s,t) = 16\pi^2(2\alpha(t)+1)\,\beta(t)\,P_{\alpha(t)}(z_t), \quad z_t > 1$$
$$\underset{s\to\infty}{\sim} s^{\alpha(t)} \qquad (2.8.2)$$

as expected from (2.7.1). In fact if (2.8.2) is substituted in (2.6.2) we find, from (A.22),

$$A_l^{\mathscr{S}}(t) = \frac{(2\alpha(t)+1)\,\beta(t)}{(l-\alpha(t))\,(l+\alpha(t)+1)} \xrightarrow[l\to\alpha(t)]{} \frac{\beta(t)}{l-\alpha(t)} \tag{2.8.3}$$

confirming that (2.8.1) does give rise to a pole in the l plane.

If $\alpha(t)$ is a function of t, then, for a given fixed l, $A_l(t)$ will have a pole in t at the point t_r where $\alpha(t_r) = l$. We shall examine the properties of $\alpha(t)$, $\beta(t)$ in detail in section 3.2, and will find that usually $\alpha(t)$ is a real analytic function of t with a branch point at the threshold t_T. Thus for real $t > t_T$ we can separate it into its real and imaginary parts

$$\alpha(t) = \alpha_R(t) + i\alpha_I(t) \tag{2.8.4}$$

and define t_r to be the point where $\alpha_R(t) = l$. So expanding about this point gives

$$\alpha(t) = l + \alpha_R'(t_r)\,(t-t_r) + \ldots + i\alpha_I(t_r) + i\alpha_I'(t_r)\,(t-t_r) + \ldots \tag{2.8.5}$$

(where $' \equiv \mathrm{d}/\mathrm{d}t$) and so for $\alpha_R \approx l$

$$A_l^{\mathscr{S}}(t) \approx \frac{\beta(t_r)}{-\alpha_R'(t_r)\,(t-t_r) - i\alpha_I(t_r) - i\alpha_I'(t_r)\,(t-t_r)} \approx \frac{\beta(t_r)/\alpha_R'(t_r)}{t_r - t - i\,\alpha_I(t_r)/\alpha_R'(t_r)} \tag{2.8.6}$$

assuming $\alpha_I' \ll \alpha_R'$. This may be compared with the Breit–Wigner formula (2.2.15) from which we see that (2.8.6) corresponds to a t-channel resonance of mass $M_r = \sqrt{t_r}$ and total width

$$\Gamma = \frac{\alpha_I(t_r)}{\alpha_R'(t_r)\,M_r} \tag{2.8.7}$$

Below threshold $\alpha_I = 0$ and we have a bound state pole on the real t axis. This puts bound states and resonances on a very similar footing, both being Regge poles (fig. 2.9).

When such a Regge pole occurs for a physical integer value of l it will correspond to a physical particle or resonance. This is also evident from (2.8.1) in which we see that a pole in t will occur when $\alpha(t)$ passes through an integer because of the vanishing of $\sin \pi\alpha(t)$. However, (2.8.1) is the signatured amplitude, and to obtain the physical amplitude we must use (2.5.10) giving

$$A^R(s,t) = -16\pi^2\,(2\alpha(t)+1)\,\beta(t)\,\frac{P_{\alpha(t)}(-z_t) + \mathscr{S}P_{\alpha(t)}(z_t)}{\sin \pi\alpha(t)} \tag{2.8.8}$$

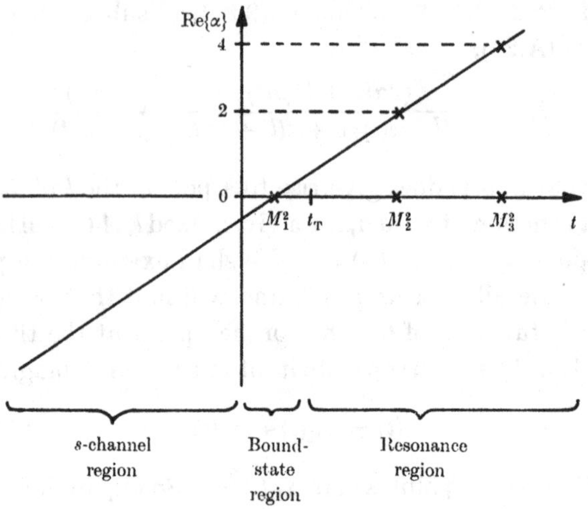

FIG. 2.9 A Regge trajectory of even signature. The trajectory has Re $\{\alpha\} = 0$ for $t < t_T$ (the threshold) giving a spin = 0 bound state of mass M_1, and then resonances of spin 2 mass M_2, and spin 4 mass M_3. For $t < 0$ the trajectory contributes to the power behaviour of the crossed s-channel amplitude, $\sim s^{\alpha(t)}$.

which with (A.10) becomes

$$A^{\mathrm{R}}(s,t) = -16\pi^2(2\alpha(t)+1)\beta(t)\left[(1+\mathscr{S}\mathrm{e}^{-\mathrm{i}\pi\alpha(t)})\frac{P_{\alpha(t)}(-z_t)}{\sin\pi\alpha(t)} - \mathscr{S}\frac{2}{\pi}\sin\pi\alpha(t)\,Q_{\alpha(t)}(-z_t)\right] \quad (2.8.9)$$

But the last term is asymptotically negligible because of (A.27), and is usually omitted giving

$$A^{\mathrm{R}}(s,t) = -16\pi^2(2\alpha(t)+1)\beta(t)(1+\mathscr{S}\mathrm{e}^{-\mathrm{i}\pi\alpha(t)})\frac{P_{\alpha(t)}(-z_t)}{\sin\pi\alpha(t)} \quad (2.8.10)$$

The factor $(1+\mathscr{S}\mathrm{e}^{-\mathrm{i}\pi\alpha})$ is called the 'signature factor', and it ensures that a trajectory of given signature $\mathscr{S} = \pm 1$ contributes a pole in t to the scattering amplitude only when $\alpha(t)$ passes through a right-signature integer (i.e. even/odd integer); see (2.5.4) *et seq.*

The Froissart bound (2.4.9) requires that $\alpha(t) < 1$ for $t < 0$, but if trajectories rise through several integers for positive t we can expect to find families of particles which lie on the same trajectory, and whose spins are separated by 2 units of angular momentum. We shall find in chapter 5 that this is indeed the case, with $\alpha(t)$ taking an approximately linear form

$$\alpha(t) = \alpha^0 + \alpha' t \quad (2.8.11)$$

as shown for example in fig. 2.9 and figs. 5.4–5.6. This provides one verification of the applicability of Regge's ideas to particle physics.

Another simple test is to look at the crossed s-channel physical region $s > s_{\mathrm{T}}$, $t < 0$. Here (2.8.10) gives, through (1.7.19) and (A.25),

$$A^{\mathrm{R}}(s,t) \underset{s\to\infty}{\sim} s^{\alpha(t)} \tag{2.8.12}$$

where now t gives the momentum transfer. Hence we expect to find that at high energy the s dependence of the s-channel scattering amplitude is a simple power behaviour, the power being a function of the momentum transfer (remember $\alpha(t)$ is real in this region). It should be an analytic continuation of the spins of the particles lying on the leading t-channel trajectory (see fig. 2.9 and fig. 6.6). Thus whereas $\alpha(t)$ is observable only at discrete points for positive t, where $\alpha(t) =$ integer and a particle occurs, it can be detected in the asymptotic s behaviour for all $t < 0$, at least in principle. In practice several trajectories may be exchanged in a given process making it hard to identify the different powers of s accurately, but it has proved possible to determine quite a lot of trajectories from the experimental data in this way – see section 6.8.

The power behaviour expected from the exchange of a Regge trajectory (sometimes called 'Reggeon') (2.8.12) may be contrasted with that from a fixed-spin (elementary) particle, (2.6.10), which corresponds to a Kronecker δ in the l plane, (2.6.11). From (A.25) we see that (2.6.10) gives $A(s,t) \sim s^{\sigma}$, where σ is always integral, and independent of t. At first sight it is rather surprising that the exchange of many particles with high spins on a trajectory like fig. 2.9 should give rise to the power $\alpha(t) < 1$ for $t < 0$ (as required by the Froissart bound) when each particle individually would give s^{σ_i}, $i = 1, 2, 3, \ldots$. The reason for this is that, in a sense, the contributions of the different partial waves cancel; but remember the partial-wave series does not converge in the s-channel region. Thus suppose we have a linear trajectory like (2.8.11), with poles at

$$t = M_l^2 \equiv \frac{l-\alpha^0}{\alpha'}, \quad l = 0, 1, 2, \ldots \quad (\alpha^0 < 0) \tag{2.8.13}$$

Then we can write the partial-wave series for these poles

$$\begin{aligned} A^{\mathscr{P}}(s,t) &= 16\pi \sum_l (2l+1) \frac{\beta(M_l^2)}{\alpha'(M_l^2-t)} P_l(z_t) \\ &= -\frac{16\pi}{2\mathrm{i}} \int_{C_1} (2l+1) \frac{\beta(t)}{l-\alpha(t)} \frac{P_l(-z_t)}{\sin \pi l} \,\mathrm{d}l \end{aligned} \tag{2.8.14}$$

and when we apply the Sommerfeld–Watson transform (2.7.8) we find of course that $A^{\mathscr{S}}(s,t) \sim s^{\alpha(t)}$.

The hypothesis of maximal analyticity of the second kind implies that all the subtractions needed in dispersion relations such as (1.10.7) are due to singularities in the angular-momentum plane like (2.7.2) and (2.7.4). If we allowed arbitrary subtractions, as in (1.10.10), the function $F_{n-1}(s,t)$ (a polynomial of degree $n-1$ in s) would contribute to all the (integer) partial waves $l = 0, 1, 2, \dots, n-1$ in the t channel, giving Kronecker δ terms in the l plane, $\delta_{l0}, \delta_{l1}, \dots, \delta_{ln-1}$, rather than singularities. But such terms are precluded by our analyticity postulate. The Froissart bound implies that the degree of $F_{n-1}(s,t)$ can be at most 1, so the higher partial waves are certainly obtainable from $D_s^{\mathscr{S}}(s,t)$; but the analyticity postulate also requires that the lowest partial waves should be obtained from the higher by analytic continuation, so they are given by $D_s^{\mathscr{S}}(s,t)$ too, and F is not arbitrary.

This closes a most important gap in the determination of the scattering amplitude by the unitarity equations. For we have seen in chapter 1 (especially section 1.10) that given all the particle poles (masses and couplings) one can, in principle, determine all the other singularities from the unitarity equations, and thence find the scattering amplitudes by using dispersion relations (apart from the subtractions). But there seemed to be no limitation on the number of particles which could occur. However, it is unlikely that one needs to put in all the particle poles *a priori*, since the composite particles which are generated by the forces should emerge as consequences of unitarity, and will lie on trajectories. For example, if one regards the deuteron as a neutron–proton bound state it should be possible to deduce its properties (mass and coupling) from a knowledge of the strong interaction forces, and it would be inconsistent to insert arbitrary values for these quantities.

Now maximal analyticity of the second kind tells us that if one knows $D_s^{\mathscr{S}}(s,t)$ one can work back, via the Froissart–Gribov projection, and determine the nature of all the poles, because they are all Regge poles. This requires a very high degree of self-consistency in strong-interaction theory. For if we were to try and invent a new particle, and insert it into the unitarity equations, it would generate further singularities, and hence further contributions to the asymptotic behaviour of the scattering amplitudes, and hence further Regge poles which would themselves have to be included in the unitarity equations – and so on.

Clearly if our postulates are correct the actual (perhaps infinite) number of different types of particles in the universe must be self-consistent, i.e. must reproduce itself, and no other particles, under the combined processes of unitarization and analytic continuation in l. But whether it is the unique set with this property, so that the self-consistency requirement determines the theory completely, is not clear. The proposal that all the strongly interacting particles are self generating in this way is called the 'bootstrap hypothesis' (see Chew 1962) and we shall examine it further below. Intuitively, it seems clear that if all the hadrons are to be composites of each other, and all the forces are due to the exchange of particles, then some form of self-consistency is necessary, and by invoking Regge theory it is possible to give a more rigorous formulation of this idea. Since this proposal eliminates elementary particles, and puts all the observed particles on an equal footing as composite Reggeons, it is sometimes referred to as 'nuclear democracy' (Chew 1965).

Alternatively, it may be that there are some basic elementary particles, for example quarks (see chapter 5), which do not lie on Regge trajectories, and whose properties one needs to know before one can predict the particle spectrum. If so, Regge theory will not be sufficient by itself to tell us everything about strong-interaction physics, but it will still provide important consistency constraints on scattering amplitudes. We shall return to these more philosophical problems in chapter 11.

2.9 The Mandelstam–Sommerfeld–Watson transform*

In (2.7.8) we chose the contour for the background integral, C_3, along $\mathrm{Re}\{l\} = -\frac{1}{2}$ because (see (A.25), (A.26)) this gives the most convergent behaviour of $P_l(z)$ ($\sim z^{-\frac{1}{2}}$ for $\mathrm{Re}\{l\} = -\frac{1}{2}$). However, this line is not a natural boundary of analytic continuation, and Mandelstam (1962) has shown how it may be crossed.

We begin by rewriting (2.7.7) as

$$A^{\mathscr{S}}(s,t) = 16\pi \sum_{l=0}^{\infty} \left\{(2l+1)\, A_l(t)\, P_l(z_t) + \frac{1}{\pi}(-1)^{l-1}\,(2l)\, A^{\mathscr{S}}_{l-\frac{1}{2}}(t)\, Q_{l-\frac{1}{2}}(z_t)\right\}$$
$$-16\pi \sum_{l=1}^{\infty} \frac{1}{\pi}(-1)^{l-1}\,(2l)\, A^{\mathscr{S}}_{l-\frac{1}{2}}(t)\, Q_{l-\frac{1}{2}}(z_t) \qquad (2.9.1)$$

We then make a Sommerfeld–Watson transform of the two terms in

* This section may be omitted at first reading.

brackets { } in (2.9.1), the first giving (2.7.5) and the second involving $Q_l(-z_t)(\cos\pi l)^{-1}$ which has the required poles at half-integer values of l. Then using (A.18) these two integrals can be combined giving, when we open up the contour as in (2.7.6),

$$\begin{aligned} A^{\mathscr{S}}(s,t) = {} & \frac{16}{2\mathrm{i}}\int_{-\frac{1}{2}+\epsilon-\mathrm{i}\infty}^{-\frac{1}{2}+\epsilon+\mathrm{i}\infty}(2l+1)\,A_l^{\mathscr{S}}(t)\,\frac{Q_{-l-1}(z_t)}{\cos\pi l}\,\mathrm{d}l \\ & +16\pi(2\alpha(t)+1)\,\beta(t)\,\frac{Q_{-\alpha(t)-1}(-z_t)}{\cos\pi\alpha(t)} \\ & +\frac{16}{2\mathrm{i}}\int^{\alpha_c(t)}(2l+1)\,\Delta_l^{\mathscr{S}}(t)\,\frac{Q_{-l-1}(-z_t)}{\cos\pi l}\,\mathrm{d}l \\ & -16\pi\sum_{l=1}^{\infty}\frac{1}{\pi}(-1)^{l-1}(2l)\,A_{l-\frac{1}{2}}^{\mathscr{S}}(t)\,Q_{l-\frac{1}{2}}(z_t) \end{aligned} \tag{2.9.2}$$

The contour of the background integral has been put at $\frac{1}{2}+\epsilon$ $(\epsilon>0)$ to avoid the pole of $(\cos\pi l)^{-1}$ at $l=-\frac{1}{2}$ (fig. 2.10). If we now displace this contour to $\mathrm{Re}\{l\}=-l$ we pick up contributions from the poles at $l\equiv l'$ (say) $=-\frac{1}{2},\,-\frac{3}{2},\ldots,\,-L'$, where $-L'$ is the first half-integer above $-L$, giving

$$\begin{aligned} A^{\mathscr{S}}(s,t) = {} & \frac{16}{2\mathrm{i}}\int_{-L-\mathrm{i}\infty}^{-L+\mathrm{i}\infty}(2l+1)\,A_l^{\mathscr{S}}(t)\,\frac{Q_{-l-1}(-z_t)}{\cos\pi l}\,\mathrm{d}l+\text{poles}+\text{cuts} \\ & -16\pi\sum_{l'=-L'}^{-\frac{1}{2}}(2l'+1)\,A_{l'}^{\mathscr{S}}(t)\,Q_{-l'-1}(-z_t)\,\frac{(-1)^{l'-\frac{1}{2}}}{\pi} \\ & -16\pi\sum_{l=1}^{\infty}\frac{(-1)^{l-1}}{\pi}(2l)\,A_{l-\frac{1}{2}}^{\mathscr{S}}(t)\,Q_{l-\frac{1}{2}}(z_t) \end{aligned} \tag{2.9.3}$$

If we now replace the summation index l' in the second line by $l=-l'-\frac{1}{2}$, this line becomes

$$16\pi\sum_{l=0}^{L'-\frac{1}{2}}\frac{(-1)^{-l-1}}{\pi}(-2l)\,A_{-l-\frac{1}{2}}^{\mathscr{S}}(t)\,Q_{l-\frac{1}{2}}(-z_t) \tag{2.9.4}$$

which will cancel with the first $L'-\frac{1}{2}$ terms of the last summation in (2.9.3) provided

$$A_{l-\frac{1}{2}}^{\mathscr{S}}(t)=A_{-l-\frac{1}{2}}^{\mathscr{S}}(t)\quad\text{for}\quad l=\text{integer} \tag{2.9.5}$$

This symmetry of partial-wave amplitudes about $l=-\frac{1}{2}$, the so-called 'Mandelstam symmetry', follows from the Froissart–Gribov projection (2.6.2) and the corresponding symmetry (A.19) of $Q_l(z)$ (except that of course the projection does not converge without subtractions), and as we shall see in the next chapter it is true in

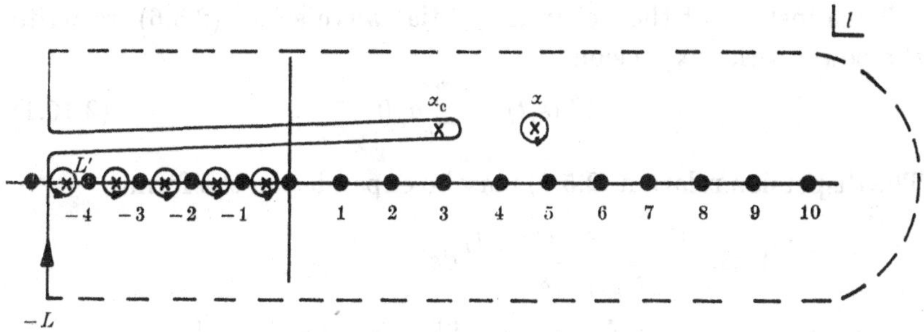

FIG. 2.10 The integration contour in (2.9.2) with the same singularities as fig. 2.8, but we also pick up extra poles at the negative half-integers.

potential scattering, so it seems reasonable to suppose that it will also hold in strong interactions. If so we end up with

$$A^{\mathscr{S}}(s,t) = \frac{16}{2i}\int_{-L-i\infty}^{-L+i\infty} (2l+1)\,A_l^{\mathscr{S}}(t)\,\frac{Q_{-l-1}(-z_t)}{\cos\pi l}\,dl + \text{poles} + \text{cuts}$$
$$-16\sum_{l=L'+\frac{1}{2}}^{\infty}(-1)^{l-1}(2l)\,A^{\mathscr{S}}_{-l-\frac{1}{2}}(t)\,Q_{l-\frac{1}{2}}(z_t) \qquad (2.9.6)$$

Since from (A.27) $Q_l(z) \sim z^{-l-1}$, the Regge pole and cut terms (given explicitly in (2.9.2)) still have the asymptotic behaviour $\sim s^{\alpha(t)}$, but the first and last terms of (2.9.6) $\sim s^{-L}$ where L can be made as large as we like. Of course in displacing the contour in this way we can expect to expose more poles and cuts, and the magnitude of the background integral at fixed z may increase.

The actual pole in the Regge term in (2.9.2) has been absorbed into $Q_{-\alpha-1}$, which has poles for α = a non-negative integer (see (A.32)). The apparent poles from $(\cos\pi\alpha)^{-1}$, at positive half-integer values of α, cancel with the zeros of $Q_{-\alpha-1}$ which contains $(\Gamma(-\alpha+\frac{1}{2}))^{-1}$ (see (A.8)) while the symmetry (2.9.5) ensures that the residues of these poles vanish for negative half-integers.

2.10 The Mellin transform*

Frequently we shall be concerned only with the leading asymptotic s behaviour of the scattering amplitude, in which case many of our equations can be greatly simplified by including only the asymptotic behaviour of the Legendre functions, (A.25), (A.27), and making the replacement $z_t \to s$ for $s \to \infty$.

* This section may be omitted at first reading.

Thus instead of the t-channel partial-wave series (2.5.6) we write the power series expansion

$$A^{\mathscr{S}}(s,t) = \sum_{n=0}^{\infty} a_n(t)\, s^n \tag{2.10.1}$$

The dispersion relation (2.5.8) may be expanded in the form

$$\begin{aligned} A^{\mathscr{S}}(s,t) &= \frac{1}{\pi}\int_{s_{\mathrm{T}}}^{\infty} \frac{D_s^{\mathscr{S}}(s',t)}{s'-s}\,\mathrm{d}s' \\ &= \frac{1}{\pi}\int_{s_{\mathrm{T}}}^{\infty} D_s^{\mathscr{S}}(s',t)\frac{1}{s'}\left[1+\frac{s}{s'}+\left(\frac{s}{s'}\right)^2+\dots\right]\mathrm{d}s' \end{aligned} \tag{2.10.2}$$

and on comparing with (2.10.1) for each power of s we find

$$a_n(t) = \frac{1}{\pi}\int_{s_{\mathrm{T}}}^{\infty} D_s^{\mathscr{S}}(s',t)\, s'^{-(n+1)}\,\mathrm{d}s' \tag{2.10.3}$$

which corresponds to taking the leading s term of the Legendre function in the Froissart–Gribov projection (2.5.3). However, the position of the threshold is irrelevant as far as the leading behaviour is concerned, and so it will not make much difference if we write instead of (2.10.3)

$$a_n(t) = \frac{1}{\pi}\int_{0}^{\infty} D_s^{\mathscr{S}}(s',t)\, s'^{-(n+1)}\,\mathrm{d}s' \tag{2.10.4}$$

This is the Mellin transform of $D_s^{\mathscr{S}}(s',t)$ (see Titchmarsh (1937) p. 7), and its inverse is

$$D_s^{\mathscr{S}}(s,t) = \frac{1}{2\mathrm{i}}\int_{-\mathrm{i}\infty+\gamma}^{\mathrm{i}\infty+\gamma} a_n(t)\, s^n\,\mathrm{d}n \tag{2.10.5}$$

where the contour of integration is along a line parallel to the imaginary axis to the right of all the singularities in n of $a_n(t)$.

Now if we take the leading power of the Legendre function in the Sommerfeld–Watson transform (2.7.5) we get

$$A^{\mathscr{S}}(s,t) = -\frac{16\pi}{2\mathrm{i}}\int_{C_1}(2l+1)\,A_l^{\mathscr{S}}(t)\frac{(-s)^l}{\sin\pi l}\,\mathrm{d}l \tag{2.10.6}$$

which agrees with (2.10.5) if we remember that

$$\mathrm{Disc}_s\{(-s)^l\} = -s^l\sin\pi l, \quad s>0,$$

and if we incorporate the factor $16\pi(2l+1)$ into $a_n(t)$. The contour C_1 in (2.10.6) can be expanded to that in (2.10.5), but if $D_s^{\mathscr{S}}(s,t) \sim s^{\alpha(t)}$ then $a_n(t)$ will obtain a pole at $n = \alpha(t)$ from (2.10.3) (see (2.7.2)), whose contribution will have to be added to (2.10.5) similar to (2.7.8). Hence Regge poles in the l plane give rise to poles in the n plane. However, since the Legendre function can be expanded as a power series in z_t, of which (A.25) is only the first term, a given Regge pole will produce a series of poles in the n plane at $n = \alpha(t) - m$, $m = 0, 1, 2, \ldots$; and vice versa. But as long as we are only concerned with the leading behaviour this many-to-one correspondence between poles in the l and n planes will not matter.

The dispersion properties are somewhat different in that (2.10.6) is cut for $0 \leqslant s \leqslant \infty$ while the pole in (2.7.8) is cut for $z_t > 1$ (see (A.13)), i.e. $-4q_t^2 \leqslant s \leqslant \infty$ for equal-mass kinematics, from (1.7.22). Of course neither of these is correct because the s cuts of the amplitude should start at the threshold $s = s_{\mathrm{T}}$. So there must be a cancellation between the discontinuities of the pole terms and the background integral in the regions $0 \leqslant s \leqslant s_{\mathrm{T}}$ and $-4q_t^2 \leqslant s \leqslant s_{\mathrm{T}}$, respectively. Also we shall find in chapter 6 that the replacement of z_t by s is not always trivial with unequal-mass kinematics. But provided these points are borne in mind it is frequently convenient to use (2.10.4) and (2.10.5) instead of the more exact expressions.

3
Some models containing Regge poles*

3.1 Introduction

In the previous chapter we showed how, by analytically continuing the partial-wave amplitudes in angular momentum, one can represent the scattering amplitude as a sum of pole and cut contributions in the complex l plane. Cuts do not occur in potential scattering, or in some of the simpler models for strong interactions, and they will not be introduced until chapter 8. But Regge poles correspond to bound-state or resonance particles, and in this chapter we shall examine their occurrence in non-relativistic potential-scattering amplitudes, in Feynman perturbation field theory, and in various models of strong-interaction dynamics.

Though clearly none of these examples can prove that Regge poles will actually occur in hadronic processes, they do help to make it plausible. They also give some indication of the properties which Regge trajectories may be expected to possess.

We begin by discussing some of the more general results which are independent of particular models.

3.2 Properties of Regge trajectories

The analyticity and unitarity properties of the partial-wave amplitudes imply certain general features of the Regge trajectories.

For example the occurrence of a pole at $l = \alpha(t)$ implies that

$$(B_l(t))^{-1} \to 0 \quad \text{as} \quad l \to \alpha(t) \tag{3.2.1}$$

which may be used implicitly to define the function $\alpha(t)$, and hence tells us about the analyticity of $\alpha(t)$. It is more useful however to begin by writing, from (2.6.2) and (2.6.8) (Oehme and Tiktopoulos

* This chapter may be ommitted at first reading.

1962, Barut and Zwanziger 1962),

$$B_l(t) = \int_{s_T}^{s_1} + \int_{s_1}^{\infty} \left[\frac{1}{16\pi^2} Q_l(z_t)\, D_s(s,t)\, \mathrm{d}z_t\, (q_{t13}q_{t24})^{-l}\right]$$
$$\equiv E_l(t) + F_l(t) \qquad (3.2.2)$$

where to define $E_l(t)$ and $F_l(t)$ we have split the region of integration at some arbitrary point s_1. Then if $D_s(s,t) \sim s^{\alpha(t)}$, since from (A.27)

$$Q_l(z_t) \sim s^{-l-1}$$

we find that $$F_l(t) \sim \int_{s_1}^{\infty} s^{-l-1+\alpha(t)}\mathrm{d}s = -\frac{\mathrm{e}^{(\alpha(t)-l)\log s_1}}{\alpha(t)-l} \qquad (3.2.3)$$

and so contains the pole. $E_l(t)$ involves only a finite integration in s and so has no pole. Thus instead of (3.2.1) we can define $\alpha(t)$ by

$$(F_l(t))^{-1} \to 0 \quad \text{as} \quad l \to \alpha(t) \qquad (3.2.4)$$

It is evident from (3.2.2) that $F_l(t)$ has similar singularities to $B_l(t)$, i.e. the same dynamical right-hand cut starting at the threshold t_T, and a similar left-hand cut due to the s-singularities, but with the branch point pushed further to the left in the t plane as its position is determined by s_1 not s_T (substituted for M^2 in (2.6.16), see section 2.6). The kinematical threshold singularity has of course been removed from $B_l(t)$, and hence $F_l(t)$ in (3.2.2).

The implicit function theorem (Titchmarsh (1939) p. 198) tells us that if $(F_l(t))^{-1}$ is regular in the neighbourhood of some point $t = t_p$, say, and if

$$\frac{\partial}{\partial l}(F_l(t_p))^{-1}\big|_{l=\alpha(t_p)} \neq 0 \qquad (3.2.5)$$

then $\alpha(t_p)$ is also a regular function in the neighbourhood of t_p. This is easily demonstrated by expanding $(F_l(t_p))^{-1}$ in a Taylor series about $t = t_p$. $l = \alpha(t_p)$, i.e.

$$(F_l(t))^{-1} = a_1(l-\alpha(t_p)) + a_2(l-\alpha(t_p))^2 + \ldots + b_1(t-t_p)$$
$$+ b_2(t-t_p)^2 + \ldots + c_2(t-t_p)(l-\alpha(t_p)) + \ldots \qquad (3.2.6)$$

Then setting $(F_l(t))^{-1} = 0$ at $l = \alpha(t)$ gives

$$\alpha(t) = \alpha(t_p) - \frac{b_1}{a_1}(t-t_p) + \ldots \qquad (3.2.7)$$

a Taylor series for $\alpha(t)$, so α must be regular in the neighbourhood of t_p. However, if (3.2.5) does not hold, i.e. if $a_1 = 0$, then

$$\alpha(t) = \alpha(t_p) \pm \left(-\frac{b_1}{a_2}\right)^{\frac{1}{2}} (t-t_p)^{\frac{1}{2}} + \ldots \qquad (3.2.8)$$

and so there are two trajectories which cross at $t = t_p$, each with a square-root branch point, such that their imaginary parts for $t < t_p$ are equal and opposite, to preserve the analyticity of F_l^{-1}. Of course if b_1 also vanishes at this point there will not be a branch point.

Thus we conclude that $\alpha(t)$ will be analytic where F_l^{-1} is analytic unless two (or more) trajectories cross each other, in which case there may, but need not, be a branch point in each trajectory function. So unless trajectories cross we can expect $\alpha(t)$ to have the same singularities as $(F_l(t))^{-1}$. However, the position of the left-hand cut in $F_l(t)$ is arbitrary as it depends on s_1. We can make s_1 as large as we like and still obtain a pole in (3.2.3) from the divergence of the integrand in (3.2.2) as $s \to \infty$, so it is evident that $\alpha(t)$ cannot contain the left-hand cut of $(F_l(t))^{-1}$. Hence $\alpha(t)$ has just the dynamical right-hand cut from $t_T \to \infty$, unless two trajectories collide.

Such collisions must in fact occur at $t = 0$ for fermion trajectories in order to satisfy the generalized MacDowell symmetry (see section 6.5 below). Also they have been observed to occur in various potential-scattering calculations, but this can only happen for $\text{Re}\{l\} < -\frac{1}{2}$ (see the next section). There is no direct evidence that complex trajectories occur in hadron physics for $t < 0$ (see however section 8.6), and it is usually assumed that the trajectory functions are real for $t < t_T$.

Then since $\alpha(t)$ is real analytic we can write a dispersion relation

$$\alpha(t) = \frac{1}{\pi}\int_{t_T}^{\infty} \frac{\text{Im}\{\alpha(t')\}}{t'-t}\, dt' \tag{3.2.9}$$

However, subtractions will usually be needed. For example if

$$\text{Re}\{\alpha(t)\} \underset{t\to\infty}{\to} A(t),$$

a polynomial in t, we may have

$$\alpha(t) = A(t) + \frac{1}{\pi}\int_{t_T}^{\infty} \frac{\text{Im}\{\alpha(t')\}}{t'-t}\, dt' \tag{3.2.10}$$

We shall find in the next section that with well behaved potentials like the Yukawa the trajectories tend to negative integers as $t \to \infty$, giving

$$\alpha(t) = -n + \frac{1}{\pi}\int_{t_T}^{\infty} \frac{\text{Im}\{\alpha(t')\}}{t'-t}\, dt', \quad n = 1, 2, 3, \ldots \tag{3.2.11}$$

On the other hand in particle physics trajectories seem to be approximately linear, with rather small imaginary parts (see section 5.3) suggesting instead

$$\alpha(t) = \alpha_0 + \alpha_1 t + \frac{1}{\pi}\int_{t_T}^{\infty} \frac{\text{Im}\{\alpha(t')\}}{t'-t}\, dt' \tag{3.2.12}$$

Or the integral in (3.2.12) may not converge, in which case subtractions will be needed as in (1.10.10), and if for example two subtractions are sufficient we get

$$\alpha(t) = \alpha_0 + \alpha_1 t + \frac{t^2}{\pi}\int_{t_T}^{\infty} \frac{\operatorname{Im}\{\alpha(t')\}}{t'^2(t'-t)}\,dt' \tag{3.2.13}$$

We have chosen to make the subtractions at $t = 0$ so that $\alpha_0 = \alpha(0)$ and $\alpha_1 = \alpha'(0) \equiv (d\alpha/dt)_{t=0}$.

We shall find (see section 5.4) that $\operatorname{Im}\{\alpha(t)\} > 0$ for $t > t_T$, so if we take the nth derivative of (3.2.11) or (3.2.12) or (3.2.13)

$$\frac{d^n\alpha}{dt^n} = \frac{n!}{\pi}\int_{t_T}^{\infty} \frac{\operatorname{Im}\{\alpha(t')\}}{(t'-t)^{n+1}}\,dt' \tag{3.2.14}$$

we find that all the derivatives are positive for $t < t_T$. A function with this property is called a Herglotz function (Herglotz 1911).

If the pole takes the form (2.8.3), we have from (2.6.8)

$$B_l(t) \xrightarrow[l\to\alpha(t)]{} \frac{\gamma(t)}{l-\alpha(t)} \quad \text{where} \quad \gamma(t) \equiv \beta(t)\,(q_{t13}\,q_{t24})^{-\alpha(t)} \tag{3.2.15}$$

The function $\gamma(t)$, the Regge residue with the threshold behaviour removed, is often referred to as the 'reduced residue'. We can use Cauchy's residue theorem to write (from (3.2.2))

$$\gamma(t) = \frac{1}{2\pi i}\oint dl\, F_l(t) \tag{3.2.16}$$

where the integration contour is a closed path encircling the point $l = \alpha(t)$, but no other singularities of F_l. This equation together with the implicit function theorem tells us that $\gamma(t)$ will have similar analyticity properties to $\alpha(t)$, i.e. just the dynamical right-hand cut of $F_l(t)$ unless two or more trajectories cross. So as with (3.2.9) we can write

$$\gamma(t) = \frac{1}{\pi}\int_{t_T}^{\infty} \frac{\operatorname{Im}\{\gamma(t')\}}{t'-t}\,dt' \tag{3.2.17}$$

again making subtractions if necessary.

It is also possible to deduce the nature of the branch point in the trajectory function at t_T from the unitarity equation. If we consider the elastic scattering process $1+3 \to 1+3$ below the inelastic threshold in the t channel, t_I, we have, from (2.6.23),

$$\operatorname{Im}\{(B_l(t))^{-1}\} = -\rho(t)\,(q_{t13})^{2l}, \quad t_T < t < t_I \tag{3.2.18}$$

where $\rho(t) \equiv 2q_{t13}t^{-\frac{1}{2}}$. Now the function

$$-\frac{\mathrm{i}\rho(t)\,(-q_{t13})^{2l}}{\cos \pi l} = -\frac{\mathrm{i}\rho(t)\,(q_{t13})^{2l}\,\mathrm{e}^{\pm \mathrm{i}\pi l}}{\cos \pi l} \tag{3.2.19}$$

has the same discontinuity as $(B_l(t))^{-1}$ for $t_{\mathrm{T}} < t < t_{\mathrm{I}}$, so that

$$Y(t, l) \equiv \cos \pi l (B_l(t))^{-1} + \mathrm{i}\rho(t)\,(-q_{t13})^{2l} \tag{3.2.20}$$

is analytic in this region. From (3.2.1.) we have

$$Y(t, l) \to \mathrm{i}\rho(t)\,(-q_{t13})^{2l}, \quad \text{for} \quad l \to \alpha(t) \tag{3.2.21}$$

If we define $\alpha_{\mathrm{T}} \equiv \alpha(t_{\mathrm{T}})$ we have (using (1.7.15), $t_{\mathrm{T}} = (m_1 + m_3)^2$)

$$Y(t, \alpha_{\mathrm{T}}) \approx -\frac{2}{\sqrt{t_{\mathrm{T}}}}\left(\frac{t_{\mathrm{T}} - t}{4}\right)^{\alpha_{\mathrm{T}}+\frac{1}{2}} \quad \text{for} \quad t \to t_{\mathrm{T}} \tag{3.2.22}$$

so $Y(t_{\mathrm{T}}, \alpha_{\mathrm{T}}) = 0$ if $\alpha_{\mathrm{T}} > -\frac{1}{2}$. We can expand Y in a Taylor series about the threshold values of t and α, giving

$$Y(t, \alpha(t)) = Y(t_{\mathrm{T}}, a_{\mathrm{T}}) + Y'_l(\alpha(t) - \alpha_{\mathrm{T}}) + Y'_t(t - t_{\mathrm{T}}) + \ldots \tag{3.2.23}$$

where

$$Y'_l \equiv \left.\frac{\partial Y}{\partial l}\right|_{\substack{l=\alpha_{\mathrm{T}},\\ t=t_{\mathrm{T}}}} \quad Y'_t \equiv \left.\frac{\partial Y}{\partial t}\right|_{\substack{l=\alpha_{\mathrm{T}}\\ t=t_{\mathrm{T}}}} \tag{3.2.24}$$

and so

$$\alpha(t) = \alpha_{\mathrm{T}} - \frac{2}{\sqrt{t_{\mathrm{T}}}}\,Y_l'^{-1}\left(\frac{t - t_{\mathrm{T}}}{4}\right)^{\alpha_{\mathrm{T}}+\frac{1}{2}} \mathrm{e}^{-\mathrm{i}\pi(\alpha_{\mathrm{T}}+\frac{1}{2})}$$
$$- (t - t_{\mathrm{T}})\left(\frac{Y'_t}{Y'_l}\right) + \ldots, \quad \alpha_{\mathrm{T}} > -\tfrac{1}{2} \tag{3.2.25}$$

Hence the trajectory has a threshold cusp for $-\frac{1}{2} < \alpha_{\mathrm{T}} < \frac{1}{2}$ and above threshold

$$\mathrm{Im}\,\{\alpha(t)\} \sim (t - t_{\mathrm{T}})^{\alpha+\frac{1}{2}}, \quad t \approx t_{\mathrm{T}} \tag{3.2.26}$$

However in potential scattering these cusp effects seem to be small (Warburton 1964).

Since

$$Y(t, l) \to -\frac{2}{\sqrt{t_{\mathrm{T}}}}, \quad l \to -\tfrac{1}{2}, \quad t \to t_{\mathrm{T}}, \tag{3.2.27}$$

the condition for a pole (3.2.1) becomes, from (3.2.20),

$$\frac{2}{\sqrt{t_{\mathrm{T}}}} = \frac{2}{\sqrt{t_{\mathrm{T}}}}\,(q_{t13})^{2(l+\frac{1}{2})}\,\mathrm{e}^{-\mathrm{i}\pi(l+\frac{1}{2})} \tag{3.2.28}$$

which can be satisfied by $l = \alpha_n$ for any α_n such that

$$(\log(q_{t13}^2) - \mathrm{i}\pi)(\alpha_n + \tfrac{1}{2}) = 2\pi n\mathrm{i}, \quad n = 0, \pm 1, \pm 2, \ldots \tag{3.2.29}$$

that is

$$\alpha_n = \frac{2\pi n}{\pi + \mathrm{i}\log(q_{t13}^2)} - \tfrac{1}{2} \tag{3.2.30}$$

So an infinite number of trajectories converge on $\alpha = -\frac{1}{2}$ as $t \to t_T$ ($q_{t13} \to 0$). This is sometimes called the Gribov–Pomeranchuk phenomenon (Gribov and Pomeranchuk 1962). Their occurrence should serve as a warning against supposing that the left-half angular-momentum plane is likely to have a simple singularity structure.

3.3 Potential scattering

In this section we shall briefly review the behaviour of solutions of the Schroedinger equation for non-relativistic potential scattering as a function of l. As we have already mentioned this is how Regge poles were first discovered (Regge 1959) and there is the great advantage that all the results can be proved rigorously. But as potential scattering is only of limited relevance to particle physics our discussion will be rather cursory, and we refer the interested reader to more complete studies, where the required proofs are given in detail (Squires 1963, Newton 1964, de Alfaro and Regge 1965).

a. Solutions of the Schroedinger equation

If the interaction potential $V(r)$ is a function of the r only, the solutions of the Schroedinger equation (1.13.3)

$$\nabla^2\psi - U(r)\psi + k^2\psi = 0 \tag{3.3.1}$$

can be decomposed into partial waves (see for example Schiff (1968) p. 81)

$$\psi(r,\theta,\phi) = \sum_{l=0}^{\infty} \frac{1}{r}\phi_l(r) P_l(\cos\theta) \tag{3.3.2}$$

The cylindrical symmetry removes any dependence on the azimuthal angle ϕ, and the radial wave function $\phi_l(r)$ satisfies the radial Schroedinger equation (2.1.1)

$$\frac{d^2\phi_l(r)}{dr^2} + \left(k^2 - \frac{l(l+1)}{r^2} - U(r)\right)\phi_l(r) = 0 \tag{3.3.3}$$

The quantization of angular momentum, which restricts l to integer values, stems from the requirement that angular dependence of (3.3.2) be finite for all values of θ. But in (3.3.3) l appears as a free parameter, and the equation can be solved for any value of l. Poincaré's theorem (see below) tells us that the solutions of such a differential equation are usually analytic functions of such parameters, so we may expect $\phi_l(r)$ to be analytic in l. It is also useful to note the symmetry of (3.3.3) under the replacements $l \to -(l+1)$, and $k \to -k$.

As long as the potential is 'regular', i.e. $r^2 U(r) \to 0$ as $r \to 0$, the small-r solutions of (3.3.3) are controlled by the centrifugal barrier term $l(l+1)r^{-2}$. This constitutes a repulsive addition to the effective potential (for $l > 0$), and physically of course it represents the increased difficulty of holding particles together if they have a high relative angular momentum due to the centrifugal force. As $r \to 0$ we can neglect k^2 and U in (3.3.3). Evidently there are two independent solutions which behave like r^{-l} and r^{l+1}, respectively, as $r \to 0$. The physical solution must be finite at the origin, however, and we denote it by $\phi_l(r) = \phi(l, k, r) \sim r^{l+1}$.

It satisfies the integral equation (Newton (1964) p. 21, de Alfaro and Regge (1965), p. 21)

$$\phi(l,k,r) = \phi_0(l,k,r) + \int_0^r \mathrm{d}r'\, G(r,r')\, U(r')\, \phi(l,k,r')\, \mathrm{d}r' \quad (3.3.4)$$

where G is the Green's function, which may be written in terms of Hankel functions as

$$G(r,r') = \mathrm{i}\,\frac{\pi}{4}\,(rr')^{\frac{1}{2}}\,(H^{(2)}_{l+\frac{1}{2}}(kr)\, H^{(1)}_{l+\frac{1}{2}}(kr') - H^{(1)}_{l+\frac{1}{2}}(kr)\, H^{(2)}_{l+\frac{1}{2}}(kr')) \quad (3.3.5)$$

and where ϕ_0 is a solution of (3.3.3) with $U(r) = 0$, i.e.

$$\phi_0(l,k,r) = r^{\frac{1}{2}}\,\Gamma(l+\tfrac{3}{2})\left(\frac{k}{2}\right)^{-l-\frac{1}{2}} J_{l+\frac{1}{2}}(kr) \quad (3.3.6)$$

J being a Bessel function. It can be checked by direct substitution that (3.3.4) satisfies (3.3.3), and the boundary condition at $r = 0$.

As long as $rU(r) \to 0$ as $r \to \infty$, both $U(r)$ and the centrifugal barrier term become irrelevant in (3.3.3) as $r \to \infty$, and in this limit it is more convenient to consider the 'irregular' solutions $\chi(l, \pm k, r)$ whose boundary conditions are $\chi(l, \pm k, r) \sim \mathrm{e}^{\mp ikr}$ as $r \to \infty$, because these give the incoming and outgoing plane waves, in terms of which the scattering amplitude is defined. They satisfy the integral equation (Newton (1964) p. 14, de Alfaro and Regge (1965) p. 23)

$$\chi(l,k,r) = \chi_0(l,k,r) - \int_r^\infty G(r,r')\, U(r')\, \chi(l,k,r')\, \mathrm{d}r' \quad (3.3.7)$$

where again G is given by (3.3.5) and χ_0 is a solution of (3.3.3) with $U(r) = 0$, i.e.

$$\chi_0(l, k, r) = e^{-i\frac{1}{2}\pi(l+1)} \left(\frac{\pi kr}{2}\right)^{\frac{1}{2}} H^{(2)}_{l+\frac{1}{2}}(kr) \tag{3.3.8}$$

The other independent solution is obtained by letting $k \to -k$.

Since any solution of (3.3.3) can be expressed in terms of these independent solutions, we can relate the physical solution (3.3.4) to the asymptotic plane-wave solutions (3.3.7), viz.

$$\phi(l, k, r) = \frac{1}{2ik}(f(l, k)\chi(l, -k, r) - f(l, -k)\chi(l, k, r)) \tag{3.3.9}$$

where the f's are called Jost functions and satisfy (de Alfaro and Regge (1965) p. 39)

$$f(l, k) = f_0(l, k) + \int_0^\infty U(r')\chi(l, k, r')\phi(l, k, r')\,dr' \tag{3.3.10}$$

$$f_0(l, k) = \frac{2}{\pi^{\frac{1}{2}}}\Gamma(l+\tfrac{3}{2})\left(\frac{k}{2}\right)^{-l} e^{-\frac{1}{2}i\pi l} \tag{3.3.11}$$

Hence as $r \to \infty$

$$\phi(l, k, r) \to \frac{1}{2ik}(f(l, k)\, e^{ikr} - f(l, -k)\, e^{-ikr}) \tag{3.3.12}$$

But the partial-wave S-matrix is $S(l, k) = e^{2i\delta_l(k)}$, where $\delta_l(k)$ is the phase shift (see (2.2.10)), and is related to the asymptotic form of the regular solution by

$$\phi(l, k, r) \sim (e^{-ikr} - e^{-i(\pi l - kr)} S(l, k)) \tag{3.3.13}$$

i.e. $S(l, k)$ gives the ratio of the outgoing flux ($\chi \sim e^{ikr}$) to the incoming flux ($\chi \sim e^{-ikr}$) for the given partial wave, So in terms of the Jost functions

$$S(l, k) = \frac{f(l, k)}{f(l, -k)} e^{i\pi l} \tag{3.3.14}$$

and the partial-wave scattering amplitude is obtained from this S-matrix by

$$A_l(k) = \frac{S(l, k) - 1}{2ik} \tag{3.3.15}$$

(See (2.2.10). With non-relativistic kinematics $\rho(s) \to k$.)

b. *Analyticity properties of the solutions*

The analyticity properties of $A_l(k)$ are readily deduced from those of $f(l, k)$ obtained from (3.3.10).

Poincaré's theorem (Poincaré 1884) states that if a given parameter occurs in a differential equation only in functions which are holomorphic in that parameter, and if the boundary conditions are independent of the parameter, then the solutions to the equation will be holomorphic in the given parameter.

Thus since (3.3.3) is analytic in l, and since if we consider the function $r^{-l-1}\phi(l,k,r)$ the boundary conditions become independent of l, the regular solution $\phi(l,k,r)$ must be analytic in l for $\mathrm{Re}\{l\} > -\frac{1}{2}$. However, for $\mathrm{Re}\{l\} < -\frac{1}{2}$ the regular solution $\to\infty$ as $r\to 0$, because $r = 0$ is not a regular point of (3.3.3).

To continue to $\mathrm{Re}\{l\} < -\frac{1}{2}$ we have to analytically continue the integral equation (3.3.4), and the possibility of doing this depends on the nature of the potential. If the potential is singular, i.e. $rU(r)\to\infty$ for $r\to 0$, then for a repulsive potential the boundary condition becomes independent of l, since the potential provides the most singular term. So we can simply use the symmetry of (3.3.3) under $l\to-(l+1)$ to obtain the S-matrix for $\mathrm{Re}\{l\} < -\frac{1}{2}$, i.e. from (3.3.14)

$$S(l,k) = -\mathrm{e}^{-2\pi i l}S(-l-1,k) \tag{3.3.16}$$

This exhibits the Mandelstam symmetry (2.9.5). However, for an attractive singular potential the S-matrix cannot be defined as there will be an infinite number of bound states (see Frank, Land and Spector 1971).

But we are mainly concerned with potentials which are regular at the origin, like the generalized Yukawa potential (1.13.17). For such we can make the expansions

$$\left.\begin{aligned} rU(r)-k^2r &= \sum_{n=0}^{\infty} a_n r^n \\ \phi(l,k,r) &= r^{l+1}\sum_{n=0}^{\infty} b_n r^n \end{aligned}\right\} \tag{3.3.17}$$

and on substituting in (3.3.3), and equating coefficients of the various powers of r, one finds

$$\left.\begin{aligned} b_n &= \frac{1}{(2l+n+1)n}\sum_{m=0}^{n-1} a_m b_{n-1-m}, \quad n\geqslant 1 \\ b_0 &= 1 \end{aligned}\right\} \tag{3.3.18}$$

So ϕ is meromorphic in l with poles at $2l = -(n+1)$, i.e. $2l$ = negative integers, provided that the series (3.3.17) converges for r near zero. The same will be true of the Jost functions in (3.3.9) except that the poles at half-integer l values vanish due to the Mandelstam symmetry.

And since the positions of the poles at negative integer l are independent of r, these fixed poles will cancel in the ratio (3.3.14), and so will be absent from the S-matrix.

If the potential vanishes at the origin, so that $rU(r) \sim r^{p+1}$, which in (1.13.17) implies (expanding the exponential) that

$$\int_m^\infty \rho(\mu)\,\mu^n\,\mathrm{d}\mu = 0 \quad \text{for} \quad n = 0, 1, \ldots, p, \tag{3.3.19}$$

then there are no poles of ϕ_l for integer $\mathrm{Re}\{l\} > -1-p/2$.

A special intermediate case is potentials which contain a singular term V_0/r^2. This may be combined with the centrifugal barrier term in (3.3.3) to give an effective angular momentum L, where

$$L(L+1) \equiv l(l+1)+V_0.$$

Thus the poles in L at $L = n$ give rise to branch points in l at

$$l = \tfrac{1}{2}\{-1 \pm [1-4V_0+4n(n+1)]\}^{\frac{1}{2}} \tag{3.3.20}$$

whose positions depend on V_0.

In strong interactions the very-short-distance behaviour of the interaction is the part we know least well, and so the applicability of the above analysis is uncertain. But the fact that the Yukawa potential and its generalizations, which are so analogous to particle exchange forces, do give rise to meromorphic Jost functions for $\mathrm{Re}\{l\} > -1$ suggests that the same may be true in particle physics too.

By precisely similar arguments to the above it can be shown that $\phi(l, k, r)$ is also holomorphic in k for all k ($\mathrm{Re}\, l > -\frac{1}{2}$), since k appears analytically in (3.3.3) and does not affect the boundary conditions. Similarly $\chi(l, k, r)\,\mathrm{e}^{\mathrm{i}kr}$ is holomorphic in k for $\mathrm{Re}\{k\} > 0$, $\mathrm{Im}\{k\} < 0$. But at $k = 0$ χ has a branch point which can be seen directly in the expression (3.3.8) for χ_0. The solution for $\mathrm{Re}\{k\} < 0$ can be obtained by continuing round this singularity replacing χ by $\chi(l, k\,\mathrm{e}^{-\mathrm{i}\pi}, r)$. Continuation to $\mathrm{Im}\{k\} > 0$ can be achieved by series methods, and it is found that the Jost functions have the Hermitian analyticity property

$$f(l, k) = f^*(l^*, k^*) \tag{3.3.21}$$

However if the potential has the Yukawa form, say, and behaves like e^{-mr} as $r \to \infty$, then the asymptotic form of the outgoing wave function $\chi \sim \mathrm{e}^{\mathrm{i}kr}$ is damped away faster than $U(r)\,\mathrm{e}^{-\mathrm{i}kr}$ as $r \to \infty$ if $\mathrm{Im}\{k\} > m/2$ and the series solution breaks down at this point. This is because the partial-wave amplitude has a left-hand cut in k^2 beginning at

$k^2 = -m^2/4$, as one would expect from the analyticity properties discussed in sections 1.13 and 2.6.

Having obtained the singularities of the Jost functions in k and l we can now discuss those of the scattering amplitude, which from (3.3.14) and (3.3.15) may be written

$$A_l(k) = \frac{1}{2ik}\left[\frac{e^{i\pi l}f(l,k) - f(l,-k)}{f(l,-k)}\right] \tag{3.3.22}$$

Clearly its singularities in k^2 will be the same as those of the f's, namely a left-hand cut starting at $k^2 = -m^2/4$, and a right-hand cut along the positive k^2 axis starting at $k^2 = 0$, as we found in section 1.13. In fact these partial-wave methods can be used to prove that Yukawa potential scattering satisfies the Mandelstam representation (Blankenbecler *et al.* 1960). The right-hand cut is of course a consequence of the unitarity condition $SS^* = 1$, and for integral l, from (3.3.15) and (3.3.21), this becomes

$$A_l(k_+) - A_l(k_-) = 2ikA_l(k_+)\,A_l(k_-) \tag{3.3.23}$$

where $k_{+,-}$ are evaluated above and below the cut (cf. (2.2.7)). But for non-integral l it is necessary to take out the threshold behaviour first (as in (2.6.8)) so we define

$$B_l(k) = \frac{A_l(k)}{k^{2l}} \tag{3.3.24}$$

which is Hermitian analytic and along the right-hand cut, $k^2 > 0$, satisfies the unitarity equation

$$2i\,\mathrm{Im}\{B_l(k)\} \equiv B_l(k_+) - B_l(k_-) = 2ik^{2l+1}B_l(k_+)\,B_l(k_-)$$
$$= 2ik^{2l+1}\,|B_l(k)|^2 \tag{3.3.25}$$

(cf. (2.6.23)).

c. *Regge poles*

In addition to these branch points there is the possibility that pole singularities may appear in (3.3.22) due to the vanishing of $f(l,-k)$. If this happens for a given l at say $k = ik_b$, $k_b > 0$, then it is evident from (3.3.12) that as $r \to \infty$ the wave function is damped exponentially like $e^{-k_b r}$, corresponding to a bound-state pole on the real negative k^2 axis. Since f is an analytic function of l the position of this pole at $l = \alpha(k_b^2)$, say, where the function α is defined by

$$f(\alpha(k_b^2), -k_b) = 0, \tag{3.3.26}$$

will also be an analytic function of l. On the other hand if there is a zero of $f(l, -k)$ at some $\operatorname{Im}\{k\} < 0$, say $k = k_\mathrm{R} - \mathrm{i}k_\mathrm{I}$, we may write, in this neighbourhood of k,

$$f(l, -k) \approx C(k - k_\mathrm{R} + \mathrm{i}k_\mathrm{I})$$

so
$$f(l, k) = f^*(l^*, -k^*) = C^*(k - k_\mathrm{R} - \mathrm{i}k_\mathrm{I}) \tag{3.3.27}$$

where C is some constant, producing a resonance pole in the S-matrix (3.3.14) of the form

$$S(l, k) \approx \mathrm{e}^{\mathrm{i}(\pi l - 2 \arg C)} \left(\frac{k - k_\mathrm{R} - \mathrm{i}k_\mathrm{I}}{k - k_\mathrm{R} + \mathrm{i}k_\mathrm{I}}\right) \tag{3.3.28}$$

(Note that we cannot have $k_\mathrm{I} = 0$ since then both $f(l, k)$ and $f(l, -k)$ would vanish at the same place and so ϕ would vanish.) So resonances will also lie on Regge trajectories, like bound states.

To find the Regge trajectories produced by a given potential one must search for the zeros of $f(l, -k)$. One potential which has particularly simple trajectories is the Coulomb potential $V(r) = e^2/r$. Though this violates the convergence requirements as $r \to \infty$ $(rU(r) \nrightarrow 0)$, it is well known (see for example Schiff (1968) p. 138) that the phase shift $\delta_l(k)$ can still be defined if one first removes the infinite part $\exp[(\mathrm{i}\, e^2 \log r)/2k]$ stemming from the infinite range of the interaction. The S-matrix is then (Singh 1962)

$$S(l, k) = \frac{\Gamma(l + 1 - \mathrm{i}e^2/2k)}{\Gamma(l + 1 + \mathrm{i}e^2/2k)} \tag{3.3.29}$$

This has poles where the argument of the numerator Γ-function passes through negative integers, i.e. at

$$l = \alpha_n(s) \equiv -m - 1 + \frac{\mathrm{i}e^2}{2k}, \quad m = 0, 1, 2, \ldots \tag{3.3.30}$$

giving bound states at

$$s = E = k^2 = -\frac{e^4}{4(l + m + 1)^2} \tag{3.3.31}$$

which is the usual Rydberg formula for the hydrogen atom (see fig. 3.1). Note how the trajectories tend to infinity at $E = 0$, which is a characteristic of the zero-mass photon exchange.

With Yukawa-like potentials the Schroedinger equation can be solved numerically using the series method (3.3.17) and some examples are shown in fig. 3.2. A sufficiently attractive potential will produce a bound state for low l, which will become less bound as l increases due to the centrifugal repulsion, and perhaps manifest itself as a higher

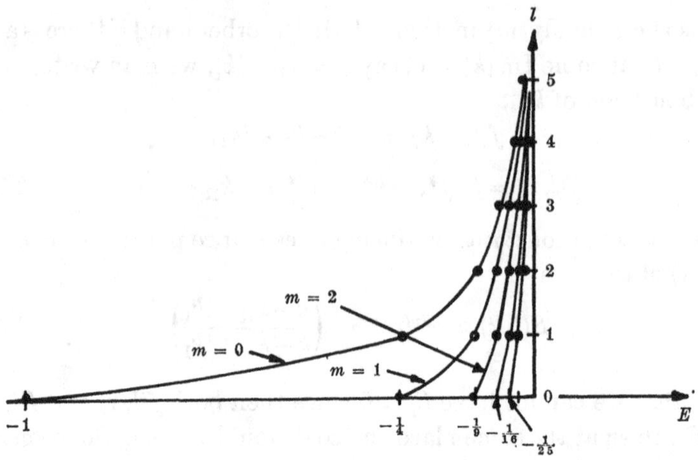

FIG. 3.1 Regge trajectories for the Coulomb potential from (3.2.29). For integer l we have the degenerate hydrogen-atom levels of principle quantum number $n = l+m+1$ ($m = 0, 1, 2$), where m is the radial quantum number. (E is measured in units of $e^4/4 = 1$ rydberg.)

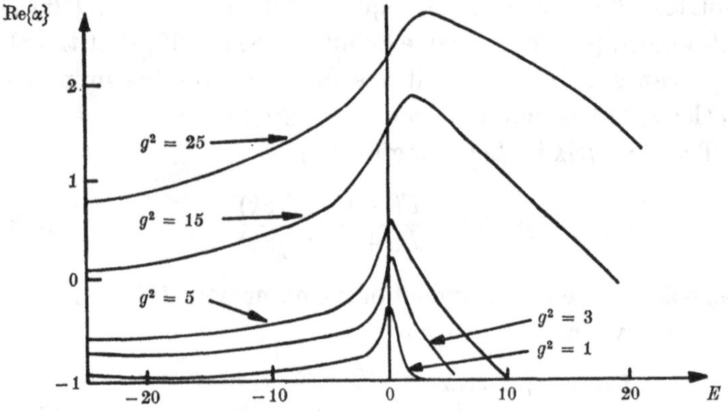

FIG. 3.2 Regge trajectories for an attractive Yukawa potential

$$V(r) = -g^2 e^{-r}/r$$

for various values of g^2, from Lovelace and Masson (1962). See also Ahmadzadeh, Burke and Tate (1963).

spin resonance. The trajectory turns down again once the effective potential, $U(r) - l(l+1)r^{-2}$, becomes too weak to produce a pole for the given l value. It will also be seen that as $g^2 \to 0$ the leading trajectory remains near $l = -1$ for all k, i.e. near the position of the highest fixed pole in the Jost function. This is because the Born approximation (1.13.16) or (1.13.18), which behaves like t^{-1} for all s, is a good approximation to the scattering amplitude in this limit.

In fact the leading trajectory asymptotes to -1 for $s \to \pm\infty$ even for large g^2 because the first Born approximation dominates for large s. However, if the potential vanishes at the origin, $rU(r) \sim r^{p+1}$ as $r \to 0$, then the trajectory asymptotes to the highest integer $l \leqslant -1-(p+1)/2$. This follows from (**1.13.18**) since if the denominator is expanded for large t

$$A^{\mathrm{B}}(s,t) = -\int_m^\infty \mathrm{d}\mu\, \rho(\mu) \left(\frac{1}{t} + \frac{\mu^2}{t^2} + \dots\right) \qquad (3.3.32)$$

it is clear from (**3.3.19**) that coefficients of $t^{-1}, t^{-2}, \dots, t^{-\frac{1}{2}p-1}$ all vanish.

Other potentials for which the trajectories have been calculated include the square well (see Newton 1964) and the three-dimensional harmonic oscillator, $V(r) = \frac{1}{2}M\omega^2r^2$, where ω is the classical frequency. The eigenstates are (Morse and Feshbach (1953) p. 1662)

$$E = k^2 = \hbar\omega(n+\tfrac{1}{2}) = \hbar\omega(2m+l+\tfrac{3}{2}) \qquad (3.3.33)$$

giving trajectories with $l \propto E$. This is particularly interesting because with relativistic kinematics $E^2 = k^2+m^2$ one might expect to get $l \propto E^2$ instead, which corresponds to the behaviour found in particle physics (see chapter 5). Various quark models for meson trajectories have been proposed based on this observation (see Dalitz (1965), and chapter 5) using a static version of the relativistic Bethe–Salpeter equations (see (**3.4.11**) below) instead of the Schroedinger equation, with a harmonic oscillator potential between the quarks. However, such potentials do not satisfy the convergence requirement that $rV(r) \to 0$ as $r \to \infty$ so there are no quark–quark scattering solutions. The quarks can never get out of the potential which, since they have not been observed, may not be a bad thing!

For well behaved potentials it is possible to determine the slope of the trajectory below threshold from the 'size' of the bound state. The Schroedinger equation (**3.3.3**) may be written

$$D\phi = 0 \quad \text{where} \quad D \equiv \left(\frac{\mathrm{d}^2}{\mathrm{d}r^2} + E - \frac{l(l+1)}{r^2} - U(r)\right) \qquad (3.3.34)$$

We seek a solution $\phi(l,k,r)$ for $l = \alpha(E)$ where $E = k^2$. Differentiating with respect to E gives

$$\frac{\mathrm{d}D}{\mathrm{d}E}\phi + D\frac{\mathrm{d}\phi}{\mathrm{d}E} = 0 \quad \text{where} \quad \frac{\mathrm{d}D}{\mathrm{d}E} = 1 - \frac{2\alpha+1}{r^2}\frac{\mathrm{d}\alpha}{\mathrm{d}E} \qquad (3.3.35)$$

Multiplying (**3.3.34**) by $\mathrm{d}\phi/\mathrm{d}E$ and (**3.3.25**) by ϕ and subtracting gives

$$\frac{\mathrm{d}\phi}{\mathrm{d}E}D\phi - \phi D\frac{\mathrm{d}\phi}{\mathrm{d}E} = \phi\frac{\mathrm{d}D}{\mathrm{d}E}\phi \qquad (3.3.36)$$

But
$$D\phi - \phi D = \frac{\mathrm{d}^2}{\mathrm{d}r^2}\phi - \phi\frac{\mathrm{d}^2}{\mathrm{d}r^2} \tag{3.3.37}$$

so the left-hand side of (3.3.36) may be written

$$\frac{\mathrm{d}}{\mathrm{d}r}\left[\frac{\mathrm{d}\phi}{\mathrm{d}E}\frac{\mathrm{d}\phi}{\mathrm{d}r} - \phi\frac{\mathrm{d}^2\phi}{\mathrm{d}E\,\mathrm{d}r}\right] \tag{3.3.38}$$

and integrating both sides from $r = 0$ to ∞ we get

$$\left[\frac{\mathrm{d}\phi}{\mathrm{d}E}\frac{\mathrm{d}\phi}{\mathrm{d}r} - \phi\frac{\mathrm{d}^2\phi}{\mathrm{d}E\,\mathrm{d}r}\right]_0^\infty = \int_0^\infty \frac{\mathrm{d}D}{\mathrm{d}E}\phi^2\,dr \tag{3.3.39}$$

Since $\phi \sim r^{l+1}$ for $r \to 0$ and $\sim \mathrm{e}^{-|k|r}$ for $r \to \infty$ (for a bound state) the left-hand side vanishes at both limits for $l > -\frac{1}{2}$ and $E < 0$. Then substituting (**3.3.35**) in the right-hand side of (**3.3.39**) we end up with

$$\frac{\mathrm{d}\alpha}{\mathrm{d}E} = \frac{1}{2\alpha+1}\frac{\int_0^\infty \phi^2 \mathrm{d}r}{\int_0^\infty (1/r^2)\,\phi^2\,\mathrm{d}r} \equiv \frac{R^2}{2\alpha+1} > 0 \tag{3.3.40}$$

where R^2 defined by (3.3.40) is the mean-square radius of the state described by the wave function ϕ. It shows that $\mathrm{d}\alpha/\mathrm{d}E$ is positive for $\alpha > -\frac{1}{2}$, $E < 0$.

d. The N/D method

In obtaining the scattering amplitude from the potential one is seeking a function whose left-hand cut in $E = k^2$ is given by the potential, and whose right-hand cut satisfies the unitarity condition (3.3.25). An alternative to solving the Schroedinger equation which exploits these analyticity properties is the so-called N/D method (Blankenbecler *et al.* 1960). This is of some interest because, unlike the Schroedinger equation, it is readily generalized to particle physics provided the scattering amplitudes have the expected analyticity properties.

From (**3.3.22**) and (**3.3.24**) we can write

$$B_l(E) = \frac{f(l,k)\,\mathrm{e}^{\mathrm{i}\pi l} - f(l,-k)}{2(\mathrm{i}k)^{l+1}} \cdot \frac{1}{(-\mathrm{i}k)^l f(l,-k)} \equiv \frac{N_l(E)}{D_l(E)} \tag{3.3.41}$$

Now from (**3.3.21**) we find that $N_l(k) = N_l(k\mathrm{e}^{-\mathrm{i}\pi})$ (for real l) so that $N_l(E)$ has no right-hand cut in E but just the left-hand cut stemming from the potential beginning at $E = -m^2/4$, and $N \to 0$ as $|E| \to \infty$.

Similarly $D_l(E)$ has no left-hand cut, but just the right-hand unitarity cut, and $D_l(E) \to 1$ as $|E| \to \infty$. Both N and D are real analytic.

Hence we can write dispersion relations

$$N_l(E) = \frac{1}{\pi}\int_{-\infty}^{-m^2/4} \frac{\mathrm{Im}\{N_l(E')\}}{E'-E}\,\mathrm{d}E' \tag{3.3.42}$$

$$D_l(E) = 1+\frac{1}{\pi}\int_0^{\infty} \frac{\mathrm{Im}\{D(E')\}}{E'-E}\,\mathrm{d}E' \tag{3.3.43}$$

If we define the discontinuity of $B_l(E)$ across the left-hand cut as $b_l(E)$ we have

$$\mathrm{Im}\{N_l(E)\} = D_l(E)\,b_l(E), \quad E < -\frac{m^2}{4} \tag{3.3.44}$$

while on the right-hand cut

$$\mathrm{Im}\{D_l(E)\} = N_l(E)\,\mathrm{Im}\left\{\frac{1}{B_l(E)}\right\} = -N_l(E)\frac{\mathrm{Im}\{B_l(E)\}}{|B_l(E)|^2} = -N_l(E)k^{2l+1} \tag{3.3.45}$$

from (3.3.25), and hence we obtain the simultaneous equations

$$N_l(E) = \frac{1}{\pi}\int_{-\infty}^{-m^2/4} \frac{D_l(E')\,b_l(E')}{E'-E}\,\mathrm{d}E' \tag{3.3.46}$$

$$D_l(E) = 1-\frac{1}{\pi}\int_0^{\infty} \frac{N_l(E')\,E'^{l+\frac{1}{2}}}{E'-E}\,\mathrm{d}E' \tag{3.3.47}$$

The solution of these equations, given $b_l(E)$, corresponds to the solution of the Schroedinger equation with the given potential. The problem of course is to find $b_l(E)$. This is easy for the first Born approximation (1.13.16) whose t-discontinuity is just

$$D_t(E,t) = \pi g^2\,\delta(t-\mu^2)$$

which substituted in (2.6.19) (interchanging s and t and putting $q = k$) gives

$$b_l^1(E) = \frac{g^2}{64}P_l\left(1+\frac{\mu^2}{2k^2}\right)\frac{1}{k^{2l+2}}, \quad E < -\frac{\mu^2}{4} \tag{3.3.48}$$

If this is substituted in (3.3.46) and (3.3.47) we get quite a good approximation to the exact solution for small g^2. The second Born approximation can also be calculated fairly easily (see Collins and Johnson 1968), but higher order terms are more difficult.

The Regge poles appear as zeros of the D function, i.e. $D_{\alpha(E)}(E) = 0$ implicitly defines $\alpha(E)$, and so a trajectory $\alpha(E)$ can be followed by observing the movement of this zero with l. This tells us that $\alpha(E)$ will have just the singularities of $D_l(E)$, i.e. just the right-hand cut, in agreement with our conclusions of the previous section.

3.4 Regge poles in perturbation field theory

It is important to check that Regge singularities also occur in perturbation field theory, because this has a much more realistic singularity structure in s and t than potential scattering. We shall find in chapter 8 that more complicated l-plane singularities, Regge cuts, which are absent from potential scattering, also arise in such field theories. But in this chapter we restrict our attention to the poles.

Perhaps the first thing to note is that the theory will include not only Regge poles but also the input elementary particles which correspond to Kronecker-δ functions in the l plane. We are concerned only with scalar mesons, and the partial-wave projection of a t-channel propagator like (1.12.1) is, from (2.2.18) and (A.20),

$$A_l(t) = \frac{g^2}{16\pi} \frac{1}{t-m^2} \delta_{l0} \tag{3.4.1}$$

that is a contribution to the S wave only. Such elementary particles do not seem to exist so we can be fairly sure from the beginning that not all aspects of the l-plane structure of the field theory will correspond to that of particle physics. (However we shall show in chapter 12 that in some circumstances these input δ's may be cancelled away.) We shall only be interested in the composite particles which may arise as bound or resonant states formed by the interaction between the elementary particles. These should occur on trajectories in analogy with potential scattering.

Such composite particles involve infinite sets of Feynman diagrams, and we shall have to assume that the asymptotic behaviour of such sets of diagrams can be obtained by summing the leading behaviours of the individual diagrams. This certainly need not be true mathematically, of course, but, at least for weak couplings where the perturbation series may make some sense, it has a certain plausibility.

A much more complete review of this subject may be found in Eden *et al.* (1966, chapter 3). Here we are mainly concerned to obtain (3.4.11) below.

For a general Feynman integral like (1.12.5), with n internal lines and l closed loops, conservation of four-momentum at each vertex can be used to express all the q_i in terms of the loop momenta k_l and the external momenta p_j. Then after judicious changes of variables $k_l \to k'_l$ the denominator can be rearranged so that the k' integrations can be performed using

$$\int \mathrm{d}^4k' \frac{1}{(k'^2+U)^3} = \frac{\mathrm{i}\pi^2}{2U} \tag{3.4.2}$$

and its derivatives with respect to U, and (see Eden *et al.* 1966) one ends up with

$$A = \frac{\int_0^1 \prod_{i=1}^{n} \mathrm{d}\alpha_i \,\delta(1-\Sigma\alpha_i)\, C(\alpha)^{n-2l-2}}{(D(p,\alpha)+\mathrm{i}\epsilon\, C(\alpha))^{n-2l}} \tag{3.4.3}$$

where D is a function of the p's and α's and C a function of the α's only. Thus for the $2 \to 2$ scattering amplitude where there are just the two independent invariants s and t and D is linear in s we can rewrite this (dropping the iϵ term) as

$$A = \frac{\int_0^1 \prod_{i=1}^{n} \mathrm{d}\alpha_i \,\delta(1-\Sigma\alpha_i)\, C(\alpha)^{n-2l-2}}{(g(\alpha)\, s+d(t,\alpha))^{n-2l}} \tag{3.4.4}$$

where g and d are some functions. We are interested in the limit $s \to \infty$, t fixed, and clearly the integrand $\sim s^{-n+2l}$ unless $g(\alpha) = 0$. So this will also be the behaviour of the integral unless somewhere on the contour of integration $g(\alpha) = 0$, and it is impossible to distort the contour round this point because either (i) $g(\alpha) = 0$ at one of the end points of integration (giving a so-called 'end-point' contribution) or (ii) the point $g(\alpha) = 0$ is 'pinched' by two or more singularities of the integrand as $s \to \infty$ (see section 1.12).

It can be shown that as long as we stick to just planar diagrams (i.e. diagrams which can be drawn on a sheet of paper without any lines crossing) there will be no pinch contributions on the physical sheet. We shall have to consider non-planar diagrams in chapter 8, but here we shall only be concerned with the end-point contributions of planar diagrams.

Obviously the pole diagram, fig. **3.3** (*a*), gives

$$A_1 = \frac{g^2}{m^2-s} \sim \frac{1}{s} \tag{3.4.5}$$

which is just the Born approximation for the t-channel scattering

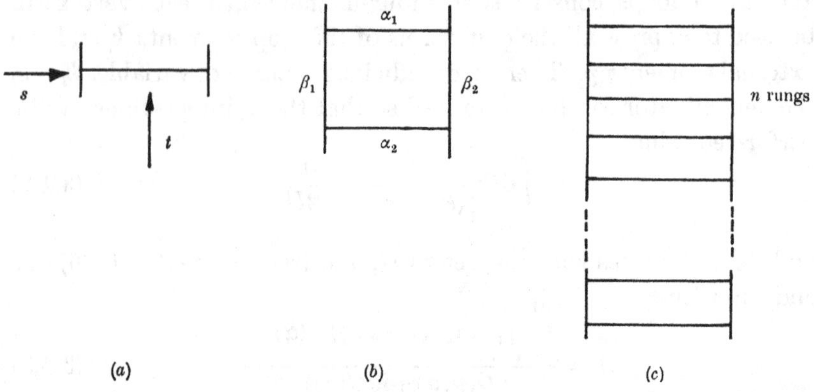

FIG. 3.3 A sequence of t-channel ladder Feynman diagrams: (a) the single particle exchange Born approximation, (b) the box diagram with its associated Feynman parameters, (c) an n-rung ladder.

process. Then there is the box diagram, fig. 3.3 (b), whose amplitude is

$$A_2 = g^2\left(-\frac{g^2}{16\pi^2}\right)\frac{\int_0^1 \prod_{i=1}^{2} \mathrm{d}\alpha_1\,\mathrm{d}\beta_1\,\delta(1-\Sigma\alpha_i-\Sigma\beta_i)}{(\alpha_1\alpha_2 s+d_2(\alpha,\beta,t))^2} \tag{3.4.6}$$

As $s\to\infty$ only the behaviour near $\alpha_1, \alpha_2 = 0$ need be considered, and defining $d_2' = d_2(0,0,\beta_1,\beta_2,t)$ we need

$$\int_0^\epsilon \mathrm{d}\alpha_1\,\mathrm{d}\alpha_2\,\frac{1}{(\alpha_1\alpha_2 s+d_2')^2} = \int_0^\epsilon \mathrm{d}\alpha_2\,\frac{\epsilon}{d_2'(\epsilon\alpha_2 s+d_2')}$$

$$= \frac{1}{\mathrm{d}_2' s}\log\left(\frac{\epsilon^2 s+d_2'}{\mathrm{d}_2'}\right) \sim \frac{\log s}{\mathrm{d}_2' s} \tag{3.4.7}$$

so
$$A_2 \to g^2 K(t)\,\frac{\log s}{s} \tag{3.4.8}$$

where
$$K(t) = \frac{-g^2}{16\pi^2}\int_0^1 \frac{\mathrm{d}\beta_1\,\mathrm{d}\beta_2\,\delta(1-\beta_1-\beta_2)}{d_2(0,0,\beta_1,\beta_2,t)}$$

$$= \frac{g^2}{16\pi^3}\int \frac{\mathrm{d}^2 K}{(K^2+m^2)\,[(K+q)^2+m^2]}, \quad t=-q^2 \tag{3.4.9}$$

is the loop integral corresponding to the Feynman diagram fig. 3.4 (a) in which the sides have been contracted out (since $\alpha_1 = \alpha_2 = 0$), which is evaluated only with two-dimensional momentum $\boldsymbol{K}$ rather than four-dimensional k (because d_2 appears only in the first power, unlike in (3.4.6)).

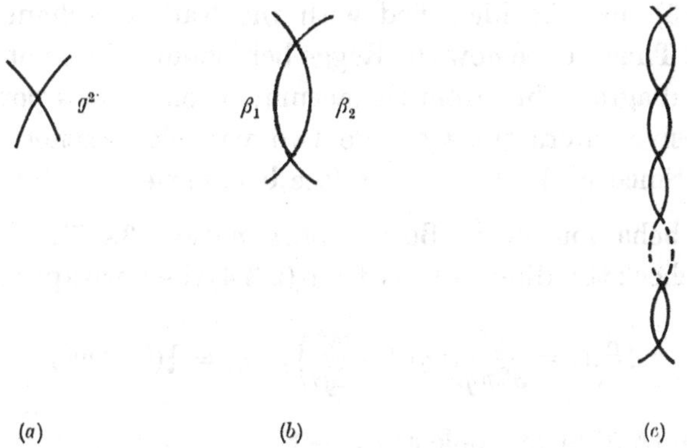

FIG. 3.4 The contracted diagrams corresponding to fig. 3.3 which give the coefficient of the $\sim s^{-1}$ asymptotic behaviour.

For the n-rung ladder diagram, fig. 3.3 (c), it is found similarly that

$$A_n = g^2\left(\frac{-g^2}{16\pi}\right)^{n-1}(n-1)!\frac{\int_0^1 \prod_{i=1}^{n} \mathrm{d}\alpha_i \pi \mathrm{d}b\, \delta(1-\Sigma\alpha_i-\Sigma\beta)\, C(\alpha,\beta)^{n-2}}{[\alpha_1\ldots\alpha_n s+d_n(\alpha,\beta,t)]^n} \tag{3.4.10}$$

and again, since the leading behaviour comes from the region where the α's vanish (Fig. 3.4 (b)) the α integrations can be performed to give

$$A_n \sim \frac{g^2}{s}\frac{(\log sK(t))^{n-1}}{(n-1)!} \tag{3.4.11}$$

The power behaviour of all the diagrams in fig. 3.3 is thus s^{-1} like (3.4.5). This is because just a single-particle propagator is needed to get across the diagram. But the power of $\log s$ which appears depends on the number of such propagators.

The next step is to take the asymptotic behaviour of the sum of all such ladder diagrams with any number of rungs, assuming, as mentioned above, that the asymptotic behaviour of the sum is the sum of the asymptotic behaviours. The similarity of figs. 3.3 to figs. 1.14 indicates why this may be rather like solving the Schroedinger equation with a 'potential' given by the Born approximation (3.4.5). From (3.4.11) we get

$$A(s,t) \equiv \sum_n A_n \sim \sum_{n=1}^{\infty}\frac{g^2}{s}\frac{(\log sK(t))^{n-1}}{(n-1)!} \sim \frac{g^2}{s}\mathrm{e}^{K(t)\log s} \tag{3.4.12}$$

$$\sim g^2 s^{\alpha(t)} \quad \text{where} \quad \alpha(t) \equiv -1+K(t) \tag{3.4.13}$$

Clearly, through the Froissart–Gribov projection (2.6.2), the power of

s in (3.4.13) may be identified with the leading t-channel Regge trajectory. Thus we see how the Regge behaviour comes not from any individual diagram, but from the accumulation of $\log s$ powers from the successive interactions of the two particles scattering in the t channel. Since $K(t) \to 0$ for $t \to \infty$ (see below) we have $\alpha(t) \xrightarrow[t\to\infty]{} -1$, due to the behaviour of the Born approximation (3.4.5).

We can check this directly since from (2.3.4) the Born approximation gives

$$A_l^{\mathrm{B}}(t) = \frac{g^2}{32\pi q_t^2} Q_l\left(1+\frac{m^2}{2q_t^2}\right), \quad q_t^2 = \tfrac{1}{4}(t-4m^2) \tag{3.4.14}$$

which from (A.32) has a pole at $l = -1$

$$A_l^{\mathrm{B}}(t) \sim \frac{g^2}{32\pi q_t^2(l+1)} \tag{3.4.15}$$

When this fixed pole is inserted in the unitarity equations it is Reggeized. The partial-wave amplitude must tend to (3.4.14) as $g^2 \to 0$, and it must satisfy the unitarity equation (2.2.8) which it does if we write it as a series in g^2

$$A_l(t) = \frac{g^2}{32\pi q_t^2}\left[\frac{1}{\alpha(t)-l} - \frac{1}{l+1}\left(1+\frac{g^2\alpha_1(t)}{l+1}+\ldots\right)\right] \tag{3.4.16}$$

where we have expanded the trajectory function in g^2

$$\alpha(t) = -1 + \frac{g^2}{16\pi}\alpha_1(t) + \ldots \tag{3.4.17}$$

and

$$\operatorname{Im}\{\alpha(t)\} = \frac{g^2}{16\pi q_t \sqrt{t}} \tag{3.4.18}$$

Since $\alpha(t)$ is an analytic function satisfying the dispersion relation (3.2.11) with $n = 1$ we have

$$\alpha(t) = -1 + \frac{g^2}{16\pi^2}\int_{4m^2}^{\infty} \frac{dt'}{q_t'\sqrt{t'}\,(t'-t)} = -1 + \frac{g^2}{16\pi^2}\frac{1}{q_t\sqrt{t}}\log\left[\frac{2q_t + t^{\frac{1}{2}}}{2q_t - t^{\frac{1}{2}}}\right] \tag{3.4.19}$$

in agreement with (3.4.13). So as expected $\alpha(t) \to -1$ as $t \to \pm\infty$ for all g^2, and for all t as $g^2 \to 0$. This is almost certainly unrealistic for strong interactions because it stems from the elementary nature of the exchanged scalar meson. But the way in which the trajectory is built up from this basic interaction is so similar to potential scattering that it seems very plausible that a similar mechanism will operate in hadronic physics too. In fact, summing the ladders corresponds to

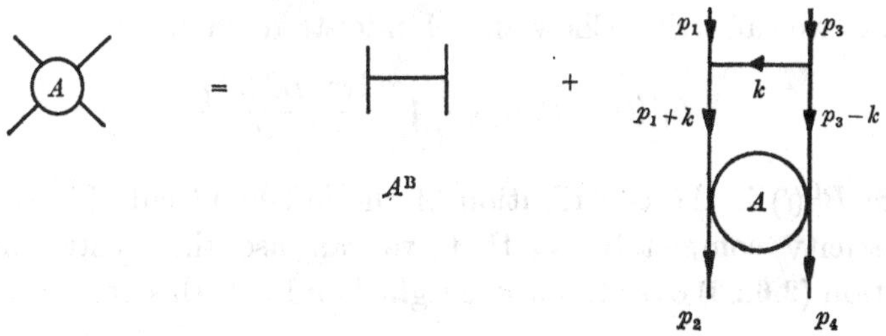

FIG. 3.5 The Bethe–Salpeter equation (3.4.20) for summing ladder diagrams.

solving the t-channel Bethe–Salpeter equation (see fig. 3.5) (Bethe and Salpeter 1951, see Polkinghorne, 1964)

$$A(s,t) = A^{\mathrm{B}}(s,t) + \frac{g^2}{(2\pi)^4}\int \frac{\mathrm{d}^4k}{[(p_1+k)^2-m^2][(p_3-k)^2-m^2]}$$
$$\times A^{\mathrm{B}}(p_1,p_3,p_1+k,p_3-k)\,A(p_1+k,p_3-k,p_2,p_4) \quad (3.4.20)$$

which is the relativistic version of the Lippman–Schwinger equation (1.13.27). Trajectories generated by solving the Bethe–Salpeter equation with various potentials have been published by Swift and Tucker (1970, 1971).

3.5 Bootstraps

In section 2.8 we introduced the bootstrap hypothesis that the only set of particles whose existence is compatible with unitarity, analyticity in s and t, and analyticity in l, is the actual set of hadrons found in the real world. If this is so it should be possible to deduce the properties of the particles just by implementing the unitarity equations together with the constraints of crossing. Attempts to achieve this are called 'bootstrap calculations'.

The complexity of many-body unitarity has made it impossible to test this hypothesis properly so far. We shall examine some of the progress made in this direction in section 11.7, but here we want to illustrate the application of two-body unitarity, to complement our discussion of the previous sections. We review briefly the three main techniques which have been employed.

a. N/D equations

These are based on partial-wave dispersion relations, and their development closely parallels the discussion in section 3.3d. From

(2.6.20) we can write (Chew and Mandelstam 1960)

$$B_l^{\mathscr{S}}(t) = B_l^{\mathrm{L}}(t) + \frac{1}{\pi}\int_{t_\mathrm{T}}^{\infty} \frac{\operatorname{Im}\{B_l^{\mathscr{S}}(t')\}}{t'-t}\,\mathrm{d}t' \tag{3.5.1}$$

where $B_l^{\mathrm{L}}(t)$ is the contribution of the left-hand cut. If we neglect inelasticity completely, so that we can use the elastic unitarity equation (2.6.23) over the whole right-hand cut, this becomes

$$B_l^{\mathscr{S}}(t) = B_l^{\mathrm{L}}(t) + \frac{1}{\pi}\int_{t_\mathrm{T}}^{\infty} \frac{\rho_l(t')\,|B_l^{\mathscr{S}}(t')|^2}{t'-t}\,\mathrm{d}t' \tag{3.5.2}$$

And if we suppose that all the crossed channel singularities are known, i.e. $B_l^{\mathrm{L}}(t)$ is given, then (3.5.2) is an integral equation for the scattering amplitude. To solve it we linearize by writing (cf. (3.3.41))

$$B_l^{\mathscr{S}}(t) = \frac{N_l(t)}{D_l(t)} \tag{3.5.3}$$

where, by definition, the numerator function $N_l(t)$ has the left-hand cut of $B_l^{\mathscr{S}}(t)$, and $D_l(t)$ the right-hand cut. So

$$\operatorname{Im}\{N_l(t)\} = \operatorname{Im}\{B_l^{\mathscr{S}}(t)\}\,D_l(t) \equiv b_l(t)\,D_l(t), \quad \text{say}, \quad t < t_\mathrm{L} \tag{3.5.4}$$

and

$$\operatorname{Im}\{D_l(t)\} = N_l(t)\operatorname{Im}\left\{\frac{1}{B_l^{\mathscr{S}}(t)}\right\}, \quad t > t_\mathrm{T}$$

$$= -N_l(t)\frac{\operatorname{Im}\{B_l^{\mathscr{S}}(t)\}}{|B_l^{\mathscr{S}}(t)|^2} = -\rho_l(t)\,N_l(t) \tag{3.5.5}$$

from (2.6.23). Since, using (2.2.10) and (2.6.8)

$$\frac{1}{B_l^{\mathscr{S}}(t)} = \frac{D_l(t)}{N_l(t)} = \frac{\mathrm{e}^{-\mathrm{i}\delta_l(t)}}{\sin\delta_l(t)}\rho_l(t) \tag{3.5.6}$$

and N is real for $t > t_\mathrm{L}$, $D_l(t)$ must have the phase $\mathrm{e}^{-\mathrm{i}\delta_l(t)}$ along the right-hand cut, $t > t_\mathrm{T}$.

The Wiener–Hopf method (see Titchmarsh (1937) p. 339) allows one to construct $D_l(t)$ knowing this phase, and the positions of the p_l poles at $t = t_{il}$, say, and the m_l zeros at $t = t_{il}$, on the physical sheet. It takes the form

$$D_l(t) = D_l(t_\mathrm{T})\prod_{i=1}^{p_l}\left(\frac{t_\mathrm{T}-t_{il}}{t-t_{il}}\right)\prod_{j=1}^{m_l}\left(\frac{t-t_{jl}}{t_\mathrm{T}-t_{jl}}\right)\exp\left\{-\frac{t-t_\mathrm{T}}{\pi}\int_{t_\mathrm{T}}^{\infty}\frac{\delta_l(t')-\delta_l(t_\mathrm{T})}{(t'-t)(t'-t_\mathrm{T})}\,\mathrm{d}t'\right\} \tag{3.5.7}$$

We have assumed that $\delta_l(t) \xrightarrow[t\to\infty]{}$ constant, so that only one subtraction at t_T is needed in the integral. We insist that (as in section 3.3*d*) all the poles of the amplitude correspond to zeros of $D_l(t)$ (not poles of $N_l(t)$). These may be either bound states on the physical sheet at $t = t_{jl}$, or resonances on unphysical sheets where $\delta_l(t) \to (2n+1)\pi/2$. Then from (3.5.7)

$$D_l(t) \sim t^{\{m_l - p_l + \pi^{-1}[\delta_l(\infty) - \delta_l(t_T)]\}} \tag{3.5.8}$$

We choose conventionally that $D_l(t) \xrightarrow[t\to\infty]{} 1$ so

$$\delta_l(\infty) - \delta_l(t_T) = \pi[p_l - m_l] \tag{3.5.9}$$

and also that $\delta_l(t_T) = m_l$ giving

$$\delta_l(\infty) = \pi p_l \tag{3.5.10}$$

This relation between the asymptotic value of the phase shift and the number of poles of the D function is known as Levinson's theorem (Levinson 1949).

From (3.5.5) and (3.5.7) we can write a dispersion relation for $D_l(t)$ in the form

$$D_l(t) = 1 - \frac{1}{\pi}\int_{t_T}^{\infty} \frac{\rho_l(t')\,N_l(t')}{t'-t}\,dt' + \sum_{i=1}^{p_l} \frac{\gamma_{il}}{t-t_{il}} \tag{3.5.11}$$

where the γ_{il} are the residues of the poles. Since the γ_{il} and t_{il} are arbitrary, $D_l(t)$ is evidently not completely determined by the input $B_l^L(t)$. This is known as the CDD ambiguity, after its discoverers Castillejo, Dalitz and Dyson (1956). An elementary (non-composite) particle like that represented by (3.4.1) would correspond to a CDD pole in the appropriate partial wave.

However for large l the result (2.5.5) implies that $B_l(t) \xrightarrow[l\to\infty]{} B_l^L(t) \xrightarrow[l\to\infty]{} 0$ so that $\delta_l(\infty) \to \delta_l(t_T)$. There will clearly be no bound states in this limit, i.e. $m_l \to 0$, and hence from (3.5.9) $p_l \to 0$ too. Thus for large l there is no CDD ambiguity and the scattering amplitudes will be completely determined by $B_l^L(t)$. However, our assumption of analyticity in l requires that the low partial waves should be obtainable from the high partial waves by analytic continuation, and so we cannot just start adding poles in (3.5.11) as l is decreased. So analyticity in l precludes CDD poles in low partial waves as well.

Hence from (3.5.4) and (3.5.5) we arrive at the pair of simultaneous N/D equations

$$N_l(t) = \frac{1}{\pi}\int_{-\infty}^{t_L} \frac{b_l(t')\,D_l(t')}{t'-t}\,dt' \tag{3.5.12}$$

$$D_l(t) = 1 - \frac{1}{\pi}\int_{t_T}^{\infty} \frac{\rho_l(t')\,N_l(t')}{t'-t}\,dt' \tag{3.5.13}$$

like (3.3.46), (3.3.47). If we introduce the function

$$C_l(t) \equiv N_l(t) - B_l^{\mathrm{L}}(t)\, D_l(t) \tag{3.5.14}$$

it will have no left-hand cut since

$$\operatorname{Im}\{B_l^{\mathrm{L}}(t)\} = \frac{\operatorname{Im}\{N_l(t)\}}{D_l(t)} \tag{3.5.15}$$

while on the right-hand cut

$$\operatorname{Im}\{C_l(t)\} = -B_l^{\mathrm{L}}(t)\operatorname{Im}\{D_l(t)\} \tag{3.5.16}$$

and so it satisfies the dispersion relation

$$C_l(t) = \frac{1}{\pi}\int_{t_{\mathrm{T}}}^{\infty} \frac{\operatorname{Im}\{C_l(t')\}}{t'-t}\,\mathrm{d}t' \tag{3.5.17}$$

or from (3.5.14)

$$N_l(t) = B_l^{\mathrm{L}}(t)\, D_l(t) - \frac{1}{\pi}\int_{t_{\mathrm{T}}}^{\infty} \frac{B_l^{\mathrm{L}}(t')\operatorname{Im}\{D_l(t')\}}{t'-t}\,\mathrm{d}t' \tag{3.5.18}$$

Then using (3.5.13) and (3.5.5) to eliminate $D_l(t)$ this becomes

$$N_l(t) = B_l^{\mathrm{L}}(t) + \frac{1}{\pi}\int_{t_{\mathrm{T}}}^{\infty} \frac{B_l^{\mathrm{L}}(t') - B_l^{\mathrm{L}}(t)}{t'-t}\,\rho_l(t')\, N_l(t')\,\mathrm{d}t' \tag{3.5.19}$$

This is an integral equation for $N_l(t)$ given $B_l^{\mathrm{L}}(t)$ which can be solved numerically. Once $N_l(t)$ is found it can be substituted in (3.5.13) to find $D_l(t)$.

These equations can be generalized to include inelastic states (for a review see Collins and Squires (1968) chapter 6). The most important change is that it is then possible for bound or resonant states of one channel to appear as CDD poles in another channel. However, such a CDD zero will emerge from the inelastic cut as l is decreased, so continuity in l is not destroyed, and such CDD poles do not correspond to elementary particles.

A zero of $D_l(t)$ at some $t = t_{\mathrm{r}}$, say, corresponds to a pole of the partial-wave amplitude. Continuing the solution in l we generate a trajectory $\alpha(t)$ such that

$$D_{\alpha(t_{\mathrm{r}})}(t_{\mathrm{r}}) = 0 \tag{3.5.20}$$

Then expanding $D_l(t)$ about $l = \alpha(t_{\mathrm{r}})$ we have (from (3.5.3))

$$B_l^{\mathscr{S}}(t) \approx \frac{N_\alpha(t_{\mathrm{r}})}{(l-\alpha(t_{\mathrm{r}}))\,(\partial D_l/\partial l)_{\substack{l=\alpha(t_{\mathrm{r}})\\ t=t_{\mathrm{r}}}}}, \quad l \approx \alpha(t_{\mathrm{r}}), \quad t \approx t_{\mathrm{r}} \tag{3.5.21}$$

so the residue of the Regge pole is given by $N(\partial D/\partial l)^{-1}$.

A simple example of the use of such equations is the ρ bootstrap

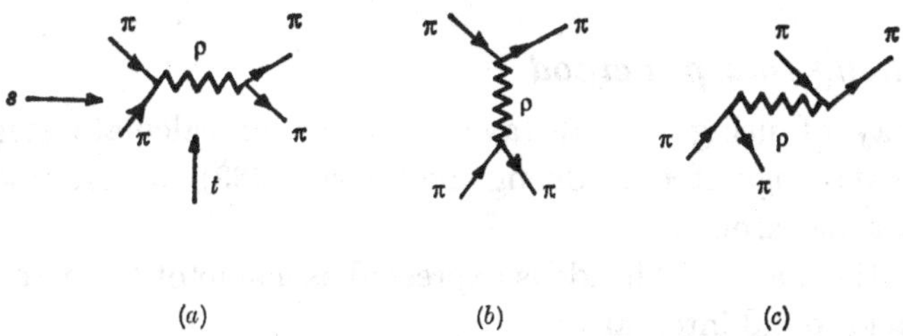

FIG. 3.6 The ρ-exchange poles in the s-, t- and u-channels of $\pi\pi$ scattering.

(Zachariasen 1961, Balazs 1962, 1963, Collins 1966). This is based on the observation that the dominant singularity in low energy elastic $\pi\pi$ scattering is the spin = 1 ρ resonance. Because $\pi\pi$ scattering is crossing symmetric this resonance will occur in all three s, t and u channels (fig. 3.6). So if we make the very drastic approximation that this is the only important singularity we can obtain the left-hand cut of the t-channel partial-wave amplitude from the ρ poles in the s and u channels. Thus from (2.6.14)

$$B_l^L(t) = \frac{g_\rho^2}{16\pi} \frac{q_\rho^2}{q_t^{2l+2}} Q_l\left(1+\frac{m_\rho^2}{2q_t^2}\right) P_1\left(1+\frac{t}{2q_\rho^2}\right) \qquad (3.5.22)$$

where $$q_\rho^2 \equiv \tfrac{1}{4}(m_\rho^2 - m_\pi^2)$$

The mass of ρ, m_ρ, and its coupling strength to $\pi\pi$, g_ρ, can be regarded as free parameters. Then if we insert (3.5.22) in (3.5.19), solve the equation, and insert the solution for $N_l(t)$ in (3.5.13) we obtain an output t-channel trajectory and residue from (3.5.20) and (3.5.21). Crossing symmetry requires that $D_l(t)$ should have a zero for $l = 1$ at $t = m_\rho^2$, and that the residue should be g_ρ^2. Hence one can try and adjust these parameters until self-consistency under crossing and unitarity is achieved, and thereby deduce the mass and coupling of the ρ from self-consistency requirements only.

Unfortunately there are several technical problems concerning the divergence of the integral in (3.5.19) which requires a cut-off, but a qualitative success may be claimed (see Collins and Squires (1968) chapter 6). This is probably the most we can expect given that we have neglected all the other singularities and inelastic unitarity. But the most important point is that this method of generating trajectories in particle physics is based on methods which we know can be employed successfully in potential scattering.

b. *The Cheng–Sharp method*

Another way of using partial-wave unitarity to calculate Regge trajectories was suggested by Cheng and Sharp (1963) and Frautschi, Kaus and Zachariasen (1964).

If the partial-wave amplitude is expressed as a sum of Regge poles plus the background integral

$$B_l^{\mathscr{S}}(t) = \sum_{i=1}^{n} \frac{\gamma_i(t)}{l-\alpha_i(t)} + \bar{B}_l^{\mathscr{S}}(t) \tag{3.5.23}$$

and substituted in the unitarity equation (2.6.23), or (4.7.4) below, for $l \to \alpha_j(t)$ we get

$$\frac{1}{2\mathrm{i}\rho_{\alpha_j}(t)} = \sum_i \frac{\gamma_i^*(t)}{\alpha_j(t)-\alpha_i^*(t)} + \bar{B}_{\alpha_j^*}^{\mathscr{S}*}(t), \quad \text{for} \quad j = 1, 2, \ldots, n \tag{3.5.24}$$

a set of simultaneous equations for the Regge parameters given the background integral (which contains the crossed-channel singularities, i.e. the 'potential'). If one supposes that just a single pole α_j dominates with $\operatorname{Im}\{\alpha_j\}$ small, then $\bar{B}$ can be neglected and (3.5.24) becomes

$$\operatorname{Im}\{\alpha_j(t)\} = \rho_{\alpha_j}(t)\,\gamma_j(t), \quad \operatorname{Im}\{\gamma_j(t)\} = 0 \tag{3.5.25}$$

which has the correct threshold behaviour (3.2.26).

To proceed further it is necessary to modify the Regge pole terms so that they have the correct Mandelstam analyticity. (The s discontinuity in (2.8.10) starts at $z_t = -1$, from (A.13), i.e. at $s = -4q_t^2$ for equal-mass kinematics, rather than at the threshold s_{T} (see Collins and Squires (1968) chapter 3). One must also add the crossed-channel poles, which provide the potential, in $\bar{B}_l^{\mathscr{S}}$. This method has been applied successfully in calculating trajectories in potential-scattering problems (Hankins, Kaus and Pearson 1965), and, with many necessary modifications, for some bootstrap calculations (Abbe *et al.* 1967).

c. *The Mandelstam iteration*

This method makes direct use of the Mandelstam representation discussed in section 1.11. Elastic unitarity is used to obtain the double spectral functions, ρ_{st}, in those regions of the s–t plane where elastic unitarity holds, and the asymptotic behaviour of ρ_{st} gives the trajectory.

From (1.5.7) the discontinuity across the elastic cut for $t_{\mathrm{T}} < t < t_{\mathrm{I}}$ in the t channel is

$$D_t(s,t) = \frac{q_t}{32\pi^2\sqrt{t}}\int \mathrm{d}\Omega_t A^+(s',t)\,A^-(s'',t) \qquad (3.5.26)$$

where (see (2.2.3) with $s \leftrightarrow t$) $s' = s(z',t)$, $z' \equiv \cos\theta_{in}$ being the cosine of the scattering angle between the direction of motion of the particles in the initial and intermediate states, and where $s'' = s(z'',t)$ and $z'' \equiv \cos\theta_{nf}$, the cosine of the angle between the intermediate and final states, in the t-channel centre-of-mass system. Similarly $s = s(z_t,t)$ where $z_t = \cos\theta_{if}$ (see fig. 2.1) and $\mathrm{d}\Omega_t \equiv \mathrm{d}z''\,\mathrm{d}\phi$. These angles are related by the addition theorem (2.2.4), i.e.

$$z' = z_t z'' + \sqrt{(1-z_t^2)}\sqrt{(1-z''^2)}\cos\phi \qquad (3.5.27)$$

Formally we can substitute the dispersion relation (1.10.7) for A^+ and A^- into (3.5.26) and obtain at fixed t (neglecting the pole terms for simplicity)

$$D_t(s,t) = \frac{q_t}{32\pi^2\sqrt{t}}\int \mathrm{d}\Omega_t \left[\frac{1}{\pi}\int_{s_{\mathrm{T}}}^{\infty}\frac{D_s(s_1,t_+)}{s_1-s'}\mathrm{d}s_1 + \frac{1}{\pi}\int_{u_{\mathrm{T}}}^{\infty}\frac{D_u(u_1,t_+)}{u_1-u'}\mathrm{d}u_1\right]$$
$$\times\left[\frac{1}{\pi}\int_{s_{\mathrm{T}}}^{\infty}\frac{D_s(s_2,t_-)}{s_2-s''}\mathrm{d}s_2 + \frac{1}{\pi}\int_{u_{\mathrm{T}}}^{\infty}\frac{D_u(u_2,t_-)}{u_2-u''}\mathrm{d}u_2\right] \qquad (3.5.28)$$

with (from (1.7.21))

$$s+t+u = s'+t+u' = s''+t+u'' = s_1+t+u_1 = s_2+t+u_2 = \Sigma \qquad (3.5.29)$$

If then we replace the s's and u's by z's using (2.3.2) and change the order of integration we find terms of the form

$$\int_{-1}^{1}\mathrm{d}z''\int_0^{2\pi}\mathrm{d}\phi\,\frac{1}{(z_1-z')(z_2-z'')} = \frac{2\pi}{\Delta^{\frac{1}{2}}}\log\left(\frac{z-z_1z_2+\Delta^{\frac{1}{2}}}{z-z_1z_2-\Delta^{\frac{1}{2}}}\right) \qquad (3.5.30)$$

using (3.5.27), where

$$\Delta(z_t,z_1,z_2) \equiv -1+z_t^2+z_1^2+z_2^2-2z_tz_1z_2 \qquad (3.5.31)$$

and we must take the branch of the logarithm which is real for $-1 < z < 1$. So converting back from z's to s's we get

$$D_t(s,t) = \frac{q_t}{16\pi^3\sqrt{t}}\int_{s_{\mathrm{T}}}^{\infty}\frac{\mathrm{d}s_1}{2q_t^2}\int_{s_{\mathrm{T}}}^{\infty}\frac{\mathrm{d}s_2}{2q_t^2}(D_s(s_1,t_+)+D_u(s_1,t_+))\,(D_s(s_2,t_-)$$
$$+D_u(s_2,t_-))\,2q_t^2K^{-\frac{1}{2}}\log\left(\frac{s-s_1-s_2-(s_1s_2/2q_t^2)+K^{\frac{1}{2}}}{s-s_1-s_2-(s_1s_2/2q_t^2)-K^{\frac{1}{2}}}\right) \qquad (3.5.32)$$

where

$$K(s, s_1, s_2, t) \equiv [s^2+s_1^2+s_2^2-2(ss_1+ss_2+s_1s_2)-ss_1s_2/q_t^2] \quad (3.5.33)$$

Now from (1.11.11) the double spectral function $\rho_{st}(s,t)$ is just the discontinuity of $D_t(s,t)$ across its cuts in s. This discontinuity arises from the vanishing of K. But $K \to 0$ makes the logarithm tend to $\log 1 = 2\pi n\mathrm{i}$, where n depends on the branch of the logarithm which is chosen. So the discontinuity in going round the threshold branch point in s for $K > 0$ is just 2π. Hence

$$\rho_{st}(s,t) = \frac{1}{8\pi^2}\frac{q_t}{\sqrt{t}}\overbrace{\int_{s_T}\frac{\mathrm{d}s_1}{2q_t^2}\int_{s_T}}^{K=0}\frac{\mathrm{d}s_2}{2q_t^2}\frac{D_s(s_1,t_+)\,D_s(s_2,t_-)}{K^{\frac{1}{2}}(s,s_1,s_2,t)}2q_t^2 \quad (3.5.34)$$

The region of integration is over s_1, $s_2 > s_T$ but with $K > 0$, since there is no discontinuity for $K < 0$. The boundary in s of $\rho_{st}(s,t)$ is given by the lowest values of s_1, s_2, i.e. where

$$K(s, s_T, s_T, t) = s\left(s-4s_T-\frac{s_T^2}{q_t^2}\right) = 0 \quad (3.5.35)$$

But $s = 0$ is not a singular point of (3.5.32) so the boundary is

$$s = 4s_T+\frac{s_T^2}{q_t^2} \equiv b(t) \quad (3.5.36)$$

From (1.11.4) we have

$$D_s(s,t) = \frac{1}{\pi}\int_{b(s)}^{\infty}\frac{\rho_{st}(s,t'')}{t''-t}\mathrm{d}t'' + \text{other terms} \quad (3.5.37)$$

The most important 'other term' is the s-channel bound-state pole from the Born approximation (2.6.13)

$$D_s^{\mathrm{B}} = \pi g^2\,\delta(s-m^2) \quad (3.5.38)$$

If this is substituted in (3.5.34) we get

$$\rho_{st}(s,t) = \frac{g^4}{16q_t(s-4m^2-m^4/q_t^2)^{\frac{1}{2}}\sqrt{t}\sqrt{s}} \quad (3.5.39)$$

whose boundary is at $K(s, m^2, m^2, t) = 0$, i.e. (1.12.10). Then if (2.5.39) is substituted into (3.5.37) we get an additional contribution to D_s (over and above (3.5.38)), which may in turn be substituted in (3.5.34) to give a further contribution to $\rho_{st}(s,t)$ with a boundary at $K(s, 4m^2, 4m^2, t) = 0$; and so on. Hence we can find $D_s(s,t)$ by iteration, the successive contributions to the double spectral function having boundaries at higher and higher s, as shown in fig. 3.7. This is just

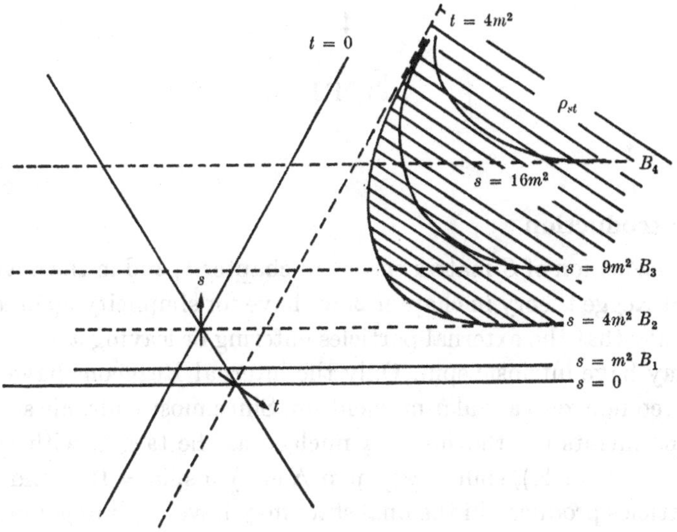

FIG. 3.7 The boundaries of the successive contributions to the double spectral function ($B_2, B_3, B_4, \ldots$) obtained by iterating the input s-channel pole B_1 with t-channel unitarity. The asymptotic s behaviour will be $\rho(s,t) \sim s^{\alpha(t)}$ for fixed t, which enables the trajectory to be found.

another way of summing the ladders corresponding to multiple exchange of the Born approximation (3.5.38). Indeed (3.5.39) gives us the behaviour (3.4.8), and the various iterations agree with (3.4.11).

Of course (3.5.38) is unrealistic as a Born approximation for particle physics. Attempts have been made to incorporate crossing symmetry by taking s-channel Regge poles as the input, and generating t-channel Regge poles as output, and seeking bootstrap self-consistency as described in section 3.5a, but so far with only modest success (see Collins and Johnson 1969, Webber 1971). We shall explore other similar dynamical schemes in chapter 11. However, it seems likely that the restriction to just planar diagrams with elastic unitarity precludes a proper self-consistent answer. Our purpose in discussing this method here has been to show that the Mandelstam iteration gives yet another procedure for generating Regge trajectories by summing ladder diagrams.

4
Spin

4.1 Introduction

In our discussion of S-matrix theory in chapter 1, and in the development of Regge theory in chapter 2, we have for simplicity ignored the possibility that the external particles entering or leaving a given process may have intrinsic spin. Only the internal Reggeons have been permitted non-zero angular momentum. Since most hadronic scattering experiments use the spin $= \frac{1}{2}$ nucleon as the target, with beams of spin $= 0$ (π or K), spin $= \frac{1}{2}$ (p, n, $\bar{\text{p}}$, Λ etc.) or spin $= 1(\gamma)$, and since the particles produced in the final state may have any integer or half-integer spin, it is essential to rectify this deficiency before we can confront the predictions of Regge theory with the real world.

There are three important points to bear in mind while doing this. First, an experiment may include in the initial state particles whose spin orientations have been predetermined (polarization experiments), or may involve detection of the spin direction of some of the final-state particles, by secondary scattering or by observing their subsequent decay. So there are further experimental observables (in addition to σ^{tot} and $\text{d}\sigma/\text{d}t$) which show how the scattering probability depends on these spin directions. Secondly, the dependence of the scattering process on the spin vectors means that the Lorentz invariance and crossing properties of the scattering amplitudes will generally be more complicated than those for spinless particles. And finally, and most important for Regge theory, the total angular momentum of a given state, $\boldsymbol{J}$, will no longer be just the orbital angular momentum $\boldsymbol{l}$, as in chapter 2, but the vector sum of $\boldsymbol{l}$ and the spins of the particles, $\boldsymbol{\sigma}_j$, so that for the initial state for example

$$\boldsymbol{J} = \boldsymbol{l}+\boldsymbol{\sigma}_1+\boldsymbol{\sigma}_2 \tag{4.1.1}$$

and care is needed in making an analytic continuation in J rather than l.

The two most commonly employed methods for discussing spin problems are invariant amplitudes, and centre-of-mass helicity amplitudes.

To obtain the invariant amplitudes each particle of spin σ_j is represented by a wave function $\psi(\sigma_z)$, the spin being quantized along a chosen z axis. For spin $= \frac{1}{2}$ particles these wave functions are just the usual four-component Dirac spinors $u(\sigma_z)$, $\sigma_z = \pm\frac{1}{2}$, while for spin $= 1$ we use the polarization vectors $\epsilon_\mu(\sigma_z)$, and higher-spin wave functions can be constructed by taking products of these with suitable Clebsh–Gordan coefficients. The transition amplitude for the scattering process $1+2 \to 3+4$ between these spin states is then written in the form (see for example Barut (1967), Pilkuhn (1967))

$$A_{fi} = \chi_f M_{fi} \chi_i \tag{4.1.2}$$

where the χ's are the spinor wave functions of the particles in the initial and final states ($\chi_i = \psi_1 \otimes \psi_2, \chi_f = \psi_3 \otimes \psi_4$), and the M-functions are matrices. Because of Lorentz invariance they may be decomposed in the form

$$M_{fi} = \Sigma_\alpha A_\alpha(s,t) Y_\alpha \tag{4.1.3}$$

where the A_α's are scalar functions of the invariants, and the Y_α's are all the different independent Lorentz invariant matrices which can be constructed from the spin operators (Dirac matrices, polarization vectors etc.) and the momentum vectors of the particles (see Scadron and Jones, 1968, and Cohen-Tannoudji *et al.*, 1968). For example in pseudo-scalar-meson–baryon scattering (spins $0+\frac{1}{2} \to 0+\frac{1}{2}$) it is found that there are only two independent terms in (4.1.3) (paying due regard to TCP invariance and the algebra of Dirac matrices), and in the now conventional notation of Chew *et al.* (1957) one writes

$$M = A(s,t) + B(s,t)\,\tfrac{1}{2}(p_1+p_3)_\mu \gamma^\mu \tag{4.1.4}$$

where p_1 and p_3 are the four-momenta of the pions in the initial and final states respectively, γ_μ is the Dirac matrix, and A, B are the required invariant amplitudes for the process.

This method has the advantage that, if the Y's are suitably chosen, the invariant amplitudes $A_\alpha(s,t)$ are free of kinematical singularities, and so have just the dynamical singularities generated by the unitarity equations. Also they can be crossed directly from one channel to another ($s \to t$ etc.) as the spin rotations etc. involved in going from one channel to another are taken care of by the Y's. So these invariant amplitudes are completely analogous to the spinless particle amplitudes of chapter 1. Their disadvantages are that the determination of a complete independent set of Y_α which satisfy TCP invariance and have no arbitrary zeros (which would introduce compensating kinematical poles in the A_α) is quite difficult for high spins, and their unitarity equations are complicated by the occurrence of spinors in the intermediate states, which necessitates the evaluation of the trace of a matrix product. Also the relation of invariant amplitudes to experimentally observable quantities is somewhat complicated, and, perhaps most serious for us, the angular-momentum decomposition of these amplitudes is non-trivial (see for example Durand (1967), Jones and Scadron (1967), Taylor (1967), for a discussion of covariant Reggeization).

For all these reasons the helicity representation of Jacob and Wick (1959) has become more popular. (A full discussion of helicity amplitudes may be found in Martin and Spearman (1970).)

As described in chapter 1 a helicity state for a particle of four-momentum p and spin σ is denoted by $|p, \sigma, \lambda\rangle$, where the helicity, λ, is the spin component along the direction of motion of the particle ((1.2.4): $\lambda \equiv \boldsymbol{\sigma}.\mathbf{p}/|\mathbf{p}|$), and has $2\sigma+1$ possible values, $\sigma, \sigma-1, \ldots, -\sigma$. These states are irreducible representations of the Lorentz group, and are invariant under rotations. A state containing two non-interacting particles is described by the direct product

$$|p_1, \sigma_1, \lambda_1\rangle \otimes |p_2, \sigma_2, \lambda_2\rangle \equiv |p_1, \sigma_1, \lambda_1, p_2; \lambda_2, \sigma_2\rangle \qquad (4.1.5)$$

We work in the centre-of-mass system where $\mathbf{p}_1 = -\mathbf{p}_2$, and $s \equiv (p_1+p_2)^2$ is the square of the total energy (see (1.7.5)), and in this system, to avoid possible confusion, we shall denote the helicities by μ (λ will be used subsequently for helicities in the t-channel centre-of-mass system).

Thus for the scattering process $1+2 \to 3+4$, the s-channel centre-of-mass scattering amplitude may be written

$$\langle p_3, \sigma_3, \mu_3; p_4, \sigma_4, \mu_4| A |p_1, \sigma_1, \mu_1; p_2, \sigma_2, \mu_2\rangle$$
$$\equiv \langle \mu_3, \mu_4| A(s,t) |\mu_1 \mu_2\rangle \equiv A_{H_s}(s,t) \qquad (4.1.6)$$

where the dependence on the p_i has been expressed in terms of the invariants s and t, as in chapter 1, and the spins σ_i, being internal quantum numbers (like Q, B, I, Y etc.), have been suppressed. For brevity we use

$$H_s \equiv \{\mu_1, \mu_2, \mu_3, \mu_4\} \qquad (4.1.7)$$

for the helicities of the particles in the s-channel centre-of-mass system. These amplitudes are Lorentz invariant, except under reversal of the directions of the $\mathbf{p}_i$ (see below).

They have the advantage of being immediately applicable for particles of any spin, their unitarity equations are quite simple, requiring just a summation over intermediate-state helicity labels (see section 4.7), and, as we shall find below, they are directly related to experimental observables. Also their angular-momentum decomposition is comparatively easy. This is because for a two-particle state the orbital angular momentum is perpendicular to the direction of relative motion of the particles. So in the centre-of-mass frame the component of the total angular momentum in the direction of motion is just the difference of the helicities, which is fixed. Thus for the initial state $J_z = \mu_1 - \mu_2$ (the minus sign occurring because particle 2 is travelling in the $-z$ direction).

The disadvantage of these helicity amplitudes is that they are not free of kinematical singularities, so we must learn how to extract the necessary kinematical factors before we can write dispersion relations like (1.10.7), integrating just over the dynamical singularities. Also their crossing properties are non-trivial because the directions of motion of the particles are different in the s- and t-channel centre-of-mass systems, and so a given s-channel helicity amplitude crosses into a sum of t-channel amplitudes, and vice versa (see (4.3.7) below).

However, both of these problems have been solved for arbitrary spins, and so helicity amplitudes are now widely used for discussing spin problems and we shall employ them throughout this book. However, invariant amplitudes were invented first, and are still quite often invoked for pseudo-scalar-meson–baryon scattering and photo-production.

In the next section we shall briefly discuss the relation between helicity amplitudes and experimental observables, and then go on to consider their crossing properties. We then repeat the procedures of partial-wave decomposition and analytic continuation in angular momentum which we followed in chapter 2, showing the extra complications which spin introduces into Regge theory. We conclude the chapter with a review of the restrictions which unitarity places on the Regge singularities.

4.2 Helicity amplitudes and observables

For a given scattering process $1+2 \to 3+4$ there are $\prod_{i=1}^{4}(2\sigma_i+1)$ different helicity amplitudes, the different possible combinations of μ_i in (4.1.6). However, not all of these are independent because strong interactions are invariant under parity inversion and time reversal.

Under a parity inversion $((x, y, z) \overset{\mathrm{P}}{\to} (-x, -y, -z))$ the momentum vector $\boldsymbol{p} \overset{\mathrm{P}}{\rightarrow} -\boldsymbol{p}$, but since the spin, vector $\boldsymbol{\sigma}$ is an axial vector (i.e. transforms like a vector product $\boldsymbol{r} \times \boldsymbol{p} \overset{\mathrm{P}}{\rightarrow} (-\boldsymbol{r}) \times (-\boldsymbol{p}) = \boldsymbol{r} \times \boldsymbol{p}$) $\boldsymbol{\sigma} \overset{\mathrm{P}}{\rightarrow} \boldsymbol{\sigma}$. Hence the sign of the helicity (1.2.4) is reversed under a parity transformation, i.e. $\mu \overset{\mathrm{P}}{\rightarrow} -\mu$. Since the scattering process is invariant under P we have

$$\langle \mu_3, \mu_4 | A | \mu_1, \mu_2 \rangle = \eta \langle -\mu_3, -\mu_4 | A | -\mu_1, -\mu_2 \rangle \qquad (4.2.1)$$

where η is a phase factor ($= \pm 1$). The phase convention usually adopted for helicity amplitudes, following Jacob and Wick (1959), is obtained by representing the parity inversion operator, P, as a reflection in the x–z plane, Y, followed by a rotation by π about the y axis. Also by convention the particle is travelling along the $\pm z$ axis, so for example

$$\mathrm{P}|p_1, \sigma_1, \mu_1\rangle = \mathrm{e}^{\mathrm{i}\pi J_y} Y |p_1, \sigma_1, \mu_1\rangle = P_1(-1)^{\sigma_1 - \mu_1} \mathrm{e}^{\mathrm{i}\pi J_y} |p_1, \sigma_1, -\mu_1\rangle \qquad (4.2.2)$$

where P_1 is the intrinsic parity of the particle, and the factor $(-1)^{\sigma_1 - \mu_1}$ appears because the reflection is achieved by the rotation matrix, $d^{\sigma}_{\mu', \mu}(\pi) = (-1)^{\sigma - \mu} \delta_{\mu', -\mu}$, from (B.7) and (B.8). Since the scattering plane is taken to be the x–z plane ($\phi = 0$) the phase factor in (4.2.1) is, remembering that 2 is travelling in the opposite direction to 1, etc.,

$$\eta = P_1 P_2 P_3 P_4 (-1)^{\sigma_1 - \mu_1 + \sigma_2 + \mu_2 - \sigma_3 + \mu_3 - \sigma_4 + \mu_4} \qquad (4.2.3)$$

(see Martin and Spearman (1970) p. 227).

Similarly time-reversal invariance implies that the amplitudes for $1+2 \rightarrow 3+4$ must equal those for $3+4 \rightarrow 1+2$, again apart from a phase factor, and with this convention

$$\langle \mu_3 \mu_4 | A | \mu_1 \mu_2 \rangle = (-1)^{\mu_3 - \mu_4 - \mu_1 + \mu_2} \langle \mu_1 \mu_2 | A | \mu_3 \mu_4 \rangle \qquad (4.2.4)$$

(Martin and Spearman (1970) p. 232).

These relations greatly reduce the number of amplitudes which we have to consider. Thus for a process with spins $0 + \frac{1}{2} \rightarrow 0 + \frac{1}{2}$, of the 4 possible helicity amplitudes only 2 are independent, while for $\frac{1}{2} + \frac{1}{2} \rightarrow \frac{1}{2} + \frac{1}{2}$ only 6 of the 16 possible amplitudes are independent. Further restrictions may follow in some cases from the identity of the particles (depending on whether they obey Fermi or Bose statistics).

In general in a scattering experiment it is impossible to determine completely the spin orientations of all the particles. This means that one is not able to deal with pure helicity states in which each particle has a well defined spin projection, but must consider mixed states (statistical ensembles) which are incoherent sums of the different helicity states, occurring with various probabilities (see for example Schiff (1968) p. 378).

The simplest experiment is one in which no attempt is made to determine any of the spin directions, so that all the $2\sigma_i + 1$ helicity states for each particle are equally probable. In this case we simply have to average over all the possible helicity states which could occur in the initial state, and sum over all those which may occur in the final state, so instead of (1.8.16) the unpolarized differential cross-section in terms of the amplitudes (4.1.6) is

$$\frac{d\sigma}{dt} = \frac{1}{64\pi s q_{s12}^2} \frac{1}{(2\sigma_1+1)(2\sigma_2+1)} \sum_{H_s} |A_{H_s}(s,t)|^2 \qquad (4.2.5)$$

where the sum over H_s is over all the $2\sigma_i + 1$ values of each μ_i ($i = 1, \ldots, 4$). Similarly the total cross-section, $1+2 \to$ all, for scattering from an initially unpolarized state is related via the optical theorem (1.9.6) to the forward elastic scattering amplitudes $1+2 \to 1+2$ by

$$\sigma_{12}^{tot} = \frac{1}{2q_{s12}\sqrt{s}} \frac{1}{(2\sigma_1+1)(2\sigma_2+1)} \sum_{\mu_1\mu_2} \operatorname{Im}\{\langle \mu_1\mu_2 | A^{el}(s,0) | \mu_1\mu_2 \rangle\} \qquad (4.2.6)$$

It is possible to obtain information about the spin dependence of the scattering process by doing experiments with polarized particles, that is to say particles for which the average spin projection in some chosen direction is different from zero. This can be achieved for example by a polarization experiment in which the target proton is placed in a strong magnetic field along a chosen y axis at very low temperatures giving, say, a more than 50% probability that $\sigma_y = +\frac{1}{2}$ rather than $-\frac{1}{2}$. Or, if one of the final-state particles is unstable we can determine the average spin orientation of that particle from the angular distribution of its decay products.

We describe such a mixed-spin state for a given particle, i, by a spin density matrix, $\rho_{mm'}$, a $(2\sigma_i+1)$ by $(2\sigma_i+1)$ Hermitian matrix of unit trace, such that the expectation value (or average value) of some spin-dependent observable, O, in this state is given by

$$\langle O \rangle = \operatorname{tr}(O\rho) \qquad (4.2.7)$$

(tr = trace). Thus suppose we observe the angular distribution (θ, ϕ) of the two-body decay of one of the final-state particles (4 say), so that the full process is $1+2 \to 3+4$, $4 \to a+b$. Then the scattering amplitude will take the form (Jackson 1965)

$$\sum_m A_{\mu_1\mu_2\mu_3 m} A(m \to ab; \theta, \phi) \tag{4.2.8}$$

where $A_{\mu_1\mu_2\mu_3 m}$ is the probability amplitude for producing particle 4 with helicity $\mu_4 = m$, and $A(m \to ab; \theta, \phi)$ is the probability amplitude for the decay of 4 from this helicity state into $a+b$, with particle a travelling in the direction specified by the polar angles θ, ϕ relative to the direction of motion of particle 4. (These angles are measured in the rest frame of particle 4.) So the production angular distribution for this process will be

$$W(\theta, \phi) \propto \sum_{\mu_1\mu_2\mu_3} \left|\sum_m A_{\mu_1\mu_2\mu_3 m} A(m \to ab; \theta, \phi)\right|^2 \tag{4.2.9}$$

Hence if we define the production spin density matrix for particle 4 by

$$\rho_{mm'} \equiv \frac{\sum_{\mu_1\mu_2\mu_3} A_{\mu_1\mu_2\mu_3 m} A^*_{\mu_1\mu_2\mu_3 m'}}{\sum_{\mu_1\mu_2\mu_3\mu_4} \left|A_{\mu_1\mu_2\mu_3\mu_4}\right|^2} \tag{4.2.10}$$

which is normalized so that $\operatorname{tr}(\boldsymbol{\rho}) = 1$, and define the decay density matrix by

$$R_{mm'} = A(m \to ab)\, A^*(m' \to ab) \tag{4.2.11}$$

then the angular distribution (4.2.9) will be given by

$$W(\theta, \phi) = \operatorname{tr}(\boldsymbol{\rho} \mathbf{R}^*) \tag{4.2.12}$$

Thus if we know $\mathbf{R}$, $\boldsymbol{\rho}$ can be determined directly from $W(\theta, \phi)$ and this gives further information about the A_{H_s} in addition to (4.2.5).

To obtain R we let $\boldsymbol{q}$ and $-\boldsymbol{q}$ be the momenta of a and b, respectively, in the rest frame of particle 4, and $\hat{\boldsymbol{q}}$ a unit vector in the direction of $\boldsymbol{q}$. The final state after the decay is then $|\hat{\boldsymbol{q}}, \mu_a, \mu_b\rangle$. For a parity conserving decay the decay amplitude takes the form (when suitably normalized)

$$A(m \to \mu_a \mu_b) = \left(\frac{2\sigma_4 + 1}{4\pi}\right)^{\frac{1}{2}} \mathscr{D}^{\sigma_4 *}_{m\mu}(\phi, \theta, 0) \tag{4.2.13}$$

where $\mathscr{D}$ is the rotation matrix (B.3) corresponding to the rotation of a system having angular momentum σ_4 from the direction of motion of particle 4 (in which m is its spin projection) to the direction $\hat{\boldsymbol{q}}$ (in which $\mu \equiv \mu_a - \mu_b$ is its spin projection) θ is the angle between $\hat{\boldsymbol{q}}$ and $\boldsymbol{p}_4$, and ϕ the azimuthal angle about $\hat{\boldsymbol{q}}$. Using the representation (B.4)

$$\mathscr{D}^{\sigma_4}_{m\mu}(\phi, \theta, 0) = e^{im\phi} d^{\sigma_4}_{m\mu}(\theta) \tag{4.2.14}$$

and summing over the helicities, μ_a, μ_b, we find that the normalized angular distribution is

$$W(\theta, \phi) = \frac{2\sigma_4+1}{4\pi} \sum_{mm'} \sum_{\mu_a\mu_b} \rho_{mm'} \mathrm{e}^{\mathrm{i}(m-m')\phi} d^{\sigma_4}_{m\mu}(\theta)\, d^{\sigma_4}_{m'\mu}(\theta) \quad (4.2.15)$$

Thus for the decay of a spin = 1 particle into two spin = 0 particles (e.g. $\rho \to \pi\pi$) we have

$$W_1(\theta, \phi) = \frac{3}{4\pi}\Big[\cos^2\theta\,\rho_{00} + \tfrac{1}{2}\sin^2\theta\,(\rho_{11} - \rho_{-1-1}) - \sin^2\theta\,\mathrm{Re}\{\rho_{1-1}\mathrm{e}^{2\mathrm{i}\phi}\} - \frac{1}{\sqrt{2}}\sin 2\theta\,\mathrm{Re}\{\rho_{10}\,\mathrm{e}^{\mathrm{i}\phi} - \rho_{-1,0}\,\mathrm{e}^{-\mathrm{i}\phi}\}\Big] \quad (4.2.16)$$

It is then quite easy to take suitable moments of the observed experimental distributions to invert (4.2.16) to give the ρ's directly, e.g.

$$\rho_{00} = \tfrac{1}{2}\int \mathrm{d}\Omega\,(5\cos^2\theta - 1)\,W_1(\theta, \phi)$$
$$\rho_{11} + \rho_{-1-1} = \tfrac{1}{2}\int \mathrm{d}\Omega\,(3 - 5\cos^2\theta)\,W_1(\theta, \phi) \quad (4.2.17)$$

Similar, but slightly more complicated expressions are obtained for parity-violating weak decays such as $\Lambda \to p\pi^-$ since the decay amplitude corresponding to (4.1.23) will then involve two terms, one even under parity reflections and the other odd (see Jackson 1965).

Because of the parity relation (4.2.1) not all the production density matrix elements are independent, but

$$\rho_{-m-m'} = (-1)^{m-m'}\rho_{mm'} \quad (4.2.18)$$

Also the Hermitian nature of the density matrix implies that ρ_{mm} is real, which, together with the normalization condition that $\mathrm{tr}(\rho) = \sum_m \rho_{mm} = 1$, leaves only the following independent real observables

$$\left.\begin{array}{ll} \rho_{mm} & 0 \leqslant m \leqslant \sigma_4 \\ \mathrm{Re}\{\rho_{mm'}\} & |m'| < m \leqslant \sigma_4 \\ \rho_{m-m} & \text{for (integral } \sigma_4) \end{array}\right\} \quad (4.2.19)$$

If both the final-state particles decay there are similar joint production density matrices

$$\rho^{mm'}_{nn'} \equiv \frac{\sum_{\mu_1\mu_2} A_{\mu_1\mu_2 mn} A_{\mu_1\mu_2 m'n'}}{\sum_{\mu_1\mu_2\mu_3\mu_4} |A_{\mu_1\mu_2\mu_3\mu_4}|^2} \quad (4.2.20)$$

which can be obtained from the joint decay distribution

$$W(\theta_3\phi_3;\, \theta_4\phi_4).$$

For spin $= \frac{1}{2}$ particles it is more usual to re-express the density matrix in terms of the polarization vector $\boldsymbol{P}$ defined by

$$\rho_{mm'}(\tfrac{1}{2}) = \tfrac{1}{2}(1+\boldsymbol{P}.\boldsymbol{\sigma}) = \frac{1}{2}\begin{pmatrix} 1+P_z & P_x - \mathrm{i}P_y \\ P_x+\mathrm{i}P_y & 1-P_z \end{pmatrix} \tag{4.2.21}$$

where $\boldsymbol{\sigma}$ is the Pauli matrix, and where as usual the z axis is along the direction of motion, and $\mathbf{y}$ is perpendicular to the production plane. Parity conservation (4.2.18) requires $P_x = P_z = 0$. Thus for example for $\pi+\mathrm{p}\rightarrow\pi+\mathrm{p}$, with a polarized proton target,

$$P_y = \langle\sigma_y\rangle = \mathrm{tr}\,(\rho\sigma_y) = -2\,\mathrm{Im}\{\rho_{\frac{1}{2}-\frac{1}{2}}\} = -2\frac{A_{++}A^*_{+-}}{|A_{++}|^2+|A_{+-}|^2} \tag{4.2.22}$$

where $\pm \equiv \pm\frac{1}{2}$ for the nucleon helicities, and the pion helicity label $(=0)$ is omitted. This can be determined directly from the left–right asymmetry of the scattering cross-section about the y–z plane.

4.3 Crossing of helicity amplitudes

To discuss the Regge pole exchange contributions to a scattering process it is necessary to be able to cross from the t-channel centre-of-mass scattering amplitude $A^t(s,t)$, for the process $1+\overline{3}\rightarrow\overline{2}+4$ in which the Reggeon appears as a physical particle, to the s-channel centre-of-mass amplitude $A^s(s,t)$, which describes the process $1+2\rightarrow 3+4$. For spinless-particle scattering the crossing relation is simply

$$A^s(s,t) = A^t(s,t) \tag{4.3.1}$$

from the crossing postulate (section 1.6).

However, for helicity amplitudes things are not quite so simple because the helicities are defined in terms of the spin projections in the directions of motion of the various particles, so if we change the directions of motion the helicities will change too. Moreover, we have to make not just a physical Lorentz transformation, but a complex Lorentz transformation in which we pass from the values of the momenta appropriate for a physical process in the t channel, to those appropriate for the s channel, where the four-momenta of particles 2 and 3 are reversed. Thus great care is needed in following the path of continuation of the kinematical factors involved in the Lorentz transformation. However, it can be shown (Trueman and Wick 1964) that with a suitable choice of path the helicities are unchanged by crossing so that (apart from a possible phase factor)

$$\langle\lambda_3\lambda_4|\,A^s(s,t)\,|\lambda_1\lambda_2\rangle = \langle\lambda_2\lambda_4|\,A^t(s,t)\,|\lambda_1\lambda_3\rangle \tag{4.3.2}$$

where the λ's are t-channel centre-of-mass-frame helicities (i.e. the spin projections of the particles in their directions of motion in that frame). It is then necessary to re-express (4.3.2) in terms of s-channel helicities, and to achieve this we use the fact that under a general Lorentz transformation a helicity state is transformed as

$$|\mathrm{p},\sigma,\lambda\rangle \to \sum_{\lambda'} \mathscr{D}^{\sigma}_{\lambda'\lambda}(R)\,|p',\sigma,\lambda'\rangle \tag{4.3.3}$$

where $\mathscr{D}$ is the rotation matrix (B.3) and p' is the Lorentz transformed four-momentum. But the momenta appear only in the Lorentz scalars s and t, and so

$$\langle\mu_3\mu_4|\,A^s(s,t)\,|\mu_1\mu_2\rangle = \sum_{\lambda_i} d^{\sigma_1}_{\lambda_1\mu_1}(\chi_1)\,d^{\sigma_2}_{\lambda_2\mu_2}(\chi_2)$$
$$\times\, d^{\sigma_3}_{\lambda_3\mu_3}(\chi_3)\,d^{\sigma_4}_{\lambda_4\mu_4}(\chi_4)\langle\lambda_2\lambda_4|\,A^t(s,t)|\,\lambda_1\lambda_3\rangle \tag{4.3.4}$$

where we have used (B.4) to express the rotation matrices in terms of the rotation functions $d^{\sigma}_{\lambda\mu}$, and χ_i is the angle of rotation for particle i between its direction of motion in the s- and t-channel centre-of-mass frames. In terms of s and t these angles are given by (see for example Martin and Spearman (1970) p. 337)

$$\left.\begin{aligned}
\cos\chi_1 &= \frac{-(s+m_1^2-m_2^2)(t+m_1^2-m_3^2)-2m_1^2\Delta}{(\lambda(s,m_1,m_2)\,\lambda(t,m_1,m_3))^{\frac{1}{2}}}\\
\cos\chi_2 &= \frac{(s+m_2^2-m_1^2)(t+m_2^2-m_4^2)-2m_2^2\Delta}{(\lambda(s,m_1,m_2)\,\lambda(t,m_2,m_4))^{\frac{1}{2}}}\\
\cos\chi_3 &= \frac{(s+m_3^2-m_4^2)(t-m_3^2-m_1^2)-2m_3^2\Delta}{(\lambda(s,m_3,m_4)\,\lambda(t,m_1,m_3))^{\frac{1}{2}}}\\
\cos\chi_4 &= \frac{-(s+m_4^2-m_3^2)(t+m_4^2-m_2^2)-2m_4^2\Delta}{(\lambda(s,m_3,m_4)\,\lambda(t,m_2,m_4))^{\frac{1}{2}}}\\
\sin\chi_i &= \frac{2m_i\phi^{\frac{1}{2}}}{(\lambda(s,m_i,m_j)\,\lambda(t,m_i,m_k))^{\frac{1}{2}}} \quad \begin{array}{c}(j,k\text{ chosen as for}\\ \cos\chi_i\text{ above})\end{array}
\end{aligned}\right\} \tag{4.3.5}$$

where
$$\Delta \equiv m_2^2-m_4^2-m_1^2+m_3^2 \tag{4.3.6}$$

and ϕ and λ are defined in (1.7.23) and (1.7.11).

It is often convenient to rewrite (4.3.4) as

$$A_{H_s}(s,t) = \sum_{H_t} M(H_s,H_t)\,A_{H_t}(s,t) \tag{4.3.7}$$

where $H_s \equiv \{\mu_1, \mu_2, \mu_3, \mu_4\}$, $H_t \equiv \{\lambda_1, \lambda_2, \lambda_3, \lambda_4\}$, and M is the helicity crossing matrix given in (4.3.4). It is of course a square matrix with $\prod_{i=1}^{4}(2\sigma_i+1)$ rows and columns, but the number of elements can often be reduced because of the parity and time-reversal relations (4.2.1) and (4.2.4).

As an example we consider $\pi+\mathrm{p}\to\pi+\mathrm{p}$ elastic scattering in the s channel, for which the t channel is $\pi\pi\to\mathrm{p}\bar{\mathrm{p}}$. The crossing relation reads

$$A_{H_s}(s,t) = \sum_{\lambda_2\lambda_4} \mathrm{d}^{\frac{1}{2}}_{\lambda_2\mu_2}(\chi_2)\, d^{\frac{1}{2}}_{\lambda_4\mu_4}(\chi_4)\, A_{H_t}(s,t) \tag{4.3.8}$$

with $\chi_2 = \pi - \chi_4$ given by substituting the appropriate masses in (4.3.5). So using (B.19) and the relations $A_{++} = A_{--}$, $A_{+-} = -A_{-+}$ from (4.2.1) (where $\pm \equiv \pm\frac{1}{2}$ as in (4.2.22)) we find the crossing relation becomes

$$\left.\begin{aligned} A^s_{++}(s,t) &= \sin\chi_4\, A^t_{++}(s,t) - \cos\chi_4\, A^t_{+-}(s,t) \\ A^s_{+-}(s,t) &= \cos\chi_4\, A^t_{++}(s,t) + \sin\chi_4\, A^t_{+-}(s,t) \end{aligned}\right\} \tag{4.3.9}$$

These amplitudes are related to the invariant amplitudes $A(s,t)$ and $B(s,t)$ of (4.1.4) by (Cohen-Tannoudji, Salin and Morel 1968)

$$\left.\begin{aligned} A^s_{++} &= \left(\frac{1+z_s}{2}\right)^{\frac{1}{2}} [2m_{\mathrm{N}} A(s,t) + (s - m^2_{\mathrm{N}} - m^2_\pi)\, B(s,t)] \\ A^s_{+-} &= -\left(\frac{1-z_s}{2}\right)^{\frac{1}{2}} s^{-\frac{1}{2}}[(s + m^2_{\mathrm{N}} - m^2_\pi)\, A(s,t) + (s - m^2_{\mathrm{N}} + m^2_\pi)\, m_{\mathrm{N}} B(s,t)] \end{aligned}\right\} \tag{4.3.10}$$

and

$$\left.\begin{aligned} A^t_{++} &= -(t-4m^2_{\mathrm{N}})^{\frac{1}{2}} A(s,t) + m_{\mathrm{N}}(t-4m^2_\pi)^{\frac{1}{2}} z_t B(s,t) \\ &\equiv -(t-4m^2_{\mathrm{N}})^{\frac{1}{2}} A'(s,t) \\ A^t_{+-} &= \tfrac{1}{2}(t-4m^2_\pi)^{\frac{1}{2}}\, t^{\frac{1}{2}}(1-z^2_t)^{\frac{1}{2}} B(s,t) \end{aligned}\right\} \tag{4.3.11}$$

Since the invariant amplitudes are free of kinematical singularities these equations directly exhibit the kinematical singularities of the helicity amplitudes. ($A'(s,t)$ defined in (4.3.11) will be used below.)

The rotation matrices $d^\sigma_{\lambda\mu}$ are orthogonal, and so the crossing matrix is too. Hence, we can also write the differential cross-section as

$$\frac{\mathrm{d}\sigma}{\mathrm{d}t} = \frac{1}{64\pi s q^2_{s12}} \frac{1}{(2\sigma_1+1)(2\sigma_2+1)} \sum_{H_t} |A_{H_t}(s,t)|^2 \tag{4.3.12}$$

Equations (4.2.5) and (4.3.12) are equivalent in both the s- and t-channel physical regions so it does not matter whether one uses s- or t-channel helicity amplitudes. However, outside the physical regions the crossing matrix has singularities so care is needed in interpreting the equivalence of these two equations. The density matrices (4.2.10) are obviously not the same with the two sets of amplitudes, though both frames are quite commonly used. Equation (4.2.10) gives what are called the s-channel or 'helicity frame' density matrices, while the similar expressions with λ's substituted for the μ's gives the t-channel or 'Gottfried–Jackson' density matrices (named after their originators Gottfried and Jackson (1964)). The crossing matrix of (4.3.7) enables one to transform from one set of density matrices to the other.

4.4 Partial-wave amplitudes with spin

Our main motive for introducing helicity amplitudes has been to provide a basis for defining partial-wave amplitudes, so that we can make an analytic continuation in the total angular momentum, J, similar to that made in chapter 2.

The initial state, $|p_1, \sigma_1, \mu_1; p_2, \sigma_2, \mu_2\rangle$, has the two particles travelling in opposite directions along the z axis in the s-channel centre-of-mass system. It can be decomposed into partial waves of angular momentum J by

$$|p_1, \sigma_1, \mu_1; p_2, \sigma_2, \mu_2\rangle = (16\pi)^{\frac{1}{2}} \sum_{J=|\mu|}^{\infty} (2J+1)^{\frac{1}{2}} |s, J, \mu, \mu_1, \mu_2\rangle \quad (4.4.1)$$

where

$$\mu \equiv \mu_1 - \mu_2 \quad (4.4.2)$$

is the z component of J, $s = (p_1+p_2)^2$ as usual, and the factor $[16\pi(2J+1)]^{\frac{1}{2}}$ gives a convenient normalization. We have absorbed the spin labels, $\sigma_{1,2}$, into the implicit particle-type label on the right-hand side of (4.4.1) (see section 1.2).

Similarly, in the final state the particles are travelling in opposite directions at polar angles, θ, ϕ, relative to the z axis (see fig. 2.1 (c)), and the corresponding decomposition is

$$|p_3, \sigma_3, \mu_3; p_4, \sigma_4, \mu_4\rangle = (16\pi)^{\frac{1}{2}} \sum_{J=|\mu'|}^{\infty} \sum_{\mu''=-J}^{J} (2J+1)^{\frac{1}{2}}$$
$$\times \mathscr{D}^J_{\mu''\mu'}(\phi, \theta, -\phi)|s, J.\mu'', \mu_3, \mu_4\rangle \quad (4.4.3)$$

using (4.4.1), (B.1) and (B.3), where

$$\mu' \equiv \mu_3 - \mu_4 \quad (4.4.4)$$

is the component of J along the direction of motion, and μ'' is the component of J along the z axis. $\mathscr{D}^J_{\mu''\mu'}(\phi, \theta, -\phi)$ is the rotation matrix defined in (B.3) corresponding to the rotation from the θ, ϕ direction to the z axis.

Because of angular-momentum conservation we can define a partial-wave scattering amplitude for scattering in each J, i.e.

$$A_{HJ}(s) \equiv \langle s, J, \mu'', \mu_3, \mu_4 | A | s, J, \mu, \mu_1, \mu_2 \rangle \tag{4.4.5}$$

$$H \equiv \{\mu_1, \mu_2, \mu_3, \mu_4\} \tag{4.4.6}$$

where $\mu'' = \mu$ to conserve the z component of J, and so the full scattering amplitude (4.1.6) may be written (using (4.4.1), (4.4.3) and (4.4.5)) as

$$A_{H_s}(s,t) = 16\pi \sum_{J=M}^{\infty} (2J+1) A_{HJ}(s) \mathscr{D}^{J*}_{\mu\mu'})\phi, \theta, -\phi) \tag{4.4.7}$$

where
$$M \equiv \max\{|\mu|, |\mu'|\} \tag{4.4.8}$$

If we take the scattering plane to be the x–z plane $\phi = 0$, so, from (B.4), (4.4.7) simplifies to

$$A_{H_s}(s,t) = 16\pi \sum_{J=M}^{\infty} (2J+1) A_{HJ}(s) d^J_{\mu\mu'}(z_s) \tag{4.4.9}$$

which may be compared to (2.2.2) for spinless scattering.

The partial-wave amplitudes can be obtained from (4.4.9) using the orthogonality relation (B.14), viz.

$$A_{HJ}(s) = \frac{1}{32\pi} \int_{-1}^{1} A_{H_s}(s,t) d^J_{\mu\mu'}(z_s) \, dz_s \tag{4.4.10}$$

It is evident that for spinless scattering where $\mu_i = 0$, $i = 1, \ldots, 4$, (4.4.10) reduces to (2.2.1) because of (B.18).

The values of J in the series (4.4.9) are either integer or half-odd-integer depending on whether the number of fermions in the s channel is even or odd (i.e. J is integer for boson–boson and fermion–fermion scattering, but half-odd-integer for boson–fermion scattering). The sum starts at $J = M$ (defined in (4.4.8)) not 0 or $\frac{1}{2}$, because, as we noted in section 4.1, there is no component of $\boldsymbol{l}$ in the direction of motion of the particles, so for the initial state

$$J_z = \sigma_{1z} + \sigma_{2z} = \mu_1 - \mu_2 \equiv \mu$$

(with a similar expression for the final-state particles in their direction of motion) and obviously one must have $J \geqslant |J_z|$.

Following similar arguments to those in section 2.2 we find that the unitarity relation for these partial-wave amplitudes is

$$A^{if}_{HJ}(s_+) - A^{if}_{HJ}(s_-) = \frac{4iq_{sn}}{\sqrt{s}} \sum_{H_n} A^{in}_{HJ}(s_+) A^{nf}_{HJ}(s_-) \tag{4.4.11}$$

like (2.2.7), but where the sum runs over all the possible helicities of the intermediate state $|n\rangle$.

Like (2.2.2), the series (4.4.9) is valid only until we reach the nearest dynamical t-singularity (i.e. only inside the small Lehmann ellipse) and to continue outside the neighbourhood of the s-channel physical region it is necessary to make an analytic continuation. However, unlike the $P_l(z_s)$, the $d^J_{\mu\mu'}(z_s)$ are not in general entire functions of z_s, and so there are additional 'kinematical' singularities which we must also take into account. They can be read off directly from (B.9), for since the Jacobi polynomials are entire functions of z, the singularities of the $d^J_{\mu\mu'}(z_s)$ stem just from the half-angle factor

$$\xi_{\mu\mu'}(z_s) \equiv \left(\frac{1-z_s}{2}\right)^{\frac{1}{2}|\mu-\mu'|}\left(\frac{1+z_s}{2}\right)^{\frac{1}{2}|\mu+\mu'|} = \left(\sin\frac{\theta_s}{2}\right)^{|\mu-\mu'|}\left(\cos\frac{\theta_s}{2}\right)^{|\mu+\mu'|} \tag{4.4.12}$$

and so occur at $z_s = \pm 1$. They have a rather simple physical interpretation in that for forward scattering, $z_s = 1$, μ and μ' are the projections of J along the z axis in the initial and final states, respectively. Since angular momentum is to be conserved the scattering amplitude must obviously vanish as $z_s \to 1$ unless $\mu = \mu'$. The same applies for backward scattering ($z_s = -1$) where μ and $-\mu'$ are the corresponding z-components of J.

It is thus convenient to define s-channel helicity amplitudes free of these kinematical singularities in t by

$$\hat{A}_{H_s}(s,t) \equiv A_{H_s}(s,t)\,[\xi_{\mu\mu'}(z_s)]^{-1} \tag{4.4.13}$$

These amplitudes will satisfy the same sort of fixed-s dispersion relations, involving integrals over the dynamical singularities in t, as do spinless-particle scattering amplitudes. Note, however, that (4.4.13) still has kinematical s-singularities, which we shall discuss later (see section 6.2).

We could of course repeat the discussion of this section for t-channel helicity amplitudes to obtain the partial-wave series

$$A_{H_t}(s,t) = 16\pi \sum_{J=M}^{\infty} (2J+1)\,A_{HJ}(t)\,d^J_{\lambda\lambda'}(z_t) \tag{4.4.14}$$

where $\lambda \equiv \lambda_1 - \lambda_3, \quad \lambda' \equiv \lambda_2 - \lambda_4, \quad M \equiv \max\{|\lambda|, |\lambda'|\}$ (4.4.15)

and $\hat{A}_{H_t}(s,t) \equiv A_{H_t}(s,t)\,[\xi_{\lambda\lambda'}(z_t)]^{-1}$ (4.4.16)

will be free of kinematical singularities in s. The inverse of (4.4.14) is (like (4.4.10))

$$A_{HJ}(t) = \frac{1}{32\pi}\int_{-1}^{1} A_{H_t}(s,t)\,d^J_{\lambda\lambda'}(z_t)\,dz_t \tag{4.4.17}$$

(We have for simplicity dropped the channel label for the helicities of the partial-wave amplitudes in (4.4.10) and (4.4.17) as they are always implied by the channel invariants.)

4.5 The Froissart–Gribov projection

Since $\hat{A}_{H_t}(s,t)$ defined in (4.4.16) has no kinematical s-singularities it satisfies a dispersion relation in s at fixed t like (1.10.7), i.e.

$$\hat{A}_{H_t}(s,t) = \frac{1}{\pi}\int_{s_T}^{\infty} \frac{D_{sH}(s',t)}{s'-s}\,ds' + \frac{1}{\pi}\int_{u_T}^{\infty} \frac{D_{uH}(u',t)}{u'-u}\,du' \qquad (4.5.1)$$

where D_{sH} is the discontinuity of $\hat{A}_H$ across the dynamical s-cuts above the threshold s_T (and correspondingly for D_{uH}). Bound-state poles, if they occur, can be added as in (1.10.7).

This expression can be employed, following the method of section 2.3, to define partial-wave amplitudes even outside the region of convergence of the partial-wave series. Substituting (4.5.1) into (4.4.17), remembering (4.4.16) and (2.3.2), we obtain (Calogero, Charap and Squires 1963*b*, Drechsler 1968)

$$A_{HJ}(t) = \frac{1}{32\pi}\int_{-1}^{1} dz_t\, d^J_{\lambda\lambda'}(z_t)\,\xi_{\lambda\lambda'}(z_t)\left\{\frac{1}{\pi}\int_{z_T}^{\infty} \frac{D_{sH}(s',t)}{z'-z_t}\,dz' + \frac{1}{\pi}\int_{-z_T}^{-\infty} \frac{D_{uH}(u',t)}{z'-z_t}\,dz'\right\} \qquad (4.5.2)$$

($z_T \equiv z_s(s_T,t)$), which, with the generalized Neumann relation (B.21), gives the Froissart–Gribov projection (cf. (2.3.4))

$$A_{HJ}(t) = \frac{1}{16\pi^2}\int_{z_T}^{\infty} dz_t \{D_{sH}(s,t)\, e^J_{\lambda\lambda'}(z_t)\,\xi_{\lambda\lambda'}(z_t) + (-1)^{J-\lambda} D_{uH}(s,t)\, e^J_{\lambda-\lambda'}(z_t)\,\xi_{\lambda-\lambda'}(z_t)\} \qquad (4.5.3)$$

where (B.23) has been used for the second term.

If the asymptotic behaviour is $A_{H_t} \underset{s\to\infty}{\sim} s^{\alpha}$, then $\hat{A}_{H_t} \sim s^{\alpha-M}$ from (4.4.16) since $\xi_{\lambda\lambda'}(z_t) \sim s^M$, and since from (B.25) $e^J_{\lambda\lambda'}(z) \sim s^{-J-1}$, the criterion for the convergence of (4.5.3) is the same as for (2.3.4), i.e. $J > \alpha$.

As $J \to \infty$ we find from (B.26) that the first term in (4.5.3) tends to zero like

$$\underset{J\to\infty}{\sim} J^{-\frac{1}{2}} e^{-(J+\frac{1}{2})\zeta(z_T)} \qquad (4.5.4)$$

but the second term behaves like

$$\underset{J\to\infty}{\sim} J^{-\frac{1}{2}} e^{-(J+\frac{1}{2})\zeta(z_T)} e^{-i\pi(J-\lambda)} \qquad (4.5.5)$$

and so diverges as $J \to i\infty$. So (4.5.3) does not satisfy the conditions for Carlson's theorem, and (as in section 2.5) before we can make an analytic continuation in J we have to introduce amplitudes of definite signature. These are defined by replacing $(-1)^{J-v}$ by the signature $\mathscr{S} = \pm 1$, where

$$\left.\begin{aligned} v &\equiv 0 \quad \text{for physical } J = \text{integer} \\ v &\equiv \tfrac{1}{2} \quad \text{for physical } J = \text{half-odd-integer} \end{aligned}\right\} \qquad (4.5.6)$$

(Note that whether λ is integral or half-integral depends on the physical J values.) Hence

$$A^{\mathscr{S}}_{HJ}(t) \equiv \frac{1}{16\pi^2}\int_{z_{\mathrm{T}}}^{\infty} \mathrm{d}z_t \{D_{sH}(s,t)\, e^J_{\lambda\lambda'}(z_t)\, \xi_{\lambda\lambda'}(z_t) + \mathscr{S}(-1)^{\lambda-v} D_{uH}(s,t)\, e^J_{\lambda-\lambda'}(z_t)\, \xi_{\lambda-\lambda'}(z_t)\} \qquad (4.5.7)$$

For $\mathscr{S} = \pm 1$ these amplitudes coincide with the physical $A_{HJ}(t)$ for $J - v$ = even/odd, so instead of (4.4.14) we can write

$$A_{H_t}(s,t) = 16\pi \sum_{J=M}^{\infty} (2J+1)\,(A^{+}_{HJ}(t)\, d^{+}_{\lambda\lambda'}(J,z) + A^{-}_{HJ}(t)\, d^{-}_{\lambda\lambda'}(J,z)) \qquad (4.5.8)$$

if we define

$$d^{\mathscr{S}}_{\lambda\lambda'}(J,z) \equiv \tfrac{1}{2}[d^J_{\lambda\lambda'}(z) + \mathscr{S}(-1)^{\lambda-v} d^J_{\lambda-\lambda'}(-z)] \qquad (4.5.9)$$

Note that $d^{\pm}_{\lambda\lambda'}(J,z)$ vanishes for $J - v$ = odd/even because of the symmetry relation (B.7).

Scattering amplitudes of definite signature are defined by

$$A^{\mathscr{S}}_{H_t}(s,t) = 16\pi \sum_{J=M}^{\infty} (2J+1)\, A^{\mathscr{S}}_{HJ}(t)\, d^{\mathscr{S}}_{\lambda\lambda'}(J,z) \qquad (4.5.10)$$

Equation (4.5.7) may be used to define definite-signature partial-wave amplitudes for all J. The physical J values are of course those having integer $J - v$, with $J \geqslant |\lambda|$ for the initial state ($1+\bar{3}$ in the t channel) and $J \geqslant |\lambda'|$ for the final state. So $J \geqslant M$ defined in (4.4.15). Because these are the values of J which make physical sense, they are known as 'sense–sense' or ss values, and the amplitudes for these values of J are called ss amplitudes. When we continue in J we may arrive at integer values of $J - v$ with $J < M$, but $J \geqslant N$ where

$$N \equiv \min\{|\lambda|, |\lambda'|\} \qquad (4.5.11)$$

If say $|\lambda| > |\lambda'|$ then this J value makes physical sense for the final state, but not for the initial state (and vice versa if $|\lambda| < |\lambda'|$). These are called 'sense–nonsense' or sn values of J. And of course for integer $J-v$, $J<N$, we have nonsense–nonsense or nn amplitudes which do not make physical sense for either the incoming or outgoing states (Gell-Mann 1962). It is sometimes convenient to refer to all integer $J-v$ with $J<M$ as 'nonsense' values of J.

4.6 The Sommerfeld–Watson representation

The partial-wave series (4.5.10) can be rewritten as a contour integral in J, like (2.7.5), viz

$$A^{\mathscr{S}}_{H_t}(s,t) = -\frac{16\pi}{2\mathrm{i}}\int_{C_1}\frac{2J+1}{\sin\pi(J+\lambda')}A^{\mathscr{S}}_{HJ}(t)\,d^{\mathscr{S}}_{-\lambda\lambda'}(J,-z_t)\,\mathrm{d}J \quad (4.6.1)$$

where the contour C_1 encloses the physical values $J \geqslant M$, but avoids any singularities of the $A_{HJ}(t)$ as in fig. 4.1. The $(-1)^{J+\lambda'}$ from the residues of the poles of $(\sin\pi(J+\lambda'))^{-1}$ is cancelled by the use of $d^{\mathscr{S}}_{-\lambda\lambda'}(J,-z)$ instead of $d^{\mathscr{S}}_{\lambda\lambda'}(J,z)$ because of the symmetry relation (B.7).

Then when we open up the contour to C_2 of fig. 4.1 we reveal any Regge poles and cuts of $A_{HJ}(t)$, and also obtain contributions from integer values of $J-v$ in the region $-\frac{1}{2}<J<M$, i.e. from the sn and nn values defined above, so we have (substituting the integrand of (4.6.1) where indicated)

$$A^{\mathscr{S}}_{H_t}(s,t) = -\frac{16\pi}{2\mathrm{i}}\int_{C_2}[(4.6.1)] - 16\pi^2\frac{2\alpha_i(t)+1}{\sin\pi(\alpha(t)+\lambda')}\beta_H(t)d^{\mathscr{S}}_{-\lambda\lambda'}(\alpha(t),-z_t)$$
$$-\frac{16\pi}{2\mathrm{i}}\int^{\alpha_c(t)}\frac{2J+1}{\sin\pi(J+\lambda')}\Delta^{\mathscr{S}}(J,t)\,d^{\mathscr{S}}_{-\lambda\lambda'}(J,-z_t)\,\mathrm{d}J$$
$$-\sum_{J=N}^{M-1}-\sum_{J=v}^{N-1}16\pi(2J+1)A^{\mathscr{S}}_{HJ}(t)\,d^{\mathscr{S}}_{-\lambda\lambda'}(J,-z_t) \quad (4.6.2)$$

The first term is the usual background integral, $\sim s^{-\frac{1}{2}}$. For simplicity we have assumed that there is just one pole at $J=\alpha(t)$, and one branch point at $J=\alpha_c(t)$, in $\mathrm{Re}\{J\}>-\frac{1}{2}$, and evidently these terms have the usual asymptotic behaviour $\sim s^{\alpha(t)}$, and $\sim s^{\alpha_c(t)}$ respectively, from (B.14). The final terms contain the sn and nn contributions.

At a sn point $J=J_0$ say, where $J_0=v$ is an integer with $N \leqslant J_0 < M$, we can see from (B.12) that $d^J_{\lambda\lambda'}(z)$ (and hence $d^{\mathscr{S}}_{-\lambda\lambda'}(J,-z)$) vanishes like $(J-J_0)^{\frac{1}{2}}$, and so there will be no contribution from these terms unless $A_{HJ}\sim(J-J_0)^{-\frac{1}{2}}$. We shall discuss this possibility further in section 4.8, but if for the moment we assume that this does not happen

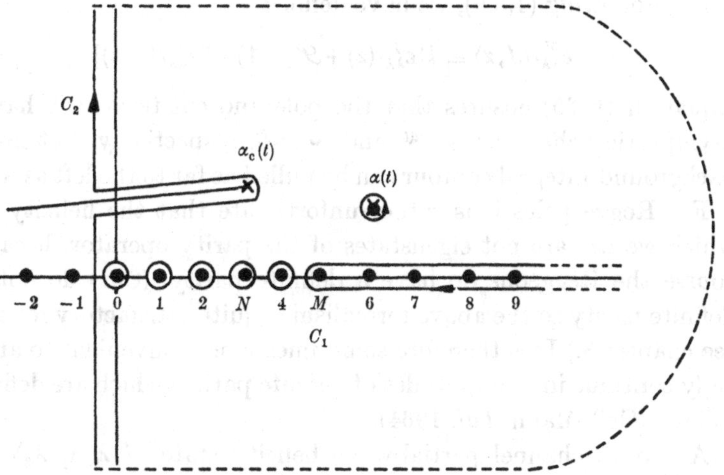

FIG. 4.1 The Sommerfeld–Watson transform for a helicity amplitude with $M = 5$ and $N = 3$. The contour C_1 encloses the integers $J \geqslant M$. When it is opened up to C_2 we get contributions from the Regge pole at $\alpha(t)$, from the branch cut starting at $\alpha_c(t)$, and from the integer values $-\frac{1}{2} < J < M$.

the first summation can be neglected. Similarly from (B.12) we find that at the nn points $J = J_0$ with $J_0 - v$ integer, $v \leqslant J_0 < N$, $d^J_{\lambda\lambda'} \sim (J-J_0)$ and so these terms also vanish unless there are fixed poles,

$$A_{HJ} \sim (J-J_0)^{-1},$$

a possibility which we shall also reconsider in section 4.8.

If we wish to explore the region $\mathrm{Re}\{J\} < -\frac{1}{2}$ we can again employ the Mandelstam method described in section 2.9, using the relation (B.28) instead of (A.18). The symmetry of the rotation functions (B.27) ensures that, from (4.5.7),

$$A_{HJ}(t) = (-1)^{\lambda-\lambda'} A^{\mathscr{S}'}_{H\,-J-1}(t), \quad J - v = \text{half-odd-integer} \quad (4.6.3)$$

(where $\mathscr{S}' \equiv \mathscr{S}$ for $v = 0$ and $\mathscr{S}' \equiv -\mathscr{S}$ for $v = \frac{1}{2}$) as long as (4.5.7) converges, and so the contribution of the poles of $[\cos \pi(J+\lambda')]^{-1}$ in the two terms of (B.28) cancel pairwise for $J < M$. So we get

$$\begin{aligned} A^{\mathscr{S}}_{H_t}(s,t) = {} & 16\pi(2\alpha(t)+1)\,\beta_H(t)\,\frac{e^{\mathscr{S}'}_{\lambda-\lambda'}(-\alpha-1,-z_t)}{\cos\pi(\alpha+\lambda')} \\ & + \frac{16\pi}{2\mathrm{i}}\int^{\alpha_c(t)} \frac{2J+1}{\cos\pi(J+\lambda')}\,\Delta^{\mathscr{S}}(J,\lambda')\,e^{\mathscr{S}'}_{\lambda-\lambda'}(-J-1,-z_t)\,\mathrm{d}J \\ & + \text{possible fixed poles or cuts} \\ & + \text{background integral} \end{aligned} \qquad (4.6.4)$$

where, following (4.5.9), we have defined

$$e^{\mathscr{S}}_{\lambda\lambda'}(J,z) \equiv \tfrac{1}{2}[e^{J}_{\lambda\lambda'}(z) + \mathscr{S}(-1)^{\lambda-v} e^{J}_{\lambda\lambda'}(-z)] \tag{4.6.5}$$

Equation (B.25) ensures that the pole and cut terms will have the asymptotic behaviour $\sim s^{\alpha(t)}$ and $\sim s^{\alpha_c(t)}$, respectively, but now the background-integral contour can be pulled as far to the left as we like.

For Regge poles it is rather unfortunate that the helicity states which we use are not eigenstates of the parity operator, because of course the Reggeons do have a definite parity. (Cuts do not have definite parity so the above formalism is quite satisfactory for them – see chapter 8.) It is therefore sometimes more convenient to analytically continue in J amplitudes of definite parity, which are defined as follows (Gell–Mann *et al.* 1964).

A given t-channel partial-wave helicity state $|J, \lambda, \lambda_1, \lambda_2\rangle$ transforms under the parity operator as

$$P|J,\lambda,\lambda_1,\lambda_3\rangle = P_1 P_3(-1)^{J-\sigma_1-\sigma_3}|J,\lambda,-\lambda_1,-\lambda_3\rangle \tag{4.6.6}$$

where P_1, P_3 are the intrinsic parities of the particles, and, as discussed in section 4.2, the helicities change sign. The phase factor $(-1)^{J-\sigma_1-\sigma_3}$ corresponds to the Condon and Shortley phase conventions for the relative phases of the helicity states as used in (4.2.2) and in the reflection properties of the rotation matrices (B.7) (see Jacob and Wick 1959). Thus we may define definite parity states by

$$|J,\lambda,\lambda_1,\lambda_3,\eta\rangle \equiv \frac{1}{\sqrt{2}}\{|J,\lambda,\lambda_1,\lambda_3\rangle + \eta P_1 P_3(-1)^{\sigma_1+\sigma_3-v}|J,\lambda,-\lambda_1,-\lambda_3\rangle\} \tag{4.6.7}$$

where $\eta = \pm 1$ for natural/unnatural parity. A state is said to have natural parity if $P = (-1)^{J-v}$ and unnatural parity if $P = (-1)^{J-v-1}$, results which are readily obtained from (4.6.7) using (4.6.6). These states are physical for $J-v$ even/odd depending on the signature, and so we have the relation

$$P = \eta\mathscr{S} \tag{4.6.8}$$

Since parity is conserved in strong interactions, scattering amplitudes occur only between states of the same parity, and a definite-parity partial-wave amplitude is given by

$$\begin{aligned}\langle J,\lambda',\lambda_2,\lambda_4,\eta|\,A^{\mathscr{S}}(t)\,|J,\lambda,\lambda_1,\lambda_3,\eta\rangle &\equiv A^{\mathscr{S}\eta}_{HJ}(t)\\ &\equiv \langle\lambda_2,\lambda_4|\,A^{\mathscr{S}}_J(t)\,|\lambda_1\lambda_3\rangle + \eta P_1 P_3(-1)^{\sigma_1+\sigma_3-v}\langle\lambda_2,\lambda_4|\,A_J(t)\,|-\lambda_1,-\lambda_3\rangle\end{aligned} \tag{4.6.9}$$

Hence we can define the so-called 'parity-conserving helicity amplitudes' free of kinematical singularities in s by

$$\hat{A}^{\mathscr{S}\eta}_{H'_t}(s,t) \equiv \langle\lambda_2\lambda_4\eta|\,\hat{A}^{\mathscr{S}}(s,t)\,|\lambda_1\lambda_3\eta\rangle = \langle\lambda_2\lambda_4|\,A^{\mathscr{S}}(s,t)\,|\lambda_1\lambda_3\rangle\xi^{-1}_{\lambda\lambda'}(z_t)$$
$$+\eta P_1 P_3(-1)^{\lambda'+M+\sigma_1+\sigma_3-v}\langle\lambda_2\lambda_4|\,\hat{A}^{\mathscr{S}}(s,t)\,|-\lambda_1-\lambda_3\rangle\,\xi^{-1}_{-\lambda\lambda'}(z_t) \tag{4.6.10}$$

The partial-wave series for this amplitude is

$$\hat{A}^{\mathscr{S}\eta}_{H'_t}(s,t) = 16\pi\sum_{J=M}^{\infty}(2J+1)\,A^{\mathscr{S}}_{HJ}(t)\,\frac{d^{\mathscr{S}}_{\lambda\lambda'}(J,z_t)}{\xi_{\lambda\lambda'}(z_t)}$$
$$+\eta P_1 P_3(-1)^{\lambda'+M+\sigma_1+\sigma_3-v}A^{\mathscr{S}}_{\bar{H}J}(t)\,\frac{d^{\mathscr{S}}_{-\lambda\lambda'}(z_t)}{\xi_{-\lambda\lambda'}(z_t)} \tag{4.6.11}$$

where we have introduced $\bar{H} \equiv \{-\lambda_1, -\lambda_3, \lambda_2, \lambda_4\}$. Or, using (4.6.9),

$$\hat{A}^{\mathscr{S}\eta}_{H'_t}(s,t) = 16\pi\sum_{J=N}^{\infty}(2J+1)\,(A^{\mathscr{S}\eta}_{HJ}(t)\,\hat{d}^{\mathscr{S}+}_{\lambda\lambda'}(J,z_t) + A^{\mathscr{S}\bar{\eta}}_{HJ}(t)\,\hat{d}^{\mathscr{S}-}_{\lambda\lambda'}(J,z_t)) \tag{4.6.12}$$

with $\bar{\eta} \equiv -\eta$ and

$$\hat{d}^{\mathscr{S}\eta}_{\lambda\lambda'}(J,z) \equiv \frac{1}{2}\left[\frac{d^{\mathscr{S}}_{\lambda\lambda'}(J,z)}{\xi_{\lambda\lambda'}(z)} + \eta(-1)^{\lambda'+M}\frac{d^{\mathscr{S}}_{-\lambda\lambda'}(J,z)}{\xi_{-\lambda\lambda'}(z)}\right] \tag{4.6.13}$$

Thus we see that the total amplitude contains contributions from partial-wave amplitudes of both parities, but asymptotically, from (4.6.12), (4.5.9), (B.17) and (B.13),

$$\hat{d}^{\mathscr{S}\eta}_{\lambda\lambda'}(J,z) \sim \left(\frac{z}{2}\right)^{J-M}\left(\frac{1+\eta}{2}\right) + O(z^{J-M-1}), \quad \mathrm{Re}\{J\} > -\tfrac{1}{2} \tag{4.6.14}$$

so to leading order $\hat{d}^{\mathscr{S}+}_{\lambda\lambda'}$ dominates over $\hat{d}^{\mathscr{S}-}_{\lambda\lambda'}$. It is only in this asymptotic sense that (4.6.12) can be regarded as a definite-parity amplitude.

If we now make a Sommerfeld–Watson transform of (4.6.12), and use the Mandelstam method like (4.6.4), we find that a Regge contribution is given by

$$\hat{A}^{\mathscr{S}\eta}_{H'_t}(s,t) = 16\pi(2\alpha(t)+1)\,\beta_H(t)\,\frac{\hat{e}^{\mathscr{S}+}_{\lambda-\lambda'}(-\alpha-1,z_t)}{\cos\pi(\alpha+\lambda')} \tag{4.6.15}$$

where, in analogy with (4.6.13), we have introduced

$$\hat{e}^{\mathscr{S}\eta}_{\lambda\lambda'}(J,z) \equiv \tfrac{1}{4}(1+\mathscr{S}e^{\pm i\pi(J-v)})\left[\frac{e^{J}_{\lambda\lambda'}(z)}{\xi_{\lambda\lambda'}(z)} + \eta(-1)^{\lambda'+M}\frac{e^{J}_{-\lambda\lambda'}(J,z)}{\xi_{\lambda\lambda'}(z)}\right] \tag{4.6.16}$$

But to leading order there is no difference between (4.6.15) and the Regge pole contribution in (4.6.4).

4.7 Restrictions on Regge singularities from unitarity

We have already noted in section 2.4 how the application of s-channel unitarity leads to the Froissart bound, and hence to the restriction that the t-channel Regge singularities cannot be above 1 for $t \leqslant 0$. This applies also in the presence of spin, since the Regge power behaviours are unchanged.

There are also some important restrictions which stem from t-channel unitarity. For spinless-particle elastic scattering in the t channel, $1+\bar{3}\rightarrow 1+\bar{3}$, the unitarity condition reads (from (2.2.7) and (2.6.8), with $s\rightarrow t$)

$$B_l^{\mathscr{S}}(t_+) - B_l^{\mathscr{S}}(t_-) = 2\mathrm{i}\rho_l(t)\, B_l^{\mathscr{S}}(t_+)\, B_l^{\mathscr{S}}(t_-) \tag{4.7.1}$$

$$\rho_l(t) \equiv (q_{t13})^{2l+1}\frac{2}{\sqrt{t}} \tag{4.7.2}$$

for $t_{\mathrm{T}} < t < t_{\mathrm{I}}$, where t_{T} is the elastic threshold, and t_{I} the inelastic threshold. Since $B_l(t)$ is a real analytic function we have

$$(B_{l^*}^{\mathscr{S}}(t+\mathrm{i}\epsilon))^* = B_l^{\mathscr{S}}(t-\mathrm{i}\epsilon) \tag{4.7.3}$$

for real t (where $* \equiv$ complex conjugate), and so we can rewrite (4.7.1) as

$$B_l^{\mathscr{S}}(t) - (B_{l^*}^{\mathscr{S}}(t))^* = 2\mathrm{i}\rho_l(t)\, B_l^{\mathscr{S}}(t)\, (B_{l^*}^{\mathscr{S}}(t))^* \tag{4.7.4}$$

To start with we only know that this equation is valid for right-signature integer values of l, but both sides of (4.7.4) satisfy the boundedness condition for Carlson's theorem (section 2.7) and hence the equation remains true if we continue in l. Note that, from the discussion in section 2.6, (4.7.4) is true for non-integer l only because we removed the kinematical threshold singularities in defining $B_l^{\mathscr{S}}(t)$ in (2.6.8).

It is evident that (4.7.1) cannot be satisfied by a fixed l-plane pole of the form

$$B_l^{\mathscr{S}}(t) \approx \frac{\beta(l,t)}{l-l_0}, \quad l\rightarrow l_0 \tag{4.7.5}$$

for if we inserted (4.7.5) into (4.7.1) we would have a single pole at l_0 on the left-hand side equated to a double pole on the right-hand side. A pole whose position changes with t, say at $l = \alpha(t)$, can satisfy (4.7.1) as long as $\alpha(t_+) \neq \alpha(t_-)$, i.e. as long as $\mathrm{Im}\,\{\alpha(t)\} \neq 0$ (for $t > t_{\mathrm{T}}$). We have seen examples of this in section 3.4 where unitarity has converted the fixed pole of the Born term into a moving pole with a right-hand cut.

The only way in which (4.7.4) can be satisfied with a fixed pole is if there is also an l-plane cut passing through l_0 for all $t_T < t < t_I$. Then one approaches $l = l_0$ on different sides of this cut in B_l and $B^*_{l^*}$, and the pole can be present on one side of the cut but not the other, in which case there is no problem (see section 8.3). But in the absence of cuts all poles must be moving poles, i.e. their positions must be functions of t.

For particles with spin we define corresponding partial-wave helicity amplitudes

$$B^{\mathscr{S}}_{HJ}(t) = A^{\mathscr{S}}_{HJ}(t)\,(q_{t13})^{-2L} \tag{4.7.6}$$

where L is the lowest possible orbital angular momentum at threshold for the given J ($L = J - Y^+_{13}$ where $Y^+_{13} = \sigma_1 + \sigma_3$ or $\sigma_1 + \sigma_3 - 1$ depending on the parity – this will be discussed in section 6.2.) Then the unitarity condition can be written in the form

$$\mathbf{B}^{\mathscr{S}}_J(t) - (\mathbf{B}^{\mathscr{S}}_{J^*}(t))^\dagger = 2\mathrm{i}(\mathbf{B}^{\mathscr{S}}_{J^*}(t))^\dagger\,\boldsymbol{\rho}_J(t)\,\mathbf{B}^{\mathscr{S}}_J(t) \tag{4.7.7}$$

where the B's have been expressed as matrices, the various initial- and final-state helicities labelling the rows and columns ($\dagger \equiv$ Hermitian conjugate = complex conjugate transposed matrix, i.e. $B^\dagger_{ij} = B^*_{ji}$). Here $\boldsymbol{\rho}_J(t)$ is a diagonal matrix of kinematical factors

$$(\rho_{HJ}(t))_{nn} = (q_{tn})^{2L_n+1}\frac{2}{\sqrt{t}} \tag{4.7.8}$$

So in (4.7.7) the sum over intermediate-state helicities is represented as a matrix product. Above the inelastic threshold, two-body inelastic processes can similarly be incorporated by increasing the numbers of rows and columns to represent the unitarity equation (2.2.11).

A fixed pole at $J = J_0$ in (4.7.7) implies that

$$\boldsymbol{\beta}(J_0, t)\,\boldsymbol{\beta}^\dagger(J_0, t) = 0, \quad \text{i.e. } \boldsymbol{\beta} = 0 \tag{4.7.9}$$

so again fixed poles on the real J axis are forbidden, but if J_0 has an imaginary part (4.7.7) simply gives

$$\boldsymbol{\beta}(J_0, t_+)\,\boldsymbol{\beta}(J_0, t_-) = 0 \tag{4.7.10}$$

which does not require $\boldsymbol{\beta} = 0$. So in principle there could be fixed poles even in the absence of cuts, but not on the real axis. However, there does not seem to be any reason why such fixed poles at complex values of J should occur. We shall find in the next section that fixed poles do occur on the real axis at wrong-signature nonsense points, and these clearly must have shielding Regge cuts.

If we define the partial-wave S-matrix by

$$\mathbf{S}(J,t) = \mathbf{1} + 2\mathrm{i}\boldsymbol{\rho}_J(t)\,\mathbf{B}(J,t) \tag{4.7.11}$$

where $\mathbf{1}$ is the unit matrix, the unitarity relation (4.7.7) reads

$$\mathbf{S}(J,t)\,\mathbf{S}^{\dagger}(J,t) = \mathbf{1} \quad \text{or} \quad \mathbf{S}(J,t)\,\frac{\operatorname{cof}(\mathbf{S}^{\dagger})}{\det(\mathbf{S}^{\dagger})} \tag{4.7.12}$$

(where cof ≡ cofactor matrix and det ≡ determinant). Thus for a two-channel process this becomes

$$\begin{pmatrix} S_{11} & S_{12} \\ S_{21} & S_{22} \end{pmatrix} = \frac{\begin{pmatrix} S_{22}^* & -S_{21}^* \\ -S_{12}^* & S_{11}^* \end{pmatrix}}{S_{11}^* S_{22}^* - S_{12}^* S_{21}^*} \tag{4.7.13}$$

so if $\mathbf{S}$ has a simple pole of the form $\boldsymbol{\beta}(J-\alpha)^{-1}$, the vanishing of the denominator on the right-hand side requires that

$$\beta_{22}\beta_{11} = \beta_{12}\beta_{21} \tag{4.7.14}$$

so that one can write

$$\beta_{ij} = \beta_i \beta_j \tag{4.7.15}$$

i.e. the Regge pole residue must factorize, as could have been anticipated from our discussion in section 1.5. This result has been proved for an arbitrary number of channels by Charap and Squires (1962).

4.8 Fixed singularities and SCR

The rotation functions, $e^J_{\lambda\lambda'}$, used in (4.5.7) to define partial-wave amplitudes of any J, have fixed J singularities stemming from the square bracket in (B.24) at unphysical values of J. ($F(a,b,c,d)$ is an entire function of its arguments.) Since $x!$ has poles at $x = -1, -2, -3, \ldots$, we see that for $J = J_0$ (where $J_0 - v$) is an integer

$$\left.\begin{aligned} e^J_{\lambda\lambda'}(z) &\sim (J-J_0)^{-\frac{1}{2}}, \quad N \leqslant J_0 < M \quad \text{and} \quad -M \leqslant J_0 < -N \\ &\sim (J-J_0)^{-1}, \quad -N \leqslant J_0 < N \quad \text{and} \quad J_0 < -M \end{aligned}\right\} \tag{4.8.1}$$

Thus for $J < -M$ the pole residue is just $d^{J_0}_{\lambda\lambda'}(z)$ (see (B.29)) and so for $J \to J_0 < -M$

$$\begin{aligned} A^{\mathscr{S}}_{HJ}(t) \to \frac{1}{J-J_0}\frac{1}{16\pi^2}\int_{z_T}^{\infty} \mathrm{d}z_t \{ & D_{sH}(s,t)\,\xi_{\lambda\lambda'}(z_t)\,d^{J_0}_{\lambda\lambda'}(z_t) \\ & + \mathscr{S}(-1)^{\lambda-v} D_{uH}(s,t)\,\xi_{\lambda-\lambda'}(z_t)\,d^{J_0}_{\lambda-\lambda'}(z_t)\} \end{aligned} \tag{4.8.2}$$

But such a real-axis fixed pole is incompatible with unitarity, as we found in the previous section, and so the integral in (4.8.2) must vanish, i.e.

$$\int_{z_T}^{\infty} dz_t\{D_{sH}(s,t)\,\xi_{\lambda\lambda'}(z_t)\,d^{J_0}_{\lambda\lambda'}(z_t)+\mathscr{S}(-1)^{\lambda-v}D_{uH}(s,t)\,\xi_{\lambda-\lambda'}(z_t)\,d^{J_0}_{\lambda-\lambda'}(z_t)\}=0 \tag{4.8.3}$$

or taking the asymptotic limit of the rotation functions (for $J<-\frac{1}{2}$ from (B.13*a*))

$$\int_{z_T}^{\infty} ds\,\{D_{sH}(s,t)+\mathscr{S}(-1)^{M-v}D_{uH}(s,t)\}\,s^{-J_0-1+M}=0 \tag{4.8.4}$$

which is needed for all $J_0<-M$.

Such integrals are known as 'superconvergence relations', or SCR for short. For example, with spinless particle scattering (N, $M=0$) $Q_l(z)$ in (2.5.3) has poles for all negative integers, $J_0=-1,-2,\ldots$, from (A.32), and the SCR becomes

$$\int_{s_T}^{\infty} D_s^{\mathscr{S}}(s,t)\,s^n\,ds=0,\quad n=0,1,2,\ldots \tag{4.8.5}$$

Similar SCR must hold in potential scattering if a trajectory is to pass below $l=-(1+n)$ (see section 3.3*b*).

Of course the integral (4.5.7) will diverge for $J>J_0$ if there are Regge poles and cuts in $\mathrm{Re}\{J\}>J_0$, and it is only after all such pole and cut contributions have been removed that the SCR obtain. Since the Froissart bound requires that poles and cuts must not be above 1 for $t\leqslant 0$, we find from (4.8.3) and (B.14) that it is essential for

$$\int_{s_T}^{\infty} ds\,\{D_{sH}(s,t)+\mathscr{S}(-1)^{M+v}(D_{uH}(s,t)\}\,s^n=0,\quad n=M,M-1,\ldots,1 \tag{4.8.6}$$

whatever Regge singularities occur, otherwise the fixed singularities (4.8.1) would give contributions to the asymptotic behaviour which violate this bound.

But there will still be $(J-J_0)^{\frac{1}{2}}$ branch points in the partial-wave amplitudes for $N\leqslant J_0<M$ and $-M\leqslant J_0<-N$ from the cancellation of the SCR zero with (4.8.1). These can conveniently be joined pairwise by kinematical cuts running from $J=M-1-k$ to $-M+k$, $k=0,1,\ldots,M-1$. They do not contribute to the asymptotic behaviour because the $d^J_{\lambda\lambda'}$ also vanish like $(J-J_0)^{\frac{1}{2}}$ at these points, as we noted when discussing (4.6.2).

However, Gribov and Pomeranchuk (1962) demonstrated that in fact these SCR cannot hold at wrong-signature nonsense values of J, and that $A^{\mathscr{S}}_{HJ}(t)$ will therefore have fixed poles (or infinite square-root branch points) at these points. This is because, from (2.6.19), the imaginary part of the partial-wave helicity amplitude contains a contribution from the 'third' double spectral function of the form

$$\operatorname{Im}\{A^{\mathscr{S}}_{HJ}(t)\} = \frac{1}{16\pi^2}\int_{a(t)}^{b(t)} \mathrm{d}z'\rho^{su}_{H}(s',u')\,e^{J}_{\lambda\lambda'}(z'_t)\,(1-\mathscr{S}\,\mathrm{e}^{-\mathrm{i}\pi(J-v)}) \tag{4.8.7}$$

This vanishes for physical J-values, i.e. at right-signature points, and is obviously absent from situations like potential scattering (without Majorana exchange forces) which have no third double spectral function. But at the wrong-signature nonsense points of hadronic scattering amplitudes the fixed singularities of (4.8.1) will occur, and this time their residues will certainly not vanish due to SCRs because, at least for some regions of t where the integral in s runs over the elastic part of the double spectral function (see fig. 2.6), we can be sure from (3.5.34) that the integrand is always positive. So the SCRs (4.8.3), (4.8.4), (4.8.5) hold only for J_0 such that $(-1)^{J_0-v} = \mathscr{S}$. (We shall return to this point in section 7.2.)

Because of the unitarity equation (4.7.7), each helicity amplitude will acquire the singularities of the others, so fixed singularities will in fact occur at all wrong-signature $J_0 = \sigma_{\mathrm{T}} - k$, $k = 2, 4, 6, \ldots$ or $1, 3, 5, \ldots$ since σ_{T} ($\equiv \max\{\sigma_1+\sigma_3, \sigma_2+\sigma_4\}$) gives the largest possible value of M. Of course the occurrence of wrong-signature fixed poles for $J_0 > 1$ does not violate the Froissart bound since the vanishing of the signature factor ensures that they will not contribute to the asymptotic behaviour. But these real-axis fixed poles are incompatible with the unitarity equation, and so the occurrence of Gribov–Pomeranchuk poles proves that Regge cuts must exist, as we shall find in chapter 8.

5
Regge trajectories and resonances

5.1 Introduction

One of the most important conclusions of chapters 2 and 4 was that whenever a Regge trajectory, $\alpha(t)$, passes through a right-signature integral value of $J-v$ a t-plane pole will occur in the scattering amplitude because of the vanishing of the factor $\sin[\pi(\alpha(t)+\lambda')]$ in (4.6.2). And, as we found in section 1.5, such poles correspond to physical particles; to a particle which is stable against strong-interaction decays if the pole occurs below the t-channel threshold, or to a resonance which can decay into other lighter hadrons if it occurs above threshold. If a given trajectory passes through several such integers it will contain several particles of increasing spin, and so it is possible to classify the observed particles and resonances into families, each family lying on a given Regge trajectory. Some examples are given in figs. 5.5 and 5.6 below.

This chapter is mainly devoted to presenting the evidence for this Regge classification, but as there will be a different trajectory for each different set of internal quantum numbers such as B, I, S, etc. it will be useful for us first to examine briefly the way in which the particles have been classified according to their internal quantum numbers using SU(3) symmetry and the quark model. Readers requiring a more complete discussion than we have space for here will find the books by Carruthers (1966), Gourdin (1967), and Kokkedee (1969) very helpful.

The complete specification of a hadron requires, in addition to its mass m, and spin σ, the values of the internal quantum numbers; i.e. baryon number B, charge Q, intrinsic parity $P = \eta\mathscr{S}$ from (4.6.8), strangeness S, and isospin I, and in some cases the charge conjugation C_n, and G-parity G, as well. All of these are good, conserved quantum numbers for strong interactions, though only B and Q are conserved in all interactions (to the best of our knowledge).

By definition $B = 0$ for mesons, $+1$ for baryons, and -1 for anti-baryons. These are the only values which occur for what are often loosely referred to as the 'elementary' particles (though see section 2.8 for a discussion of the more strict use of this terminology which we employ). But baryon number is an additive quantum number, which means that a two-particle state $|1,2\rangle$ will have baryon number $B_{12} = B_1 + B_2$, and so complex nuclei have $B = A$, the atomic mass number.

The intrinsic parity of a particle is $P = \pm 1$ depending on how its wave function transforms under the parity reflection operator in the particle's rest frame, i.e. $\mathrm{P}_{\mathrm{op}}\psi(\boldsymbol{r}) = \psi(-\boldsymbol{r}) = P\psi(\boldsymbol{r})$. This is a multiplicative quantum number, and so for a two-particle state $P_{12} = P_1 P_2(-1)^l$, where l is the relative orbital angular momentum of the two particles (see (4.6.6)).

The charge-conjugation operator C, has the effect of turning a particle into its anti-particle, i.e. a particle which has the opposite sign for all the additive quantum numbers. So under C, $B \to -B$, $Q \to -Q$ and $S \to -S$. Since strong interactions are invariant under C, particles which have $B = Q = S = 0$, i.e. non-strange, neutral mesons, are eigenstates of C with eigenvalue $C_{\mathrm{n}} = +1$ ($\mathrm{n} \equiv$ neutral). It is found (see for example Bernstein (1968)) that $C_{\mathrm{n}} = \pm 1$ for π^0 and η^0, and $C_{\mathrm{n}} = -1$ for ρ^0, ω, ϕ and the photon γ. These assignments are consistent with the observed decays $\pi^0, \eta^0 \to \gamma\gamma$ and $\rho^0, \omega, \phi \to \gamma_v \to e^+e^-$ (where γ_v is a virtual photon).

For other non-strange mesons ($B = S = 0$, $Q \neq 0$) it is useful to invoke the isospin invariance of strong interactions to define an extended particle–anti-particle conjugation operator called the G-parity operator. For such particles the z component of the isospin (see (5.2.1) below) is equal to the charge, i.e. $Q = I_z$, and so rotation of the particle state by an angle π about the y axis in 'isospin space' takes us to the charge-conjugate particle, i.e. $I_z \to -I_z$, up to a phase factor. The Condon and Shortley phase convention for isospin multiplets gives (cf. (B.7))

$$e^{i\pi I_y}|I, I_z\rangle = (-1)^{I-I_z}|I, -I_z\rangle \tag{5.1.1}$$

So for non-strange mesons the combined operation

$$G \equiv C\,e^{i\pi I_y} \tag{5.1.2}$$

will have an eigenvalue $G = \pm 1$. Thus for the pion multiplet, π^+, π^0, π^-, with $I = 1$, $I_z = 1, 0, -1$, we have $G_\pi = -1$ since $C_{\pi^0} = +1$. This is obviously also a multiplicative quantum number, and hence a state consisting of n pions will have $G|n\rangle = (-1)^n|n\rangle$. This allows one to determine the G-parity of other non-strange mesons from their hadronic decays into pions; for example the fact that the decay $\rho \to \pi\pi$

occurs indicates that ρ has $G = +1$. And of course the decays $\rho \to 3\pi, 5\pi$ etc. are forbidden by G-parity conservation.

The remaining quantum numbers I and S require a brief discussion of unitary symmetry, which we give in the next section.

5.2 Unitary symmetry

a. Isospin

It is well known from nuclear physics that the strong interaction is approximately invariant under the transformations of the isotopic spin (or isospin) group SU(2), at least to an accuracy of a few per cent. This group is isomorphic to the rotation group, the isospin vector $\boldsymbol{I}$ corresponding to $\boldsymbol{J}$, while its z component in isospin space I_z corresponds to J_z. This isospin invariance manifests itself in two related ways.

(i) All the hadrons may be grouped conventionally into multiplets of a given isospin I (such that $I(I+1)$ is the eigenvalue of $\boldsymbol{I}^2$) which are approximately degenerate in mass, and are identical in all their other quantum numbers except the charge. Well known examples are

Nucleon, N	p, n	$I = \frac{1}{2}, I_z = \pm\frac{1}{2}$
Pion, π	π^+, π^0, π^-	$I = 1, I_z = 1, 0, -1.$
3–3 resonance, Δ	$\Delta^{++}, \Delta^+, \Delta^0, \Delta^-$	$I = \frac{3}{2}, I_z = \frac{3}{2}, \frac{1}{2}, -\frac{1}{2}, -\frac{3}{2}$

The isospin is assigned according to the multiplicity of charge states exhibited by the particle, so that I_z spans the range $I, I-1, \ldots, -I$, and the z component is associated with the charge according to the relation

$$Q = I_z + \tfrac{1}{2}B \tag{5.2.1}$$

(for non-strange particles only). A particle may thus be represented by the isotopic state vector $|I, I_z\rangle$.

The mass differences within a given multiplet are rather small (for example $m_p = 938.3\,\text{MeV}$, $m_n = 939.6\,\text{MeV}$) and are believed to be caused by the differing electromagnetic interactions of the particles. As far as strong interactions are concerned such differences can be ignored, and so we use a single symbol for all the members of a multiplet (for example N ≡ {p, n}), and regard them all as lying on the same Regge trajectory, which carries a definite isospin. For example $\alpha_N(t)$ has $I = \frac{1}{2}$, and only if we want to discuss electromagnetic interactions need we take account of the fact that this is really two trajectories, with $I_z = \pm\frac{1}{2}$, which are very slightly split.

(ii) The various scattering amplitudes involving these particles are related by isospin invariance, being dependent on the value of I but not on I_z, i.e. strong interactions exhibit charge independence. This property will be examined in section 6.7.

It is sometimes convenient to regard the iso-doublet

$$(\mathrm{p},\mathrm{n}) \quad |I=\tfrac{1}{2}, I_z=\pm\tfrac{1}{2}\rangle$$

as the fundamental isotopic spinor, out of which all other multiplets can be constructed (just as all possible angular momenta can be obtained by adding different numbers of spin $=\frac{1}{2}$ particles). This doublet iso-spinor can be represented by a column matrix.

$$\{2\} \equiv \eta \equiv \begin{pmatrix} \mathrm{p} \\ \mathrm{n} \end{pmatrix} \tag{5.2.2}$$

which transforms under SU(2) as

$$\eta \to \eta' = \boldsymbol{U}\eta \tag{5.2.3}$$

where $\boldsymbol{U}$ is any 2×2 unitary matrix with $\det(\boldsymbol{U})=1$. Any such matrix can be written in the form

$$\boldsymbol{U} = \mathrm{e}^{\frac{1}{2}\mathrm{i}\theta \boldsymbol{n}\cdot\boldsymbol{\tau}} \tag{5.2.4}$$

where θ is an arbitrary parameter, $\boldsymbol{n}$ is a unit three-vector, and the components of $\boldsymbol{\tau}$ are the Pauli matrices

$$\tau_x = \begin{pmatrix} 0 & 1 \\ 1 & 0 \end{pmatrix}, \quad \tau_y = \begin{pmatrix} 0 & -\mathrm{i} \\ \mathrm{i} & 0 \end{pmatrix}, \quad \tau_z = \begin{pmatrix} 1 & 0 \\ 0 & -1 \end{pmatrix} \tag{5.2.5}$$

The corresponding 'anti-particles' are given by the row-matrix iso-spinor

$$\{\bar{2}\} \equiv \bar{\eta} \equiv (\bar{\mathrm{p}},\bar{\mathrm{n}}) \tag{5.2.6}$$

Formally all the other iso-multiplets can be constructed by combining η's and $\bar{\eta}$'s. Thus for example

$$\frac{1}{\sqrt{2}}(\bar{\mathrm{p}}\mathrm{p}+\bar{\mathrm{n}}\mathrm{n}) \tag{5.2.7}$$

gives an $I=0$ singlet, like the η meson, $\{1\}$, while

$$\mathrm{p}\bar{\mathrm{n}}, \quad \frac{1}{\sqrt{2}}(\bar{\mathrm{p}}\mathrm{p}-\bar{\mathrm{n}}\mathrm{n}), \quad \text{and} \quad \bar{\mathrm{p}}\mathrm{n} \tag{5.2.8}$$

form the triplet, $\{3\}$, $I=1$, $I_z=1, 0, -1$ respectively, like the π meson. So at least in this formal sense we can regard the η and π mesons as bound states of the nucleon-antinucleon system, with

$$\{2\}\otimes\{2\} = \{1\}\oplus\{3\} \tag{5.2.9}$$

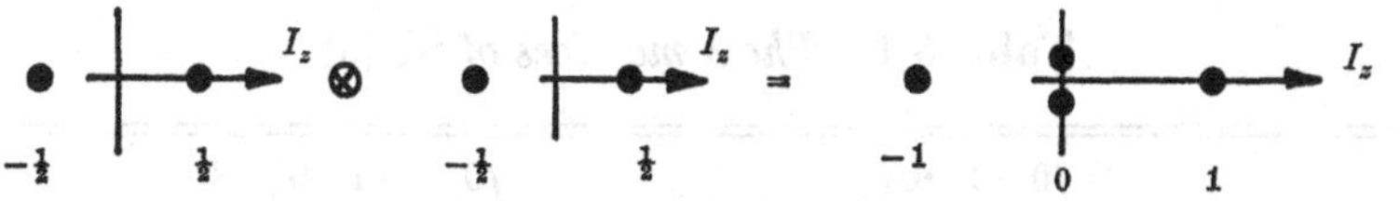

FIG. 5.1 The superposition of two iso-doublets, $I_z = \pm\frac{1}{2}$ to give four states, one with $I_z = -1$, two with $I_z = 0$ and one with $I_z = 1$.

as shown in fig. 5.1, completely in analogy with the construction of spin = 0 and spin = 1 helium atom states from two electrons of spin $\frac{1}{2}$.

b. SU(3)

The above scheme can be extended to include strange particles as well as by taking the fundamental representation to be the three-component spinor

$$\{3\} \equiv q \equiv \begin{pmatrix} p \\ n \\ \lambda \end{pmatrix} \tag{5.2.10}$$

transforming under SU(3) as

$$q \to q' = \boldsymbol{U}q \tag{5.2.11}$$

where now $\boldsymbol{U}$ is any unitary 3×3 matrix with $\det(\boldsymbol{U}) = 1$, which can be written

$$U = e^{\frac{1}{2}i\theta\boldsymbol{\alpha}\cdot\boldsymbol{\lambda}} \tag{5.2.12}$$

where $\boldsymbol{\alpha}$ is an 8-dimensional unit vector, and the λ matrices are given in table 5.1. They correspond to the three τ matrices of SU(2), (5.2.5).

The three particles p, n, λ were introduced by Gell–Mann (1964) and Zweig (1964), and are called 'quarks'. They are assigned the quantum numbers shown in table 5.2. Clearly, the p and n quarks are not to be identified with the proton and neutron of (5.2.2) as they have, *inter alia*, $B = \frac{1}{3}$. We also need a triplet of anti-quarks

$$\{\bar{3}\} \equiv \bar{q} = (\bar{p}, \bar{n}, \bar{\lambda}) \tag{5.2.13}$$

There is no evidence that such quarks actually exist, but at the very least they provide a very convenient mnemonic for the group-theory of SU(3). Also the observed hadrons frequently behave as though they were actually composed of quarks as we shall discuss particularly in chapter 7. (An extensive review of the evidence for the quark structure of hadrons in electromagnetic and weak interactions is given in Feynman (1972).)

A baryon is made up of three quarks (to give $B = 1$), while mesons are composed of quark–antiquark pairs. The hypercharge, Y, is defined

Table 5.1 *The λ matrices of* SU(3)

$$\lambda_1 = \begin{pmatrix} 0 & 1 & 0 \\ 1 & 0 & 0 \\ 0 & 0 & 0 \end{pmatrix} \qquad \lambda_2 = \begin{pmatrix} 0 & -i & 0 \\ i & 0 & 0 \\ 0 & 0 & 0 \end{pmatrix}$$

$$\lambda_3 = \begin{pmatrix} 1 & 0 & 0 \\ 0 & -1 & 0 \\ 0 & 0 & 0 \end{pmatrix} \qquad \lambda_4 = \begin{pmatrix} 0 & 0 & 1 \\ 0 & 0 & 0 \\ 1 & 0 & 0 \end{pmatrix}$$

$$\lambda_5 = \begin{pmatrix} 0 & 0 & -i \\ 0 & 0 & 0 \\ i & 0 & 0 \end{pmatrix} \qquad \lambda_6 = \begin{pmatrix} 0 & 0 & 0 \\ 0 & 0 & 1 \\ 0 & 1 & 0 \end{pmatrix}$$

$$\lambda_7 = \begin{pmatrix} 0 & 0 & 0 \\ 0 & 0 & -i \\ 0 & i & 0 \end{pmatrix} \qquad \lambda_8 = \begin{pmatrix} 1/\sqrt{3} & 0 & 0 \\ 0 & 1/\sqrt{3} & 0 \\ 0 & 0 & -2/\sqrt{3} \end{pmatrix}$$

Table 5.2 *The quantum numbers of the quarks*

	B	I	I_z	Q	S	Y
p	$\frac{1}{3}$	$\frac{1}{2}$	$\frac{1}{2}$	$\frac{2}{3}$	0	$\frac{1}{3}$
n	$\frac{1}{3}$	$\frac{1}{2}$	$-\frac{1}{2}$	$-\frac{1}{3}$	0	$\frac{1}{3}$
λ	$\frac{1}{3}$	0	0	$-\frac{1}{3}$	-1	$-\frac{2}{3}$

in terms of the strangeness S by

$$Y = S+B \tag{5.2.14}$$

and the charge is then given by the Gell-Mann–Nishijima relation

$$Q = I_z+\tfrac{1}{2}Y = I_z+\tfrac{1}{2}(S+B) \tag{5.2.15}$$

instead of (5.2.1).

Taking all possible combinations of a quark and an antiquark, as shown in fig. 5.2, we get

$$q\bar{q} = \{3\}\otimes\{\bar{3}\} = \{1\}\oplus\{8\} \tag{5.2.16}$$

so we can expect that mesons will occur in nonets, each nonet consisting of a singlet and an octet with the quantum numbers shown in fig. 5.2. Table 5.3 gives the well established mesons grouped into such multiplets. It is evident that the symmetry is very badly broken for the masses of the particles, the SU(3) mass-splitting in Y being very much greater than the isospin mass-splitting in I_z.

Also it is not clear how one should distinguish the singlet states such as ω_1 from the octet state with the same quantum numbers, ω_8. With a broken symmetry the observed ω and ϕ particles can be

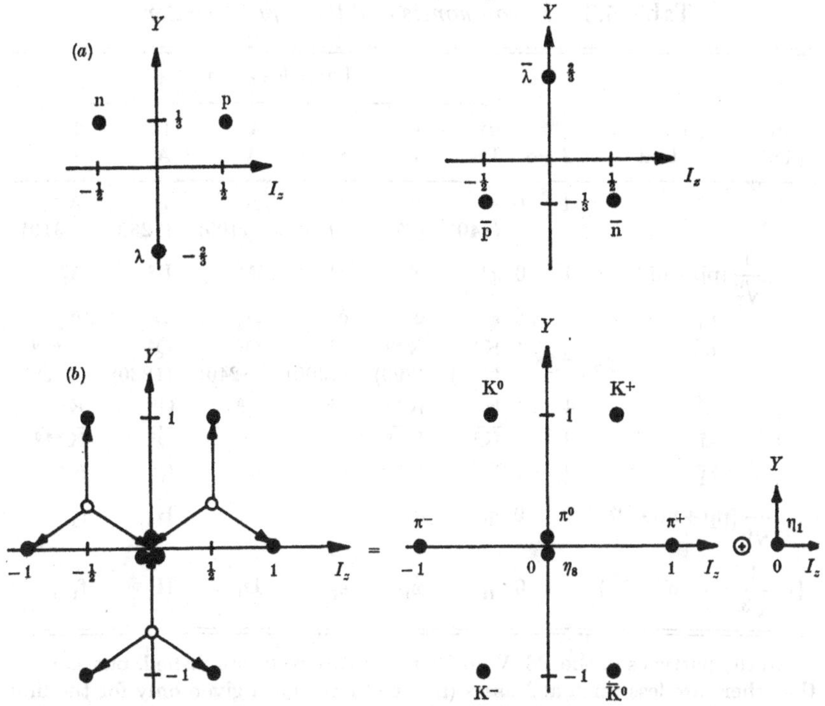

FIG. 5.2 (a) Triplets of quarks $\{p, n, \lambda\}$ and anti quarks $\{\bar{p}, \bar{n}, \bar{\lambda}\}$. (b) The decomposition $q \otimes \bar{q} = \{8\} + \{1\}$. On each quark represented by ○ is imposed an anti-quark triplet to give the nine states which are identified with pseudo-scalar mesons on the right-hand side.

mixtures of these pure SU(3) states, say

$$\left.\begin{aligned} \phi &= \omega_8 \cos\theta - \omega_1 \sin\theta \\ \omega &= \omega_8 \sin\theta + \omega_1 \cos\theta \end{aligned}\right\} \tag{5.2.17}$$

where θ is the 'mixing angle'. The so-called 'ideal' value is

$$\theta = \tan^{-1}\left(\frac{1}{\sqrt{2}}\right) \approx 38^\circ \tag{5.2.18}$$

in which case from table 5.3 we find that

$$\omega = \frac{1}{\sqrt{2}}(p\bar{p} + n\bar{n}), \quad \phi = -\lambda\bar{\lambda} \tag{5.2.19}$$

so that ω contains no strange quarks. This ideal mixing seems to hold for the vector and tensor mesons, but not the pseudo-scalars.

The mass separations within a given multiplet are assumed to be

Table 5.3 *Meson nonets and their quark content*

Multiplet	Quark content	I	S	Particles J^{PC_n}: 0^{-+} PS	1^{--} V	0^{++} S	1^{++} A^+	1^{+-} A^-	2^{++} T
{8}	$p\bar{n}$	1	0	π^+ (140)	ρ^+ (770)	δ^+ (970)	A_1^+ (1100)	B^+ (1235)	A_2^+ (1310)
	$\frac{1}{\sqrt{2}}(p\bar{p}-n\bar{n})$	1	0	π^0	ρ^0	δ^0	A_1^0	B^0	A_2^0
	$n\bar{p}$	1	0	π^-	ρ^-	δ^-	A_1^-	B^-	A_2^-
	$n\bar{\lambda}$	$\frac{1}{2}$	1	K^0 (498)	K^{*0} (890)	κ^0 (1300)	Q^0 (1240)	Q^0 (1280)	K^{**0} (1420)
	$p\bar{\lambda}$	$\frac{1}{2}$	1	K^+	K^{*+}	κ^+	Q^0	Q^0	K^{**+}
	$\lambda\bar{n}$	$\frac{1}{2}$	-1	$\overline{K^0}$	$\overline{K^{*0}}$	$\overline{\kappa^0}$	$\overline{Q^0}$	$\overline{Q^0}$	$\overline{K^{**0}}$
	$\lambda\bar{p}$	$\frac{1}{2}$	-1	K^-	K^{*-}	κ^-	Q^-	Q^-	K^{**-}
	$\frac{1}{\sqrt{6}}(p\bar{p}+n\bar{n}-2\lambda\bar{\lambda})$	0	0	η_8	ω_8	ε_8	D_8	H_8	f_8
{1}	$\frac{1}{\sqrt{3}}(p\bar{p}+n\bar{n}+\lambda\bar{\lambda})$	0	0	η_1	ω_1	ε_1	D_1	H_1	f_1

All the particles in the PS, V and T nonets are well established, but some of the others are less certain. Masses (in MeV) have been given only for the first member of each isospin multiplet. C_n is not a good quantum number for strange mesons so the assignment in the Q region is particularly uncertain. The iso-singlet mixtures are $\eta_8+\eta_1 = \eta(549)+\eta'(958)$, $\omega_8+\omega_1 = \omega(783)+\phi(1019)$, $\varepsilon_8+\varepsilon_1 = \varepsilon(600)+S^*(993)$, $D_8+D_1 = D(1285)+E(1420)$, $H_8+H_1 = H(990)+?$, $f_8+f_1 = f(1270)+f'(1514)$, mixed as in (5.2.17).

due to the λ quark having a different mass from that of the p and n quarks. So with ideal mixing, if we set $m_n = m_p = m$ and $m_\lambda = m+\Delta m$, we find that for the vector mesons

$$m_\omega = m_\rho = 2m, \quad m_{K^*} = 2m+\Delta m, \quad m_\varphi = 2(m+\Delta m) \tag{5.2.20}$$

giving

$$m_\omega + m_\varphi = 2m_{K^*} \tag{5.2.21}$$

However for mesons it is generally supposed (for no very compelling reason) that these relations should actually be written for the squares of the masses, i.e. $m_\omega^2 = m_\rho^2$, $m_\omega^2+m_\varphi^2 = 2m_{K^*}^2$, which hold equally well because the masses are much larger than the mass differences. The lighter pseudo-scalar mesons do not obey the corresponding mass formulae either for m or m^2, which is generally taken as evidence that the mixing between η and η' is far from ideal (see Kokkedee 1969).

Both the pseudo-scalar (PS) and vector (V) meson nonets can be obtained with the spin $= \frac{1}{2}$ quarks in an $l = 0$ orbital state, since they correspond to quark spins being anti-parallel (total quark spin $s = 0$) or parallel ($s = 1$) respectively. Higher spin mesons can be obtained by orbital excitation of the $q\bar{q}$ pair. Since q and $\bar{q}$, being fermions, have opposite intrinsic parity, the parity of a $q\bar{q}$ state is

$$P = (-1)^{l+1} \tag{5.2.22}$$

and for $B = S = 0$ states the charge conjugation and G-parity are

$$C_n = (-1)^{l+s}, \quad G = (-1)^{l+s+I} \tag{5.2.23}$$

Since the spin of the meson is $\boldsymbol{J} = \boldsymbol{l} + \boldsymbol{s}$ we have for $l = 0$ just the PS and V nonets with $J^{PC} = 0^{-+}$ and 1^{--} respectively, while for $l = 1$ there are four possible nonets, scalar $S \equiv 0^{++}$, two axial vectors, $A^+ \equiv 1^{++}$ and $A^- \equiv 1^{+-}$, and tensor $T \equiv 2^{++}$. A possible assignment of meson states according to this classification is given in table 5.3.

Regge theory suggests that one may expect to see recurrences of each of these six nonets at J values spaced by 2 units from the above. In the next section we shall find that only a few of these excited states have been observed. This is hardly surprising, however, because mesons can usually only be observed in production experiments such as

$$1+2 \to 3+4, \quad 4 \to a+b$$

The resonance 4 will be seen as a peak of the cross-section in the invariant mass of its decay products at $m_4^2 = (p_a + p_b)^2$, a and b having an angular distribution corresponding to the spin of 4 (see section 4.2). But at high values of m_4^2 many partial waves can be expected to contribute to the ab system and so the analysis of this decay within the three-body final state $3+a+b$ becomes difficult. Un-natural parity mesons are even more difficult to find as they only have three (or more) body decays.

The situation is more favourable for baryon resonances which can be formed in meson–baryon scattering experiments such as

$$\text{MB} \to \text{B}^* \to \text{MB}$$

where a partial-wave analysis of the two-body final state is sufficient to find the resonance. So a lot more baryon resonances are known.

They are built from three quarks

$$q \otimes q \otimes q = \{3\} \otimes \{3\} \otimes \{3\} = \{1\} \oplus \{8\} \oplus \{8\} + \{10\} \tag{5.2.24}$$

(see Carruthers 1966), and so baryons should occur in singlets, octets and decuplets, with the quantum numbers shown in fig. 5.3

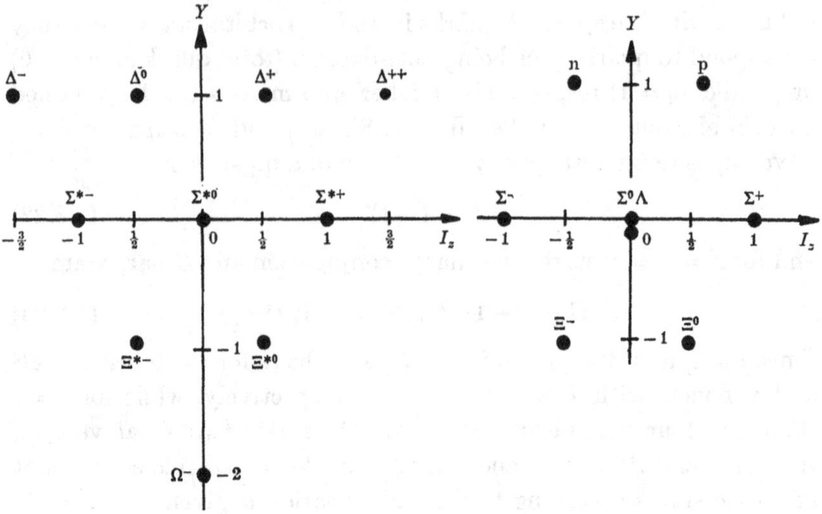

FIG. 5.3 The $J^P = \frac{3}{2}^+$ decouplet and the $\frac{1}{2}^+$ octet of baryons.

The lowest mass states, with $l = 0$ may have $J = \frac{1}{2}$ or $\frac{3}{2}$, and are given in table 5.4, and again one may expect higher l states at higher masses. (We shall ignore the difficulty that since the quarks are fermions with presumably anti-symmetric wave functions the increase of mass with J is far from obvious – see Kokkedee (1969).) By the same method as before we find that the mass-splitting in the decuplet should obey the equal spacing rule

$$m_{\Omega^-} - m_{\Xi^*} = m_{\Xi^*} - m_{\Sigma^*} = m_{\Sigma^*} - m_{\Delta} = \Delta m \qquad (5.2.25)$$

which is well satisfied. For the octet we obtain the Gell–Mann–Okubo mass formula

$$m_{\mathrm{p}} + m_{\Xi} = \tfrac{1}{2}(m_{\Sigma} + 3m_{\Lambda}) \qquad (5.2.26)$$

but the relations $m_{\Lambda} = m_{\Sigma}$ and $m_{\Lambda} - m_{\mathrm{p}} = m_{\Sigma^*} - m_{\Delta}$ are not obeyed, so there must be symmetry-breaking effects in the potential between the quarks as well.

In addition to these predictions about the masses of the particles SU(3) invariance also gives relations between scattering amplitudes, and these will be explored in section 6.7.

The scheme outlined above is only the most elementary version of the quark model. The discovery of two long-lived vector mesons, ψ_1 (3100) and ψ_2 (3700) (see Particle Data Group (1975) for references) has increased the interest in more elaborate structures based on the inclusion of a fourth quark, c, having the quantum numbers

$$B, Q, I, S, C = \tfrac{1}{3}, \tfrac{2}{3}, 0, 0, 1,$$

Table 5.4 *The lowest mass octet and decuplet of baryons and their quark content*

Multiplet	Quark content	I	S	Particles
$\{8\}, J^P = \frac{1}{2}^+$	ppn	$\frac{1}{2}$	0	p(938.3)
	pnn	$\frac{1}{2}$	0	n(939.6)
	ppλ	1	-1	$\Sigma^+(1189.5)$
	pnλ	1	-1	$\Sigma^0(1192.6)$
		0	-1	$\Lambda(1115.6)$
	nnλ	1	-1	$\Sigma^-(1197.4)$
	p$\lambda\lambda$	$\frac{1}{2}$	-2	$\Xi^0(1314.7)$
	n$\lambda\lambda$	$\frac{1}{2}$	-2	$\Xi^-(1321.2)$
$\{10\}, J^P = \frac{3}{2}^+$	ppp	$\frac{3}{2}$	0	$\Delta^{++}(1236)$
	ppn	$\frac{3}{2}$	0	Δ^+
	pnn	$\frac{3}{2}$	0	Δ^0
	nnn	$\frac{3}{2}$	0	Δ^-
	ppλ	1	-1	$\Sigma^{*+}(1383)$
	pnλ	1	-1	Σ^{*0}
	nnλ	1	-1	Σ^{*-}
	p$\lambda\lambda$	$\frac{1}{2}$	-2	$\Xi^{*0}(1532)$
	n$\lambda\lambda$	$\frac{1}{2}$	-2	Ξ^{*-}
	$\lambda\lambda\lambda$	0	-3	$\Omega^-(1672)$

where C is a new quantum number called 'charm', which has eigenvalue 0 for the p, n and λ quarks. The particles ψ_1 and ψ_2 are taken to be c$\bar{c}$ bound states, and the basic meson SU(3) nonets from $\{3\} \otimes \{\bar{3}\}$ are increased to SU(4) 16-plets formed from $\{4\} \otimes \{\bar{4}\}$. However this fourth quark must be much heavier than the others so that the predicted charmed particles (formed from c$\bar{p}$, c$\bar{n}$, c$\bar{\lambda}$, $\bar{c}$p, $\bar{c}$n, $\bar{c}\lambda$) are heavier than the nonet mesons, whose SU(3) symmetry and mixing are approximately preserved. The discovery of charmed particles has greatly increased the interest of this model, and of the related schemes based on 'coloured' quarks (see Weinberg (1974), de Rujula *et al.* (1974), Gaillard, Lee and Rosner (1975) for reviews).

An important test of the quark model is that all the observed mesons have quantum numbers which can be formed from q $\otimes$ $\bar{q}$ as in fig. 5.2, and all the baryons have quantum numbers that can be formed from q $\otimes$ q $\otimes$ q as in fig. 5.3. Channels which have quantum numbers outside these patterns, like $\pi^+\pi^+$ which has $I = 2$, or K^+p which as $S = 1$, are called 'exotic' channels, and do not seem to contain resonances. All the well established resonances have non-exotic quantum numbers.

5.3 The Regge trajectories

An authoritative survey of the experimental properties of particles and resonances is published at frequent intervals by the Particle Data Group. Their 1974 edition (Particle Data Group 1974) contains information on over 50 possible mesons and 90 baryons, though the evidence for some of these is fairly weak. In this section we shall try to group all the particles for which there is reasonably strong evidence on Regge trajectories. Of course this cannot be done with complete certainty because there are few *a priori* rules to direct which particles should be associated together on the same trajectory. But, as we shall see, this problem is greatly simplified by the fact that the trajectories seem to be straight parallel lines when $\text{Re}\{\alpha(t)\}$ is plotted against t.

a. Mesons

All the well established mesons are shown in fig. 5.4 in a Chew–Frautschi plot (Chew and Frautschi 1962) of the spin $\sigma(=\text{Re}\{\alpha\})$ versus mass$^2 = t$. It should be noted that the only well verified particle with $\sigma > 2$ is the spin $= 3$, $I = 1$, g meson which has the same internal quantum numbers as the $\rho(\sigma = 1)$ and so presumably lies on the same trajectory. Strictly this is the only trajectory on which we can put even two points! However, in drawing fig. 5.4 we have taken into account that there is also evidence for spin $= 3$ ω and K* resonances and spin $= 4$ h and A_2^* resonances, and have made some use of information about the behaviour of the trajectories in the region $t < 0$ obtained from Regge fits (see fig. 6.6. below).

Also it is found that the $\sigma = 2$ A_2 meson, which has similar quantum numbers to the ρ apart from its signature (note from (4.6.8), (5.2.22) and (5.2.23) that this in fact means opposite values of P, C_n and G), lies very close to the straight line joining ρ and g, and (fig. 6.6) the A_2 trajectory is close to that of the ρ for $t < 0$ as well. Such an identity between trajectories of opposite signature is called 'exchange degeneracy'. It seems to imply (from (2.5.3) or (4.5.7)) that, rather surprisingly, the exchange forces, i.e. the u-channel singularities, are not making much contribution to the trajectories. Similarly the ω and f, which because of ideal mixing are almost degenerate in mass with the ρ and A_2 respectively (see (5.2.20)), seem to lie on a single $I = 0$ exchange-degenerate trajectory which almost coincides with that of ρ, A_2, g while the $I = 0$, ϕ, f′ trajectory appears to be parallel with these.

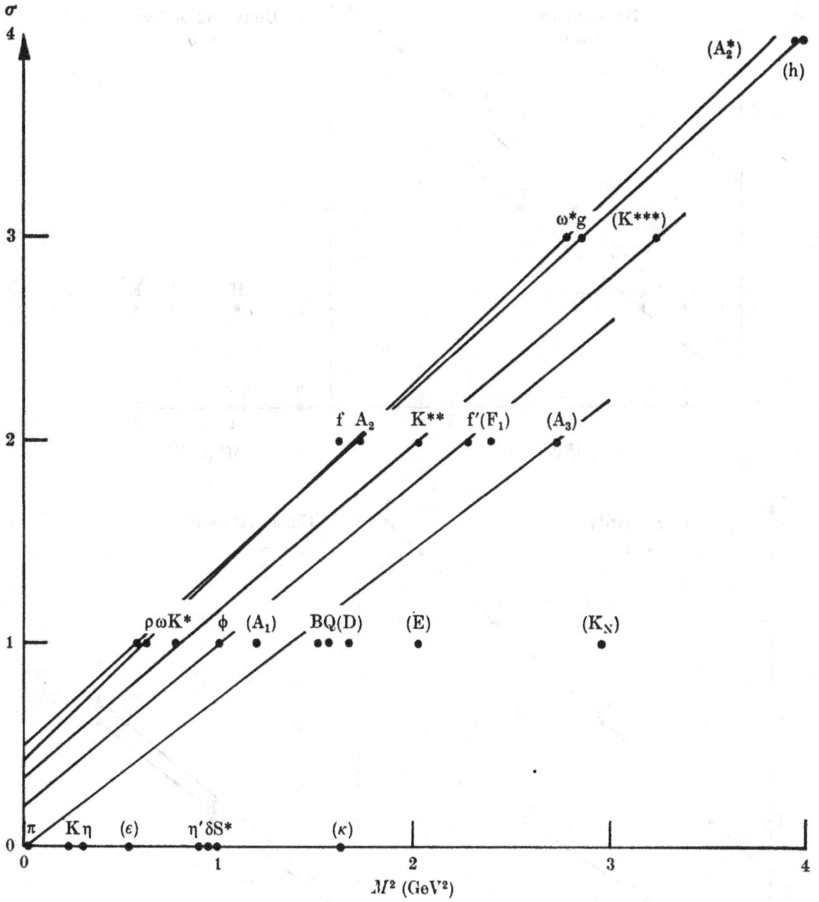

FIG. 5.4 Chew–Frautschi plot of Re $\{\alpha(t)\}$ versus t for the well established mesons. Less well verified states appear in brackets.

If we then make the rather bold assumption that all the mesons lie on approximately straight, parallel, exchange-degenerate trajectories we can associate most of the states listed by the Particle Data Group with trajectories as shown in fig. 5.5. They give leading trajectories which are very approximately

$$\left.\begin{array}{lll} \alpha_\rho(t) \approx 0.5+0.9t & \rho, \omega, A_2, f, g, \omega^*, A_2^*, h & I=0,1 \\ \alpha_{K^*}(t) \approx 0.3+0.9t & K^*, K^{**}, K^{***} & I=\frac{1}{2} \\ \alpha_\varphi(t) \approx 0.1+0.9t & \phi, f' & I=0 \\ \alpha_\pi(t) \approx 0.0+0.8t & \pi, B, A_3 & I=1 \\ \alpha_K(t) \approx -0.2+0.8t & K, Q, L & I=\frac{1}{2} \end{array}\right\} \quad (5.3.1)$$

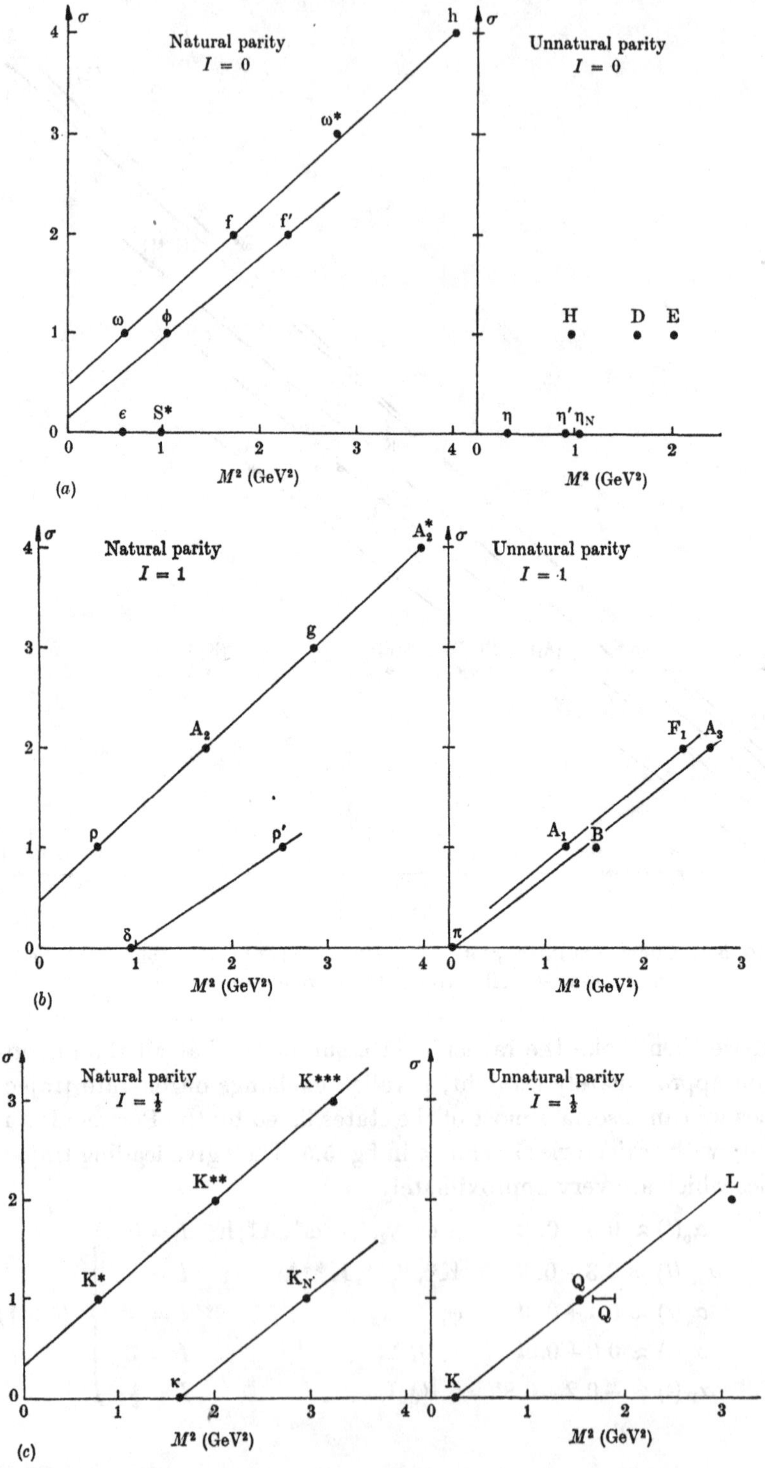

FIG. 5.5 Meson trajectories for (a) $I = 0$, (b) $I = 1$ and (c) $I = \frac{1}{2}$ mesons, including less well established states.

These straight lines are suggestive of a harmonic oscillator type of effective potential between the quarks, as mentioned in equation (3.3.33) *et seq.* An additional motivation for these figures, to be discussed in sections 6.5 and 7.4, is that there are theoretical reasons for expecting that trajectories may occur in integrally spaced sequences, with a 'parent' trajectory $\alpha(t)$, and an infinite sequence of 'daughters' $\alpha_n(t) \equiv \alpha(t)-n$, $n = 1, 2, \ldots$. Thus the $\rho'(1600)$, if it really is a resonance, may lie on the $n = 2$ daughter of the ρ.

b. Baryons

There are many more baryon states with high spin whose quantum numbers have been fairly well determined, and so the Chew–Frautschi plots of figs. 5.6 are more highly populated.

Again the trajectories seem to be straight and parallel, with similar slopes to the meson trajectories, but exchange degeneracy is badly broken in many cases. The leading trajectories are approximately given by

$$\left.\begin{array}{lll} \alpha_{\mathrm{N}}(t) \approx & -0.3+0.9t & \mathrm{N}(939), \mathrm{N}(1688), \mathrm{N}(2220) \\ \alpha_{\Delta}(t) \approx & 0.0+0.9t & \Delta(1232), \Delta(1950), \Delta(2420), \Delta(2850), \Delta(3230) \\ \alpha_{\Lambda}(t) \approx & -0.6+0.9t & \Lambda(1116), \Lambda(1520), \Lambda(1815), \Lambda(2100), \Lambda(2350), \\ & & \Lambda(2585) \\ \alpha_{\Sigma}(t) \approx & -0.8+0.9t & \Sigma(1190), \Sigma(1915) \end{array}\right\} \tag{5.3.2}$$

We have plotted the natural and unnatural parity trajectories back to back because the generalized MacDowell symmetry (see section 6.5) requires that odd-baryon-number trajectories should satisfy the relation

$$\alpha^{+}(\sqrt{t}) = \alpha^{-}(-\sqrt{t}), \quad \text{for} \quad t > 0 \tag{5.3.3}$$

where the superscripts $\pm$ refer to the parity. Since the trajectories (5.3.2) are approximately even in $\sqrt{t}$ this gives

$$\alpha^{\pm}(\sqrt{t}) = \alpha^{0}+\alpha' t \tag{5.3.4}$$

for both parities, and so the resonances should appear in exchange-degenerate pairs. It is evident from fig. 5.6 that this relation is not in fact satisfied, It is discussed further in section 6.5.

It is clear from the above figures that the Chew–Frautchi plot provides a very useful way of classifying resonances in addition to SU(3).

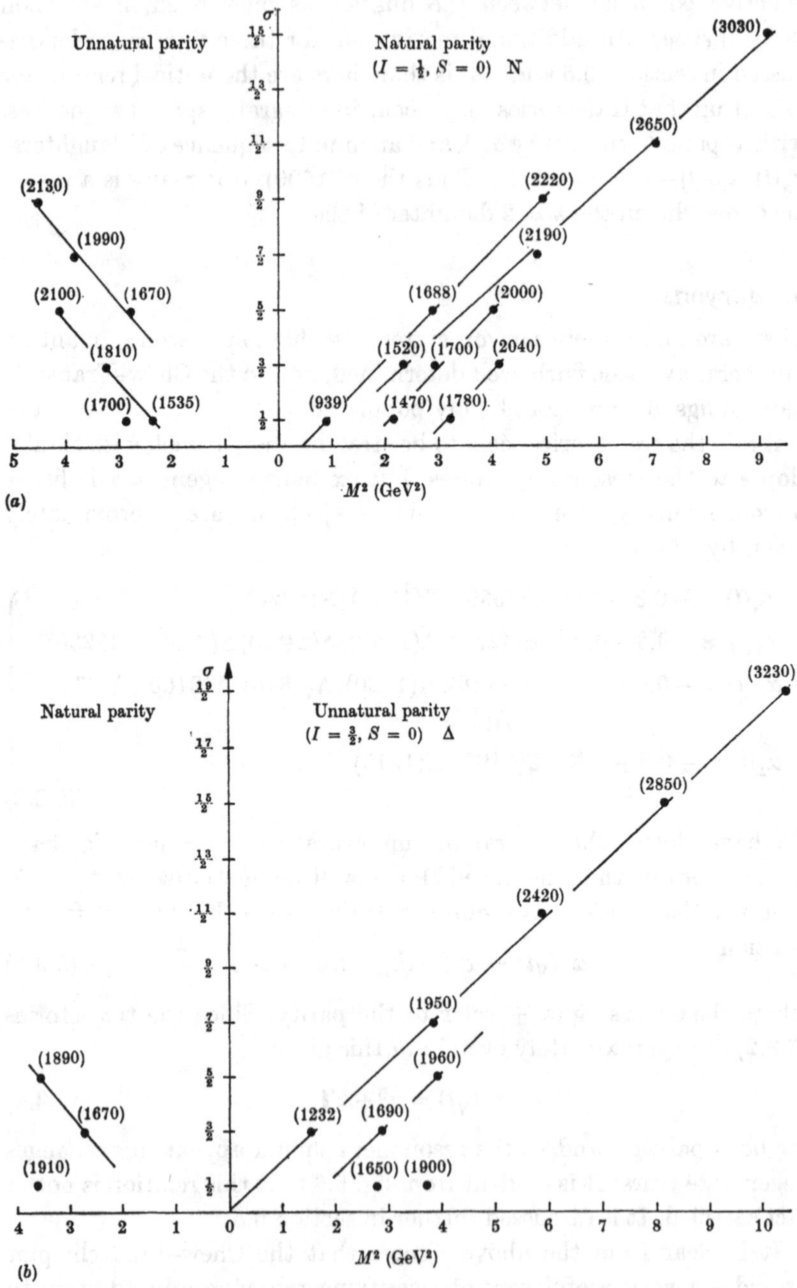

FIG. 5.6 Baryon trajectories for $(I, S) = (a)\ (\frac{1}{2}, 0)$, $(b)\ (\frac{3}{2}, 0)$.

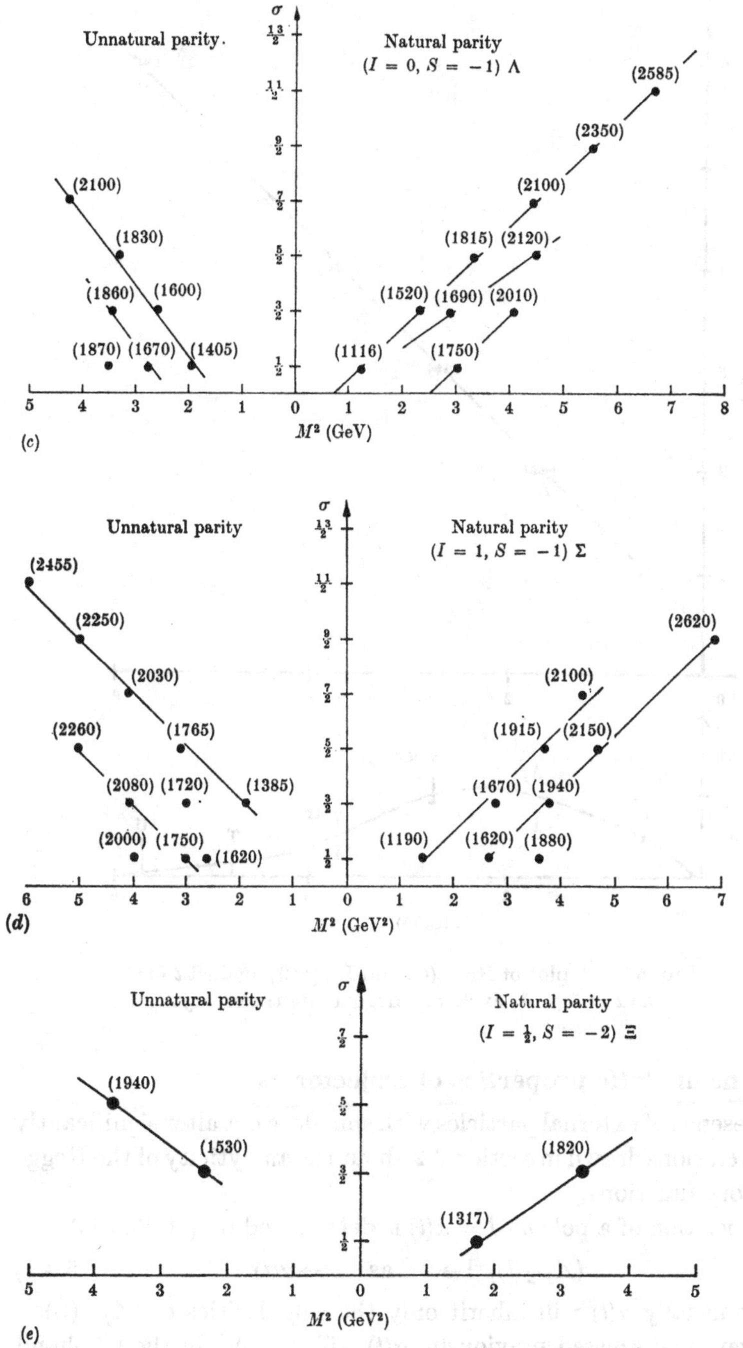

(c) (0, −1), (d) (1, −1) and (e) (½, −2).

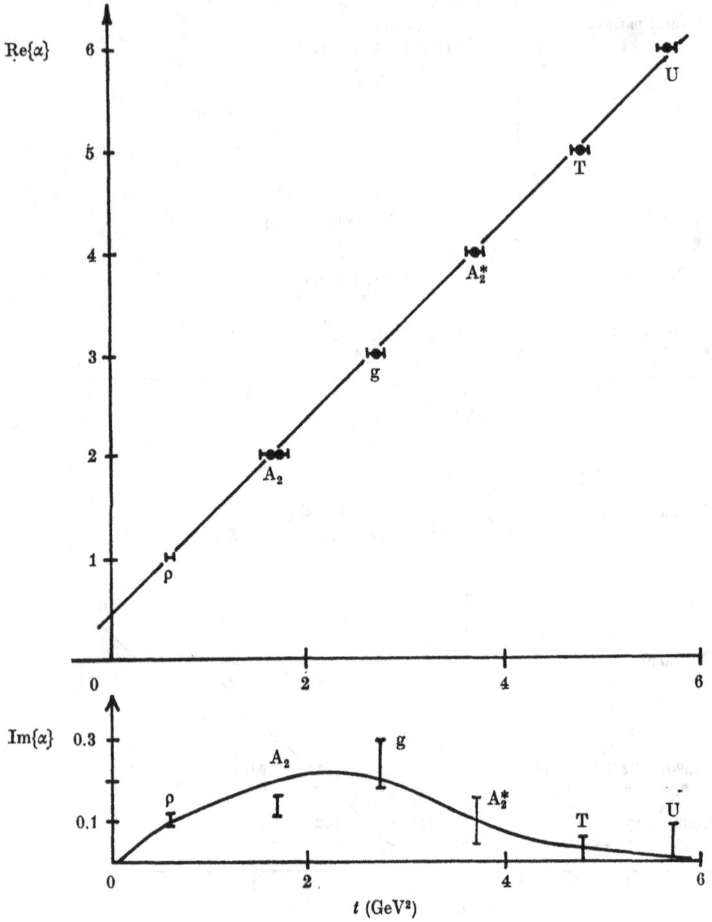

FIG. 5.7 A plot of Re $\{\alpha(t)\}$ and Im $\{\alpha(t)\}$ against t for the $I = 1$ ρ, A_2 exchange-degenerate trajectory.

5.4 The analytic properties of trajectories

The presence of external particles with spin does not alter significantly the conclusions drawn in section 3.2 about the analyticity of the Regge trajectory functions.

The position of a pole at $J = \alpha(t)$ is determined by (cf. (3.2.1))

$$(A_{HJ}(t))^{-1} \to 0 \quad \text{as} \quad J \to \alpha(t) \tag{5.4.1}$$

so that usually $\alpha(t)$ will inherit only the singularities of $(A_{HJ}(t))^{-1}$. However, as discussed previously, $\alpha(t)$ will not obtain the left-hand cuts of the partial-wave amplitude. Also since the same trajectory

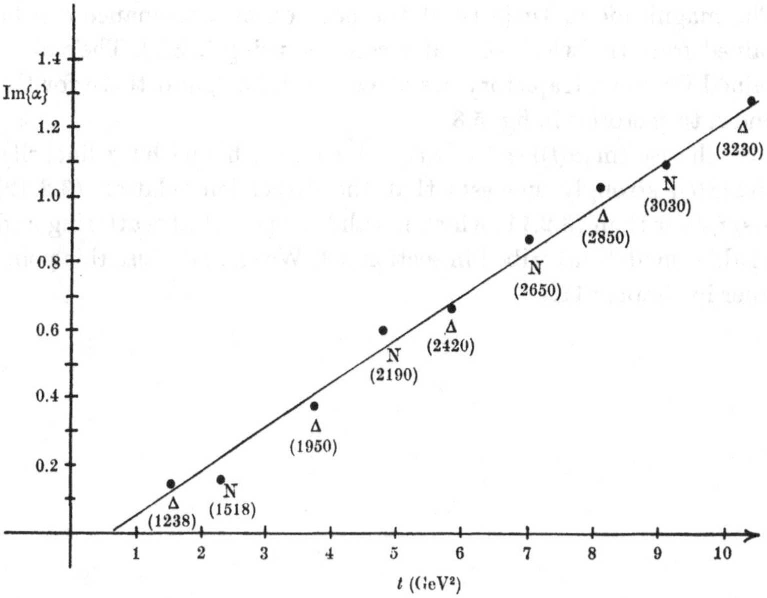

FIG. 5.8 A plot of Im $\{\alpha(t)\}$ against t for the N and Δ trajectories.

function occurs in all the different helicity amplitudes for a given process which are connected by the unitarity relation like (4.4.11), the various kinematical singularities of $A_{HJ}(t)$ which depend upon the helicities will not occur in $\alpha(t)$, though they are present in the Regge residue (see section 6.2).

So, unless trajectories cross each other, $\alpha(t)$ will have just the dynamical right-hand cut of $A_{HJ}(t)$ beginning at the t-channel threshold branch point, t_T. The unitarity relation (4.4.11) with (4.7.6) leads to the threshold behaviour

$$\operatorname{Im}\{\alpha(t)\} \propto (t-t_T)^{\alpha(t_T)-Y_{13}^{+}+\frac{1}{2}}, \quad \alpha(t_T)-Y_{13}^{+} > -\tfrac{1}{2} \tag{5.4.2}$$

instead of (3.2.26), and an infinite number of trajectories will accumulate at threshold at the point $J = Y_{13}^{+}-\frac{1}{2}$, as in (3.2.29).

For mesons one can expect that the trajectory functions will satisfy dispersion relations like (3.2.12) or (3.2.13). But for baryons the MacDowell symmetry (5.3.3) implies that the dispersion relation must be written in terms of $\sqrt{t}$ rather than t, so in unsubtracted form it reads

$$\alpha(\sqrt{t}) = \frac{1}{\pi}\int_{\sqrt{t_T}}^{\infty} \frac{\operatorname{Im}\{\alpha(\sqrt{t'})\}}{\sqrt{t'}-\sqrt{t}}\,\mathrm{d}\sqrt{t'} + \frac{1}{\pi}\int_{-\sqrt{t_T}}^{-\infty} \frac{\operatorname{Im}\{\alpha(\sqrt{t'})\}}{\sqrt{t'}-\sqrt{t}}\,\mathrm{d}\sqrt{t'} \tag{5.4.3}$$

where we have integrated over both the physical regions of $\alpha(\sqrt{t})$. Of course subtractions will in fact be necessary.

The magnitude of $\text{Im}\{\alpha(t)\}$ at the position of a resonance can be obtained from the width of that resonance using (2.8.7). The values obtained for the ρ trajectory are shown in fig. 5.7, and those for the N and Δ trajectories in fig. 5.8.

In each case $\text{Im}\{\alpha(t)\} \ll \text{Re}\{\alpha(t)\}$, which, together with the linearity of $\text{Re}\{\alpha(t)\}$ strongly suggests that the dispersion relation (3.2.12) holds, rather than (3.2.11) which is valid for potential scattering and the ladder models described in section 3.4. We shall discuss this point further in chapter 11.

6
Regge poles and high energy scattering

6.1 Introduction

Having identified, in the previous chapter, some of the leading Regge trajectories from the resonance spectrum, we next want to look more closely at the other main aspect of Regge theory, the way in which Regge poles in the crossed t channel control the high energy behaviour of scattering amplitudes in the direct s channel.

For spinless-particle scattering this presents few problems; we would simply use the expression (2.8.10) in the region where t is small and negative, and s is large. However, for real experiments with spinning particles it is a bit more difficult because, as we shall find in the next section, the t-channel helicity amplitudes contain various kinematical factors, and are subject to various constraints, which must also be incorporated in the Regge residues. Also we shall need to look closely at the behaviour of the residue function when a trajectory passes through the nonsense points discussed in section 4.5. Only when we have clarified these kinematical requirements can we write down correct expressions for the Regge pole contribution to a scattering amplitude based on (4.6.15).

In exploring these kinematical problems we shall discover that some of the difficulties at $t = 0$ may imply the occurrence of additional trajectories called 'daughters' and 'conspirators', and we shall briefly review the application of group theoretical techniques to such problems. Also we examine the way in which the internal SU(2) and SU(3) symmetries constrain Regge pole exchange models.

We are thus led to (6.8.1) below for the parameterization of a Reggeon exchange amplitude, and in the extended final section of this chapter we discuss the comparison of this expression with the experimental data on high energy scattering processes. A reader who is mainly interested in the phenomenology could start at section 6.8 and refer back as necessary.

6.2 Kinematical singularities of Regge residues*

We noted in section 4.1 that though helicity amplitudes have many advantages for Regge theory they suffer from the defect that they are not generally free of kinematical singularities. Since the residue of a t-channel Regge pole is given by (see (4.6.1) and cf. (3.2.16))

$$\beta_H(t) = \frac{1}{2\pi i}\oint \mathrm{d}J\, A^{\eta}_{HJ}(t) \tag{6.2.1}$$

the integration contour being taken round the pole at $J = \alpha(t)$, it is clear from our discussion in section 3.2 that $\beta_H(t)$ will inherit the singularities of $A^{\eta}_{HJ}(t)$, i.e. the kinematical singularities as well as the dynamical right-hand cut beginning at the t-channel threshold. But it will not, of course, contain the pole, nor, in view of the argument of section 3.2, the left-hand cut of $A^{\eta}_{HJ}(t)$.

Various methods have been devised for obtaining the kinematical singularities. One way is to make use of the relationship between helicity amplitudes and the invariant amplitudes of (4.1.3) which are free of kinematical singularities (Cohen-Tannoudji, Salin and Morel 1968), but this becomes difficult for high spins. Another technique, devised by Hara (1964), and worked out fully by Wang (1966), makes use of the fact that the only kinematical t-singularities of an s-channel helicity amplitude occur in the half-angle factors (4.4.12). And in view of the crossing relation (4.3.7) it is evident that the only kinematical singularities in t of the t-channel helicity amplitudes are either those of the s-channel amplitudes, or singularities which are present in the crossing matrix (4.3.4), which is known. A very complete account of this method is given in Martin and Spearman (1970, chapter 6).

But with both methods the physical reasons for the occurrence of the kinematical factors are rather obscure, and instead we shall employ a less rigorous method based on Jackson and Hite (1968) which makes the physics clearer.

The s-singularities of a t-channel helicity amplitude stem entirely from the half-angle factors of (4.4.16), and their occurrence is readily explained by the fact that angular-momentum conservation in the forward and backward directions requires the vanishing of helicity-flip amplitudes (see section 4.4). Similarly we shall find that the t kinematical factors, which for the processes $1+\bar{3}\rightarrow\bar{2}+4$ may occur at the thresholds $t = (m_1+m_3)^2$ and $t = (m_2+m_4)^2$, pseudo-thresholds

* This section may be omitted at first reading.

$t = (m_1 - m_3)^2$ and $t = (m_2 - m_4)^2$, or at $t = 0$, also have a simple physical explanation. We begin by assuming that $m_1 > m_3$ and $m_2 > m_4$, but will consider equal masses, for which the pseudo-threshold moves to $t = 0$, later.

We have found both in non-relativistic potential scattering, in (3.3.24), and for spinless particle scattering, in (2.6.8), that at the threshold $t = (m_1 + m_3)^2$ the partial-wave amplitude has the behaviour

$$A_l(t) \sim (q_{t13})^l \sim (T_{13}^+(t))^l \tag{6.2.2}$$

in the notation of (2.6.6), due to the opening of the partial-wave phase space. Since scattering near threshold is non-relativistic we may expect that even for particles which have spin the threshold behaviour will similarly be

$$A^\eta_{HJ}(t) \sim (T_{13}^+(t))^L \tag{6.2.3}$$

where L is the lowest value of l that can occur for the given J. This will generally be $L = J - \sigma_1 - \sigma_3$ (i.e. $\boldsymbol{\sigma}_1, \boldsymbol{\sigma}_3$ and $\boldsymbol{l}$ all parallel) unless this value of l has the wrong parity, in which case $L = J - (\sigma_1 + \sigma_3) + 1$. This may be incorporated in the expression

$$\left.\begin{aligned} L &= J - \sigma_1 - \sigma_3 + \tfrac{1}{2}[1 - \eta P_1 P_3 (-1)^{\sigma_1 + \sigma_3 - v}] \\ &\equiv J - Y_{13}^+ \quad \text{(say)} \end{aligned}\right\} \tag{6.2.4}$$

where P_1, P_3 $(= \pm 1)$ are the intrinsic parities of the particles, and v is defined in (4.5.6).

We found in section 2.6 that the behaviour (6.2.2) is guaranteed for spinless particle scattering by the Froissart–Gribov projection (2.6.2) (where it converges). However, in (4.5.7) $e^J_{\lambda\lambda'}(z_t) \sim (T_{13}^+)^{J+1}$ (from (1.7.19) and (B.25)), $\xi_{\lambda\lambda'}(z_t) \sim (T_{13}^+)^{-M}$ (from (B.11)) where

$$M \equiv \max\{|\lambda|, |\lambda'|\}, \quad \text{and} \quad dz_t \sim ds(T_{13}^+)^{-1}$$

giving instead

$$A^\eta_{HJ}(t) \sim (T_{13}^+)^{J-M}, \quad t \to (m_1 + m_3)^2 \tag{6.2.5}$$

So the only way in which (6.2.3) can be obtained from (4.5.7) is if the extra factors are already present as kinematical factors in $A_{H_t}(s,t)$, and hence in $D_{sH}(s,t)$ etc. So we must have

$$A_{H_t}(s,t) \sim (T_{13}^+)^{M - Y_{13}^+} \quad \text{as} \quad t \to (m_1 + m_3) \tag{6.2.6}$$

A similar result holds at the $\overline{2}4$ threshold. But the pseudo-threshold corresponds to the threshold for a process in which the lighter particle (say m_3) has the rest energy $E = -m_3$. Such negative energy states (or 'holes') correspond to anti-particles, and for fermions (but not bosons) the anti-particle has the opposite parity to its particle, so we must replace P_3 by $P_3(-1)^{2\sigma_3}$. So we end up with the threshold behaviour

$$A_{H_t}(s,t) \propto (T_{13}^{+})^{M-Y_{13}^{+}}(T_{13}^{-})^{M-Y_{13}^{-}}(T_{24}^{-})^{M-Y_{24}^{+}}(T_{24}^{-})^{M-Y_{24}^{-}} \tag{6.2.7}$$

where $\quad Y_{ij}^{\pm} \equiv \sigma_i + \sigma_j - \frac{1}{2}[1-\eta P_i P_j(-1)^{\sigma_i \pm \sigma_j - v}]$

for $m_i > m_j$. Of course if, say, $m_1 = m_3$, the pseudo-threshold moves to $t = 0$, while if $m_3 = m_4$ also both pseudo-thresholds will be at $t = 0$. These cases will be considered below. So after the partial-wave projection (4.5.7) has been performed, because of (6.2.5) we find

$$A_{HJ}^{\eta}(t) \propto \bar{K}_{\lambda\lambda'}(t)(q_{t13}\,q_{t24})^{J-M} \tag{6.2.8}$$

where $\bar{K}_{\lambda\lambda'}(t)$ is the kinematical factor defined in table 6.1 on p. 160, and so from (6.2.1)

$$\beta_H(t) = \bar{K}_{\lambda\lambda'}(t)\left(\frac{q_{t13}\,q_{t24}}{s_0}\right)^{\alpha(t)-M} \bar{\beta}_H(t) \tag{6.2.9}$$

where $\bar{\beta}_H(t)$ is free of kinematical singularities at the thresholds and pseudo-thresholds (but not necessarily at $t = 0$). We have introduced an arbitrary scale factor s_0, with the same units as t, so that the units in which $\bar{\beta}_H$ is measured will not vary with $\alpha(t)$. It will be discussed further in section 6.8*a*.

There is an additional problem at the thresholds, however, that in general the various helicity amplitudes for a given process are not all independent (see Jackson and Hite 1968, Trueman 1968). This is because at threshold, in view of (6.2.2), only the $l = 0$ state survives, and, to keep $l = 0$, J is restricted to the range $|\sigma_1 - \sigma_3| \leqslant J \leqslant \sigma_1 + \sigma_3$, so only these values of J appear in the partial-wave series (4.4.14). So if we define $\mathbf{s} \equiv \boldsymbol{\sigma}_1 + \boldsymbol{\sigma}_3$ and expand our partial-wave helicity states $|J, \lambda; \lambda_1, \lambda_3\rangle$ $(\lambda \equiv \lambda_1 - \lambda_3)$ in terms of $l - s$ states $|J, \lambda; l, s\rangle$, at threshold we find, since $l = 0$, $s = J$,

$$|J, \lambda; \lambda_1, \lambda_3\rangle = N_J \langle \sigma_1, \lambda_1, \sigma_3, \lambda_3 | J, \lambda\rangle\, |J, \lambda; 0, J\rangle \tag{6.2.10}$$

where N_J is a normalization factor and $\langle \sigma_1, \lambda_1, \sigma_3, \lambda_3 | J, \lambda\rangle$ is the Clebsch–Gordan coefficient. So at threshold a partial-wave helicity amplitude can be written in the form

$$A_{HJ}^{\eta}(t) = \langle \sigma_1, \lambda_1, \sigma_3, \lambda_3 | J, \lambda\rangle\, a_{\lambda_2\lambda_4}(J, t) \tag{6.2.11}$$

where $a_{\lambda_2\lambda_4}(J, t)$ is independent of λ_1 and λ_3. So on summing over J, $|\sigma_1 - \sigma_3| \leqslant J \leqslant \sigma_1 + \sigma_3$, the various $A_{H_t}(s, t)$ with the same values of λ_2, λ_4 but different λ_1, λ_3, are all related at the $1\bar{3}$ threshold by a sum over the Clebsch–Gordan coefficients appearing in (6.2.11).

This is best illustrated by an example. Thus if we consider elastic πN scattering for which the t-channel process is $\pi\pi \to N\overline{N}$ we find that at the $N\overline{N}$ threshold, $t = 4m_N^2$, the relation between the amplitudes of (4.3.11) reads

$$A_{++}(s,t) \to -\mathrm{i}A_{+-}(s,t), \quad t \to 4m_N^2 \tag{6.2.12}$$

the factor $(-\mathrm{i})$ coming from the half-angle factor (see (6.2.15) below). Then if we take out all the kinematical factors we have (cf. (4.3.11))

$$\left.\begin{aligned} A_{++}(s,t) &= \hat{A}_{++}(s,t)\,(t-4m_N^2)^{-\frac{1}{2}} \\ A_{+-}(s,t) &= \hat{A}_{+-}(s,t)\,t^{\frac{1}{2}}(t-4m_\pi^2)^{\frac{1}{2}}\,(1-z_t^2)^{\frac{1}{2}} \end{aligned}\right\} \tag{6.2.13}$$

where the $\hat{A}$'s are free of kinematical singularities in both s and t. If we express each of these amplitudes in terms of a single Regge pole $\alpha(t)$, we have (from (6.8.1) below)

$$\left.\begin{aligned} A_{++}(s,t) &= \gamma_1(t)\,(t-4m_N^2)^{-\frac{1}{2}}\left(\frac{s}{s_0}\right)^{\alpha(t)} \\ A_{+-}(s,t) &= \gamma_2(t)\,t^{\frac{1}{2}}(t-4m_\pi^2)^{\frac{1}{2}}\,(1-z_t^2)^{\frac{1}{2}}\left(\frac{s}{s_0}\right)^{\alpha(t)-1} \end{aligned}\right\} \tag{6.2.14}$$

$$\xrightarrow[s\to\infty]{} \mathrm{i}\gamma_2(t)\,t^{\frac{1}{2}}(t-4m_N^2)^{-\frac{1}{2}}\left(\frac{s}{s_0}\right)^{\alpha(t)} \tag{6.2.15}$$

where the γ's are kinematical-singularity-free residues. The relation (6.2.12) then becomes

$$\gamma_1(4m_N^2) = 2m_N\,\gamma_2(4m_N^2) \tag{6.2.16}$$

and we can always ensure that this will be satisfied by writing

$$2m_N\,\gamma_2(t) = \gamma_1(t) + \gamma_3(t)\left(\frac{4m_N^2-t}{4m_N^2}\right) \tag{6.2.17}$$

where now $\gamma_1(t)$ and $\gamma_3(t)$ are free of constraints as well as singularities. Putting (6.2.14) and (6.2.15) in (4.3.12) gives

$$\frac{\mathrm{d}\sigma}{\mathrm{d}t} = \frac{1}{64\pi s q_{s12}^2}\frac{1}{4m_N^2}\left(\frac{s}{s_0}\right)^{2\alpha(t)}\left\{\gamma_1^2(t) - \frac{t}{4m_N^2}\left[2\gamma_1(t)\,\gamma_3(t) + \gamma_3^2(t)\left(1-\frac{t}{4m_N^2}\right)\right]\right\}. \tag{6.2.18}$$

This expression has no singularity at $t = 4m_N^2$, but had we used (6.2.14) and (6.2.15) directly, ignoring the constraint (6.2.16), there would have been a spurious pole at this point.

This is a rather cumbersome procedure, and it is therefore fortunate that usually the thresholds are sufficiently far from the s-channel physical region ($t < 0$) for it not to matter much in practice if we ignore the constraint. It is only really important in cases like $\pi N \to \pi\Delta$ where the pseudo-threshold at $t = (m_\Delta - m_N)^2$ is not so far from $t = 0$.

We must next consider the point $t = 0$. If the masses are unequal, i.e. $m_1 \neq m_3$, $m_2 \neq m_4$, then from (1.7.19)

$$z_t \xrightarrow[t\to\infty]{} \epsilon \equiv \pm 1 \quad \text{for} \quad (m_1 - m_3)(m_2 - m_4) \gtrless 0 \tag{6.2.19}$$

So the half-angle factor (4.4.12) has the behaviour

$$\xi_{\lambda\lambda'}(z_t) \underset{t\to 0}{\sim} t^{\frac{1}{2}|\lambda - \epsilon\lambda'|} \tag{6.2.20}$$

and so from (4.4.16) $\quad \hat{A}_{H_t}(s,t) \sim t^{-\frac{1}{2}|\lambda-\epsilon\lambda'|} \quad$ (6.2.21)

Hence the definite parity amplitudes (4.6.10) have the behaviour

$$\hat{A}^{\eta}_{H_t}(s,t) \sim t^{-\frac{1}{2}|\lambda-\epsilon\lambda'|} a_1(s,t) \pm \eta t^{-\frac{1}{2}|\lambda+\epsilon\lambda'|} a_2(s,t) \tag{6.2.22}$$

where a_1 and a_2 are regular at $t = 0$. So $\hat{A}^{\eta}_{H_t}$ has a singularity of the form

$$\hat{A}^{\eta}_{H_t}(s,t) \sim \frac{a^{\eta}}{t^{\frac{1}{2}\max\{|\lambda+\lambda'|,\,|\lambda-\lambda'|\}}} = \frac{a^{\eta}}{t^{\frac{1}{2}(M+N)}} \tag{6.2.23}$$

where a^{η} is one of a_1, a_2 and M, N are defined in (4.4.15), (4.5.11). But a Regge pole, which has a definite parity, cannot have such a singular behaviour as this, because if it did we would find

$$A_{H_t}(s,t) \equiv \xi_{\lambda\lambda'}(z_t)\,\hat{A}_{H_t}(s,t) \underset{t\to 0}{\sim} t^{\frac{1}{2}|\lambda-\epsilon\lambda'|}\,\tfrac{1}{2}\left(a^{\eta} t^{-\frac{1}{2}(M+N)} \mp \eta a^{-\eta} t^{-\frac{1}{2}(M+N)}\right) \tag{6.2.24}$$

(where $-\eta = (-1)\eta$) which is singular unless $a^{\eta} = \pm\eta a^{-\eta}$, except when $\lambda = \lambda' = 0$. This equality of a^{η} and $a^{-\eta}$ in fact follows directly from (6.2.22), (6.2.23), but obviously it cannot be satisfied by a Regge pole with a definite parity. So instead of (6.2.23) we must choose the less singular behaviour

$$\hat{A}^{\eta}_{H_t}(s,t) \sim \frac{a^{\eta}}{t^{\frac{1}{2}\min\{|\lambda+\lambda'|,\,|\lambda-\lambda'|\}}} = \frac{a^{\eta}}{t^{\frac{1}{2}(M-N)}} \tag{6.2.25}$$

i.e. we multiply (6.2.23) by t^N. (However, for channels with odd Fermion number N is a half-integer, so this would introduce a spurious square-root branch point – see section 6.5 for this case.)

To obtain the $t = 0$ behaviour of the residue from (6.2.25) we note that (6.2.9) has a singularity of the form $t^{-(\alpha(t)-M)}$ from (1.7.15). The $t^{-\alpha}$ will cancel with the corresponding singularity in the asymptotic behaviour of the rotation function in (4.6.4),

$$e^{-\alpha-1}_{\lambda\lambda'}(z_t) \sim \left(\frac{z_t}{2}\right)^{\alpha} \sim \left(\frac{s-u}{8q_{t13}q_{t24}}\right)^{\alpha} \tag{6.2.26}$$

from (B.25), but the t^M remains, so we end up with

$$\beta_H(t) = t^{-\frac{1}{2}(M+N)}\bar{K}_{\lambda\lambda'}(t)\left(\frac{q_{t13}q_{t24}}{s_0}\right)^{\alpha(t)-M}\bar{\gamma}_H(t) \tag{6.2.27}$$

where $\bar{\gamma}_H(t)$ is free of kinematical singularities. Unfortunately this will not do either, because its behaviour for $t \to 0$, $\beta_H(t) \sim t^{\frac{1}{2}(M-N)-\alpha}$, is not factorizable between the initial and final states. We must be able to write

$$\beta_H(t) = \beta_\lambda(t)\,\beta_{\lambda'}(t) \tag{6.2.28}$$

which is possible only if we change the $t = 0$ behaviour to $t^{\frac{1}{2}(M+N)-\alpha}$, so we finally obtain

$$\beta_H(t) = t^{-\frac{1}{2}(M-N)}\bar{K}_{\lambda\lambda'}(t)\left(\frac{q_{t13}q_{t24}}{s_0}\right)^{\alpha(t)-M}\bar{\gamma}_H(t) \tag{6.2.29}$$

where the $\bar{\gamma}_H(t)$ are free of kinematical singularities, but may have to satisfy threshold constraints like (6.2.16).

If one pair of masses is equal, say $m_1 = m_3$, then $z_t \sim t^{\frac{1}{2}}$, while if $m_2 = m_4$ also then z_t is finite at $t = 0$, and in both cases the pseudo-thresholds move to $t = 0$. The minimum kinematical behaviour can be deduced by repeating the above argument. It is also necessary to ensure factorization like (6.2.28) for amplitudes which have equal masses in one state but not the other, and we find

$$\begin{aligned}\beta_H(t) &= t^{\delta}\bar{K}_{\lambda\lambda'}(t)\left(\frac{q_{t13}q_{t24}}{s_0}\right)^{\alpha(t)-M}\bar{\gamma}_H(t)\\ &\equiv K_{\lambda\lambda'}(t)\left(\frac{q_{t13}q_{t24}}{s_0}\right)^{\alpha(t)-M}\bar{\gamma}_H(t)\end{aligned} \tag{6.2.30}$$

where $K_{\lambda\lambda'}(t)$ is given in table 6.1 (for evasion – see section 6.5).

When (6.2.30) is substituted in (4.6.4) and we use the asymptotic form (B.25), (6.2.26), for the rotation function, the Regge pole contribution to a scattering amplitude becomes

$$\begin{aligned}A^{\mathrm{R}}_{H_t}(s,t) = &-16\pi(-1)^{\Lambda}K_{\lambda\lambda'}(t)\,\bar{\gamma}_H(t)\,(\mathrm{e}^{-\mathrm{i}\pi\alpha}+\mathscr{S})\\ &\times\left\{\frac{1}{2\sin\pi(\alpha-v)}\,\frac{(2\alpha)!\,(2\alpha+1)}{[(\alpha+M)!\,(\alpha-M)!\,(\alpha+N)!\,(\alpha-N)!]^{\frac{1}{2}}}\right\}\\ &\times\left(\frac{s-u}{8s_0}\right)^{\alpha(t)-M}\xi_{\lambda\lambda'}(z_t)\end{aligned} \tag{6.2.31}$$

(where Λ is defined in (B.10)) after some use of the relation

$$(-\alpha)! = \frac{\pi}{\sin\pi\alpha(\alpha-1)!} \tag{6.2.32}$$

The same result is obtained from (4.6.2) using (B.12) for $\mathrm{Re}\{\alpha\} > -\frac{1}{2}$.

Table 6.1 *Kinematical factors for a t-channel helicity amplitude*

The factors introduced in (6.2.9) and (6.2.30) are:

$$K_{\lambda\lambda'}(t) \equiv t^{\delta}\bar{K}_{\lambda\lambda'}(t)$$

$$\bar{K}_{\lambda\lambda'}(t) \equiv (T_{13}^{+})^{M-Y_{13}^{+}}(T_{13}^{-})^{M-Y_{13}^{-}}(T_{24}^{+})^{M-Y_{24}^{+}}(T_{24}^{-})^{M-Y_{24}^{-}}$$

where

$M \equiv \max\{|\lambda|, |\lambda'|\}, \quad N \equiv \min\{|\lambda|, |\lambda'|\}, \quad \lambda \equiv \lambda_1 - \lambda_3, \quad \lambda' \equiv \lambda_2 - \lambda_4$

$T_{ij}^{\pm} \equiv [t-(m_i \pm m_j)^2]^{\frac{1}{2}}$

$Y_{ij}^{\pm} \equiv \sigma_i + \sigma_j - \frac{1}{2}[1-\eta P_i P_j(-1)^{\sigma_i \pm \sigma_j - v}]$

$v = 0/\frac{1}{2}$ for even|odd fermion number

Evasion

UU $\delta = -\frac{1}{2}(M-N)$
EU $\delta = \frac{1}{2}[|\lambda'|-M]+\frac{1}{4}[1-\eta(-1)^{\lambda}]$
EE $\delta = \frac{1}{4}[1-\eta(-1)^{\lambda}]+\frac{1}{4}[1-\eta(-1)^{\lambda'}]$

Conspiracy of Toller number Λ (see (6.5.10))

UU $\delta = \frac{1}{2}\{|\Lambda - M| + |\Lambda - N|\} - M$

EU $\delta = \frac{1}{2}\{|\Lambda - |\lambda'|| - M\} + \frac{1}{4}\{1 - \eta\bar{\eta}(-1)^{\lambda} + \epsilon(\Lambda - 2\sigma_1)\}$

EE $\delta = \frac{1}{2}\{2 + \eta\bar{\eta}(-1)^{\lambda} + \eta\bar{\eta}(-1)^{\lambda'} + \epsilon(\Lambda - 2\sigma_1) + \epsilon(\Lambda - 2\sigma_3)\}$

where $\bar{\eta} = (-1)^{\Lambda+1}$ or $(-1)^{2\sigma+1}$ for $2\sigma \gtrless \Lambda$

$\epsilon(\Lambda - 2\sigma) = \Lambda - 2\sigma$ for $\Lambda - 2\sigma \geqslant 0$

$= 0$ for $\Lambda - 2\sigma \leqslant 0$

U ≡ unequal-mass vertex, E ≡ equal-mass vertex. For EU we take $m_1 = m_3$, $m_2 \neq m_4$ so that $\lambda \equiv \lambda_1 - \lambda_3$ is the helicity change at the equal-mass end. In this section we have discussed the evasive case – see section 6.5 for conspiracies.

6.3 Nonsense factors

Equation (6.2.31) is still not satisfactory, however, because the various factorials which appear would introduce singularities at the nonsense values of α (see section 4.5) which cannot be present in the scattering amplitude. So $\bar{\gamma}_H(t)$ must contain suitable factors to cancel them.

Since (Magnus and Oberhettinger 1949, p. 1)

$$(2\alpha)! = \frac{2^{2\alpha+1}(\alpha)!(\alpha+\frac{1}{2})!}{\pi^{\frac{1}{2}}(2\alpha+1)} \tag{6.3.1}$$

we can re-express the factor in braces { } in (6.2.31) in the form

$$\bar{f}_H(\alpha) \equiv \frac{2^{2\alpha+1}}{\pi^{\frac{1}{2}}} \frac{(\alpha)!(\alpha+\frac{1}{2})!}{[(\alpha+M)!(\alpha-M)!(\alpha+N)!(\alpha-N)!]^{\frac{1}{2}}} \frac{1}{\sin\pi(\alpha-v)} \tag{6.3.2}$$

Now $(\alpha+\frac{1}{2})!$ has simple poles at $\alpha = -\frac{3}{2}, -\frac{5}{2}, \ldots$, while $\alpha!$ has poles at $\alpha = -1, -2, \ldots$. But one of these sets of singularities will be cancelled by the denominator, depending on whether M, N are integers of half-integers (i.e. on whether the channel has even or odd fermion number). So we require that $\bar{\gamma}_H(t) \sim [(\alpha+\frac{1}{2}-v)!]^{-1}$ to cancel the others (v is defined in (4.5.6)). In fact such a behaviour of the residue is guaranteed by the Froissart–Gribov projection (4.5.7) because of (B.24).

The remainder has the form

$$\frac{(\alpha+v)!}{[(\alpha+M)!\,(\alpha-M)!\,(\alpha+N)!\,(\alpha-N)!]^{\frac{1}{2}}\sin\pi(\alpha-v)} \tag{6.3.3}$$

which when $\alpha \to J_0$, where $J_0 - v$ is an integer, has the behaviour

$$\begin{aligned}
&(\alpha-J_0)^{-1} \quad \text{for} \quad J_0 \geqslant M \quad \text{and} \quad v > J_0 > -N\\
&(\alpha-J_0)^{-\frac{1}{2}} \quad \text{for} \quad M > J_0 \geqslant N \quad \text{and} \quad -N > J_0 \geqslant -M\\
&\text{Finite for} \quad N > J_0 \geqslant v \quad \text{and} \quad J_0 < -M
\end{aligned}$$

We remember that only the points $J_0 \geqslant M$ make any physical sense, i.e. are sense–sense (ss) points in the terminology of section 4.5, and so the poles in this region correspond to physical particles. (Note that they are cancelled for alternate J_0 by the signature factor.) At the sense–nonsense (sn) points (6.3.3) behaves like $(\alpha-J_0)^{-\frac{1}{2}}(\alpha+J_0+1)^{-\frac{1}{2}}$, but these branch points (which since α is a function of t give branch points in t) cannot be present in the scattering amplitude, so either

$$\bar{\gamma}_H(t) \sim (\alpha-J_0)^{-\frac{1}{2}}(\alpha+J_0+1)^{-\frac{1}{2}} \quad \text{or} \quad \bar{\gamma}_H(t) \sim (\alpha-J_0)^{\frac{1}{2}}(\alpha+J_0+1)^{\frac{1}{2}}.$$

The Froissart–Gribov projection (4.5.7) gives the former behaviour, but, as discussed in section 4.8, we expect that SCR will hold, in which case the latter behaviour will occur (except perhaps at wrong-signature points where Gribov–Pomeranchuk fixed poles may be expected). Now factorization of the form (6.2.8) requires that

$$\beta_{ss}\beta_{nn} = (\beta_{sn})^2 \propto (\alpha-J_0)(\alpha+J_0+1) \tag{6.3.4}$$

where s and n are sense and nonsense values of λ, λ' for the given J_0. So since the ss residue is expected to be finite to give the physical pole there must be a vanishing of the nn residue. If this behaviour holds at every nonsense point we have

$$\bar{\gamma}_H(t) \sim \left(\frac{(\alpha+M)!\,(\alpha+N)!}{(\alpha-M)!\,(\alpha-N)!}\right)^{\frac{1}{2}} \tag{6.3.5}$$

Combining this with the previous requirements we can write

$$\bar{\gamma}_H(t) = \gamma_\lambda(t)\,\gamma_{\lambda'}(t)\,\frac{2^{M-1}}{\pi^{\frac{1}{2}}}\,\frac{1}{(\alpha+\frac{1}{2}-v)!}\left(\frac{(\alpha+M)!\,(\alpha+N)!}{(\alpha-M)!\,(\alpha-N)!}\right)^{\frac{1}{2}} \tag{6.3.6}$$

where $\gamma_\lambda(t)\,\gamma_{\lambda'}(t)$ is a factorized residue free of any special requirements at the nonsense points; and in (6.2.31) this gives

$$A^{\mathrm{R}}_{H_t}(s,t) = -16\pi(-1)^{\Lambda} K_{\lambda\lambda'}(t)\,\gamma_\lambda(t)\,\gamma_{\lambda'}(t) \times (\mathrm{e}^{-\mathrm{i}\pi(\alpha-v)} + \mathscr{S})\, f^{\mathrm{s}}_H(\alpha) \left(\frac{s-u}{2s_0}\right)^{\alpha-M} \xi_{\lambda\lambda'}(z_t) \quad (6.3.7)$$

where
$$f^{\mathrm{s}}_H(\alpha) \equiv \frac{(\alpha+v)!}{(\alpha-M)!\,(\alpha-N)!}\,\frac{1}{2\sin\pi(\alpha-v)} \quad (6.3.8)$$

(where s ≡ sense-choosing; see below).

At right-signature points, where the signature factor is finite, (6.3.7) has the behaviour

(i) $(\alpha-J_0)^{-1}$ for $J_0 \geqslant M$

(ii) Finite for $M > J_0 \geqslant N$ and $J_0 < 0$

(iii) $(\alpha-J_0)$ for $N > J_0 \geqslant v$

At wrong-signature points the signature factor behaves like $\mathrm{i}(\alpha-J_0)$ giving a finite behaviour for (i), zero for (ii) and double zero for (iii).

However, there are various further considerations which may cause us to modify these conclusions for $\sigma_{\mathrm{T}} \geqslant J_0$ ($\sigma_{\mathrm{T}} \equiv \max\{\sigma_1+\sigma_3, \sigma_2+\sigma_4\}$).

a. Ghost-killing factors

If the trajectory passes through a right-signature point for $t < 0$ the ss residue must vanish, otherwise there would be a 'ghost' particle of negative m^2, i.e. a 'tachyon'. Since the Froissart bound restricts trajectories to $\alpha < 1$ for $t < 0$ this difficulty only occurs for even-signature trajectories at $J - v = 0$, which we see from figs. 5.4–5.6 applies in practice only to the f, A_2 and K**(1400) trajectories (and perhaps the P – see section 6.8*b*) at $\alpha = 0$. If such a zero is inserted in the ss residue it must also appear in the sn and nn residues because of (6.3.4). This is sometimes called the 'Chew mechanism' (Chew 1966).

b. Choosing nonsense

At a given nonsense J_0 a trajectory may 'choose' to satisfy (6.3.4) by having β_{nn} finite and $\beta_{ss} = 0$ instead. This gives

$$\bar{\gamma}_H(t) \sim [(\alpha-J_0)(\alpha+J_0+1)]^{\frac{1}{2}}$$

for $M > J_0 \geqslant N$ as before, but $\overline{\gamma}_H(t) \approx (\alpha - J_0)(\alpha + J_0 + 1)$ for some sense points $J_0 \geqslant M$. If this happens say for $p > J_0 > M$, where $p - v$ is some integer $> M$, then we have

$$\overline{\gamma}_H(t) \sim \frac{(\alpha+p)!}{(\alpha-p)!}\left(\frac{(\alpha-M)!\,(\alpha-N)!}{(\alpha+M)!\,(\alpha+N)!}\right)^{\frac{1}{2}} \tag{6.3.9}$$

instead of (6.3.5). The resulting pole in the nn amplitudes cannot correspond to a physical particle of course, and so it must be cancelled (or compensated for). Since the asymptotic behaviour of $e_{\lambda\lambda'}^{-\alpha-1}(z_t)$ at a nn point is $z^{-\alpha-1}$, not z^{α}, the compensating trajectory must pass through $-J_0 - 1$. This is sometimes called the 'Gell-Mann mechanism' (Gell-Mann and Goldberger 1962, Gell-Mann *et al.* 1964).

However, the need for such a compensating trajectory can be avoided by putting a zero in the nn residue, in which case extra zeros will also appear in the sn and ss residues through (6.3.4). This is called the 'no compensation mechanism'.

c. *Wrong-signature fixed poles*

The arguments of section 4.8 have led us to expect fixed poles (or infinite square-root branch points) at wrong-signature nonsense points. They will not contribute to the asymptotic behaviour of the scattering amplitude because of the signature factor. However, if they are present in the residue of a Regge pole they will cancel the zero from the signature factor.

The fixed poles, which stem from the presence of the third double spectral function ρ_{su}, could be additional to the Regge poles, and not present in the Regge residues. Or, even if fixed poles are present in the residue, since at the point where $\alpha = J_0$ (J_0 being a wrong-signature nonsense point) the residue obtains a contribution only from ρ_{su}, while at all other values of α it receives contributions from all three double spectral functions, the residue might well behave like

$$a(t) + b(t)\,(\alpha(t) - J_0)$$

for example. So with $b \gg a$ there would still be a zero near $\alpha(t) = J_0$, but with $a \gg b$ there would not.

Table 6.2 summarizes the above possibilities for the behaviour of the residue, and the corresponding behaviour of the Regge pole amplitude.

The chief importance of these results is that in some cases the Regge

Table 6.2 *The behaviour of the residue and amplitude as a trajectory passes through a nonsense point,* J_0

	Residue				Amplitude		
	nn	sn	ss	Mechanism	nn	sn	ss
Right-signature	$\alpha - J_0$	$(\alpha - J_0)^{\frac{1}{2}}$	1	Sense-choosing	$\alpha - J_0$	1	$(\alpha - J_0)^{-1}$
	1	$(\alpha - J_0)^{\frac{1}{2}}$	$\alpha - J_0$	Nonsense-choosing-	1	1	1
	$(\alpha - J_0)^2$	$(\alpha - J_0)^{\frac{3}{2}}$	$\alpha - J_0$	Chew mechanism	$(\alpha - J_0)^2$	$\alpha - J_0$	1
	$\alpha - J_0$	$(\alpha - J_0)^{\frac{3}{2}}$	$(\alpha - J_0)^2$	No compensation	$\alpha - J_0$	$\alpha - J_0$	$\alpha - J_0$
Wrong-signature	$(\alpha - J_0)^{-1}$	$(\alpha - J_0)^{-\frac{1}{2}}$	1	Fixed pole	1	1	1

In the above we have assumed the presence of a fixed pole in the residue at the wrong-signature point. If this is absent the residue behaves in the same way as at the corresponding right-signature point, and the amplitude is the same except for an extra $\alpha - J_0$ from the signature factor.

pole amplitude is predicted to have a zero in t. A good example of this is the process $\pi^- p \to \pi^0 n$ which in the t channel ($\pi^- \pi^0 \to \bar{p} n$) contains only the ρ trajectory from our list in table 6.5. From fig. 5.5 (and see also fig. 6.6*a* below) this trajectory is approximately $\alpha(t) = 0.5 + 0.9t$, and so $\alpha(t) = 0$ for $t \approx -0.55\,\text{GeV}^2$. The t-channel helicity amplitudes for this process are A_{++} and A_{+-} (defined in (4.3.11)) and $\alpha = 0$ is a ss point for A_{++} ($\lambda = \lambda' = 0$) but a sn point for A_{+-} ($\lambda = 0$, $\lambda' = 1$), and is a wrong-signature point for the ρ trajectory since the ρ resonance has spin $= 1$. So from table 6.2 we see that if there is no fixed pole and the trajectory chooses sense then A_{++} will be finite but A_{+-} will vanish at $t = -0.55$, while both amplitudes will vanish if it chooses nonsense, or both will be finite if there is a strong fixed-pole contribution. (The nonsense–nonsense amplitude occurs in $\bar{p}n \to \bar{p}n$ and does not have to be considered here.) The data on this process (fig. 6.1) show a dip but not a zero of $d\sigma/dt$ at this point, suggesting that the ρ chooses sense. But the conclusion depends on what other singularities may be present, such as a lower lying ρ' trajectory, Regge cuts etc. We shall return to this problem in section 6.8*k*, and an alternative explanation of the structure involving cuts will be presented in section 8.7*c*.

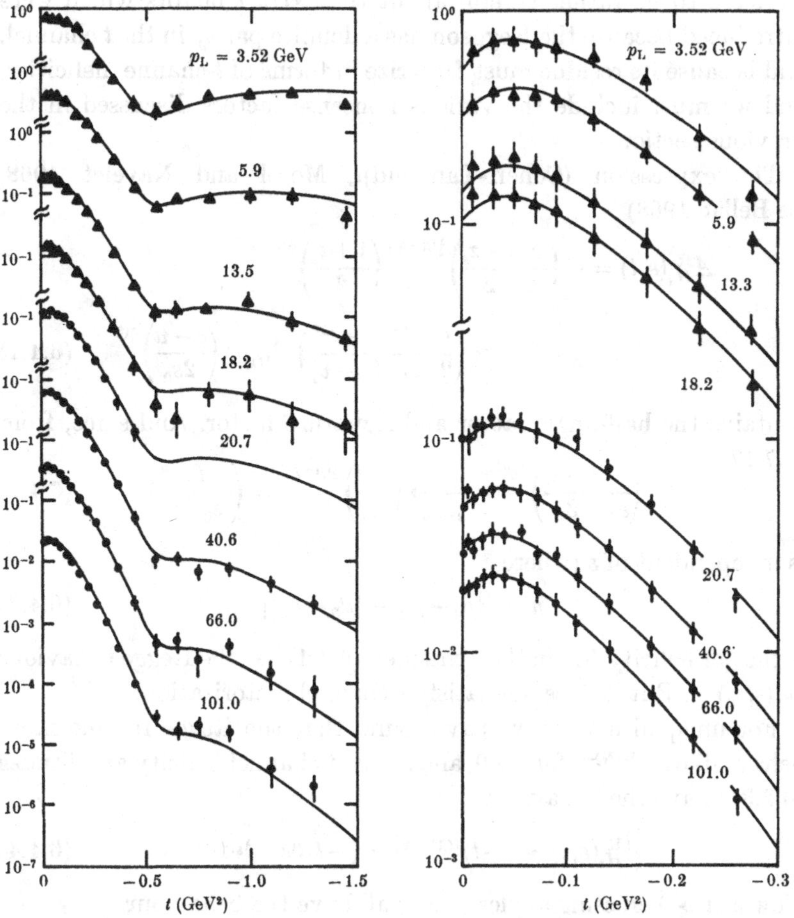

FIG. 6.1 Data for $d\sigma/dt(\pi^- p \to \pi^0 n)$ at various laboratory momenta p_L. The lines are a fit with ρ and ρ′ trajectories, from Barger and Phillips (1974).

6.4 Regge poles and s-channel amplitudes

In section 6.2 we went to a good deal of trouble to ensure that we incorporated the correct kinematical t factors into the Regge residues in the t-channel helicity amplitudes. However, many of these factors cancel out when we construct measurable quantities such as $d\sigma/dt$, density matrices etc., and the only essential t-singularities are those in the s-channel half-angle factors $\xi_{\mu\mu'}(z_s)$. It is obvious therefore that there would be many advantages to working directly with t-channel Regge poles in s-channel helicity amplitudes. But if we wish to do this we have to be rather careful about the extra t factors which were introduced because the Reggeon has a definite parity in the t channel, and because its residue must factorize in terms of t-channel helicities, and we must include the various nonsense factors discussed in the previous section.

The expression (Cohen–Tannoudji, Morel and Navelet 1968, Le Bellac 1968)

$$A^{\mathrm{R}}_{H_s}(s,t) = -\left(\frac{s}{s_0}\frac{1-z_s}{2}\right)^{\frac{1}{2}|\mu-\mu'|}\left(\frac{1+z_s}{2}\right)^{\frac{1}{2}|\mu+\mu'|}$$
$$\times\left(\frac{\mathrm{e}^{-\mathrm{i}\pi(\alpha-v)}+\mathscr{S}}{2\sin\pi(\alpha-v)}\right)\beta_{H_s}(t)\left(\frac{s-u}{2s_0}\right)^{\alpha(t)} \tag{6.4.1}$$

contains the half-angle factor and signature factor. And since, from (1.7.17),

$$\left(\frac{s}{s_0}\frac{1-z_s}{2}\right)^{\frac{1}{2}|\mu-\mu'|}\xrightarrow[s\to\infty]{}\left(\frac{-t}{s_0}\right)^{\frac{1}{2}(\mu-\mu')}=\left(\frac{-t}{s_0}\right)^{\frac{1}{2}n} \tag{6.4.2}$$

is independent of s (where

$$n\equiv\big||\mu_1-\mu_2|-|\mu_3-\mu_4|\big| \tag{6.4.3}$$

is the net helicity-flip in the s channel) (6.4.1) has the Regge behaviour $\sim(s/s_0)^{\alpha(t)}$. But it does not satisfy t-channel factorization.

For unequal masses we have found that the Regge residue must behave like $t^{\frac{1}{2}(M-N)-\alpha}$ for $t\to 0$, and so the t-channel helicity amplitudes (6.2.31) have the behaviour

$$A^{\mathrm{R}}_{H_t}(s,t)\sim(-t)^{\frac{1}{2}(M+N)}=(-t)^{\frac{1}{2}(|\lambda_1-\lambda_3|+|\lambda_2-\lambda_4|)} \tag{6.4.4}$$

Now as $t\to 0$ crossing angles (4.3.5) all have the behaviour

$$\chi_i\sim\sin\chi_i\sim(-t)^{\frac{1}{2}},\quad\text{and so}\quad d^{\sigma_i}_{\lambda_i\mu_i}(\chi_i)\sim(-t)^{\frac{1}{2}|\lambda_i-\mu_i|}$$

for $i=1,\ldots,4$. Hence the helicity crossing matrix (4.3.7)

$$M(H_s,H_t)\sim(-t)^{\frac{1}{2}(|\lambda_1-\mu_1|+|\lambda_2-\mu_2|+|\lambda_3-\mu_3|+|\lambda_4-\mu_4|)} \tag{6.4.5}$$

is diagonal to first order in t at $t=0$. Substituting (6.4.4) and (6.4.5) in (6.3.7) we deduce

$$A^{\mathrm{R}}_{H_s}(s,t)\sim\sum_{H_t}(-t)^{\frac{1}{2}(|\lambda_1-\mu_1|+|\lambda_2-\mu_2|+|\lambda_3-\mu_3|+|\lambda_4-\mu_4|+|\lambda_1-\lambda_3|+|\lambda_2-\lambda_4|)} \tag{6.4.6}$$

and the minimal kinematical behaviour is obtained from those terms in the sum over the λ_i where $\lambda_i=\mu_i$, $i=1,\ldots,4$, and so

$$A^{\mathrm{R}}_{H_s}(s,t)\sim(-t)^{\frac{1}{2}(|\mu_1-\mu_3|+|\mu_2-\mu_4|)},\quad t\to 0 \tag{6.4.7}$$

To ensure this behaviour we write instead of (6.4.1)

$$A^{\mathrm{R}}_{H_s}(s,t) = -\left(\frac{-t}{s_0}\right)^{\frac{1}{2}(|\mu_1-\mu_3|+|\mu_2-\mu_4|-|\mu'-\mu|)}\left(\frac{s}{s_0}\frac{1-z_s}{2}\right)^{\frac{1}{2}|\mu'-\mu|}$$

$$\times\left(\frac{1+z_s}{2}\right)^{\frac{1}{2}|\mu+\mu'|}\frac{\mathrm{e}^{-\mathrm{i}\pi(\alpha-v)}+\mathscr{S}}{2\sin\pi(\alpha-v)}\gamma_{H_s}(t)\left(\frac{s-u}{2s_0}\right)^{\alpha(t)} \quad (6.4.8)$$

$$\xrightarrow[s\to\infty]{} -\left(\frac{-t}{s_0}\right)^{\frac{1}{2}m}\frac{\mathrm{e}^{-\mathrm{i}\pi(\alpha-v)}+\mathscr{S}}{2\sin\pi(\alpha-v)}\gamma_{H_s}(t)\left(\frac{s}{s_0}\right)^{\alpha(t)} \quad (6.4.9)$$

where

$$m \equiv |\mu_1-\mu_3|+|\mu_2-\mu_4| \quad (6.4.10)$$

and $\gamma_{H_s}(t)$ is factorizable in terms of *s*-channel helicities

$$\gamma_{H_s}(t) = \gamma_{\mu_1\mu_3}(t)\,\gamma_{\mu_2\mu_4}(t) \quad (6.4.11)$$

and is free of kinematical singularities.

Though this deduction has been made for unequal masses, it is in fact valid for any mass combination because $A_{H_s}(s,t)$ has no *t*-singularities which depend on the masses except for those in the half-angle factor.

The only difficulty with this method is that one cannot easily incorporate the nonsense mechanisms. There is no problem with the nonsense-choosing, no-compensation or fixed-pole mechanisms which give the same behaviour for all the *t*-channel amplitudes (see table 6.2), and hence for all the *s*-channel amplitudes. But the sense-choosing and Chew mechanisms give zeros in some *t*-channel amplitudes but not others, and if a given A_{H_t} vanishes there will be constraints like

$$\sum_{H_s} M(H_s,H_t)^{-1}A_{H_s}(s,t) \sim \alpha(t)-J_0 \quad (6.4.12)$$

(where M^{-1} is the inverse matrix of M) which are difficult to parameterize. But apart from these cases (6.4.8) has much to recommend it.

6.5 Daughters and conspirators*

In obtaining (6.3.7) for the contribution of a Regge pole to a scattering amplitude we made use of (6.2.26) for the asymptotic behaviour of the rotation function. However, it is evident from (1.7.15), (1.7.19) that for unequal masses, for $t\to 0$, $q_t \sim t^{-\frac{1}{2}}$ and $z_t\to\epsilon$ ($=\pm 1$, see (6.2.19)) for all s. This might seem to imply that the unequal-mass scattering amplitude will not have Regge asymptotic behaviour at $t=0$. But in fact this cannot be true, because $t=0$ is not a singular point of the reduced scattering amplitude $\hat{A}_{H_t}$.

* This section may be ommitted at first reading.

It is easier to see what has gone wrong if we rewrite (1.7.19) as

$$z_t = \frac{s}{2q_{t13}q_{t24}}\left(1+\frac{\Delta(t)}{s}\right) \tag{6.5.1}$$

where
$$\Delta(t) \equiv \frac{1}{2t}[t^2 - t\Sigma + (m_1^2 - m_3^2)(m_2^2 - m_4^2)] \tag{6.5.2}$$

is singular at $t = 0$ for unequal masses, and then make the expansion

$$e_{\lambda\lambda'}^{-\alpha-1}(z_t) = \xi_{\lambda\lambda'}(z_t)f(\alpha)\left[\left(\frac{z_t}{2}\right)^{\alpha-M} + f_1(\alpha)\left(\frac{z_t}{2}\right)^{\alpha-M-2} + \ldots\right]$$

where $f(\alpha)$ is given by (B.25) and $f_1(\alpha)$ can be deduced from (B.24). Substituted in (4.6.4) with (6.5.1) and (6.2.30), this gives

$$A_{H_t}^{R}(s,t) \propto \left\{\left(\frac{s}{4s_0}\right)^{\alpha-M} + \Delta(t)(\alpha-M)\,4s_0\left(\frac{s}{4s_0}\right)^{\alpha-M-1}\right.$$
$$\left. + \left[\frac{(\alpha-M)(\alpha-M-1)}{2}(4s_0\Delta(t))^2 + a_1(\alpha)\left(\frac{q_{t13}q_{t24}}{s_0}\right)^2\right]\left(\frac{s}{4s_0}\right)^{\alpha-M-2} + \ldots\right\} \tag{6.5.3}$$

So each term in the expansion of order $(s/4s_0)^{\alpha-M-n}$ has a t^{-n} singularity at $t = 0$. It is these singularities which cause the problem.

However, the amplitude must be analytic at $t = 0$, since it is supposed to obey the Mandelstam representation, so there must be some other contributions which cancel them. These could be contained in the background integral (see Collins and Squires (1968), chapter 3), but a more popular suggestion (Freedman and Wang 1967) is that there are further trajectories known as 'daughters' which have singular residues which precisely cancel the singularities of the original 'parent' trajectory. So the first daughter will have

$$\alpha_1(t) \xrightarrow[t\to 0]{} \alpha(t) - 1 \tag{6.5.4}$$

and residue

$$\beta_1(t) \xrightarrow[t\to 0]{} -\beta(0)\frac{(m_1^2-m_3^2)(m_2^2-m_4^2)(\alpha(0)-M)\,2s_0}{t} + \text{non-singular terms} \tag{6.5.5}$$

to cancel the second term in (6.5.3). In fact an infinite sequence of daughters is needed with

$$\alpha_k(0) = \alpha(0) - k, \quad k = 1, 2, 3, \ldots, \quad \beta_k(0) \underset{t\to\infty}{\sim} t^{-k} \tag{6.5.6}$$

The odd-numbered daughters must have opposite signature to the parent, i.e. $\mathscr{S}_k = \mathscr{S}(-1)^k$, so that their signature factors are identical to those of the parent at $t = 0$.

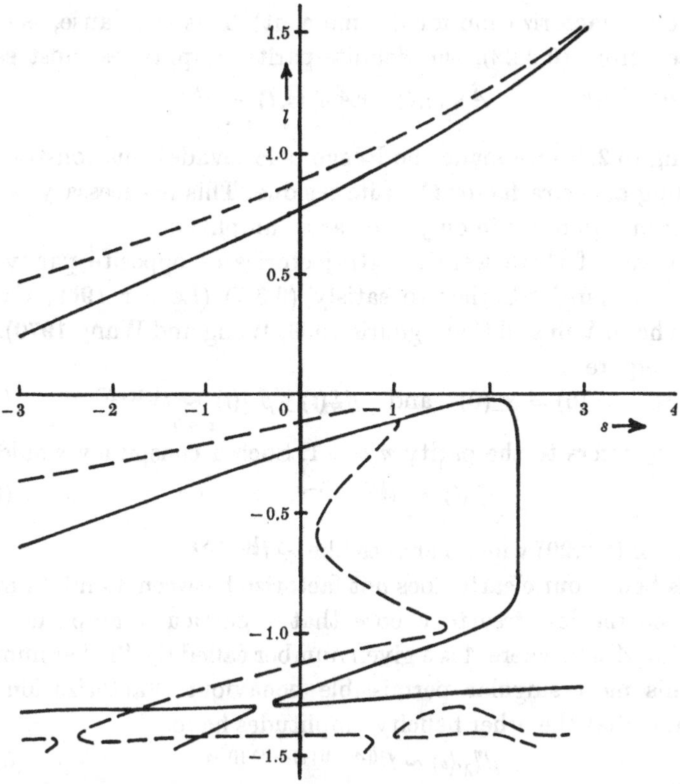

FIG. 6.2 Regge trajectories obtained by Cutkosky and Deo (1967) from the Bethe–Salpeter equation using a potential with a repulsive core. The continuous and dashed curves represent different coupling strengths. The strange behaviour of the daughters is evident.

There is not a great deal of evidence for the existence of such daughters in figs. 5.4–5.6. Indeed, calculations of trajectories using unequal-mass kinematics in the Bethe–Salpeter equation (Cutkosky and Deo 1967) produce a rather peculiar behaviour for the daughters (fig. 6.2) which do not manifest themselves as particles. Unless the non-singular terms in (6.5.5) are important, the daughters need not be visible in the s-channel energy dependence either, since their main purpose is to ensure the $s^{\alpha(t)}$ behaviour for all t, and they may be masked by other singularities (cuts etc.). But we shall discuss in the next chapter further reasons why such trajectories should exist parallel to the parent (see fig. 7.5 below).

Another problem for Regge poles at $t = 0$ is that the residues cannot have the kinematically expected behaviour (6.2.23) but only (6.2.25) (neglecting factorization for the moment). This is because, as can be deduced from (6.2.22), the definite-parity amplitudes must satisfy the constraint

$$\hat{A}^{\eta}_{H_t}(s,t) \mp \eta \hat{A}^{-\eta}_{\bar{H}_t}(s,t) \underset{t\to 0}{\sim} t^N \qquad (6.5.7)$$

In using (6.2.25) we make the Regge pole 'evade' this constraint by including an extra factor t^N in its residue. This is necessary because a Reggeon can occur in only one parity amplitude.

However, if there were two trajectories of opposite parity they could 'conspire' together to satisfy (6.5.7) (Leader 1968, Capella, Tran Thanh Van and Contogouris 1969, Wang and Wang 1970). This would require

$$\alpha_+(0) = \alpha_-(0) \quad \text{and} \quad \beta^+_H(t) \pm \beta^-_{\bar{H}}(t) \underset{t\to 0}{\sim} t^{\frac{1}{2}(M+N)-\alpha} \qquad (6.5.8)$$

where $\pm$ refers to the parity $\eta = \pm 1$. Such a conspiracy would give

$$\beta^{\eta}_H(t) \underset{t\to\infty}{\sim} t^{\frac{1}{2}(M-N)-\alpha}, \quad \eta = \pm 1 \qquad (6.5.9)$$

instead of (6.2.29) which behaves like $\sim t^{\frac{1}{2}(M+N)-\alpha}$.

This behaviour clearly does not factorize between λ and λ', but we are none the less free to choose that a particular amplitude with $\lambda = \lambda' = \Lambda$ say, where Λ is a given number called the 'Toller number', has this most singular permissible behaviour. Factorization then demands that the other helicity amplitudes have

$$\beta^{\eta}_{\lambda\lambda'}(t) \sim t^{\frac{1}{2}(|\Lambda-|\lambda||+|\Lambda-|\lambda'||)-\alpha} \qquad (6.5.10)$$

and for a conspiring trajectory the parameter δ is replaced by the values in table 6.1 (p. 160). Applying the crossing relation (4.3.7) with (6.4.5) we find

$$A^{\mathrm{R}}_{H_s}(s,t) \sim (-t)^{\frac{1}{2}(|\Lambda-|\mu_1-\mu_3||+|\Lambda-|\mu_2-\mu_4||)} \qquad (6.5.11)$$

so unlike (6.4.7) an amplitude with $|\mu_1-\mu_3| = |\mu_2-\mu_4| = \Lambda$ will not vanish at $t = 0$.

A simple example is provided by the process $\gamma p \to \pi^+ n$ which should be dominated by π-exchange near the forward direction ($t \approx 0$). Since for the photon $\mu_1 = \pm 1$ only, and the spinless pion has $\mu_3 = 0$ only, we see from (6.4.10) that $m \neq 0$ and so with the behaviour (6.4.7) all the amplitudes will vanish at $t = 0$, and hence a dip must occur in $d\sigma/dt$ at $t = 0$. In fact the data show a sharp forward spike of width $\Delta t \approx m^2_\pi$ which could be explained by a $\Lambda = 1$ conspiracy between the π and a similar natural-parity trajectory giving the behaviour (6.5.11) instead (Ball, Frazer and Jacob 1968). However, a scalar particle similar to the pion does not occur, and it has been shown (Le Bellac 1967) that such a conspiracy is incompatible with factorization in other π-exchange process, so it now seems more likely that the presence

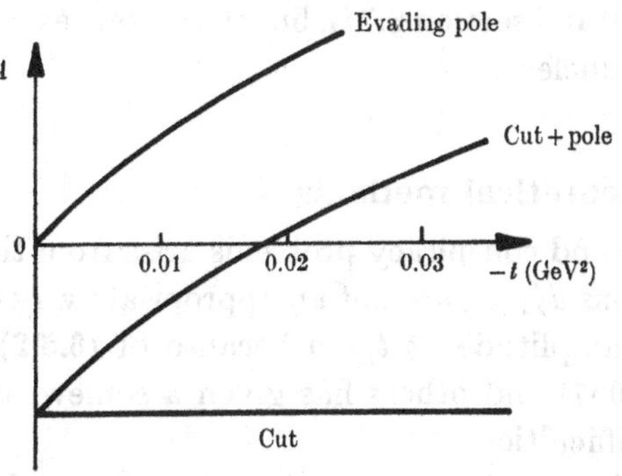

Fig. 6.3 The scattering amplitude for $\gamma p \to \pi^+ n$ showing the contributions of an evasive pion pole and a Regge cut. Cut + pole gives the sharp forward peak seen in the data.

of Regge cuts accounts for the forward peak (see fig. 6.3 and section 8.7*f*). There does not seem to be any evidence for conspiracies of meson trajectories.

A conspiracy is essential, however, if the fermion number of the exchange is odd. We mentioned after (6.2.25) that in this situation multiplying the residue by t^N would introduce a spurious square-root branch point at $t = 0$. In fact making the replacement $\sqrt{t} \to -\sqrt{t}$ in (6.2.24) we find that for half-integer λ, λ'

$$A^{\eta}_{H_t}(s, \sqrt{t}) = A^{-\eta}_{\bar{H}_t}(s, -\sqrt{t})\,(-1)^{|\lambda - \epsilon\lambda'|} \tag{6.5.12}$$

This is called the generalized MacDowell symmetry (after MacDowell 1959), and it means that for baryons there must be a conspiracy between opposite parity trajectories of Toller number $\Lambda = \frac{1}{2}$, so

$$\alpha^+(\sqrt{t}) = \alpha^-(-\sqrt{t}) \quad \text{and} \quad \beta^+_H(\sqrt{t}) = (-1)^{|\lambda - \epsilon\lambda'|}\beta^-_{\bar{H}}(-\sqrt{t}) \tag{6.5.13}$$

If such trajectories are even in $\sqrt{t}$, like the linear form (5.3.2), then the two trajectories should coincide, and one would expect baryons to occur in degenerate doublets of opposite parity. The inclusion of terms which are odd in $\sqrt{t}$, such as

$$\alpha^{\pm}(\sqrt{t}) = \alpha_0 \pm \alpha_1\sqrt{t} + \alpha_2 t + \ldots \tag{6.5.14}$$

splits the degeneracy, but makes the trajectories curved. However, we found in section 5.3 that baryon trajectories appear to be linear in t, but not parity doubled. It is possible to put zeros into the residues to make the unwanted states vanish (see for example Storrow (1972, 1975)), or to introduce a branch point at $J = \alpha_0$, and place the unwanted states on the unphysical side of the cut (see for example Carlitz and

Kisslinger (1970) and section 8.7 *i*), but the correct explanation for this problem is still unclear.

6.6 Group theoretical methods*

These daughter and conspiracy problems arise from the fact that the rotation functions $d^J_{\lambda\lambda'}(z_t)$ are not an appropriate way of representing the scattering amplitude at $t = 0$ because of (6.5.2). The work of Toller (1965, 1967) and others has given a somewhat more general view of these difficulties.

In writing the partial-wave series (4.4.14) we decomposed the scattering amplitude in terms of representation functions of the three-dimensional rotation group O(3), or more strictly, since half-integer spins may be included, its covering group SU(2). The rotation group is the so-called 'little group' of the inhomogeneous Lorentz group, or Poincaré group $\mathscr{P}$, i.e. it is the group of transformations which leaves invariant the total four-momentum of the incoming or outgoing particles (in the t channel)

$$P_\mu \equiv (p_{1\mu}+p_{3\mu}) = (p_{2\mu}+p_{4\mu}), \quad \mu = 1, \ldots, 4 \tag{6.6.1}$$

(see for example Martin and Spearman (1970) chapter 3, and Britten and Barut (1964)). The angular momentum J^2 is of course a Casimir operator of this little group, and $\sum_\mu P^2_\mu \equiv t$ is also a Casimir invariant of $\mathscr{P}$.

However, Wigner (1939) showed that although O(3) is the little group for $t > 0$, there are in fact four different classes of representations of $\mathscr{P}$ characterized by different values of t. These are

(i) Timelike, $t > 0$, little group O(3)
(ii) Spacelike, $t < 0$, little group O(2, 1)
(iii) Lightlike, $t = 0$, $P_\mu \neq 0$, little group E(2)
(iv) Null, $t = 0$, $P_\mu = 0$, little group O(3, 1)

Here O(3) is the rotation group in a space with three real dimensions, with $x^2+y^2+z^2 = R^2$ invariant; O(2, 1) is the rotation group in a space with two real dimensions and one imaginary, with $x^2+y^2-z^2$ invariant; E(2) is the group of Euclidian transformations in two dimensions; while O(3, 1) is the rotation group in a space with three real dimensions and one imaginary, with $x^2+y^2+z^2-t^2$ invariant, which is isomorphic to the Lorentz group itself.

The representation functions of O(3) are the $d^J_{\lambda\lambda'}(z_t)$, $-1 \leqslant z_t \leqslant 1$.

* This section may be omitted on first reading.

The representations of O(2, 1) are again $d^J_{\lambda\lambda'}(z_t)$, but with z_t taking the unphysical values appropriate to $t < 0$. Bargmann (1947) has shown that a function which is square-integrable on this group manifold can be expanded in terms of the principle and discrete series of representations, so that a scattering amplitude expanded in this basis takes the form (Joos 1964, Boyce 1967)

$$A_{H_t}(s,t) = -\frac{16\pi}{2\mathrm{i}}\int_{-\frac{1}{2}-\mathrm{i}\infty}^{-\frac{1}{2}+\mathrm{i}\infty} \mathrm{d}J \frac{2J+1}{\sin\pi(J+\lambda')} A_{HJ}(t)\, d^J_{\lambda\lambda'}(z_t) + \text{nonsense terms} \quad (6.6.2)$$

i.e. (4.6.2) without any Regge poles or cuts in $\mathrm{Re}\{J\} > -\frac{1}{2}$. This is because the square-integrability condition requires $A_{H_t}(s,t) = O(s^{-\frac{1}{2}})$.

So the Sommerfeld–Watson representation can be regarded as a representation on an O(2, 1) basis. However, the equivalence is incomplete in that the Sommerfeld–Watson representation is valid for all t, not just $t < 0$. Also it is valid for non-relativistic potential scattering which has E(2) rather than O(2, 1) as its little group for $t < 0$, and the E(2) representations are quite different (Inonu and Wigner 1952, Levy-Leblond 1966). And of course with Regge singularities in $\mathrm{Re}\{J\} > -\frac{1}{2}$ the Sommerfeld–Watson representation is an analytic continuation in J of (6.6.2). But if these differences are kept in mind it is possible to rephrase Regge theory as an O(2, 1) decomposition.

Because of the mass-shell conditions $p_1^2 = m_1^2$ etc.,

$$t = (p_1+p_3)^2 = (p_2+p_4)^2 = 0$$

implies that the individual components of P_μ are zero in (6.6.1) only if $m_1 = m_3$ and $m_2 = m_4$, so the little group at $t = 0$ will be O(3, 1) or E(2) depending on whether or not the masses are equal.

If the masses are equal then the amplitude can be decomposed in terms of representation functions of O(3, 1) which may be denoted by $d^{\Lambda\sigma}_{TT'}(z_t)$. They have been derived by Sciarrino and Toller (1967) and depend upon two Casimir operators, of which one is the Toller number, Λ, introduced in (6.5.10), which can take on the values $0, 1, 2, \ldots$ or $\frac{1}{2}, \frac{3}{2}, \frac{5}{2}, \ldots$ depending on the fermion number, and the other, σ, is pure imaginary, $-\infty < \mathrm{i}\sigma < \infty$. This extra Casimir operator appears because there are two degrees of freedom in satisfying $\sum_\mu P_\mu^2 = 0$ with equal masses. The other degree of freedom corresponds to variation of s. On this basis the amplitude can be expanded

$$A_{H_t}(s,t=0) = \delta_{\lambda\lambda'} \sum_{T,T'} \sum_{\Lambda=-T_M}^{T_M} \int_{-i\infty}^{i\infty} d\sigma(\Lambda^2-\sigma^2) A^{\Lambda\sigma}_{T\lambda T'}(t=0)\, d^{\Lambda\sigma}_{T\lambda T'}(z_t) \tag{6.6.3}$$

where $A^{\Lambda\sigma}_{T\lambda T'}(t)$ are O(3, 1) partial-wave amplitudes, $T_M \equiv \min\{T, T'\}$ and in the summations

$$|\sigma_1-\sigma_3| \leqslant T \leqslant \sigma_1+\sigma_3, \quad |\sigma_2-\sigma_4| \leqslant T' \leqslant \sigma_2+\sigma_4.$$

At $t = 0$ only the non-flip $\lambda = \lambda'$ amplitudes survive.

If we suppose that there is a Toller pole at $\sigma = a$ say (just as there may be a Regge pole at $J = \alpha$ in (6.6.2)) then analytic continuation in σ gives

$$A_{H_t}(s,0) = [(6.6.3)] + \delta_{\lambda\lambda'} \sum_{T,T'} A^{\Lambda a}_{T\lambda T'}(\Lambda^2-a^2)\, d^{\Lambda a}_{T\lambda T'} \tag{6.6.4}$$

where [(6.6.3)] represents the right-hand side of (6.6.3) and Λ is the Toller number of the pole. Since it is found that

$$d^{\Lambda\sigma}_{T\lambda T'}(z_t) \underset{z_t\to\infty}{\sim} (z_t)^{\sigma-1-|\Lambda-\lambda|} \tag{6.6.5}$$

we deduce from (6.6.4) that

$$A_{H_t}(s,0) \sim \delta_{\lambda\lambda'}(z_t)^{a-1-|\Lambda-\lambda|} \tag{6.6.6}$$

If this is compared with (6.3.7) (remembering (6.5.11)) it will be seen that this behaviour corresponds to a Regge pole with $\alpha(0) = a-1$ and Toller number Λ. Indeed if these O(3, 1) representation functions are decomposed in terms of $d^J_{\lambda\lambda'}(z_t)$ it is found (Sciarrino and Toller 1967) that the single Toller pole in the σ plane at $\sigma = a$ (6.6.4), corresponds to an infinite sequence of Regge poles in the J plane at $J = \alpha_k(0)$ with

$$\alpha_k(0) = a-k-1, \quad k = 0, 1, 2, \ldots \tag{6.6.7}$$

i.e. a conspiring daughter sequence of Toller number Λ. As we move away from $t = 0$ the O(3, 1) symmetry is broken so the daughter trajectories do not have to remain integrally spaced from the parent as in (6.6.7).

This argument clearly does not work for unequal masses because the $E(2)$ representations are quite different from those of O(3, 1), so continuation in the masses is needed to justify the use of Toller poles in this case (Domokos and Tindle 1968, Bitar and Tindle 1968, Kuo and Suranyi 1970). Indeed the apparent absence of conspiracies noted in the previous section leads one to suspect that nature has not in fact made use of the extra degree of freedom at $t = 0$ represented by variation of σ in (6.6.3). A single Regge pole at $t = 0$ corresponds to a counter-conspiracy consisting of an infinite sequence of Toller poles in the σ plane (just like the many-to-one relation between poles in the l and n planes in section 2.10) so the lack of conspiracies presumably reflects the primacy of the J plane over the σ plane. If so, these group-theory techniques do not appear to possess any significant advantage over the conventional Sommerfeld–Watson method which we use in this book.

6.7 Internal symmetry and crossing

a. Isospin

As we mentioned in section 5.2, the approximate invariance of strong interactions under the internal symmetries SU(2) and SU(3) leads to important relations between scattering amplitudes. We begin with the isospin group SU(2) which appears to be broken by at most a few per cent, which is often well within the errors to which scattering amplitudes can be determined. Hence it is frequently more useful to refer to scattering amplitudes for the various possible isospin states, rather than to the amplitudes for the different charge states of the particles involved.

It is convenient to consider first a particle decay such as $a \to 1+2$. The final state may be expressed in terms of the isospins of the particles (see (5.2.1)) as

$$|1,2\rangle = |I_1, I_{1z}\rangle \oplus |I_2, I_{2z}\rangle \tag{6.7.1}$$

The total isospin is the sum of the isospin vectors of the particles

$$\boldsymbol{I} = \boldsymbol{I}_1 + \boldsymbol{I}_2 \tag{6.7.2}$$

and its possible eigenvalues are

$$I = I_1 + I_2, \quad I_1 + I_2 - 1, \ldots, |I_1 - I_2| \tag{6.7.3}$$

while

$$I_z = I_{1z} + I_{2z} = I, \quad I-1, \ldots, -I \tag{6.7.4}$$

so the state (6.7.1) can be written as a superposition of the various possible total isospin states as

$$|1,2\rangle = \sum_I \langle I_1, I_2, I_{1z}, I_{2z} | I, I_z\rangle \, |I, I_z\rangle \tag{6.7.5}$$

where $\langle I_1, I_2, I_{1z}, I_{2z} | I, I_z\rangle$ are the Clebsch–Gordan coefficients (see for example Edmonds (1960) chapter 3). Since the particle a has a definite isospin, I_a, only one term in the sum (6.7.5) occurs in the decay process, and so the decay amplitude can be expressed in the form

$$A(a \to 1+2) = \langle I_1, I_2, I_{1z}, I_{2z} | I_a, I_{az}\rangle \, \bar{A}(a \to 1+2) \tag{6.7.6}$$

where $\bar{A}$ is a 'reduced' amplitude which is independent of I_{az}. Thus isospin invariance implies that the different charge states of particle a, with their different values of I_{az} (see (5.2.1)), will have decay rates which are related to each other by the Clebsch–Gordan coefficients of SU(2).

For example in the decay $\rho \to \pi\pi$ both ρ and π have $I = 1$, and $I_z = 1, 0, -1$ for the charge states $+$, 0 and $-$. So the various decay amplitudes are related according to (6.7.6) by

$$A(\rho^+ \to \pi^+\pi^0) = A(\rho^- \to \pi^0\pi^-) = A(\rho^0 \to \pi^+\pi^-) = \frac{1}{\sqrt{2}}\bar{A}(\rho \to \pi\pi) \tag{6.7.7}$$

where $\bar{A}(\rho \to \pi\pi)$ is the reduced amplitude. Such relations appear to be well satisfied in hadronic decays.

Similarly for the scattering process $1+2 \to 3+4$, both the initial and final states can be expressed as isospin states, like (6.7.5), and if the process is isospin invariant the scattering amplitude may be decomposed as

$$\langle 34| A |12\rangle = \sum_I \langle I_3, I_4, I_{3z}, I_{4z}|I, I_z\rangle^* \langle I_1, I_2, I_{1z}, I_{2z}|I, I_z\rangle A(I) \tag{6.7.8}$$

where $A(I)$ is independent of I_z. In general the number of different isospin amplitudes is smaller than the number of charged particle processes which can occur and so (6.7.8) inter-relates the amplitudes for the different processes.

For example in πN scattering the state $|\pi^+p\rangle$ has $I_z = 1+\frac{1}{2} = \frac{3}{2}$ and so $I = \frac{3}{2}$ only. Likewise $|\pi^-n\rangle$ has $I_z = -\frac{3}{2}$, $I = \frac{3}{2}$. Hence from (6.7.8)

$$\langle \pi^+p| A |\pi^+p\rangle = \langle \pi^-n| A |\pi^-n\rangle = A(\tfrac{3}{2}) \tag{6.7.9}$$

Similarly on looking up the Clebsch–Gordan coefficients we find

$$\left.\begin{aligned} \langle \pi^-p| A |\pi^-p\rangle = \langle \pi^+n| A |\pi^+n\rangle &= \tfrac{1}{3}A(\tfrac{3}{2}) + \tfrac{2}{3}A(\tfrac{1}{2}) \\ \langle \pi^0p| A |\pi^0p\rangle = \langle \pi^0n| A |\pi^0n\rangle &= \tfrac{2}{3}A(\tfrac{3}{2}) + \tfrac{1}{3}A(\tfrac{1}{2}) \\ \langle \pi^0n| A |\pi^-p\rangle = \langle \pi^0p| A |\pi^+n\rangle &= (\sqrt{2/3})A(\tfrac{3}{2}) - (\sqrt{2/3})A(\tfrac{1}{2}) \end{aligned}\right\} \tag{6.7.10}$$

So the eight different πN scattering processes are given by just two independent isospin amplitudes, $A(\frac{1}{2})$ and $A(\frac{3}{2})$.

There is at present no convincing explanation as to why nature should have chosen such a complicated symmetry structure for hadronic interactions, but it certainly works at least to a few per cent, at which level it is presumably broken by electromagnetic interactions.

We shall be particularly concerned with relations between s-channel amplitudes which arise from the exchange of particles having a definite isospin in the t channel. The t-channel process $1+\bar{3}\to\bar{2}+4$ can be decomposed as

$$\langle\bar{2}4|\,A\,|1\bar{3}\rangle = \sum_{I_t}\langle I_1, I_3, I_{1z}, I_{3z}|I_t, I_{tz}\rangle\langle I_2, I_4, I_{2z}, I_{4z}|I_t, I_{tz}\rangle^* A(I_t) \quad (6.7.11)$$

while (6.7.8) holds for s-channel isospin. The crossing relation (4.3.1) becomes for isospin amplitudes

$$A(I_s) = \sum_{I_t} M(I_s, I_t)\,A(I_t) \quad (6.7.12)$$

where the isospin crossing matrix $M(I_s, I_t)$ can be obtained from the Clebsch–Gordan coefficients in (6.7.8) and (6.7.11). However some care is needed with the phase conventions for isospin states and their behaviour under charge conjugation. These are discussed in some detail in Carruthers (1966). Some useful examples are quoted in table 6.3.

To illustrate how these matrices arise we consider $\pi\pi$ scattering. In terms of isospin states $|I, I_z\rangle$ we can write

$$\left.\begin{aligned} |\pi^+\pi^+\rangle &= |2,2\rangle \\ |\pi^+\pi^-\rangle &= \left(\frac{1}{\sqrt{3}}|0,0\rangle + \frac{1}{\sqrt{2}}|1,0\rangle + \frac{1}{\sqrt{6}}|2,0\rangle\right) \\ |\pi^-\pi^+\rangle &= \left(\frac{1}{\sqrt{3}}|0,0\rangle - \frac{1}{\sqrt{2}}|1,0\rangle + \frac{1}{\sqrt{6}}|2,0\rangle\right) \end{aligned}\right\} \quad (6.7.13)$$

etc. so for example

$$\left.\begin{aligned} \langle\pi^+\pi^+|\,A\,|\pi^+\pi^+\rangle &= A(2) \\ \langle\pi^-\pi^+|\,A\,|\pi^+\pi^-\rangle &= \tfrac{1}{3}A(0) - \tfrac{1}{2}A(1) + \tfrac{1}{6}A(2) \end{aligned}\right\} \quad (6.7.14)$$

Now under crossing the s-channel process $\pi^+\pi^+\to\pi^+\pi^+$ becomes $\pi^+\pi^-\to\pi^-\pi^+$ in the t channel, and so

$$A_s(2) = \tfrac{1}{3}A_t(0) - \tfrac{1}{2}A_t(1) + \tfrac{1}{6}A_t(2) \quad (6.7.15)$$

which gives the bottom row of the $\pi\pi$ crossing matrix in table 6.3. The remaining elements can be deduced similarly.

b. SU(3) *symmetry*

As with isospin, we expect that different scattering processes will be related by SU(3) Clebsch–Gordan coefficients if strong interactions are invariant under this symmetry (see Carruthers 1966, Gourdin 1967).

Table 6.3 *Isospin crossing matrices*

s-Channel	*t*-Channel	$M(I_s, I_t)$
$\pi\pi \to \pi\pi$	$\bar{\pi}\pi \to \pi\bar{\pi}$	$\begin{pmatrix} \frac{1}{3} & 1 & \frac{5}{3} \\ \frac{1}{3} & \frac{1}{2} & -\frac{5}{6} \\ \frac{1}{3} & -\frac{1}{2} & \frac{1}{6} \end{pmatrix}$
$\pi N \to \pi N$	$\bar{N}N \to \pi\bar{\pi}$	$\begin{pmatrix} \sqrt{\frac{1}{6}} & 1 \\ \sqrt{\frac{1}{6}} & -\frac{1}{2} \end{pmatrix}$
$KN \to KN$	$\bar{N}N \to K\bar{K}$	$\begin{pmatrix} -\frac{1}{2} & -\frac{3}{2} \\ -\frac{1}{2} & \frac{1}{2} \end{pmatrix}$
s-Channel	*u*-Channel	$M(I_s, I_u)$
$\pi N \to \pi N$	$\bar{\pi}N \to \bar{\pi}N$	$\begin{pmatrix} -\frac{1}{3} & \frac{4}{3} \\ \frac{2}{3} & \frac{1}{3} \end{pmatrix}$
$KN \to KN$	$\bar{K}N \to \bar{K}N$	$\begin{pmatrix} -\frac{1}{2} & \frac{3}{2} \\ \frac{1}{2} & \frac{1}{2} \end{pmatrix}$

The particle label matters only for the isospin so π can be replaced by any $I = 1$ particle, and K, N by any $I = \frac{1}{2}$ particles.

If we label the multiplet to which a particle belongs, i.e. {1}, {8}, {10} etc., by μ, and its quantum numbers I, I_z, Y by ν, then the state $|1, 2\rangle$ can be decomposed into irreducible representations of SU(3) by (cf. (6.7.5))

$$|\mu_1, \nu_1\rangle \otimes |\mu_2, \nu_2\rangle = \sum_{\mu,\nu} \begin{pmatrix} \mu_1 & \mu_2 & \mu \\ \nu_1 & \nu_2 & \nu \end{pmatrix} |\mu, \nu\rangle \tag{6.7.16}$$

where the bracket ( ) denotes a Clebsch–Gordan coefficient.

The cases of greatest practical importance in view of the multiplets discussed in section 5.2 are (Carruthers 1966, Gourdin 1967)

$$\left.\begin{aligned} &\{1\} \otimes \{8\} = \{8\} \\ &\{1\} \otimes \{10\} = \{10\} \\ &\{8\} \otimes \{8\} = \{1\} \oplus \{8_s\} \oplus \{8_a\} \otimes \{10\} \oplus \{\overline{10}\} \oplus \{27\} \\ &\{8\} \otimes \{10\} = \{8\} \oplus \{10\} \oplus \{27\} \oplus \{35\} \end{aligned}\right\} \tag{6.7.17}$$

(where the subscripts s and a denote symmetric '*d*-type' and anti-symmetric '*f*-type' {8} – {8}–{8} couplings respectively).

Now the SU(3) Clebsch–Gordan coefficients factorize into SU(2) Clebsch–Gordan coefficients and an iso-scalar factor in the form

$$\begin{pmatrix} \mu_1 & \mu_2 & \mu \\ \nu_1 & \nu_2 & \nu \end{pmatrix} = \langle I_1, I_2, I_{1z}, I_{2z} | I, I_z\rangle \begin{pmatrix} \mu_1 & \mu_2 & \mu \\ I_1 Y_1 & I_2 Y_2 & I Y \end{pmatrix} \tag{6.7.18}$$

These are tabulated in, for example, Particle Data Group (1974). Thus for an {8} vector meson, V, decaying into a pair of {8} pseudo-scalars, PS, we have, in the limit of exact SU(3) symmetry,

$$\left.\begin{aligned}\frac{1}{\sqrt{2}}A(\rho\to\pi\pi) &= -\frac{2}{\sqrt{2}}A(\rho\to K\bar{K}) = -\frac{2}{\sqrt{3}}A(K^*\to K\pi)\\ &= -\frac{2}{\sqrt{3}}A(K^*\to K\eta)\\ &= -\sqrt{\frac{2}{3}}\frac{1}{\cos\theta}A(\phi\to K\bar{K}) = -\sqrt{\frac{2}{3}}\frac{1}{\sin\theta}A(\omega\to K\bar{K})\\ &= A(V\to PS+PS)\end{aligned}\right\} \tag{6.7.19}$$

(where θ is the mixing angle of (5.2.17)). However, to test such relations it is essential to take account of the very different amounts of phase space available in the different decays because of the large mass splittings due to symmetry breaking. In particular $K^*\to K\eta$ and $\omega\to K\bar{K}$ are forbidden because the resonance mass is below the threshold of the decay channel. Within the considerable uncertainties as to how best to correct for this (see for example Gourdin (1967)) the relations seem to hold reasonably well.

But it is easier to test such relations for pole exchanges in scattering amplitudes. The SU(3) invariance of hadronic scattering implies that the amplitudes may depend on μ but not on ν (cf. (6.7.11)) and so for $1+2\to 3+4$ we have

$$\langle 34|\,A\,|12\rangle = \sum_{\mu,\nu}\begin{pmatrix}\mu_1 & \mu_2 & \mu\\ \nu_1 & \nu_2 & \nu\end{pmatrix}\begin{pmatrix}\mu_3 & \mu_4 & \mu\\ \nu_3 & \nu_4 & \nu\end{pmatrix}^* A(\mu) \tag{6.7.20}$$

Thus for example in processes of the type $M+B\to M'+B'$ where M, M′ and B, B′ are any members of the meson and baryon octets, respectively, there are just seven independent reduced amplitudes

$$A(1),\quad A(8_{ss})\quad A(8_{sa}),\quad A(8_{aa}),\quad A(10),\quad A(\overline{10}),\quad A(27) \tag{6.7.21}$$

from (6.7.17) ($A(8_{as}) = A(8_{sa})$ by time reversal invariance), and all the many processes of this class are related to just these seven amplitudes by the Clebsch–Gordan coefficients of (6.7.20) (analogously to (6.7.10)).

Of course the large mass splittings invalidate these relations at low energies, but at high energies, where the external particle masses become unimportant, we can expect such relations to hold provided that care is taken in dealing with the splitting of the trajectories which are exchanged – see section 6.8*i* below. If a decomposition similar

Table 6.4 *The octet crossing matrix* ($8 \otimes 8 \to 8 \otimes 8$) (*from de Swart* 1964)

	1	8_{ss}	8_{sa}	8_{as}	8_{aa}	10	$\overline{10}$	27
1	1/8	1	0	0	± 1	$\pm 5/4$	$\pm 5/4$	27/8
8_{ss}	1/8	$-3/10$	0	0	$\pm 1/2$	$\mp 1/2$	$\mp 1/2$	27/40
8_{sa}	0	0	$\pm 1/2$	1/2	0	$\sqrt{5}/4$	$-\sqrt{5}/4$	0
8_{as}	0	0	1/2	$\pm 1/2$	0	$\mp\sqrt{5}/4$	$\pm\sqrt{5}/4$	0
8_{aa}	$\pm 1/8$	$\pm 1/2$	0	0	1/2	0	0	$\mp 9/8$
10	$\pm 1/8$	$\mp 2/5$	$1/\sqrt{5}$	$\mp 1/\sqrt{5}$	0	1/4	1/4	$\mp 9/40$
$\overline{10}$	$\pm 1/8$	$\mp 2/5$	$-1/\sqrt{5}$	$\pm 1/\sqrt{5}$	0	1/4	1/4	$\mp 9/40$
27	1/8	1/5	0	0	$\mp 1/3$	$\mp 1/12$	$\mp 1/12$	7/40

The upper and lower signs refer to the s–t and s–u crossing matrices, respectively. We have changed the signs of the sa and as elements in the s–t crossing matrix to conform to the usual convention for the f-type coupling for a meson to baryon–antibaryon.

to (6.7.20) is made for the t-channel process $1+\overline{3}\to\overline{2}+4$ as well, the crossing relation may be written (cf. (6.7.12))

$$A(\mu_s) = \sum_{\mu_t} M(\mu_s, \mu_t)\, A(\mu_t) \tag{6.7.22}$$

where $M(\mu_s, \mu_t)$ is the SU(3) crossing matrix. A useful example of such a matrix is given in table 6.4. We shall make use of these results below.

6.8 Regge pole phenomenology

We have found that the Regge pole contribution to a t-channel helicity amplitude is given by (6.3.7), i.e.

$$A^{R}_{H_t}(s,t) = -16\pi(-1)^{\Lambda} K_{\lambda\lambda'}(t)\, \gamma_\lambda(t)\, \gamma_{\lambda'}(t) \times (e^{-i\pi(\alpha-v)} + \mathscr{S}) f_H(\alpha) \left(\frac{s-u}{2s_0}\right)^{\alpha-M} \xi_{\lambda\lambda'}(z_t) \tag{6.8.1}$$

Here $K_{\lambda\lambda'}(t)$ given in table 6.1 depends on whether or not there is a conspiracy, and $f_H(\alpha)$ depends on whether the trajectory chooses sense, nonsense etc., as discussed in section 6.3. Λ is defined in (B.10), and $\xi_{\lambda\lambda'}(z_t)$ in (B.11). Alternatively one can work with s-channel helicity amplitudes and use (6.4.9) instead. And since Regge poles have definite values of I, S etc., there should be SU(2) or SU(3) relations between their contributions to the various processes connected by these internal symmetries, as discussed in the previous section. This section contains a brief survey of how well these predictions compare with experiment. A bibliography of the large amount of detailed work on Regge predictions for individual process may be found in Collins and Gault (1975).

a. Regge behaviour

Equation (6.8.1) predicts that with a single Regge pole exchange all the helicity amplitudes for a process will have the asymptotic behaviour

$$A_H(s,t) \sim \left(\frac{s-u}{2s_0}\right)^{\alpha(t)} \sim \left(\frac{s}{s_0}\right)^{\alpha(t)} \tag{6.8.2}$$

for $s \to \infty$, t fixed, and so from (4.2.5) or (4.3.12)

$$\frac{\mathrm{d}\sigma}{\mathrm{d}t} \to F(t)\left(\frac{s}{s_0}\right)^{2\alpha(t)-2} \tag{6.8.3}$$

where $F(t)$ is some function of t, and from (4.2.6)

$$\sigma_{12}^{\text{tot}}(s) \sim \left(\frac{s}{s_0}\right)^{\alpha(0)-1} \tag{6.8.4}$$

so both the differential and total cross-sections should have simple power behaviours.

These expressions are valid to leading order in s/s_0 and corrections of order $(s/s_0)^{\alpha(t)-1}$ may be anticipated due to other terms in the expansion of $e_{\lambda\lambda'}^{-\alpha-1}(z_t)$, daughter trajectories, threshold corrections etc. So this prediction of Regge theory should hold for $s \gg s_0$, where s_0 is the scale factor which was introduced in (6.2.9). Obviously if s_0 were very large these predictions would be untestable. We cannot really deduce what s_0 should be (see however section 7.4 below) but empirically it seems to be about 1 GeV^2, consistent with the hadronic mass scale, and so Regge theory usually works quite well for $s > 10\,\text{GeV}^2$, or (from (1.7.30)) $p_{\text{L}} > 5\,\text{GeV}$ for a proton target, i.e. for all energies above the resonance region. Taking $s_0 = 1\,\text{GeV}^2$ has the advantage that it can be omitted from the equations, but if so its implicit occurrence should be kept in mind.

b. The Pomeron

The total cross-sections for various states are plotted in fig. 6.4 and it will be observed that though in several cases there is a fall at low energies, and a slow rise at high energies, taken over all they are remarkably constant over a large range of s. From (6.8.4), constancy

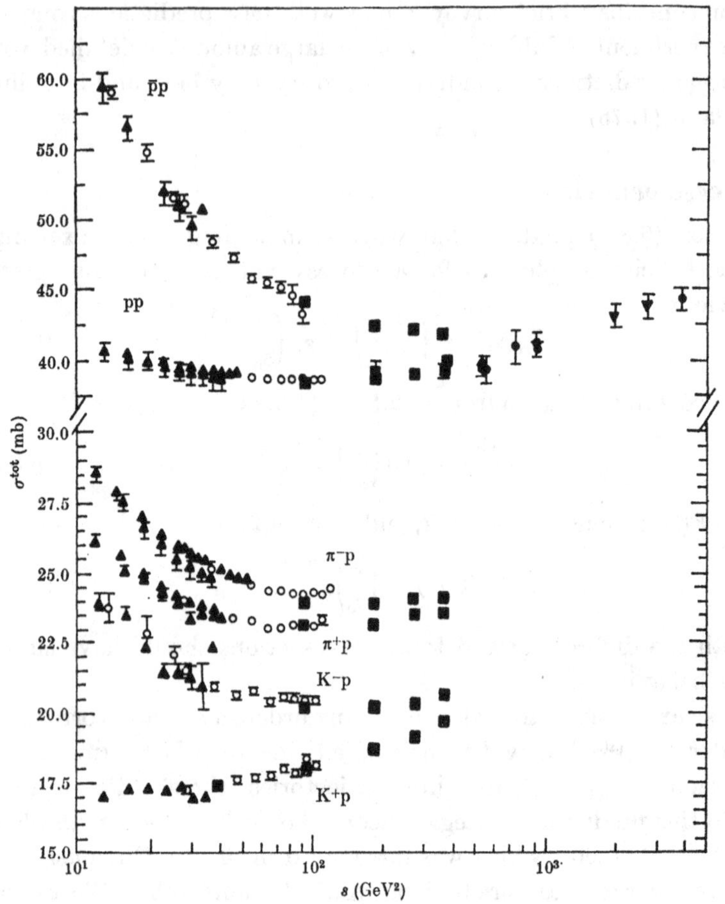

FIG. 6.4 The total cross-sections for various states as a function of s, from Barger (1974). (Note that the s scale is logarithmic.)

of $\sigma^{tot}(s)$ requires $\alpha(0) \approx 1$, but all the trajectories of figs. 5.4, 5.5 have $\alpha(0) \lesssim \frac{1}{2}$. In elastic scattering $1+2 \rightarrow 1+2$ the t channel consists of a particle and its anti-particle $(1+\bar{1} \rightarrow \bar{2}+2)$ and so the exchanged trajectories must obviously have the quantum numbers of the vacuum (i.e. $B = Q = S = I = 0$, $P = G = C_n = \mathscr{S} = +1$). The f meson has these quantum numbers, but, at least as drawn in fig. 5.5 (a), its trajectory is much too low at $t = 0$ to explain the behaviour of the total cross-sections.

This difficulty was realised rather early in the history of Regge phenomenology, and a new trajectory called the Pomeron (or Pomeranchon or Pomeranchukon by some authors), P, with $\alpha_P(0) \approx 1$ was invented (Chew and Frautschi 1961) to account for the asymptotic behaviour of the σ^{tot}'s. Since it has even signature there is no pole near $t = 0$ because $\alpha_P(0) = 1$ is a wrong-signature point. Even signature means that its contribution is symmetric under the interchange $z_t \leftrightarrow -z_t$, i.e. $s \leftrightarrow u$ at fixed t (see (2.5.3), (2.5.6)). Now the u-channel process is $\bar{1}+2 \to \bar{1}+2$, and so the P-exchange hypothesis demands that $\sigma_{12}^{tot}(s) \to \sigma_{\bar{1}2}^{tot}(s)$ as $s \to \infty$, and in fig. 6.4 we see that it is quite likely that $\sigma_{pp}^{tot} \to \sigma_{\bar{p}p}^{tot}$, $\sigma_{K^+p}^{tot} \to \sigma_{K^-p}^{tot}$, $\sigma_{\pi^+p}^{tot} \to \sigma_{\pi^-p}^{tot}$ as $s \to \infty$. Such an equality was predicted on more general grounds by Pomeranchuk (1958) which accounts for the name now given to this trajectory. (See Eden (1971) for a discussion of the status of Pomeranchuk's theorem.)

Of course $\alpha_P(0) = 1$ is the maximum value permitted by the Froissart bound (2.4.10), so to have a trajectory as high as this implies that the strong interaction is as strong as it can be under crossing – i.e. unitarity is 'saturated'. It is clearly rather unsatisfactory that we have been forced to invent a trajectory which does not seem to have any particles lying on it. However, we shall find below (fig. 6.6f) that its slope appears to be rather small, $\alpha'_P \approx 0.2\ \mathrm{GeV}^{-2}$, so that a particle at $\alpha(t) = 2$ would have a rather high mass ($m^2 \approx 5\ \mathrm{GeV}^2$). In any case the fact that the observed σ^{tot}'s are still rising at CERN-ISR energies (which would naïvely imply $\alpha_P(0) > 1$) and the complications of Pomeron cuts (see section 8.6) make one wonder if the Pomeron may not be a more complicated singularity than a pole.

The Pomeron can be exchanged not only in elastic scattering processes but also in so-called quasi-elastic processes $1+2 \to 3+4$ where 3 has the same internal quantum numbers as 1, and 4 has the same as 2 – for example $\pi N \to \pi N^*(\frac{1}{2})$ where $N^*(\frac{1}{2})$ is an $I = \frac{1}{2}$ baryon resonance – and so all such processes should have essentially constant high energy cross-sections. There are however, some empirical rules which restrict P-couplings.

In elastic scattering processes the P appears to couple only to the s-channel helicity-non-flip baryon vertex, and hence for example to A_{++}^s but not A_{+-}^s in $\pi N \to \pi N$ (see (4.3.10)). It is also found that in quasi-elastic processes such as $\gamma N \to \rho^0 N$, $\gamma N \to \omega N$, $\gamma N \to \phi N$, $\pi N \to \pi N^*(\frac{1}{2})$ and $NN \to NN^*(\frac{1}{2})$ there is at least approximate s-channel helicity conservation (i.e. $\mu_1 = \mu_3$, $\mu_2 = \mu_4$). It is of course rather odd that a t-channel exchange should have such simple s-channel helicity couplings. But the rule seems to be violated in $\pi N \to A_1 N$, $\pi N \to A_3 N$ and $KN \to QN$ (see for example Leith (1973) for a review).

Another empirical rule is the so-called Gribov–Morrison rule (Gribov 1967, Morrison 1967) that the Pomeron couples to a vertex, $1\bar{3}$, only if

$$(-1)^{\sigma_1-\sigma_3} = \eta_1\eta_{\bar{3}} \tag{6.8.5}$$

i.e. the change of spin at the vertex must be related to the change of intrinsic parity. For spinless particles ($\sigma_1 = \sigma_3 = 0$) this rule follows from parity conservation and (4.6.8), i.e. $P = \mathscr{S}\eta$. Since the Pomeron has $P = \mathscr{S} = \eta = +1$ the $1\bar{3}$ state must have

$$P = (+1) = \eta_1\eta_{\bar{3}}(-1)^l = \eta_1\eta_{\bar{3}}(-1)^J = \eta_1\eta_{\bar{3}}\mathscr{S} = \eta_1\eta_{\bar{3}} \tag{6.8.6}$$

However, for particles with spin, J is not necessarily equal to l, so there will always exist helicity states having the signature and parity of the Pomeron. But if (6.8.5) is to be violated there must be a change of helicity, and so, from (6.4.2), the Pomeron-exchange amplitudes will vanish in the forward direction.

In fact the rule often seems to apply for particles with spin (see for example Leith 1973). Thus in $\pi N \to \pi N^* \to \pi\pi N$, it is found that the N^*'s produced have $L_{2I,2S}$(I ≡ isospin, S ≡ spin) $= P_{11}, D_{13}, F_{15}$, with no sign of D_{15} which would violate (6.8.5). Similarly, while $\pi N \to A_1 N$, $KN \to QN$, $\gamma N \to \rho^0 N$ all seem to exhibit a Pomeron-like constant high energy cross-section, $\pi N \to A_2 N$, $KN \to K^*N$, $\gamma N \to BN$, which violate the rule, decrease with energy. However, the difficulty of making a clean separation of the resonances from background events, and the fact that secondary trajectories may produce a decrease of $\sigma(s)$ at low s anyway, make the rule hard to test decisively, and its status is still unclear.

c. *The leading trajectories*

If several trajectories can be exchanged in a given process then the trajectory with the highest $\text{Re}\{\alpha(t)\}$ will dominate asymptotically at any given t. How high in s one has to go before a single trajectory exchange gives a satisfactory approximation to the amplitude clearly depends on the separation of the trajectories, the relative strengths of their residues, and of course on s_0.

So for a given process all one has to do is work out the possible quantum numbers which can occur in the t channel, and look up the leading trajectory with those quantum numbers in figs. 5.4–5.6. Table 6.5 lists the leading trajectories for most of the experimentally accessible processes.

For processes where the t-channel quantum numbers are $B = S = 0$, if charge is exchanged, or if there is a change of isospin at one of the

Table 6.5 *Regge trajectory exchanges for various processes*

	Exchanged quantum numbers				
Exchanges	B	S	$(\mathrm{I})^G$	$\eta\ C_n$	Processes
π± beams					
ρ	0	0	$(1)^+$	+	$\pi^-p \rightarrow \pi^0n$
					$\pi^+p \rightarrow \pi^0\Delta^{++}$
A_2	0	0	$(1)^-$	+	$\pi^-p \rightarrow \eta n$
					$\pi^+p \rightarrow \eta\Delta^{++}$
ρ, B	0	0	$(1)^+$	±	$\pi^-p \rightarrow \omega n$
					$\pi^+p \rightarrow A_2^0\Delta^{++}$
					$\pi^+p \rightarrow \omega\Delta^{++}$
π, A_1	0	0	$(1)^-$	−	$\pi^-p \rightarrow \varepsilon n$
A_2, π, A_1	0	0	$(1)^-$	±	$\pi^-p \rightarrow \rho^0n$
					$\pi^-p \rightarrow fn$
					$\pi^+p \rightarrow \rho^0\Delta^{++}$
					$\pi^+p \rightarrow f\Delta^{++}$
ρ, f	0	0	$(0, 1)^+$	+	$\pi p \rightarrow \pi p$
					$\pi p \rightarrow \pi N^*$
ρ, B, f, n, D	0	0	$(0, 1)^+$	±	$\pi p \rightarrow A_2p$
					$\pi p \rightarrow A_1p$
A_2, π, A_1, ω, H	0	0	$(0, 1)^-$	±	$\pi p \rightarrow \rho p$
					$\pi p \rightarrow Bp$
					$\pi p \rightarrow gp$
K*, K**	0	1	$(\frac{1}{2})$	+	$\pi - p \rightarrow K^0\Lambda$
					$\pi p \rightarrow K\Sigma$
K*, K**, K, Q	0	1	$(\frac{1}{2})$	±	$\pi p \rightarrow K^*\Lambda$
					$\pi p \rightarrow K^*\Sigma$
N	1	0	$(\frac{1}{2})$	±	$\pi^-p \rightarrow n\eta$
Δ	1	0	$(\frac{3}{2})$	±	$\pi^-p \rightarrow p\pi^-$
					$\pi^-p \rightarrow p\rho^-$
N, Δ	1	0	$(\frac{1}{2}, \frac{3}{2})$	±	$\pi p \rightarrow \pi p$
					$\pi p \rightarrow p\rho$
					$\pi p \rightarrow \Delta\pi$
Σ	1	−1	(1)	±	$\pi - p \rightarrow \Lambda K^0$
Exotic					$\pi^-p \rightarrow K^+\Sigma^-$
K± beams					
ρ, A_2	0	0	(1)	+	$K^-p \rightarrow \overline{K}^0n$
					$Kp \rightarrow K\Delta$
ρ, A_2, B, π, A_1	0	0	(1)	±	$K^-p \rightarrow K^{*0}n$
					$K^-p \rightarrow K^{**0}n$
					$Kp \rightarrow K^*\Delta$
ρ, A_2, f, ω	0	0	(0, 1)	+	$Kp \rightarrow Kp$
ρ, A_2, B, π, A_1, f, ω, η, H, D	0	0	(0, 1)	±	$Kp \rightarrow K^*p$
					$Kp \rightarrow K^{**}p$
					$Kp \rightarrow Qp$
K*, K**	0	1	$(\frac{1}{2})$	+	$Kp \rightarrow \pi\Lambda$
					$Kp \rightarrow \pi\Sigma$
					$K^-p \rightarrow \eta\Lambda$
					$Kp \rightarrow \eta\Sigma$
					$K^-p \rightarrow \eta'\Lambda$
					$K^-p \rightarrow \pi^-\Sigma^{*+}$

Table 6.5 (*cont.*)

Exchanges	Exchanged quantum numbers B	S	$(I)^G$	η	C_n	Processes
K^*, K^{**}, K, Q	0	1	$(\frac{1}{2})$	+		$Kp \to \rho\Lambda$
						$K^-p \to \omega\Lambda$
						$K^-p \to \phi\Lambda$
						$Kp \to \rho\Sigma$
						$K^-p \to \omega\Sigma^0$
						$K^-p \to \phi\Sigma^0$
N	1	0	$(\frac{1}{2})$	±		$Kp \to \Lambda\pi$
						$K^-p \to \Lambda n$
N, Δ	1	0	$(\frac{1}{2}, \frac{3}{2})$	±		$K^-n \to \Sigma^0\pi^-$
Λ, Σ	1	−1	(0, 1)	±		$Kp \to pK$
Exotic						$Kp \to K\Xi$
p *beam*						
ρ, A_2, B, π, A_1	0	0	(1)	±		$pn \to np$
						$pp \to p\Delta$
						$pp \to \Delta\Delta$
$\rho, A_2, B, \pi, A_1, f, \omega, \eta, H, d$	0	0	(0, 1)	±		$pp \to pp$
N, Δ	1	0	$(\frac{1}{2}, \frac{3}{2})$	±		$pp \to \pi D$
						$pp \to \rho D$
$\bar{p}$ *beam*						
ρ, A_2, B, π, A_1	0	0	(1)	±		$\bar{p}p \to \bar{n}n$
						$\bar{p}n \to \bar{\Delta}^{++}p$
						$pp \to \bar{\Delta}\Delta$
$\rho, A_2, B, \pi, A_1, f, \omega, \eta, H, D$	0	0	(0, 1)	±		$\bar{p}p \to \bar{p}p$
K^*, K^{**}, K, Q	0	1	$(\frac{1}{2})$	±		$\bar{p}p \to \bar{\Lambda}\Lambda$
						$\bar{p}p \to \bar{\Lambda}\Sigma^0$
						$\bar{p}p \to \bar{\Sigma}\Sigma$
N, Δ	1	0	$(\frac{1}{2}, \frac{3}{2})$	±		$\bar{p}p \to \pi^-\pi^+$
Λ, Σ	1	−1	(0, 1)	±		$\bar{p}p \to K^+K^-$
Exotic						$\bar{p}p \to \bar{\Sigma}^-\Sigma^-$
Λ *beam*						
f, ω, η, H, D	0	0	(0)	±		$\Lambda p \to \Lambda p$
γ *beam*						
ρ, A_2, B, π, A_1	0	0	(1)	±		$\gamma p \to \pi^+n$
						$\gamma p \to A_1^+n$
						$\gamma p \to \pi^-\Delta^{++}$
ρ, B, ω, H	0	0	(0, 1)	±	−	$\gamma p \to \pi^0p$
						$\gamma p \to \eta p$
$A_2, \pi, A_1, f, \eta, D$	0	0	(0, 1)	±	+	$\gamma p \to \rho^0p$
						$\gamma p \to \omega p$
						$\gamma p \to \phi p$
$A_2, \pi, A_1, f, \eta, D$	0	0	(0,1)	±	+	$\gamma p \to \gamma p$
K^*, K^{**}, K, Q	0	1	$(\frac{1}{2})$	±		$\gamma p \to K^+\Lambda$
						$\gamma p \to K^{*+}\Lambda$
N	1	0	$(\frac{1}{2})$	±		$\gamma p \to \Delta\eta$
N, Δ	1	0	$(\frac{1}{2}, \frac{3}{2})$	±		$\gamma p \to n\pi^+$
						$\gamma p \to p\pi^0$
						$\gamma p \to \Delta^{++}\pi^-$
K^0 *beam*						
ρ, ω	0	0	(0, 1)	+	−	$K_L^0p \to K_S^0p$

vertices (such as $N \to \Delta$) then $I_t = 1$ only. But if there is no exchange of charge, or change of isospin at a vertex, then $I_t = 0$ or 1. If the process has been initiated by a pion beam ($G_\pi = -1$) then the t channel will have a definite G-parity (± 1) depending on the G-parity of the final-state meson. But with K, γ or baryon beams (on a baryon target) G-parity will not be a good t-channel quantum number, and so is not restricted. If the initial state contains a pseudo-scalar particle (π or K), and the final state a pseudo-scalar, then the t channel can only contain normal parity exchanges, $\eta = +1$ (see (6.8.6)). Or more rarely if the final state contains a scalar such as ϵ then we must have abnormal parity, $\eta = -1$, exchanges. But for other spin combinations the normality is not restricted. With the neutral γ or K_L^0 beams the G-parity is not restricted, and $I_t = 0$ or 1, but if the final state contains a neutral meson then the t channel has a definite value of $C_n (= \pm 1)$. Otherwise, C_n is not restricted.

With S and/or $B \neq 0$ exchanges, G and C_n are not restricted, so the rules are much easier to apply.

The simplest set of processes are meson–baryon charge–exchange scattering such as $\pi^- p \to \pi^0 n$ Since the t-channel $\pi^- \pi^0 \to \bar{p} n$ has charge, only $I = 1$ non-strange mesons can contribute, and the π–π vertex is restricted to even G-parity and normal parity. Only the ρ satisfies all these requirements. Similar remarks apply to $\pi^- p \to \eta n$ except that η has even G-parity and so only A_2 can be exchanged. However, in most processes the exchanges are not so simple. Thus in $K^- p \to \bar{K}^0 n$ the K mesons are not eigenstates of G-parity so both ρ and A_2 can be exchanged, and if the mesons have non-zero spin, as in $\pi^- p \to \rho^0 n$, the normality is not restricted so π exchange is allowed as well as A_2.

Bearing the above rules in mind the reader should have no difficulty in checking table 6.5. However, these are only the leading trajectories with the given quantum numbers, and secondary or daughter ρ′, A_2' may also occur, as well as Regge cuts. (For f read f and P.)

The appearance of a Regge pole in the t (or u channel) should result in a peak of the differential cross-section near the forward (or backward) direction. An example shown in fig. 6.5 is the data for K^+p elastic scattering. Near the forward direction we see the effect of the t-channel poles, P, f, ω, ρ and A_2, while the u channel of K^+p has the quantum numbers of the Λ and Σ baryons and so there is a smaller backward peak. However, the u channel of $K^- p \to K^- p$ is $K^+ p \to K^+ p$, which has exotic quantum numbers, and so there are no Reggeons

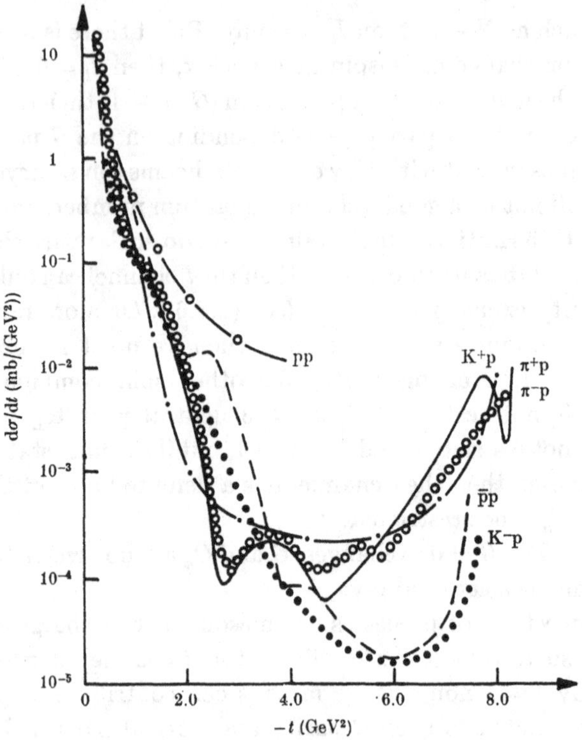

FIG. 6.5 The differential cross-sections for various elastic scattering processes at 5 GeV/C.

which can be exchanged (unless the conjectured Z particle exists – see Particle Data Group (1974)), and as expected the backward peak is strongly suppressed.

This sort of correlation between the occurrence of forward or backward peaks of $d\sigma/dt$, and the presence of non-exotic quantum numbers (and hence known trajectories) in the crossed channel, provides an excellent confirmation that particle exchange is the mediator of the strong interaction.

d. The effective trajectory

From (6.8.3)

$$\log\left(\frac{d\sigma}{dt}\right) = (2\alpha(t)-2)\log\left(\frac{s}{s_0}\right) + \log(F(t)) \tag{6.8.7}$$

and so by plotting $\log(d\sigma/dt)$ for a given process against $\log s$, at fixed t, we can determine the 'effective trajectory' for that process. At sufficiently high energy this effective trajectory should correspond to the leading trajectory for the process (apart from any complications

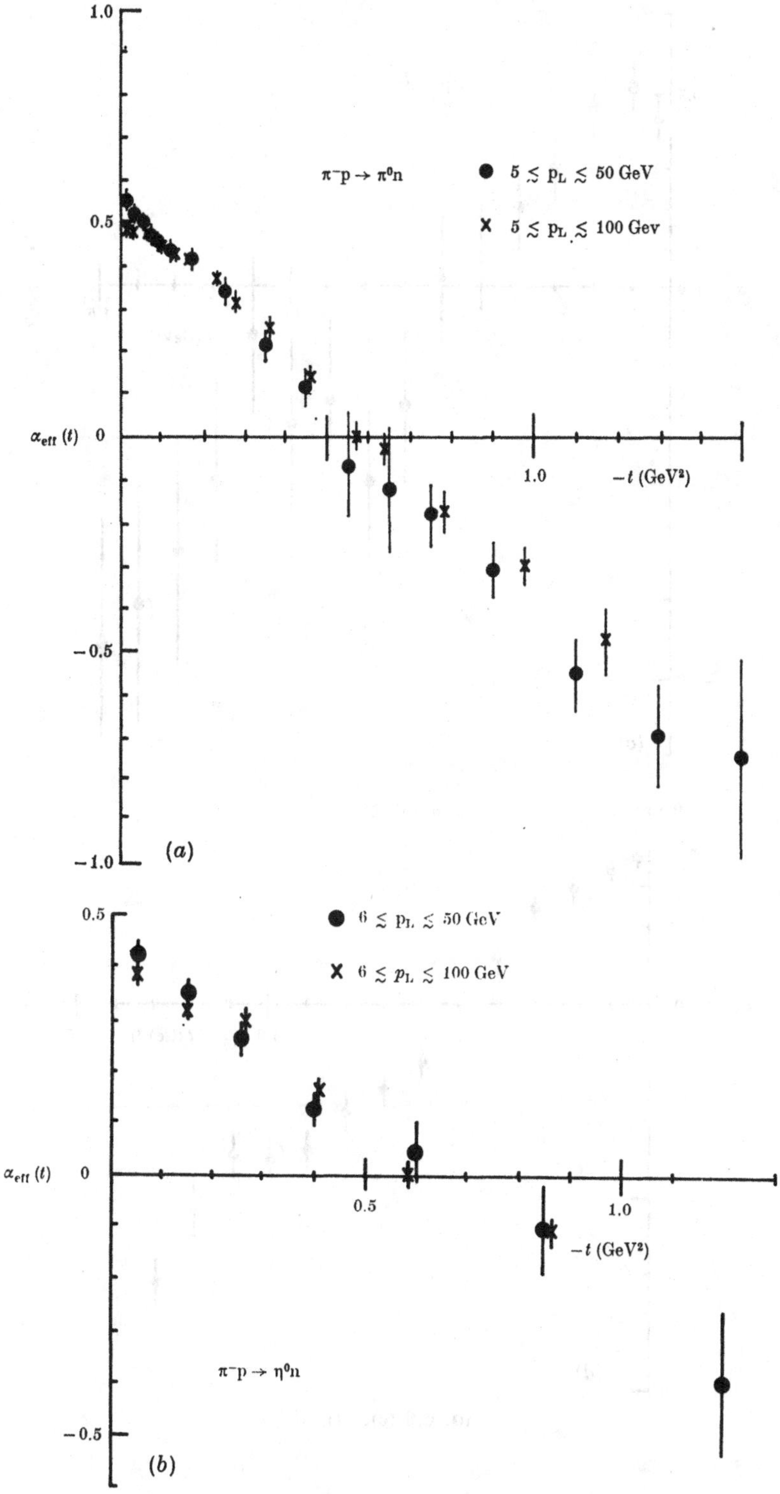
1.0
0.5
0
−0.5
−1.0
$\alpha_{eff}(t)$
$\pi^- p \to \pi^0 n$
● $5 \lesssim p_L \lesssim 50$ GeV
× $5 \lesssim p_L \lesssim 100$ Gev
1.0
$-t$ (GeV²)
(a)
0.5
0
−0.5
$\alpha_{eff}(t)$
● $6 \lesssim p_L \lesssim 50$ GeV
× $6 \lesssim p_L \lesssim 100$ GeV
0.5
1.0
$-t$ (GeV²)
$\pi^- p \to \eta^0 n$
(b)

FIG. 6.6 (*a*), (*b*)

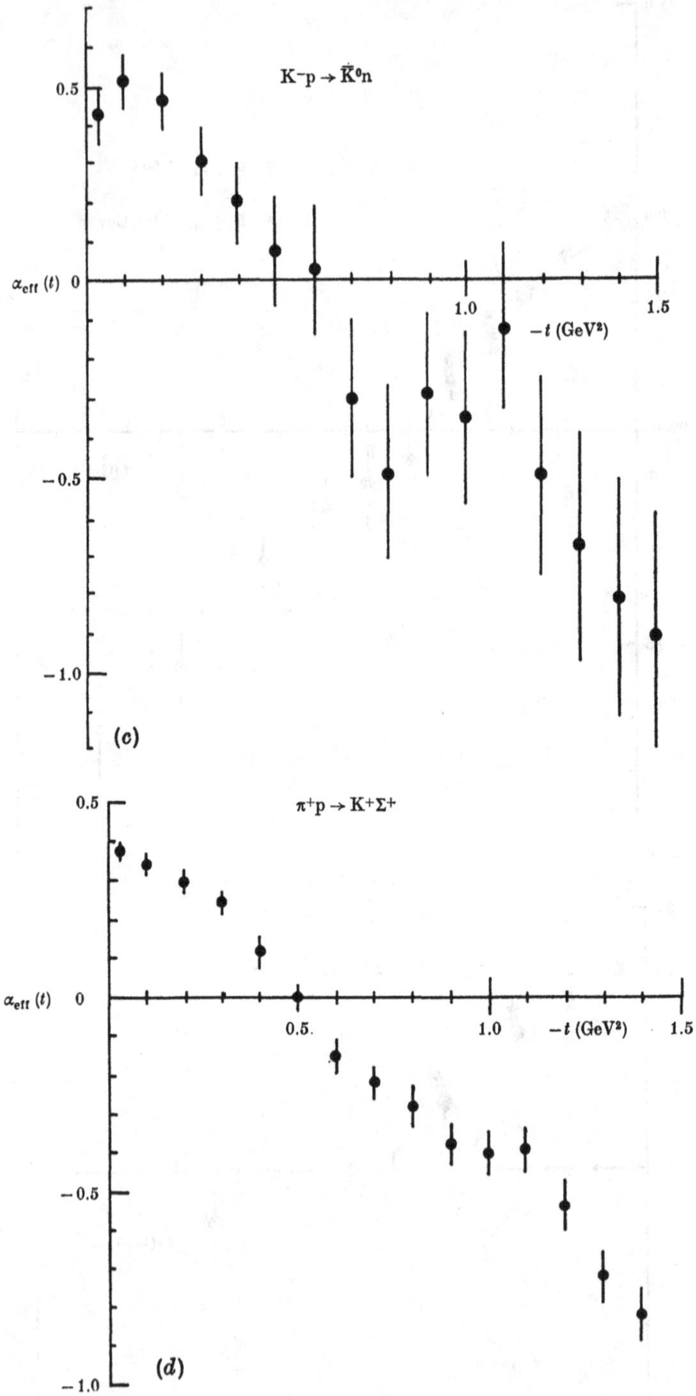

K⁻p → K̄⁰n
0.5
$\alpha_{eff}(t)$ 0
−0.5
−1.0
1.0
1.5
−t (GeV²)
(c)
π⁺p → K⁺Σ⁺
0.5
$\alpha_{eff}(t)$ 0
0.5
1.0
−t (GeV²)
1.5
−0.5
−1.0
(d)

FIG. 6.6 (c), (d)

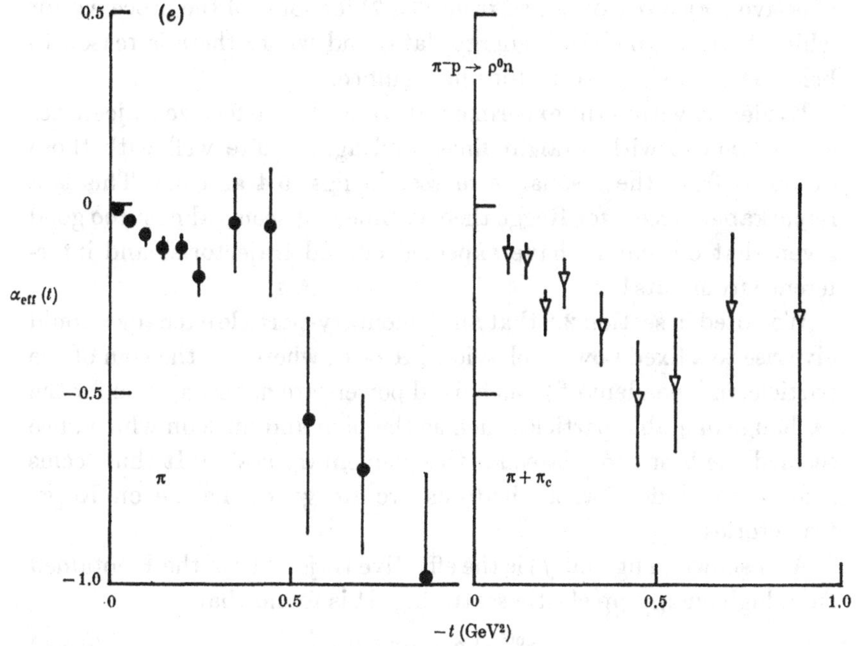

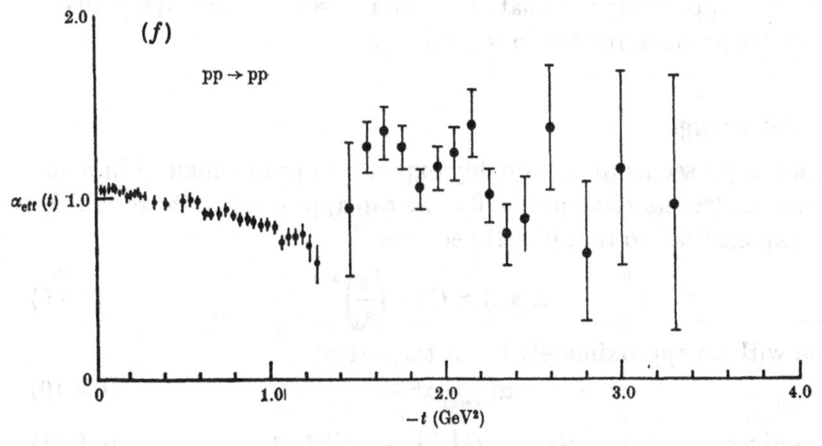

FIG. 6.6 (*a*)–(*f*) The effective trajectories for a variety of processes obtained using (6.8.7). The trajectories are: (*a*) ρ exchange, (*b*) A_2 exchange, (*c*) $\rho + A_2$ exchange, (*d*) $K^* + K^{**}$ exchange, (*e*) π exchange, (*f*) P exchange.

due to Regge cuts etc., see chapter 8 below). In fig. 6.6 we show the effective trajectory obtained from (6.8.7) for some of the processes for which there is good high energy data, and where there is reason to believe that a single trajectory may suffice.

Evidently within the experimental errors these effective trajectories are consistent with straight lines, and agree quite well with those obtained from the resonance masses in figs. 5.4 and 5.5. This is a remarkable success for Regge theory. Indeed it seems almost too good given that one might have expected curved trajectories and interference from cuts!

We noted in section 2.8 that an elementary-particle exchange would give rise to a fixed power behaviour, $A \sim s^{\sigma}$, where σ is the spin of the particle, independent of t. Such fixed powers are not seen, even for the exchange of stable particles such as the pion and nucleon which once seemed the best candidates for this elementary status. It thus seems safe to conclude that all hadrons are Reggeons, i.e. lie on Regge trajectories.

Also shown in fig. 6.6 (f) is the effective trajectory of the P obtained from high energy pp elastic scattering. It is found that

$$\alpha_{\mathrm{P}}^{\mathrm{eff}}(t) \simeq 1.08 + 0.2t \tag{6.8.8}$$

for $|t| < 1.4\,\mathrm{GeV}^2$, i.e. the trajectory has a small slope but an intercept above 1, apparently in violation of the Froissart bound. We shall discuss this problem further in section 8.7a.

e. Shrinkage

Since $\mathrm{d}\sigma/\mathrm{d}t$ seems to fall roughly exponentially for small $|t|$ in many processes (see for example fig. 6.5) we can approximate the residue by an exponential, so that (6.8.1) becomes

$$A(s,t) \approx G\,\mathrm{e}^{at}\left(\frac{s}{s_0}\right)^{\alpha(t)} \tag{6.8.9}$$

and with an approximately linear trajectory

$$\alpha(t) = \alpha^0 + \alpha' t \tag{6.8.10}$$

this gives

$$A(s,t) \approx G\left(\frac{s}{s_0}\right)^{\alpha^0} \mathrm{e}^{(a+\alpha' \log (s/s_0))t} \tag{6.8.11}$$

So if we define the 'width' of forward peak in t by

$$\Delta t \equiv \left(\frac{\mathrm{d}}{\mathrm{d}t}\left(\frac{\mathrm{d}\sigma}{\mathrm{d}t}\right)\bigg/\frac{\mathrm{d}\sigma}{\mathrm{d}t}\right)^{-1}$$

we find

$$\Delta t = [2(a+\alpha' \log s/s_0)]^{-1} \tag{6.8.12}$$

(from (6.8.11) in (4.3.12)). So Δt decreases as $\log s$ increases, i.e. Regge theory predicts that the width of the forward peak will 'shrink' as s increases. This effect can be detected by a close examination of fig. 6.1 in which the low energy data has a somewhat broader peak than the high energy (at small t). It is this shrinkage which produces the slopes of the effective trajectories in figs. 6.6.

From our discussion in section 2.4, one can interpret this shrinkage as an increase in the effective size of the target, but as the cross-section does not increase the target is evidently becoming more 'transparent' as the energy increases. Though these predictions of Regge theory once seemed rather surprising from an 'optical' point of view they are now well verified in a great variety of processes.

With $\alpha_P(0) = 1$ we obtain for the elastic differential cross-section from (6.8.11) and (1.8.16)

$$\left(\frac{d\sigma}{dt}\right)^{el} \to \frac{1}{16\pi} G_P^2 e^{2[a+\alpha_P' \log(s/s_0) t]}$$

and so
$$\sigma_{12}^{el}(s) = \int_{-\infty}^{0} \left(\frac{d\sigma}{dt}\right) dt \to \frac{1}{16\pi} \frac{G_P^2}{2[a+\alpha_P' \log(s/s_0) t]} \tag{6.8.13}$$

while from (1.9.6) $\sigma_{12}^{tot}(s) \to G_P$, and hence $\sigma_{12}^{el}/\sigma_{12}^{tot} \sim (\log s)^{-1}$. So because of the shrinkage the elastic cross-section becomes a decreasing fraction of the total cross-section as $\log s \to \infty$.

f. The phase–energy relation

As the trajectory and residue functions are expected to be real below threshold (except where trajectories collide – see section 3.2) the phase of the Regge pole amplitude (6.8.1) is given entirely by the signature factor $(e^{-i\pi(\alpha(t)-v)} + \mathscr{S})$ and so the phase angle, ϕ, is related to the energy dependence $\alpha(s)$ by

$$\cot\phi \equiv \frac{\mathrm{Re}\{A\}}{\mathrm{Im}\{A\}} \equiv \rho = -\frac{\cos\pi(\alpha(t)-v)+\mathscr{S}}{\sin\pi(\alpha(t)-v)} \tag{6.8.14}$$

It is often convenient to rewrite the signature factor as (for $v = 0$)

$$e^{-i\pi\alpha} + \mathscr{S} = e^{-i\pi\alpha/2}(e^{-i\pi\alpha/2} + \mathscr{S} e^{i\pi\alpha/2})$$
$$= e^{-i\pi\alpha/2}\, 2\cos\left(\frac{\pi\alpha}{2}\right) \quad \text{for} \quad \mathscr{S} = +1$$
$$= -i\, e^{-i\pi\alpha/2}\, 2\sin\left(\frac{\pi\alpha}{2}\right) \quad \text{for} \quad \mathscr{S} = -1 \tag{6.8.15}$$

which exhibits this phase directly.

It is possible to determine the phase of helicity-non-flip elastic scattering amplitudes at $t = 0$, either by measuring

$$\sigma_{12}^{\text{tot}}(s) \propto \text{Im}\{A_{12}^{\text{el}}(s, 0)\}$$

and
$$\text{d}\sigma/\text{d}t\,(12 \to 12) \propto \text{Re}\{A_{12}^{\text{el}}\}^2 + \text{Im}\{A_{12}^{\text{el}}\}^2$$

(but $\text{d}\sigma/\text{d}t$ has to be extrapolated to $t = 0$ from the finite negative values at which it can be measured), or by observing the interference between the hadronic scattering amplitude and the known Coulomb scattering amplitude (see for example Eden (1967)). In fig. 6.7 we show the data on the ratio ρ at $t = 0$ for pp elastic scattering compared with the predictions of a Regge pole fit (Collins, Gault and Martin 1974) to $\sigma_{\text{tot}}(\text{pp})$ and $\sigma_{\text{tot}}(\bar{\text{p}}\text{p})$ using just the dominant P, f and ω trajectories (with $\alpha_{\text{P}}(0) > 1$) and evidently the agreement is quite good.

However, this is not really a test of Regge theory so much as of the power behaviour of $\text{Im}\{A_{12}^{\text{el}}\}$ and dispersion relations. Thus, for example, if we write a once-subtracted dispersion relation for the amplitude for s above threshold (from (1.10.7))

$$\text{Re}\{A(s,t)\} = \frac{s}{\pi} P \int_{s_{\text{T}}}^{\infty} \frac{\text{Im}\{A(s',t)\}}{(s'-s)\,s'} \text{d}s' + \frac{s}{\pi} \int_{u_{\text{T}}}^{-\infty} \frac{\text{Im}\{A(s',t)\}}{(s'-s)\,s'} \text{d}s' \tag{6.8.16}$$

(where $P \equiv$ principal value) and if $\text{Im}\{A(s,t)\} \underset{s \to \infty}{\sim} s^{\alpha(t)}$ and $\underset{s \to -\infty}{\sim} (-s)^{\alpha(t)}$ then since (Erdelyi *et al.* 1953)

$$\left.\begin{aligned} \frac{P}{\pi} \int_0^{\infty} \frac{\text{d}s'}{s'-s} s'^{\alpha-1} &= -s^{\alpha-1} \cot(\pi\alpha) \\ \frac{1}{\pi} \int_0^{\infty} \frac{\text{d}s'}{s'+s} s'^{\alpha-1} &= -s^{\alpha-1} \operatorname{cosec}(\pi(\alpha-1)) \end{aligned}\right\} \tag{6.8.17}$$

(6.8.16) gives for $s \to \infty$

$$\frac{\text{Re}\{A(s,t)\}}{\text{Im}\{A(s,t)\}} \sim -(\cot(\pi\alpha) + \mathscr{S} \operatorname{cosec}(\pi\alpha)) \tag{6.8.18}$$

in agreement with (6.8.14) (for $v = 0$). This result holds for any number of subtractions. It is clear from (6.8.17) that where $\alpha_{\text{eff}} < 1$ we can expect $\rho < 0$, but where the cross-section rises, so $\alpha_{\text{eff}} > 1$, ρ should become positive which is indeed the case in fig. 6.7.

In general the absolute phases of scattering amplitudes cannot be determined experimentally, but the relative phases of different

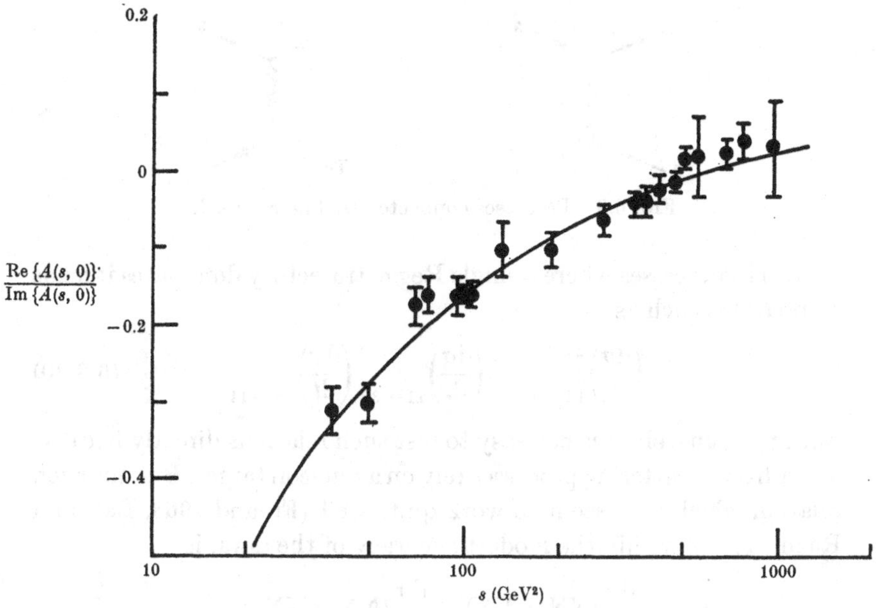

FIG. 6.7 Data on Re $\{A(s, 0)\}/$Im $\{A(s, 0)\}$ for pp scattering compared with the Regge pole fit of Collins *et al.* (1974).

amplitudes can be obtained. For example in $\pi^-p \to \pi^0 n$ the polarization is given by (4.2.22), and so depends on the phase difference between the helicity-flip and non-flip amplitudes. A single ρ pole gives the same phase (6.8.14) (with $v = 0$, $\mathcal{S} = -1$) to both amplitudes and so ρ exchange predicts that the polarization will vanish. In fact it is observed to be small but not zero ($\approx$ 10–20 per cent) at low energies ($<$ 10 GeV) indicating the need for some other contribution in addition to the ρ pole, perhaps a cut or a secondary ρ' trajectory.

We shall discuss further examples of Regge phase predictions below.

g. Factorization and line reversal

The disconnectedness of the S-matrix leads us to expect that Regge pole residues will factorize into a contribution to each vertex (see (4.7.15)) so that for a t-channel Regge pole (fig. 6.8)

$$\beta^{\text{R}}_{12\to 34}(t) = \beta^{\text{R}}_{13}(t)\,\beta^{\text{R}}_{24}(t) \tag{6.8.19}$$

We have found in sections 6.2 and 6.3 that this relation puts important constraints on the residues of helicity amplitudes, and it is built into (6.8.1).

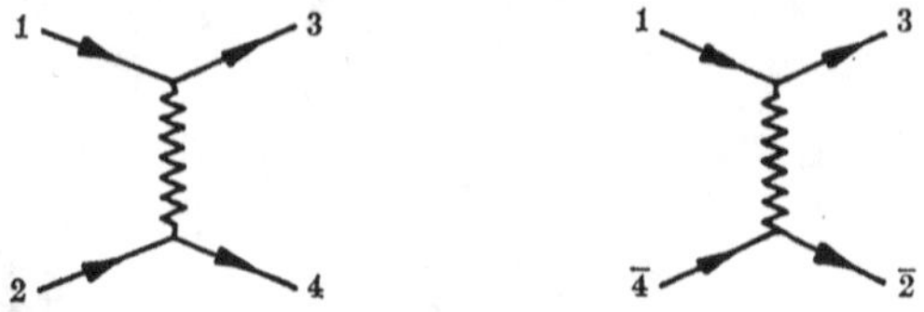

FIG. 6.8 Processes connected by line reversal.

Also in processes where a single Regge trajectory dominates it leads to relations such as

$$\left(\frac{d\sigma}{dt}\right)^2_{12\to 34} = \left(\frac{d\sigma}{dt}\right)_{11\to 33}\left(\frac{d\sigma}{dt}\right)_{22\to 44} \tag{6.8.20}$$

but unfortunately it is not easy to test such relations directly because all hadronic scattering processes rely on a nucleon target. But one such relation which does seem to work quite well (Freund 1968, Bari and Razmi 1970), within the modest accuracy of the data, is

$$\frac{\frac{d\sigma}{dt}(NN\to NN)}{\frac{d\sigma}{dt}(\pi N\to \pi N)} = \frac{\frac{d\sigma}{dt}(NN\to NN^*)}{\frac{d\sigma}{dt}(\pi N\to \pi N^*)}$$

where N^* is any $I = \frac{1}{2}$ baryon resonance so that P can be exchanged. The best direct tests of factorization can be made in inclusive reactions (chapter 10) where a greater variety of vertices is available.

Another important consequence of factorization is line reversal symmetry. Clearly if one end of the exchange diagram for $1+2\to 3+4$ is rotated as in fig. 6.8 then $s\leftrightarrow u$ and the process $1+\overline{4}\to 3+\overline{2}$ is obtained, which will thus have exactly the same Regge pole exchanges, with the same couplings, except that the sign will be changed for negative signature exchanges which are odd under $s\leftrightarrow u$ (see section 2.5).

For example the processes $K^-p\to\pi^-\Sigma^+$ and $\pi^+p\to K^+\Sigma^+$ differ only by the rotation of the K–π vertex. The only possible pole exchanges are (see table 6.5) the natural-parity strange mesons K*(890) and K**(1400), of which the first has spin = 1 and hence odd signature, while the second has spin = 2 and even signature. So the Regge exchanges for these processes can be written as $K^{**}\pm K^*$ respectively. Of course only the relative signs of the contributions are determined in this way, and the individual terms have phases given by (6.8.14). Similarly the elastic scattering process $\pi^-p\to\pi^-p$ differs from $\pi^+p\to\pi^+p$ only by line reversal, so these processes have $P+f\pm\rho$ exchanges, respectively, with the same couplings. The equality of these processes as $s\to\infty$ and P dominates is just the Pomeranchuk theorem of section 6.8*b*.

There are, however, some serious failures of factorization. For example, the zero of the ρ-exchange amplitude $A_{+-}(\pi^-p\to\pi^0n)$ at $|t|\simeq 0.55\,\text{GeV}^2$, which, as we discussed in section 6.3, could be due to a nonsense factor, should also appear in $\gamma p\to\eta p$ which is similarly dominated by ρ exchange. But there is no sign of a zero in the latter process, which makes one feel that the $\pi^-p\to\pi^0n$ dip may not be a property of the ρ pole alone, but could involve cuts as well (see section 8.7*c*). Cuts do not generally have this factorization property, so the success of factorization gives some indication of the extent to which poles dominate. But of course sums of poles do not factorize either, so it is essential to isolate a single Regge exchange in making such tests.

h. *Exchange degeneracy*

We remarked in section 5.3 that trajectories often occur in approximately exchange-degenerate pairs, so that for example the ρ and A_2 trajectories of fig. 5.4 and fig. 6.6 look rather like a single $B=S=0$, $I=1$, $\eta=+1$ trajectory, with particles having $P=(-1)^J$, $C_n=(-1)^J$, $G=(-1)^{J+1}$ at every positive integer value of J. This so-called 'weak exchange degeneracy' seems to hold quite well for the leading meson exchanges (excluding the Pomeron) and for strange baryons, though it is less good for non-strange baryons. From (4.5.7) it is evident that if amplitudes of both signature contain the same trajectory then the position of the trajectory does not depend on the u-channel (or 'exchange force') discontinuity.

If the u-channel forces do not contribute to the residues of the trajectories either then they will have degenerate residues too. This is called 'strong exchange degeneracy'. The absence of the u-channel contribution seems rather surprising, but we shall find in the next chapter theoretical reasons why this may happen. In this case the trajectories must 'choose nonsense', i.e. decouple from all amplitudes at nonsense points. This may be seen by considering for example the A_2 and f trajectories which need ghost-killing factors (see section 6.3) in all their residues at $\alpha=0$ to avoid negative m^2 particles. And if they are exchange degenerate with the ρ and ω trajectories, respectively, the latter will have zeros in their residues too, even though for them

$\alpha = 0$ is a wrong signature-point, and so they choose nonsense (see table 6.2).

This strong exchange degeneracy has the rather important consequence that if a given process is controlled by the sum of two such degenerate trajectories the amplitudes will be proportional to

$$\beta_H[(e^{-i\pi(\alpha-v)}+\mathscr{S})+(e^{-i\pi(\alpha-v)}-\mathscr{S})] = 2\beta_H e^{-i\pi(\alpha-v)} \qquad (6.8.21)$$

while if the process depends on the difference we get

$$\beta_H[(e^{-i\pi(\alpha-v)}+\mathscr{S})-(e^{-i\pi(\alpha-v)}-\mathscr{S})] = 2\beta_H \qquad (6.8.22)$$

The magnitudes in (6.8.21) and (6.8.22) are the same, but the latter is purely real, while the former has a phase which 'rotates' as $\alpha(t)$ changes.

This relation should obtain for pairs of trajectories connected by line reversal. Thus for example ($K^+n \to K^0p$, $K^-p \to \overline{K}^0n$) are controlled by $A_2 \pm \rho$, respectively, as are ($K^+p \to K^0\Delta^{++}$, $K^-n \to K^0\Delta^-$), while ($K^-p \to \pi^0\Lambda$, $\pi^-p \to K^0\Lambda$) and ($K^-p \to \pi^-\Sigma^+$, $\pi^+p \to K^+\Sigma^+$) are given by $K^{**} \mp K^*$. So if strong exchange degeneracy holds we expect in each case that the first reaction of the pair will have real phase, and the second rotating phase, and that their magnitudes will be identical. The first pair seem to achieve equality above about 5 GeV, but the situation is less clear for the others (see for example Irving, Martin and Michael (1971)), partly because of uncertainties in the normalization of the data. But these relations are not expected to be exact because there must be other contributions besides these leading trajectories to explain the non-zero polarization which is observed. According to (6.8.21) and (6.8.22) all the helicity amplitudes for a given process would have the same phase, giving zero polarization.

i. Internal symmetry relations

Since we assume that the isospin SU(2) invariance of strong interactions is exact there are a large number of relations between amplitudes involving different charge states. Thus for a process such as $\pi N \to \pi\Delta$ all the different charge combinations such as $\pi^+p \to \pi^+\Delta^+$, $\pi^+p \to \pi^0\Delta^{++}$, $\pi^-p \to \pi^0\Delta^0$, etc., share the same $I_t = 1$ ρ-exchange amplitude and are equal apart from Clebsch–Gordan coefficients. It is thus useful to analyse them all together, which is why the charges are not specified in many cases in table 6.5.

Also from (6.7.9) and (6.7.10) we find

$$\langle \pi^0 n | A | \pi^- p \rangle = \frac{1}{\sqrt{2}} (\langle \pi^+ p | A | \pi^+ p \rangle - \langle \pi^- p | A | \pi^- p \rangle) \quad (6.8.23)$$

which means that the ρ exchange, which dominates the charge-exchange process, should also, via the optical theorem (4.2.6), give the energy dependence of the difference of the total cross-sections of fig. 6.4, i.e.

$$\Delta\sigma^{\text{tot}}(\pi p) \equiv \sigma^{\text{tot}}_{\pi^- p} - \sigma^{\text{tot}}_{\pi^+ p} \sim \left(\frac{s}{s_0}\right)^{\alpha_\rho(0)-1} \quad (6.8.24)$$

which is quite well satisfied, and gives a value for $\alpha\rho(0)$ which is consistent with fig. 6.6. Similar relations, such as

$$\langle K^0 n | A | K^- p \rangle = \langle K^- p | A | K^- p \rangle - \langle K^- n | A | K^- n \rangle \quad (6.8.25)$$

can be deduced for many processes, which means that before trying to fit the elastic scattering data it is useful to obtain information about the $I_t = 1$ exchanges by analysing the charge-exchange data.

Further interesting relations stem from the approximate SU(3) invariance. Since this symmetry is badly broken for particle masses, the splitting of the exchanged trajectories often implies quite different energy dependences for SU(3) related processes. However, in some cases the trajectories are the same because of exchange degeneracy. Thus the set of charge–exchange reactions $\pi^- p \to \pi^0 n$ (ρ exchange), $\pi^- p \to \eta n$ (A_2 exchange), $K^- p \to \bar{K}^0 n$ ($A_2 + \rho$ exchange), $K^+ n \to K^0 p$ ($A_2 - \rho$) all share the same degenerate ρ–A_2 trajectory, with a common residue if strong exchange degeneracy holds. The external mesons, π, η and K, all belong to the same SU(3) octet, and so if SU(3) is exact for the residues we obtain the relation

$$\frac{d\sigma}{dt}(\pi^- p \to \pi^0 n) + 3\frac{d\sigma}{dt}(\pi^- p \to \eta n) = \frac{d\sigma}{dt}(K^- p \to \bar{K}^0 n) + \frac{d\sigma}{dt}(K^+ n \to K^0 p) \quad (6.8.26)$$

(assuming $\eta \approx \eta_8$) which is quite well satisfied experimentally (fig. 6.9). If higher spin particles are produced it is necessary to project out particular spin density matrices to test such equalities, and for example the relation

$$\rho\frac{d\sigma}{dt}(\pi^- p \to \rho^0 n) + \rho\frac{d\sigma}{dt}(\pi^- p \to \omega^0 n) = \rho\frac{d\sigma}{dt}(K^- p \to K^{*0} n) + \rho\frac{d\sigma}{dt}(K^+ n \to K^{*0} p) \quad (6.8.27)$$

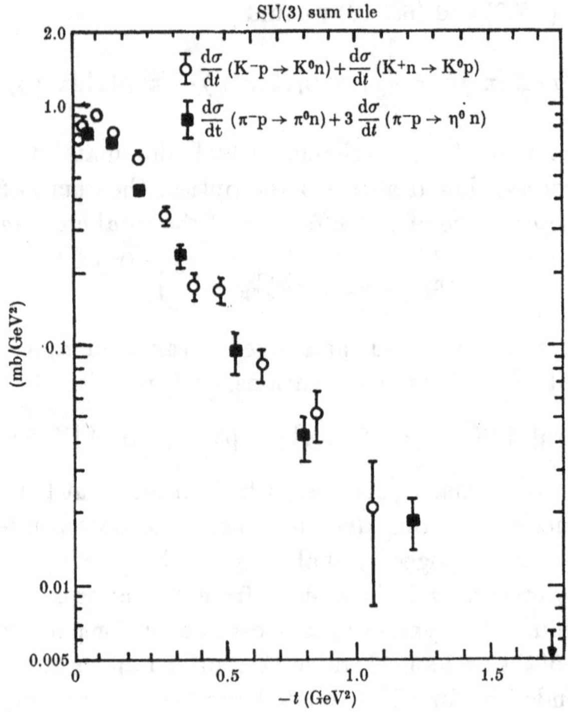

FIG. 6.9 Test of the relation (6.8.26) at 6 GeV (Barger 1974).

should work for any density matrix ρ if we assume SU(3) couplings for the vector–meson octet with ideal ω, ϕ mixing, and strong exchange degeneracy of ρ and A_2, and again it is successful experimentally (Barger 1974).

These SU(3) predictions seem to work to about 10 per cent accuracy for all helicity amplitudes and for the differences of total cross-sections (even though one expects substantial additional contributions from Regge cuts in many processes). So SU(3) appears to be a much better symmetry for Regge couplings than it is for particle masses.

j. Forward dips and peaks

In section 6.4 we found that though an s-channel helicity amplitude has the kinematical behaviour (6.4.2) at $t = 0$, i.e.

$$A_{H_s}(s,t) \sim (-t)^{n/2} \quad \text{where} \quad n \equiv \big||\mu_1 - \mu_2| - |\mu_3 - \mu_4|\big| \qquad (6.8.28)$$

a non-conspiring t-channel Regge pole, because of factorization and

Table 6.6 *Processes with dips and spikes near $t = 0$ due to π exchange*

Process	Structure
$\pi^- p \to \rho^0 n$	Dip
$\pi^+ n \to \rho^0 p$	Dip
$\pi^\pm p \to \rho^\pm p$	Dip
$\pi^+ p \to \rho^0 \Delta^{++}$	Spike
$\pi^+ p \to f^0 \Delta^{++}$	Spike
$\pi^+ n \to f^0 p$	Dip
$\gamma p \to \pi^+ n$	Spike
$\gamma n \to \pi^- p$	Spike
$\gamma p \to \pi^- \Delta^{++}$	Dip
$\gamma n \to \pi^+ \Delta^-$	Dip
$K^\pm p \to K^{*\pm} p$	Dip
$K^- p \to K^* n$	Dip
$K^\pm p \to K^* \Delta$	Spike

parity requirements, gives (6.4.7)

$$A_{H_s}(s,t) \sim (-t)^{m/2} \quad \text{where} \quad m \equiv (|\mu_1-\mu_3|+|\mu_2-\mu_4|) \quad (6.8.29)$$

So if we consider for example the process $\gamma p \to \pi^+ n$, in which, *inter alia*, the π trajectory can be exchanged, since $\mu_\gamma = \pm 1$, $\mu_\pi = 0$, all the helicity amplitudes must vanish according to (6.8.29), but according to (6.8.28) the non-flip amplitudes with $|\mu_3-\mu_4| = |\mu_1-\mu_2|$ will not. In fact, as table 6.6 indicates, the differential cross-section has a spike in the forward direction which is of width $\Delta t \approx m_\pi^2$.

One explanation for this, which we discussed in section 6.5, is that the pion engages in a $\Lambda = 1$ conspiracy with a natural-parity trajectory. But as no such particle is observed, and as such conspiracies run into difficulties with factorization, it is generally assumed that the forward peak is due to the presence of a cut which does not have a definite t-channel parity and so is not constrained to (6.8.29) (see fig. 6.3 and section 8.7*f* below). Table 6.6 implies that the minimum possible helicity-flip is favoured at each vertex, i.e. at the baryon vertex $\Delta\mu \equiv \mu_2-\mu_4 = 0$ dominates, except for the $\pi N\bar{N}$ coupling where parity conservation demands $\Delta\mu = 1$, while for meson vertices $\Delta\mu \equiv \mu_1-\mu_3 = 0$ dominates, except that obviously for $\gamma\pi$ we can only have $\Delta\mu = \pm 1$. If these rules do not allow $n = 0$ there is a forward dip, but if $n = 0$ is permitted there is a forward spike despite (6.8.29).

Table 6.7 *Processes controlled by* ρ, ω *and* A_2 *exchange*

Process	Dip seen at $t \approx -0.55$?	Trajectories	n
$\pi^- p \to \pi^0 n$	Yes	ρ	1
$\pi^- p \to \eta^0 n$	No	A_2	1
$K^- p \to \bar{K}^0 n$	No	$\rho + A_2$	1
$K^+ n \to K^0 p$	No	$\rho - A_2$	1
$\pi^+ p \to \pi^0 \Delta^{++}$	Yes?	ρ	1
$\pi^+ p \to \eta \Delta^{++}$	No	A_2	1
$K^+ p \to K^0 \Delta^{++}$	No	$\rho - A_2$	1
$K^- n \to \bar{K}^0 \Delta^-$	No	$\rho + A_2$	1
$\pi^0 p \to \rho^0 p$	Yes	ω	1
$\pi^\pm p \to \rho^\pm p$	Yes	$\omega + A_2$	1
$\pi^- p \to \omega n$	No	ρ	0, 2
$\pi^+ n \to \omega p$	No	ρ	0, 2
$\gamma p \to \pi^0 p$	Yes	ω(+ρ)	1
$\gamma n \to \pi^0 n$	Yes	ω(+ρ)	1
$\gamma p \to \eta p$	No	ρ(+ω)	0, 2
$\gamma N \to \pi^\pm N$	No	$\rho + A_2$	0, 2
$\pi^+ p \to \rho^0 \Delta^{++}$	No	A_2	0, 2
$K^+ p \to K^{*0} \Delta^{++}$	No?	$\rho - A_2$	0, 2
$K^- n \to \bar{K}^{*0} \Delta^-$	?	$\rho + A_2$	0, 2
$\pi^+ p \to \omega \Delta^{++}$	No?	ρ	0, 2

Note: (i) We have ignored π exchange which may dominate near $t = 0$ in some of these processes. The $n\,(\equiv |\mu_1 - \mu_3| \pm |\mu_2 - \mu_4|)$ given is relevant only to the natural-parity ρ, ω and A_2 exchanges.

(ii) We have assumed that ρ and A_2 have dominantly flip $N\bar{N}$ and $N\bar{\Delta}$ couplings, while ω is dominantly non-flip.

(iii) From SU(3), $\gamma_{\omega\pi\gamma} > \gamma_{\rho\pi\gamma}$ and $\gamma_{\rho\eta\gamma} > \gamma_{\omega\eta\gamma}$.

(iv) The ρ, ω couplings to $\pi\gamma$ and πV are flip.

k. Nonsense dips

Exchange-degeneracy arguments favour nonsense-choosing couplings for Reggeons, which implies that there should be dips in various differential cross-sections where trajectories pass through wrong-signature nonsense points (see table 6.2).

The trajectories of fig. 6.6 show that the wrong-signature point $\alpha(t) = 0$ occurs for the ρ and ω trajectories at $|t| \approx 0.55\,\text{GeV}^2$. However, this point is right-signature for A_2 and f, which will give a finite contribution (but not a pole) at $\alpha(t) = 0$. Similarly $\alpha(t) = -1$, which with linear trajectories is at $|t| \approx -1.6\,\text{GeV}^2$, is right-signature for ρ, ω and wrong-signature for A_2, f. Table 6.7 lists some of the processes which should be dominated by these trajectories (except that f is always overshadowed by P) and it is evident that many of the expected dips occur, but by no means all. Hence either the poles do not always choose nonsense, or there are other important contributions, probably cuts, in addition to these leading trajectories, or both. Given that factorization relates the behaviour in various processes it seems to be rather hard to salvage this nonsense decoupling idea despite its apparent success in many cases.

Similar conclusions apply to other exchanges. Some of the zeros expected from other bosons, such as K* exchange at $\alpha(t) = 0$ (i.e. $|t| \approx 0.2\,\mathrm{GeV}^2$), and from baryons, like N exchange at $\alpha(u) = -\frac{1}{2}$ (i.e. $|u| \approx 0.2\,\mathrm{GeV}^2$), are seen, but not all. It seems clear that cuts must play an important role, and we shall discuss this problem further in section 8.7*c*.

l. The cross-over problem

One rather unexpected feature of elastic differential cross-sections is that for example, $\mathrm{d}\sigma/\mathrm{d}t(\pi^-\mathrm{p}\to\pi^-\mathrm{p}) > \mathrm{d}\sigma/\mathrm{d}t(\pi^+\mathrm{p}\to\pi^+\mathrm{p})$ for t near zero, but they become equal for $|t| \approx 0.15\,\mathrm{GeV}^2$ and at larger $|t|$ the sign of the inequality is reversed (Ambats *et al.* 1974). From (6.8.3) and table 6.5 we see that the difference between these cross-sections is due to ρ exchange. So we can write

$$\frac{\mathrm{d}\sigma}{\mathrm{d}t}(\pi^\pm\mathrm{p}) = |(\mathrm{P}+\mathrm{f}\mp\rho)_{++}|^2 + |(\mathrm{P}+\mathrm{f}\mp\rho)_{+-}|^2 \qquad (6.8.30)$$

where we have dropped the kinematical factors in (4.2.5), the subscripts refer to the s-channel helicity amplitudes (4.3.10), and the Regge pole amplitudes are represented by their symbols.

It is found that the largest contribution is that of the P, which near $t = 0$ is almost purely imaginary (from $\alpha_{\mathrm{P}}(0) \approx 1, \mathscr{S} = +1$ in (6.8.14)), and that P and f have at most a very small coupling to the helicity-flip amplitude, so we have

$$\frac{\mathrm{d}\sigma}{\mathrm{d}t}(\pi^\pm\mathrm{p}) \approx |(\mathrm{P})_{++}|^2 + |(\mathrm{P})_{++}|\,\mathrm{Im}\{(\mathrm{f}\mp\rho)_{++}\} \qquad (6.8.31)$$

so that

$$\Delta\left[\frac{\mathrm{d}\sigma}{\mathrm{d}t}(\pi^\pm\mathrm{p})\right] \equiv \frac{\mathrm{d}\sigma}{\mathrm{d}t}(\pi^-\mathrm{p}\to\pi^-\mathrm{p}) - \frac{\mathrm{d}\sigma}{\mathrm{d}t}(\pi^+\mathrm{p}\to\pi^+\mathrm{p}) \propto \mathrm{Im}\{(\rho)_{++}\} \qquad (6.8.32)$$

and hence the imaginary part of the ρ non-flip amplitude must have the 'cross-over zero' at $|t| \approx 0.15\,\mathrm{GeV}^2$.

Similar cross-overs occur at about the same value of t in other elastic processes such as $\Delta[d\sigma/dt(K^{\pm}p)]$, and $d\sigma/dt(pp) - d\sigma/dt(\bar{p}p)$, as well as in some quasi-elastic processes like $\Delta[d\sigma/dt(K^{\pm}p \to Q^{\pm}p)]$, and for these processes the difference depends on $\text{Im}\{(\rho+\omega)_{++}\}$, the ω contribution being much the larger.

It is possible to fit these differential cross-sections with poles by inserting arbitrary zeros in the ρ and ω non-flip residues (see for example Barger and Phillips (1969)), but there are two difficulties. First, in other processes such as $\pi^- p \to \omega n$ (ρ exchange), $\pi^{\pm}p \to \rho^{\pm}p$ (ω and A_2 exchange) or $\gamma p \to \pi^0 p$ (ρ and ω), where ρ and ω are also coupled to the p–$\bar{p}$ vertex, no corresponding zero is seen. In other words, the residue does not factorize. Secondly, a zero of the pole residue would imply that the real and imaginary parts of the amplitude have coincident zeros. We shall find in the next section that this is not the case. It seems clear therefore that there must be some other explanation for these zeros, and again cuts seem likely to take the blame (see section 8.7*b*).

m. The phases of amplitudes and polarization

Since a Regge pole gives the same phase to all helicity amplitudes, processes in which only a single Regge trajectory (or an exchange-degenerate pair of trajectories) is exchanged are predicted to have zero polarization (from, for example, (4.2.22)).

In fact polarization in $\Delta S = 0$ meson–baryon scattering processes is generally quite small, usually < 20 per cent (although at present the crucial $\pi^- p \to \pi^0 n$ data is contradictory on this point, cf. Bonamy *et al.* (1973) and Hill *et al.* (1973)), but the fact that it is non-zero means that there must certainly be other contributions, either lower-lying poles or cuts.

It has proved possible, by judiciously combining and interpolating data on $\pi^{\pm}p$ elastic scattering and $\pi^- p \to \pi^0 n$, including polarization and spin-correlation measurements, completely to determine the structure of the $\pi N \to \pi N$, $I_t = 0, 1$, A_{++} and A_{+-} amplitudes up to a common over-all phase (Halzen and Michael 1971). Since the $I_t = 0$ A_{++} amplitude should have the almost pure-imaginary phase of the Pomeron for small $|t|$ this amounts almost to a complete phase determination.

The results for $I_t = 1$ are shown in fig. 6.10. Looking first at A_{+-}, we see the forward zero required by kinematics, and the nonsense-

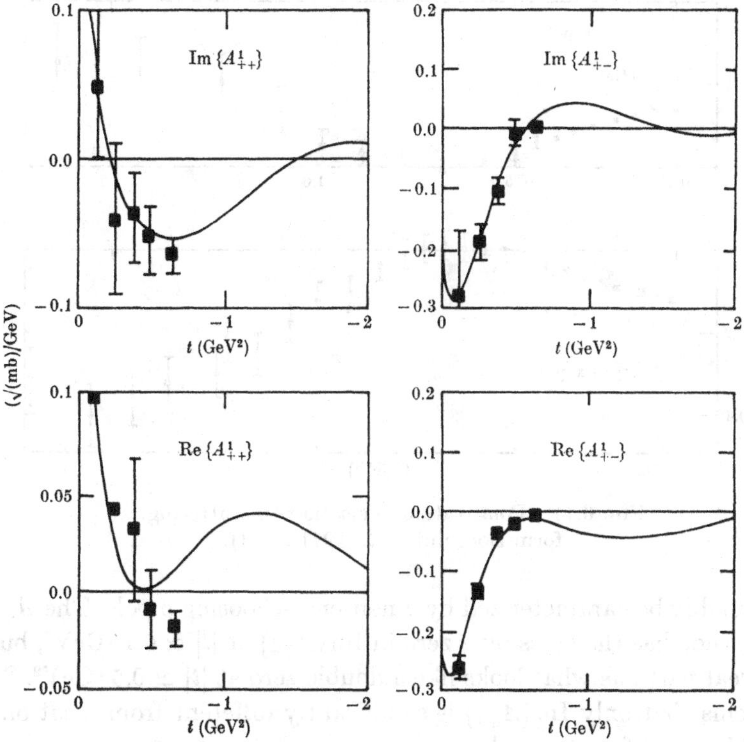

FIG. 6.10 The s-channel helicity amplitudes for $I_t = 1$ πN scattering at 6 GeV, from Halzen and Michael (1971). ■ Halzen–Michael amplitude analysis; —— Barger–Phillips FESR Regge analysis.

choosing phase given by

$$\mathrm{i}\,\mathrm{e}^{-\frac{1}{2}\mathrm{i}\pi\alpha(t)}\alpha(t), \quad \alpha(t) \approx 0.5+0.9t \tag{6.8.33}$$

(from (6.8.15) with $\mathscr{S} = -1$), so that the imaginary part has a single zero, and the real part a double zero at $\alpha(t) = 0$, i.e. at $|t| \approx 0.55\,\mathrm{GeV}^2$. This double zero can be seen directly in the elastic polarization since, from (4.2.22), using the same notation and approximations as led to (6.8.32),

$$\frac{d\sigma}{dt}P(\pi^{\pm}p\to\pi^{\pm}p) = \mp\,\mathrm{Im}\{(P+f)_{++}(\rho)^*_{+-}\} \approx \mp\,|(P)_{++}|\,\mathrm{Re}\{(\rho)_{+-}\} \tag{6.8.34}$$

since the Pomeron is nearly pure imaginary. The elastic polarization (fig. 6.11) does indeed have the mirror symmetry and double zero at $|t| \approx 0.55\,\mathrm{GeV}^2$ predicted by (6.8.33). So the $I_t = 1$, A_{+-} amplitude

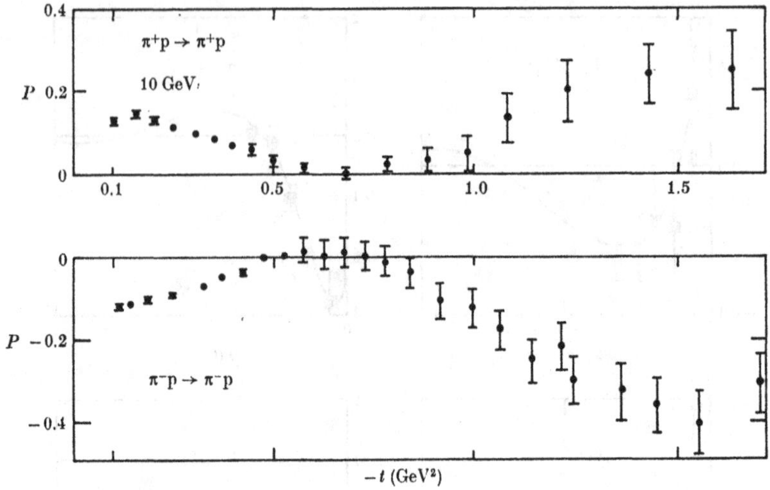

FIG. 6.11 Polarization in elastic $\pi^{\pm}p$ scattering, form Borghini *et al.* (1971, 1971).

can readily be parameterized by a nonsense-choosing ρ pole. The A_{++} amplitude has the cross-over zero in $\text{Im}\{A_{++}\}$ at $|t| \simeq 0.15\,\text{GeV}^2$, but the real part has what looks like a double zero at $|t| \simeq 0.55\,\text{GeV}^2$. So it seems that only $\text{Im}\{A_{++}\}$ is significantly different from what one would expect from a ρ pole.

Although at present we lack sufficient spin-dependent measurements to make similar complete amplitude decompositions for other processes, a careful use of the assumption that Regge pole phases hold good in some amplitudes has permitted quite a lot of information to be obtained about amplitude structures. Many amplitudes do seem to have approximate Regge phases, but certainly not all, and there is as yet no proper understanding of the successes and failures.

7
Duality

7.1 Introduction

For low energy scattering in the s channel it is often convenient to write the scattering amplitude as a partial-wave series (4.4.9)

$$A_{H_s}(s,t) = 16\pi \sum_J (2J+1) A_{HJ}(s)\, d^J_{\mu\mu'}(z_s) \qquad (7.1.1)$$

because, as we have discussed in section 2.2, if the forces are of finite range, R, then for a given s only partial waves $J \lesssim (\sqrt{s})R/\hbar$ will be important. Furthermore the various partial-wave amplitudes are frequently dominated by resonance pole contributions, so, using the Breit–Wigner formula (2.2.15), we can write

$$A_{HJ}(s) \approx \sum_r \frac{g_r}{s_r - s}, \quad s_r \equiv M_r^2 - \mathrm{i}M_r\Gamma_r \qquad (7.1.2)$$

and (7.1.2) in (7.1.1) often gives quite a good approximation to the low-energy scattering amplitude, for $s < 6\,\mathrm{GeV}^2$ say.

But as s increases the number of partial waves which must be included increases, and the density of resonances in each partial wave also seems to increase, so that it becomes harder to identify the individual resonance contributions. Hence (7.1.1) is much less useful for larger s. Also we know that it is not valid much beyond the s-channel physical region because the series diverges at the nearest t-singularity (at the boundary of the Lehmann ellipse (2.4.11)), so approximations to the scattering amplitude based on (7.1.1) are effective only in the region of the Mandelstam plot where s and $|t|$ are small, in the neighbourhood of the s-channel physical region (see fig. 1.5).

At high s on the other hand we have found it very useful to work instead with the t-channel partial-wave series, transformed via the Sommerfeld–Watson representation (4.6.4) into a sum of t-channel Regge poles and cuts. At high energies, say $s > 10\,\mathrm{GeV}^2$, only a few leading J-plane singularities need be included, but in principle this Sommerfeld–Watson representation is valid for all s and t.

The question thus arises as to how these two different viewpoints are to be married. This is an important practical problem in the inter-

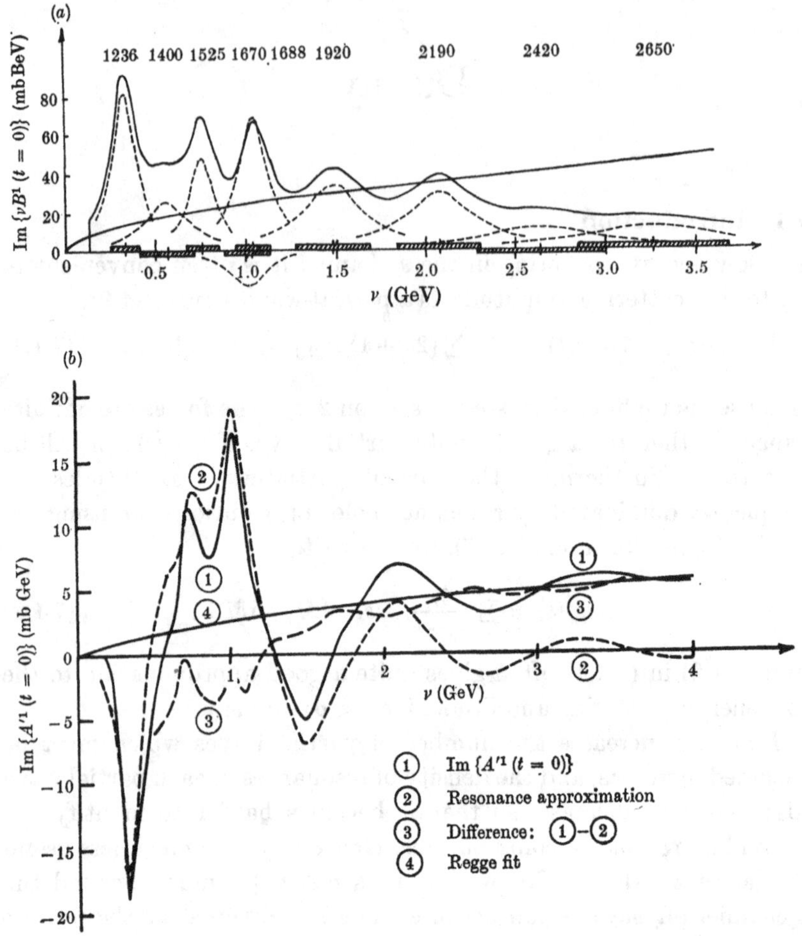

FIG. 7.1 The resonance and Regge pole contributions to (*a*) $\text{Im}\{\nu B\}$ and (*b*) $\text{Im}\{A'\}$ for $I_t = 1$ in $\pi^- p \to \pi^0 n$ at $t = 0$, from Dolen *et al.* (1968). At least at low energies the resonances almost saturate the amplitudes, while the ρ-pole Regge fit averages through the data. (For definition of ν see (7.2.3) below.)

mediate energy region, say $4 < s < 10\,\text{GeV}^2$, where the amplitudes are approaching their smooth Regge asymptotic s behaviour but some resonance bumps can still be seen (see fig. 7.1). It also poses a very important theoretical question as to how the s-channel resonances contribute to the asymptotic s behaviour, or, equivalently, where these resonances appear in the Sommerfeld–Watson representation.

Since all the residues g_r in (7.1.2) are constants, if there are only a finite number of resonances (however large), then clearly the total resonance contribution to the scattering amplitude must have the behaviour

$$A^{\mathrm{r}}_H(s,t) \sim \frac{1}{s}, \quad \text{for all fixed } t \tag{7.1.3}$$

and so would appear as a fixed pole at $J_t = -1$ in the Sommerfeld–Watson representation (from (2.7.2)). In this case one might try adding (7.1.2) and (4.6.4) giving

$$A_H(s,t) = A^{\mathrm{r}}_H(s,t) + A^{\mathrm{R}}_H(s,t) \tag{7.1.4}$$

where A^{r} includes all the s-channel resonances, and A^{R} all the t-channel Regge singularities with $\operatorname{Re}\{\alpha(t)\} > -1$. This is often called the interference model because the amplitude oscillates as a function of s because of interference between the resonances and the Regge poles (see for example Barger and Cline (1966, 1967)).

However, we have seen in chapter 3 how, in simple dynamical models like the ladder model, fig. 3.3, if the s-channel poles behave like s^{-1} then the t-channel trajectories obtain the asymptotic behaviour $\alpha(t) \xrightarrow[t\to-\infty]{} -1$ from above, from the unitarization of this fixed-pole input. And we have also found (fig. 6.6) that trajectories appear in practice to be essentially linear, $\alpha(t) \approx \alpha^0 + \alpha' t$, and seem to be descending well below -1. This could mean that somehow the fixed pole does not contribute to the leading Regge trajectories, but is to be added to them as in (7.1.4). For even-signature amplitudes, where $J = -1$ is a wrong-signature nonsense point, such an additional fixed pole is certainly possible (see sections 4.8 and 6.3), but in an odd-signature amplitude the fixed pole would be incompatible with t-channel unitarity. And a moving pole which remained in the region of $J_t = -1$ should have been observed by now in effective trajectory plots.

It seems fairly clear therefore that at least at large $-t$ the s-channel resonance poles are cancelling against each other in such a way as to produce an asymptotic behaviour $\sim s^x$, where $x \leqslant \alpha(t)$, $\alpha(t)$ being the leading t-channel singularity. The most interesting possibility is $x = \alpha(t)$, so that the s-channel resonances actually combine to produce the leading Regge pole behaviour. Of course this is only possible in the t region where $\alpha(t) > -1$ if there is an infinite number of resonances so that the series (7.1.2) diverges.

This possibility was first suggested in the now classic paper of Dolen, Horn and Schmid (1968), who noted that if one adds the contributions of all the resonances discovered by phase-shift analysts in πN scattering (for $I_t = 1$) the result not only gives almost the whole scattering amplitude but is, on the average, approximately equal to the ρ-pole exchange contribution obtained from fits to high energy data, extrapolated down to the low-s region (see fig. 7.1). There thus seems to be an equivalence, an 'average duality', between the direct channel resonances, r, and the crossed channel Regge poles, R, because, at least in the intermediate energy region, the average of the former is equal to the latter, i.e.

$$\langle A_H(s,t)\rangle \approx \langle A_H^{\mathrm{r}}(s,t)\rangle \approx \langle A_H^{\mathrm{R}}(s,t)\rangle \tag{7.1.5}$$

(this statement is made more precise in the next section). One may then hope that as s is increased the density of resonances will also increase, thus smoothing out the bumps, until eventually there is 'local duality', i.e.

$$A_H(s,t) \approx A^{\mathrm{r}}(s,t) \approx A^{\mathrm{R}}(s,t) \tag{7.1.6}$$

without any need for averaging.

Unfortunately this argument is not completely compelling for at least two reasons. First, it is always possible to re-parameterize the Regge pole terms so as to retain their asymptotic behaviour but reduce their magnitude in the intermediate energy region. For example replacing $\beta(t)\,(s/s_0)^{\alpha(t)}$ by $\beta(t)\,[(s-s_a)/s_0)]^{\alpha(t)}$ reduces the magnitude in the neighbourhood of the arbitrarily chosen point s_a. Of course the branch point at $s = s_a$ would be spurious, but so is the one at $s = 0$ in the usual parameterization, which stems from the approximation (6.2.26). Essentially these two parameterizations differ just by terms of order $s^{\alpha(t)-1}$, i.e. at the daughter level, where the predictions of Regge theory are ambiguous.

Secondly, the actual identification of inelastic resonances in phase-shift analyses is called into question by the success of (7.1.5). For as Schmid (1968) showed, if one takes a Regge pole term (6.8.1), with a linear trajectory $\alpha(t) = \alpha^0 + \alpha' t$, and uses equal-mass kinematics (1.7.22)

$$z_s = 1 + \frac{t}{2q_s^2}, \quad q_s^2 = \frac{s-4m^2}{4},$$

the s-channel partial-wave projection (2.2.1) of the Regge term depends on (Chiu and Kotanski 1968)

$$A_J(s) \propto \int_{-1}^{1} \mathrm{e}^{-\mathrm{i}\pi\alpha(t)} P_J(z_s)\,\mathrm{d}z_s = \mathrm{e}^{-\mathrm{i}\pi(\alpha^0 - 2q_s{}^2\alpha')}(\mathrm{i})^J J_J(-2q_s^2\pi\alpha') \tag{7.1.7}$$

where J_J is the spherical Bessel function of order J (see for example

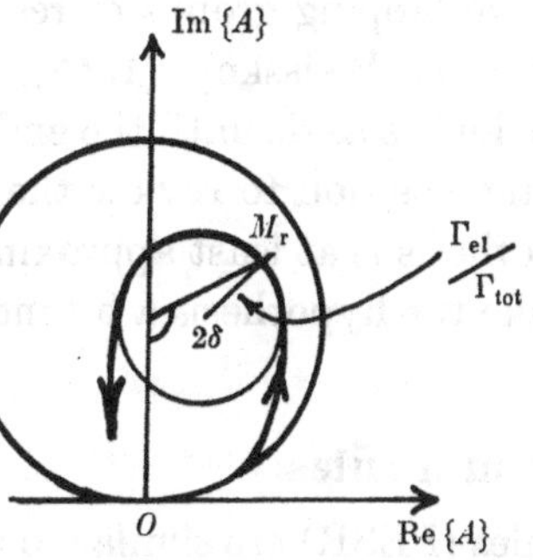

Fig. 7.2 The partial-wave Argand diagram for an inelastic resonance (see (2.2.13) *et seq.*). For a range of energies near M_r the curve follows a circle due to the Breit–Wigner formula, but it is smaller than the unitarity circle due to the inelasticity, and it is pushed off centre, and the phase may be rotated by the background.

Magnus and Oberhettinger (1949) p. 26). So as s (and hence q_s^2) increases the phase of the amplitude given by (7.1.7) will rotate anti-clockwise, giving a loop just like that predicted for an inelastic resonance by (2.2.15) (see fig. 7.2). Note that if the phase reaches $\pi/2$ at a given 'resonance' position $s = s_r$, there will be further resonances at $s_r^n = s_r + n/\alpha'$, $n = 1, 2, \ldots$ where the phase goes through $(2n+1)\pi/2$, and all the partial waves will resonate at the same s_r^n since the phase in (7.1.7) is independent of J. Thus the Regge pole terms will give rise to resonance-like loops in the partial-wave Argand plots, despite the fact that the Regge pole term does not contain any poles in s.

There are clearly two ways of interpreting this result (Collins *et al.* 1968*b*). Either one accepts the postulate of duality, in which case these loops do correspond to resonances and are a manifestation of the average equality (7.1.5) even though the Regge terms do not contain s-channel poles. Or, if one chooses to deny duality, Argand loops can no longer be regarded as sufficient evidence for the existence of resonances, and there may well be fewer actual resonances than one has been led to suppose from phase-shift analyses. If so the phenomenological case for duality crumbles. The essence of this difficulty is that there can only be experimental evidence about the behaviour of scattering amplitudes along the real s axis, and so to analytically continue to the pole on the unphysical sheet requires a model based on unitarity. The Breit–Wigner formula (2.2.14) is certainly a valid model for elastic amplitudes dominated by isolated poles, but its use for highly inelastic, overlapping groups of resonances is much more questionable; see Blatt and Weisskopf (1952), Weidenmuller (1967).

We shall put these doubts aside until the end of the chapter, where we shall be in a better position to review the quite strong evidence that the duality hypothesis is at least approximately valid. Our next step is to try and make the hypothesis a bit more precise.

7.2 Finite-energy sum rules

Finite-energy sum rules (FESR) are similar to the SCR of section 4.8, but apply also in circumstances where the amplitude is not convergent at infinity. All that is necessary is that the asymptotic behaviour be known. We shall assume for simplicity that the asymptotic behaviour is Regge-pole-like, so that, from (6.8.1)

$$\hat{A}_{H_t}(s,t) \xrightarrow[s\to\infty]{} \hat{A}^{\mathrm{R}}(s,t) = \sum_i -G_i(t)\,\frac{\mathrm{e}^{-\mathrm{i}\pi(\alpha-v)}+\mathcal{S}_i}{\sin\pi(\alpha-v)}\left(\frac{\nu}{s_0}\right)^{\alpha_i(t)-M} \tag{7.2.1}$$

where the sum is over all the leading Regge poles, say those with

$$\mathrm{Re}\{\alpha_i(t)\} > -k, \quad k>1 \tag{7.2.2}$$

We have combined all the various residue factors into $G_i(t)$, and have introduced the notation

$$\nu \equiv \frac{s-u}{2} \tag{7.2.3}$$

So asymptotically

$$\left.\begin{aligned} D_s(s,t) &\xrightarrow[s\to\infty]{} \sum_i G_i(t)\left(\frac{\nu}{s_0}\right)^{\alpha_i(t)-M} \\ D_u(s,t) &\xrightarrow[s\to-\infty]{} \sum_i -\mathcal{S}_i G_i(t)\left(\frac{\nu}{s_0}\right)^{\alpha_i(t)-M}(-1)^{M-v} \end{aligned}\right\} \tag{7.2.4}$$

The scattering amplitude is expected to obey the fixed-t dispersion relation (4.5.1), and hence

$$\begin{aligned} \hat{A}_{H_t}(\nu,t)-\hat{A}^{\mathrm{R}}(\nu,t) &= \frac{1}{\pi}\int_{\nu_{\mathrm{T}}}^{\infty}\frac{D_s(\nu',t)-\sum_i G_i(t)\,(\nu'/s_0)^{\alpha_i(t)-M}}{\nu'-\nu}\,\mathrm{d}\nu' \\ &+\frac{1}{\pi}\int_{\nu_{\mathrm{T}}}^{\infty}\frac{D_u(\nu',t)-(-1)^{M-v}\sum_i \mathcal{S}_i G_i(t)\,(\nu'/s_0)^{\alpha_i(t)-M}}{\nu'+\nu}\,\mathrm{d}\nu' \end{aligned} \tag{7.2.5}$$

where $\nu_{\mathrm{T}}\,(\equiv s_{\mathrm{T}}+\frac{1}{2}(t-\Sigma))$ is the position of the s-threshold in terms of ν (where Σ is defined in (1.7.18)), and the integrals will converge. Because of the hypothesis (7.2.2) that all the leading terms in the asymptotic behaviour of $\hat{A}$ are contained in $\hat{A}^{\mathrm{R}}$ we know that at most

$$\hat{A} - \hat{A}^{\mathrm{R}} \sim \frac{1}{\nu^k}, \quad \nu \to \infty \tag{7.2.6}$$

so that when we take $\nu \to \infty$ on the right-hand side of (7.2.5) the coefficient of the ν^{-1} term must vanish, i.e.

$$\int_{\nu_{\mathrm{T}}}^{\infty} \left\{ D_s(\nu',t) - D_u(\nu',t) - \sum_i [1 - \mathscr{S}_i(-1)^{M-v}] G_i(t) \left(\frac{\nu'}{s_0}\right)^{\alpha_i(t)-M} \right\} \mathrm{d}\nu' = 0 \tag{7.2.7}$$

Obviously among all the poles, i, only a sub-set, denoted by j, which have signature

$$\mathscr{S}_j = (-1)^{M-v+1} \tag{7.2.8}$$

will contribute to (7.2.7).

Since the poles give the asymptotic form of D_s and D_u the integrand will be negligible for $\nu' > N$, for some sufficiently large N, and so

$$\int_{\nu_{\mathrm{T}}}^{N} (D_s(\nu',t) - D_u(\nu',t))\,\mathrm{d}\nu' = \int_{\nu_{\mathrm{T}}}^{N} \sum_i 2G_i(t) \left(\frac{\nu'}{s_0}\right)^{\alpha_j(t)-M} \mathrm{d}\nu' \tag{7.2.9}$$

The integral on the right-hand side is readily performed to give the FESR

$$\int_{\nu_{\mathrm{T}}}^{N} (D_s(\nu',t) - D_u(\nu',t))\,\mathrm{d}\nu' = \sum_i \frac{2s_0 G_j(t)}{\alpha_j(t) - M + 1} \times \left[\left(\frac{N}{s_0}\right)^{\alpha_j(t)-M+1} - \left(\frac{\nu_{\mathrm{T}}}{s_0}\right)^{\alpha_j(t)-M+1}\right] \tag{7.2.10}$$

For $\alpha_j > -1 + M$ the threshold term on the right-hand side can obviously be neglected if $N \gg s_0$.

An alternative way of deriving (7.2.10) (and its generalizations below), more elegant but perhaps less instructive, is to use Cauchy's theorem to write

$$\int_C \hat{A}_H(\nu',t)\,\mathrm{d}\nu' = 0 \tag{7.2.11}$$

where C is a contour which excludes the threshold branch points, as shown in fig. 7.3. So closing the contour onto the branch cuts gives

$$2\mathrm{i} \int_{\nu_{\mathrm{T}}}^{N} (D_s(\nu,t) - D_u(\nu',t))\,\mathrm{d}\nu' = -\int_{C'} \hat{A}_{H_t}(\nu',t)\,\mathrm{d}\nu' \tag{7.2.12}$$

where C' is the circle at $|\nu| = N$. Putting $\nu = N\,\mathrm{e}^{\mathrm{i}\phi}$, replacing $\hat{A}_H$ by $\hat{A}_H^{\mathrm{R}}$ of (7.2.1), and taking proper care of the discontinuity of the Regge term across the branch cuts gives (7.2.10) without the threshold term.

The FESR (7.2.10) provides a relation between the average (i.e. the zeroth moment) of the imaginary part of the scattering amplitude at

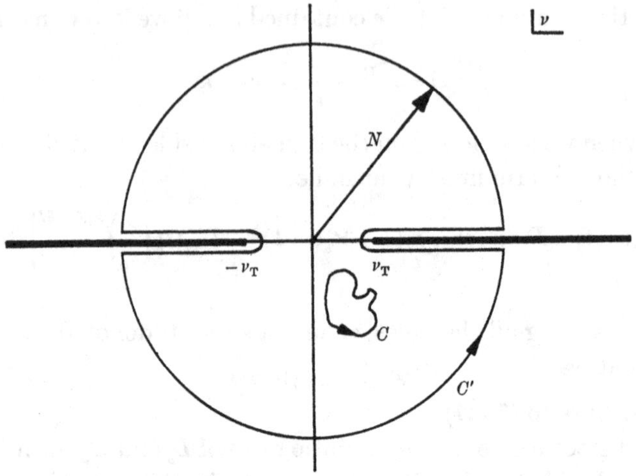

FIG. 7.3 Contours of integration in the complex ν plane for (7.2.11) and (7.2.12).

low energies and the Regge pole asymptotic behaviour at high energies, a relation which obtains because of the assumed analyticity of the amplitude and Regge pole dominance for $\nu \geqslant N$. This should clearly be helpful for understanding duality.

Several generalizations of (7.2.10) are possible. First, instead of (7.2.5) we can write the dispersion relation for

$$(\hat{A}_H(\nu,t) - \hat{A}_H^{\mathrm{R}}(\nu,t))\left(\frac{\nu}{s_0}\right)^{2n}, \quad n = 0, 1, 2, \ldots \tag{7.2.13}$$

and as long as $2n < k$ the coefficient of the ν^{-1} term must vanish giving

$$\int_{\nu_T}^{N} (D_s(\nu',t) - D_u(\nu',t))\left(\frac{\nu'}{s_0}\right)^{2n} \mathrm{d}\nu' = \sum_j \frac{2s_0 G_j(t)}{\alpha_j(t) - M + 2n + 1} \times \left[\left(\frac{N}{s_0}\right)^{\alpha_j(t)-M+2n+1} - \left(\frac{\nu_T}{s_0}\right)^{\alpha_j(t)-M+2n+1}\right] \tag{7.2.14}$$

i.e. even-moment FESR. Alternatively, if an odd power of (ν/s_0) is included, only poles k with opposite signature to (7.2.8), i.e.

$$\mathscr{S}_k = (-1)^{M-v} \tag{7.2.15}$$

contribute, giving the odd-moment FESR

$$\int_{\nu_T}^{N} (D_s(\nu',t) + D_u(\nu' t))\left(\frac{\nu'}{s_0}\right)^{2n-1} \mathrm{d}\nu' = \sum_k \frac{2s_0 G_k(t)}{\alpha_k(t) - M + 2n} \times \left[\left(\frac{N}{s_0}\right)^{\alpha_k(t)-M+2n} - \left(\frac{\nu_T}{s_0}\right)^{\alpha_k(t)-M+2n}\right] \tag{7.2.16}$$

(the + sign appearing on the left-hand side because of the odd power of ν). These integrals involve only the imaginary part of the scattering amplitude, but it is possible to include both real and imaginary parts by writing a dispersion relation for (Liu and Okubo 1967)

$$(\hat{A}_H(\nu,t)-\hat{A}_H^{\mathrm{R}}(\nu,t))\left(\frac{\nu_{\mathrm{T}}^2-\nu^2}{s_0^2}\right)^{\beta/2} \tag{7.2.17}$$

where β is an arbitrary parameter, giving

$$\int_{\nu_{\mathrm{T}}}^{N}\left[\cos\left(\frac{\pi\beta}{2}\right)\mathrm{Im}\{\hat{A}_{H_t}(\nu',t)\}-\sin\left(\frac{\pi\beta}{2}\right)\mathrm{Re}\{\hat{A}_{H_t}(\nu',t)\}\right]\left(\frac{\nu'^2-\nu_{\mathrm{T}}^2}{s_0^2}\right)^{\beta/2}\mathrm{d}\nu'$$
$$=\sum_i\frac{2s_0G_i(t)}{\alpha_i(t)+\beta+1}\left(\frac{N}{s_0}\right)^{\alpha_i(t)+\beta+1}\frac{\cos[\frac{1}{2}\pi(\alpha_i(t)+\beta)]}{\cos(\frac{1}{2}\pi\alpha_i(t))} \tag{7.2.18}$$

which for example reduces to (7.2.14) (without the ν_{T} term) for $\beta=$ even integer. These are called continuous moment sum rules (CMSR), but as information about the real parts of amplitudes is seldom available except from dispersion relations which have clearly been assumed in deriving (7.2.18) CMSR are only occasionally useful.

It is also interesting to write FESR for amplitudes of definite signature which have the fixed-t dispersion relations (like (2.5.7))

$$\hat{A}_H^{\mathscr{S}}(s,t)=\frac{1}{\pi}\int_{s_{\mathrm{T}}}^{\infty}\frac{D_s(s',t)}{s'-s}\,\mathrm{d}s'+(-1)^{M-v}\mathscr{S}\int_{u_{\mathrm{T}}}^{\infty}\frac{D_u(s',t)}{s'-s}\,\mathrm{d}s' \tag{7.2.19}$$

so if we follow the above procedure we find

$$\int_{\nu_{\mathrm{T}}}^{N}[D_s(\nu',t)+(-1)^{M-v}D_u(\nu',t)]\left(\frac{\nu'}{s_0}\right)^n\mathrm{d}\nu'$$
$$=\sum_l\frac{2s_0G_l(t)}{\alpha_l(t)-M+n+1}\left(\frac{N}{s_0}\right)^{\alpha_l(t)-M+n+1},\quad n=0,1,2,\ldots \tag{7.2.20}$$

where $l=j$ or k depending on $\mathscr{S}$ (see (7.2.8), (7.2.15)). These FESR coincide with (7.2.14) or (7.2.16) only for alternate moments. But the 'wrong moments' (i.e. n even for $\mathscr{S}=(-1)^{M-v}$ or n odd for $\mathscr{S}=(-1)^{M-v+1}$) are likely to be incorrect because we have neglected the fact that definite-signature amplitudes contain fixed poles at wrong-signature nonsense points (see section 4.8) which should also be included in the right-hand side of (7.2.20). So for wrong moments we must add

$$\sum_l\frac{G_l(t)}{J_l-M+n+1}\left(\frac{N}{s_0}\right)^{J_l-M+n+1} \tag{7.2.21}$$

to the right-hand side of (7.2.20), where J_l are the positions of the wrong-signature nonsense fixed poles, i.e. $J_l=M-1, M-3, \ldots$ or

$M-2, M-4, \ldots$ for $\mathscr{S} = \pm(-1)^{M-v}$. However, if the fixed-pole residues are small (7.2.20) will be approximately valid as it stands for all moments.

We shall discuss some of the phenomenological applications of FESR in the next section, but here our main interest is to examine the implications of duality for FESR. If the imaginary part of the scattering amplitude at low energy can be represented as a sum of resonance pole contributions (r), (7.2.10) becomes

$$\int_{\nu_{\mathrm{T}}}^{N} (D_s^{\mathrm{r}}(\nu', t) - D_u^{\mathrm{r}}(\nu', t)] \,\mathrm{d}\nu' = \sum_j \frac{2s_0 G_j(t)}{\alpha_j(t) - M + 1} \left(\frac{N}{s_0}\right)^{\alpha_j(t)\ M+1} \tag{7.2.22}$$

This gives a definite meaning to (7.1.5), that the integral of the imaginary part of the resonance contributions to the scattering amplitude is equal to that of the Regge pole contributions (fig. 7.1). Note, however, that to obtain (7.2.22) we have already made the duality assumption because the sum of a finite set of resonances $\sim s^{-1}$, but in (7.2.2) we assumed that the Regge poles include all the leading terms in asymptotic behaviour down to s^{-k}, $k > 1$. So (7.2.22) does not in any sense prove duality, but it does give it a more concrete mathematical expression than (7.1.5).

The higher-moment sum rules require a more local duality and so are less likely to work at low energies. If all the moments were the same then of course A^{r} would be identically equal to A^{R}, which is clearly impossible since the one contains poles in s and the other does not.

The constraints imposed on an amplitude by (7.2.22) are quite powerful if crossing is also incorporated. For example if we consider $\pi\pi$ scattering (Gross (1967); see also Collins and Mir (1970)), the dominant $I_t = 1$, odd-signature exchange will be the ρ trajectory (see section 3.5). However ρ poles with spin $\sigma = 1$ will also be the principal s- and u-channel resonances so (see (2.6.13))

$$D_s^{\mathrm{r}} = 16\pi^2 3 \frac{\beta(s)}{\alpha'} P_1(z_s)\,\delta(s - m_\rho^2)$$

$$= 16\pi^2 3 \frac{\beta(s)}{\alpha'} \left(1 + \frac{2t}{m_\rho^2 - 4}\right) \delta(s - m_\rho^2) \tag{7.2.23}$$

if we use units where $m_\pi = 1$. We take the residue to be

$$\beta(s) = \frac{\gamma(s)\,\alpha(s)}{\Gamma(\alpha + \frac{3}{2})} (q_s^2)^{\alpha(s)}, \quad \text{where} \quad \alpha(m_\rho^2) = 1, \quad q_s^2 = \frac{s-4}{4} \tag{7.2.24}$$

$\gamma(s)$ being the reduced residue, remaining after we have extracted explicitly the threshold behaviour (6.2.9), the nonsense factor at $\alpha = 0$, and the Mandelstam-symmetry zeros (2.9.5). This gives

$$D_s^{\mathrm{r}}(s,t) = 64(\pi)^{\frac{3}{2}} \frac{\gamma(m_\rho^2)}{\alpha'} \left(\frac{m_\rho^2-4}{4}\right)\left(1+\frac{2t}{m_\rho^2-4}\right)\delta(s-m_\rho^2) \quad (7.2.25)$$

and likewise for $D_u^{\mathrm{r}}(s,t)$. Similarly the ρ trajectory in the t channel will give, using (6.8.1),

$$D_s^{\mathrm{R}}(s,t) = 16(\pi)^{\frac{3}{2}}\gamma(t)\frac{\alpha(t)}{\Gamma(\alpha+1)}\nu^{\alpha(t)} \quad (7.2.26)$$

Substituting (7.2.25) and (7.2.26) into (7.2.22) (remembering that we are considering an amplitude for spinless particles so $M = 0$, and with $I_t = 1$ so that the left-hand side must include a crossing matrix element $\frac{1}{2}$ from table 6.3 which cancels the factor 2 from adding D_s and D_u) we obtain

$$2\frac{\gamma(m_\rho^2)}{\alpha'}\left(\frac{m_\rho^2-4}{4}\right)\left(1+\frac{2t}{m_\rho^2-4}\right) = \frac{\gamma(t)\,\alpha(t)}{\Gamma(\alpha(t)+1)\,(\alpha(t)+1)}N^{\alpha(t)+1} \quad (7.2.27)$$

If these are equated at $t = m_\rho^2$ the γ's cancel out, $\alpha(t) = 1$, and we get

$$\alpha' = \frac{3m_\rho^2-4}{N^2}$$

So taking the cut-off, N, half way between the $\rho(m_\rho^2 \approx 30m_\pi^2)$ and the next s-channel resonance, the f $(m_\mathrm{f}^2 \approx 80m_\pi^2)$, i.e. taking $N = 68m_\pi^2$ (from (7.2.3)), we get

$$\alpha' = 0.019m_\pi^{-2} = 1\,\mathrm{GeV}^2$$

in quite good agreement with (5.3.1). If we take the nth moment sum rule, and ignore the possibility of fixed poles, we get

$$\alpha' = \frac{n+2}{2^{n+1}}\frac{(3m_\rho^2-4)^{n+1}}{N^{n+2}}$$

which with $N = 68$ gives a rather slow variation of α' with n for small n, so all the low moments are quite well satisfied.

Equation (7.2.27) is an FESR consistency condition for the ρ trajectory, sometimes called an 'FESR bootstrap'. It is quite different from a proper bootstrap of the type discussed in section 3.5 (and section 11.7 below) because no attempt is made to impose unitarity, and hence the magnitude of the coupling, $\gamma(t)$, factors out. Also it is necessary to know the particle spectrum before one can fix N, so the trajectory is not determined uniquely. And we have chosen to evaluate the sum rule at $t = m_\rho^2$, but it is evident that the t-dependence of the two sides of (7.2.27) is quite different. None the less before the advent of more complete dual models (see section 7.4) a good deal of work went into showing that these consistency relations do apply quite widely (see for example Ademollo *et al.* (1958, 1969), Igi and Matsuda (1967)). Their SU(3) generalization will be discussed below.

7.3 Applications of FESR and duality

The first point to note about the duality hypothesis in the form (7.2.22) is that it is clearly invalid for Pomeron (P) exchange. For example both $pp \to pp$ and $K^+p \to K^+p$ elastic scattering amplitudes have exotic quantum numbers (see section (5.2)) and do not contain any s-channel resonances, but are controlled by the t-channel P exchange. This observation led to the hypothesis of 'two-component duality' (Harari 1968, Freund 1968) which states that where vacuum quantum numbers occur in the t channel the ordinary Reggeons, R (i.e. all except P) are dual to the resonances (r), while the P is dual to the background amplitude (b) upon which the resonances are superimposed. So such amplitudes have two components

$$A_H(s,t) = A^{\text{r}}(s,t) + A^{\text{b}}(s,t) = A^{\text{R}}(s,t) + A^{\text{P}}(s,t) \tag{7.3.1}$$

with

$$\langle \text{A}^{\text{r}} \rangle = A^{\text{R}} \quad \text{and} \quad \langle A^{\text{b}} \rangle = A^{\text{P}} \tag{7.3.2}$$

the averages being taken for the imaginary parts in the sense of (7.2.22). Of course for processes where P exchange cannot occur (7.1.6) holds, and only one component is necessary.

This hypothesis has been tested directly in πN elastic scattering (e.g. Harari and Zarmi 1969) by showing that the sum of the resonances (represented by inelastic Breit–Wigner formulae) and the P amplitude (extrapolated from high-energy fits) can reproduce the scattering amplitudes obtained in low energy phase-shift analyses. Of course, as most of these resonances were actually discovered in phase-shift analyses, the test really amounts to showing (*a*) that the Breit–Wigner formula (2.2.15) without any rotation of phase parameterizes the resonance loops satisfactorily, and (*b*) that the extrapolated P amplitude can account for all the background to these resonances. Unfortunately this is not sufficient to prove the hypothesis because by giving the Breit–Wigner formulae arbitrary phases, which is not unreasonable for highly inelastic overlapping sets of resonances, the interference model

$$A = A^{P} + A^{R} + A^{r} \tag{7.3.3}$$

can be made to fit equally well (Donnachie and Kirsopp 1969), quite apart from the uncertainty which exists in the resonance interpretation of the phase shifts mentioned in section 7.1. But the fact that it is possible to construct consistent dual models, and apply (7.3.2) in a wide variety of situations (see also section 10.7) makes it seem likely that this two-component hypothesis has at least approximate validity.

Why the P should have this exceptional status is not completely clear. We shall discuss some plausible dynamical reasons in section 11.7, but we have already noted that the slope of the P is only $\alpha'_{P} \approx 0.2\,\text{GeV}^2$, compared with $\alpha'_{R} \simeq 0.9\,\text{GeV}^2$ for all the other trajectories, so that any resonance-like loops generated by the P in (7.1.7) would have a very slow phase rotation, and would be very widely spaced.

There still remains, however, the problem that exotic channels like $pp \to pp$ and $K^+p \to K^+p$ can exchange other trajectories, $R = \rho$, A_2 ω and f (table 6.5), despite the fact that they contain no resonances. This can be accounted for by invoking strong exchange degeneracy (section 6.8*h*), and supposing that as in (6.8.22) the contributions of these trajectories cancel, $A_2 - \rho$ and $f - \omega$, leaving no imaginary part. This can occur if the signs of the different contributions are arranged as in table 7.1. Since Breit–Wigner resonances dominate $\text{Im}\{A(s,t)\}$ (see (2.2.15)) the absence of an imaginary part to A^{R} implies, via (7.2.22) and (7.3.2), that there will be no resonances. Alternatively, resonances could occur with alternating signs to give $\langle \text{Im}\,\{A^{r}\}\rangle = 0$ averaged over several resonances, but clearly this is not the solution we want for exotic elastic processes.

It is thus essential that the degeneracy pattern of Regge exchanges should be consistent with the resonance spectrum. This explains the fact that the exotic processes have rather flat $\sigma^{\text{tot}}(s)$, and only a simple exponential behaviour of $d\sigma/dt$ as a function of t from P exchange, while the non-exotic line-reversed processes $\bar{p}p \to \bar{p}p$ and $K^-p \to K^-p$, in which the sign of the odd signature ρ and ω exchanges is reversed, have falling $\sigma^{\text{tot}}(s)$, and dip structures at low energy at $|t| \approx 0.55\,\text{GeV}^2$ due to the R contribution (see for example figs. 6.4 and 6.5). We shall examine the implications of these exchange-degeneracy requirements more fully below.

FESR provide a new tool for Regge analysis, because if one knows

Table 7.1 *Signs of the trajectory contributions to the imaginary part of the elastic* NN *and* KN *scattering amplitudes*

Process	Exchanges
$\bar{p}p \to \bar{p}p$	$P+f+\rho+\omega+A_2$
$\bar{p}n \to \bar{p}n$	$P+f-\rho+\omega-A_2$
$pp \to pp$	$P+f-\rho-\omega+A_2$
$pn \to pn$	$P+f+\rho-\omega-A_2$
$K^-p \to K^-p$	$P+f+\rho+\omega+A_2$
$K^-n \to K^-n$	$P+f-\rho+\omega-A_2$
$K^+p \to K^+p$	$P+f-\rho-\omega+A_2$
$K^+n \to K^+n$	$P+f+\rho-\omega-A_2$

Under $p \leftrightarrow n$ odd-isospin ρ and A_2 change sign. Under particle $\leftrightarrow$ anti-particle the odd-C_n ρ and ω change sign.

the low energy amplitude, from, for example, a phase-shift analysis, one can use (7.2.14) and (7.2.16) to determine the Regge parameters without recourse to high energy data. This was done by Dolen *et al.* (1968) who for example used the difference of the $\pi^{\pm}p \to \pi^{\pm}p$ elastic scattering amplitudes obtained from an $E < 1.5\,\text{GeV}$ phase-shift analysis to obtain the ρ-exchange parameters from (7.2.22) (see fig. 7.1).

Since even with a single trajectory exchange there are two parameters in (7.2.14) for each value of t, $\alpha(t)$ and $G(t)$, the sum rules do not have a unique solution. But if we define for the non-flip, $M = 0$ amplitude

$$S_m(t) = \frac{1}{N^{m+1}} \int_0^N \nu^m D_s^{\mathscr{S}}(\nu', t)\,\mathrm{d}\nu' = \frac{2G(t)\,N^{\alpha(t)}}{\alpha(t)+m+1} \tag{7.3.4}$$

(using the notation of (7.2.4), and setting $s_0 = 1$) then the ratio

$$\frac{S_{m'}(t)}{S_m(t)} = \frac{\alpha(t)+m'+1}{\alpha(t)+m+1} \tag{7.3.5}$$

so $\alpha(t)$ can be obtained from the ratio of the first two right-signature moments ($m = 0$ and $m = 2$ for the $\mathscr{S} = -1$ ρ), and then re-inserted in (7.3.4) to find $G(t)$. Their results were in good agreement with the ρ parameters obtained by fitting the high energy data.

The various resonance contributions have different t dependences, being proportional to $d^{\sigma}_{\mu\mu'}(z_s)$, where σ is the spin of the resonance. These rotation functions are oscillatory functions of z_s (and hence of t at fixed s) and so it is found that at some t values the left-hand side of (7.3.4) vanishes. This occurs for $\operatorname{Im}\{A_{++}(\nu, t)\}$ at $t \approx -0.15\,\text{GeV}^2$, where the cross-over zero appears in the Regge amplitude, and in $\mathrm{Im}\{A_{+-}(s,t)\}$ at $t \approx -0.55\,\mathrm{GeV}^2$, coincident with the nonsense zero (see sections 6.8k, l). To build up the Regge behaviour with the correct t dependence for the residues there has to be a very close correlation between the contributions of the various resonances.

Of course this use of FESR suffers from the same sort of ambiguity concerning secondary trajectories, cuts etc, as do the high energy fits, but at least in principle these secondary contributions may also be identified. Thus if there is a secondary ρ′ trajectory, $\alpha_1(t)$, in addition to the ρ, we deduce from (7.3.4)

$$\frac{S_0(t) - G(t)\,N^{\alpha(t)}/(\alpha(t)+1)}{S_2(t) - G(t)\,N^{\alpha(t)}/(\alpha(t)+3)} = \frac{\alpha_1(t)+3}{\alpha_1(t)+1} \tag{7.3.6}$$

so once $\alpha(t)$, $G(t)$ have been found, it is possible to obtain $\alpha_1(t)$, and so on. In fact Dolen *et al.* obtained the very high secondary trajectory $\alpha_1(t) = 0.3 + 0.8t$, which probably mainly reflects the build-up of errors which occurs when parameters are determined successively like this.

The higher-moment sum rules weight the integrals more towards the upper limit of integration, and if N is sufficiently large use of FESR becomes essentially equivalent to making a Regge fit near N. But in practice N has to be quite low because phase-shift analyses do not extend far in energy ($< 3\,\mathrm{GeV}$). This means that the results obtained depend greatly on the assumptions which are made about the high energy behaviour, and in practice with data of finite accuracy it is not possible to predict a unique analytic extrapolation. So the predictive power of the method for determining the high energy behaviour of amplitudes from low energy data alone is very limited. Certainly it provides no substitute for high energy data. Also phase-shift analyses are available only for a few channels (πN→πN, KN→KN, γN→πN and πN→πΔ at present) so the number of processes to which the method can be applied directly, even after invoking isospin relations like (6.8.23), is somewhat limited. Quite often FESR can be employed in other processes by making extra assumptions such as resonance saturation of the low energy amplitude (which we used for the ππ amplitude in the previous section) though obviously the uncertainty of the results is increased thereby.

There is, however, one crucial advantage of the FESR method over conventional Regge fits, namely that the phase-shift analysis gives the input amplitudes A_{H_s} directly, whereas $\mathrm{d}\sigma/\mathrm{d}t$ data only give

$\sum_{H_s} |A_{H_s}|^2$. Thus with FESR one can find the Regge behaviours of the different spin amplitudes separately, and determine their phases, without recourse to polarization or other spin-dependent measurements. Thus much of the information contained in the 6 GeV πN amplitude analysis discussed in section 6.8*m* could also be obtained, at least qualitatively, by extrapolating the < 2 GeV phase-shift solutions with FESR, assuming Regge behaviour.

So FESR, especially when used in conjunction with fits to high energy data, are a very valuable aid to Regge analysis (see Barger and Phillips (1969) for examples of their use).

7.4 The Veneziano model

Much of the progress which has been made in applying and generalizing the concept of duality stems from the success of Veneziano (1968) in constructing a simple model for $2 \to 2$ scattering amplitudes which satisfies most of the requirements of duality.

We begin by considering the amplitude for $\pi^+\pi^- \to \pi^+\pi^-$, which has ρ and f poles in the s and t channels, but for which the u-channel $\pi^+\pi^+ \to \pi^+\pi^+$ is exotic, $I_u = 2$. So once the P component has been removed from this elastic scattering process we expect the approximately degenerate ρ and f trajectories to give the leading contributions in both channels, but there may be an infinite number of other resonances with these same quantum numbers.

The duality requirement (7.3.2) is that the sum over all the s-channel poles should be equal to the sum over all the t-channel poles, i.e.

$$A(s,t) = \sum_n \frac{G_n(s,t))}{s-s_n} = \sum_m \frac{G_m(t,s)}{t-t_m} \tag{7.4.1}$$

and that Regge asymptotic behaviour occur in both variables, i.e.

$$A(s,t) \underset{s\to\infty}{\sim} s^{\alpha(t)} \quad (t \text{ fixed}), \quad \text{and} \quad A(s,t) \underset{t\to\infty}{\sim} t^{\alpha(s)} \quad (s \text{ fixed}) \tag{7.4.2}$$

The simplest function which has an infinite set of s-poles lying on a trajectory $\alpha(s)$, the poles occurring when $\alpha(s) =$ positive integer, is $\Gamma(1-\alpha(s))$. Since we need an identical behaviour in t as well we might try

$$A(s,t) = \Gamma(1-\alpha(s))\,\Gamma(1-\alpha(t)) \tag{7.4.3}$$

but this would give a double pole at each s–t point where both $\alpha(s)$ and $\alpha(t)$ are positive integers (see fig. 6.4). However, these double poles can

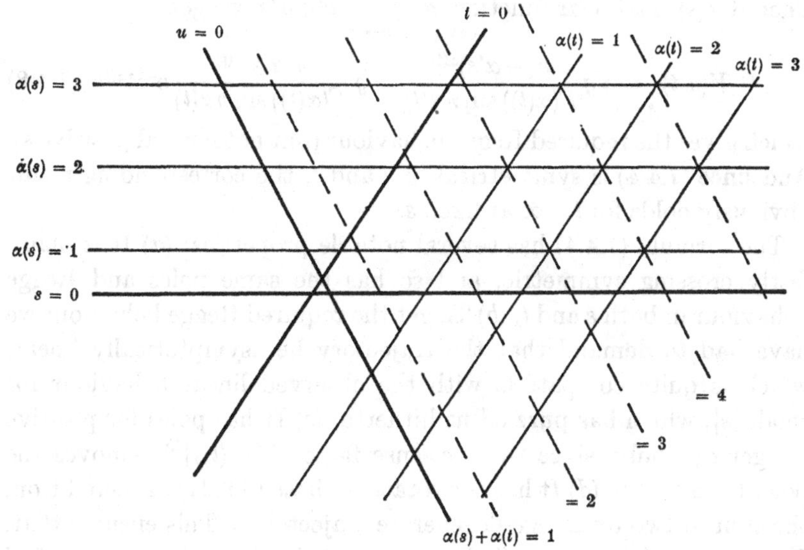

FIG. 7.4 Poles of the Veneziano amplitude in the s–t plane. The poles occur where $\alpha(s)$ and $\alpha(t)$ pass through positive integers, with lines of zeros connecting the pole intersections to prevent double poles.

easily be removed by writing

$$A(s,t) = V(s,t) \equiv g\frac{\Gamma(1-\alpha(s))\,\Gamma(1-\alpha(t))}{\Gamma(1-\alpha(s)-\alpha(t))} \tag{7.4.4}$$

which is the Veneziano formula. Here g is an arbitrary number which sets the scale of the coupling strengths as we shall see below (equation (7.4.12)).

The asymptotic behaviour of this amplitude may be deduced from Stirling's formula (see for example Magnus and Oberhettinger (1949) p. 4)

$$\Gamma(x) \xrightarrow[x\to\infty]{} (2\pi)^{\frac{1}{2}}\,e^{-x}\,x^{x-\frac{1}{2}} \tag{7.4.5}$$

(except in a wedge along the real negative x axis where poles appear for integer x) which gives

$$\frac{\Gamma(x+a)}{\Gamma(x+b)} \xrightarrow[x\to\infty]{} x^{a-b}\left(1+O\left(\frac{1}{x}\right)\right) \tag{7.4.6}$$

Hence if $\alpha(s)$ is an increasing function of s we have, for fixed t (using (6.2.32)),

$$V(s,t) \xrightarrow[s\to\infty]{} g\frac{\pi(-\alpha(s))^{\alpha(t)}}{\Gamma(\alpha(t))\sin\pi\alpha(t)} \tag{7.4.7}$$

Then if $\alpha(s)$ is a linear function, $\alpha(s) \underset{s\to\infty}{\rightarrow} \alpha^0 + \alpha' s$, we get

$$V(s,t) \underset{s\to\infty}{\longrightarrow} g\frac{\pi(-\alpha' s)^{\alpha(t)}}{\Gamma(\alpha(t))\sin\pi\alpha(t)} = g\frac{\pi(\alpha' s)^{\alpha(t)}}{\Gamma(\alpha(t))\sin\pi\alpha(t)}\mathrm{e}^{-i\pi\alpha(t)} \quad (7.4.8)$$

which gives the required Regge behaviour (but not for real positive s). And since (7.4.4) is symmetrical in s and t, the corresponding result obviously holds for $t \to \infty$ at fixed s.

The formula (7.4.4) has several notable properties: (*a*) It is manifestly crossing symmetric, and so has the same poles and Regge behaviour in both s and t. (*b*) To get the required Regge behaviour we have had to demand that the trajectory be asymptotically linear, which is quite compatible with the observed linear behaviour for small $|s|$, which has puzzled us hitherto. (*c*) It has poles for positive integer $\alpha(t)$ only, since the nonsense factor $[\Gamma(\alpha(t))]^{-1}$ removes the poles for $\alpha(t) \leqslant 0$. (*d*) It has the rotating phase (6.8.21) expected from the sum of two exchange-degenerate trajectories. This ensures that, for $s > 0$, $\mathrm{Im}\{V(s,t)\} \sim s^{\alpha(t)}$, but for $s < 0$, in the u-channel physical region $\mathrm{Im}\{V(s,t)\} = 0$, since the u-channel is exotic. However, since the poles are on the real axis the discontinuity in either the s or t channels is just a sum of δ functions, and the double spectral function is the mesh of points where the poles cross in fig. 7.4. (*e*) The scale factor in the asymptotic behaviour (7.4.8) is given by

$$s_0 = \alpha'^{-1} \quad (7.4.9)$$

and we have already noted that empirically $s_0 \approx 1\,\mathrm{GeV}^2$ and $\alpha' \approx 1\,\mathrm{GeV}^{-2}$.

To obtain the resonance spectrum in the s channel we use the result (Magnus and Oberhettinger (1949) p. 2)

$$\frac{\Gamma(x)\,\Gamma(a+1)}{\Gamma(x+a)} = \sum_{n=0}^{\infty}(-1)^n\frac{\Gamma(a+1)}{\Gamma(a-n)\,\Gamma(n+1)}\cdot\frac{1}{x+n}, \quad a \text{ real} > 0 \quad (7.4.10)$$

to write

$$V(s,t) = \sum_{n=1}^{\infty} g\frac{\Gamma(1-\alpha(t))}{\Gamma(n)\,\Gamma(1-n-\alpha(t))}\frac{(-1)^n}{\alpha(s)-n} \quad (7.4.11)$$

so that if $\alpha(s) \to n$ for $s \to s_n$ (say) there is a pole of the form

$$V(s,t) \underset{s\to s_n}{\longrightarrow} g\frac{(n-\alpha(t)-1)\,(n-\alpha(t)-2)\dots(-\alpha(t))}{(n-1)!\,\alpha'(s-s_n)} \quad (7.4.12)$$

So if $\alpha(t) = \alpha^0 + \alpha' t$ the residue of the pole is a polynomial in $t\,[= -2q_s^2(1-z_s)]$ of order n, and

$$V(s,t) \underset{s\to s_n}{\longrightarrow} \frac{g}{\alpha'(s-s_n)\,(n-1)!}[(2q_s^2\alpha' z_s)^n + O(z_s^{n-1})] \quad (7.4.13)$$

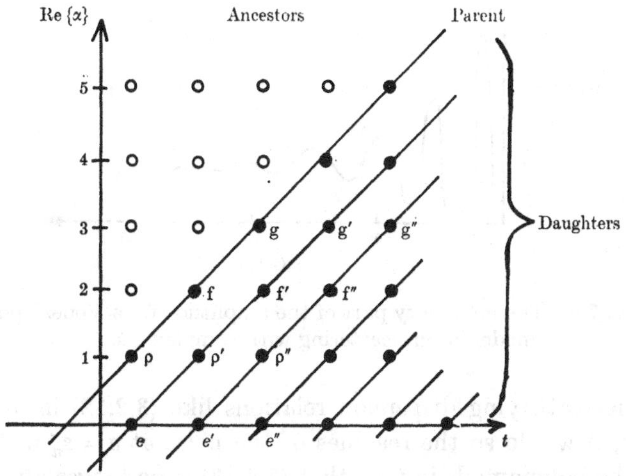

FIG. 7.5 The ε, ρ, f, g, ... states required in the Veneziano model for ππ scattering. The open circles are positions where ancestors occur if complex α's are used.

and hence the residue may be rewritten as a sum of Legendre polynomials, $P_n(z_s), P_{n-1}(z_s), \ldots, P_0(z_s)$. Thus the pole at $s = s_n$ corresponds to a degenerate sequence of $n+1$ resonances having spins $= 0, 1, \ldots, n$. The resulting resonance spectrum, an infinite sequence of integrally spaced daughters, is shown in fig. 7.5 where we have given particle names to the lowest mass states.

Since the Veneziano model is an analytic function of s and t, with just poles, and has the correct asymptotic behaviour, it clearly should provide a solution to the FESR consistency condition (7.2.22). This is not quite trivial because the Regge asymptotic behaviour does not hold along the real positive s axis. The relation between the residues in the two channels, each being proportional to g, is reminiscent of our approximate solution (7.2.27). A fairly complete review of the properties of the Veneziano formula and FESR tests can be found in Sivers and Yellin (1971).

The most obvious defect of the Veneziano model is that the poles appear on the real s axis, and so we do not get Regge behaviour where it is actually seen experimentally. This is because we have used real trajectory functions, whereas we know from section 3.2 that above the threshold in each channel unitarity requires that trajectories become complex (Im $\{\alpha\}$ being proportional to Γ, the width of the resonance – see (2.8.7)), and the poles move off the physical sheet.

It seems rather obvious therefore that one should insert complex

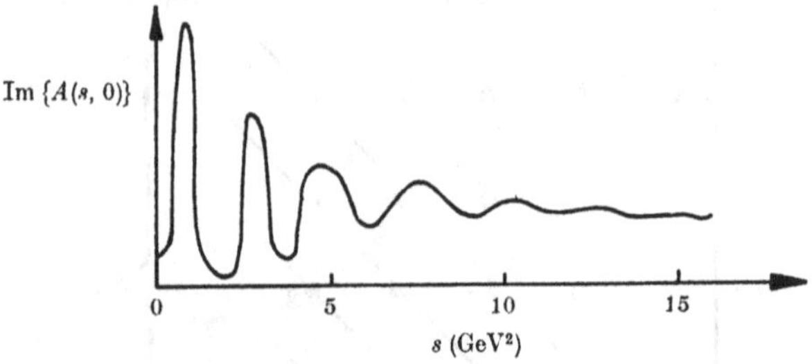

FIG. 7.6 The imaginary part of the amplitude for a Veneziano model for $\pi\pi$ scattering with complex α's.

trajectories satisfying dispersion relations like (3.2.12) into (7.4.4). However, if we do so the residues of the poles at $s = s_n$ in (7.4.12) cease to be polynomials in t, so that (7.4.13) is no longer applicable, and each pole gives rise to resonances of arbitrarily high spin. We should thus produce the so-called 'ancestor' poles of fig. 7.5. Despite the occurrence of these ancestors the asymptotic behaviour is still (7.4.8) which shows that the amplitude no longer has the convergent large-l behaviour needed for Carlson's theorem (section 2.7). Also the Argand loops are rather poorly correlated with the resonances (Collins, Ross and Squires (1969), Ringland and Phillips (1969); fig. 7.6) and the amplitude does not attain the smooth Regge asymptotic behaviour unless $\text{Im}\{\alpha\}$ grows very rapidly with s, in which case the resonances become so wide as to disappear.

Although there have been many more sophisticated attempts to insert resonances with non-zero widths into the Veneziano formula none has proved very satisfactory because the constraints of analyticity and Regge asymptotic behaviour in all directions in the complex s plane are so restrictive (see for example Bali, Coon and Dash (1969), Cohen-Tannoudji *et al.* (1971)). To use it phenomenologically it is therefore necessary to employ the asymptotic form (7.4.8) despite the fact that it is invalid on the real positive s axis. Also, for phenomenology it is essential to be able to include higher-spin external particles, especially spin $= \frac{1}{2}$. This has been done (see Neveu and Schwarz 1971) but in order to satisfy the MacDowell symmetry these models contain parity doublets. Also, because the daughter sequences of the Veneziano model do not correspond to Toller pole sequences, infinite sums of Veneziano terms are needed to satisfy the conspiracy relations (6.5.7). We shall touch on some of these generalizations of the Veneziano model in chapter 9.

It is also important to note that (7.4.4) is certainly not unique. In fact the amplitude

$$A(s,t) = \sum_{l,m,n\geqslant 0} C_{lmn} V_{lmn}(s,t), \quad n \leqslant l+m \tag{7.4.14}$$

$$V_{lmn}(s,t) \equiv g \frac{\Gamma(l-\alpha(s))\,\Gamma(m-\alpha(t))}{\Gamma(n-\alpha(s)-\alpha(t))} \tag{7.4.15}$$

where C_{lmn} are arbitrary coefficients, also satisfies all the FESR and duality requirements. The V_{lmn} are known as Veneziano 'satellite' terms. They differ from (7.4.4) in having their first pole in s at $\alpha(s) = l$, and the asymptotic behaviour $s^{\alpha(t)+n-l}$, etc. Clearly $l = 0$ is possible only if the trajectory cuts $\alpha(s) = 0$ for $s > 0$, unlike fig. 7.5. This arbitrariness demonstrates the weakness of the FESR consistency conditions compared with the full bootstrap requirements which depend on unitarity.

Despite these problems, which have greatly limited its phenomenological application, the Veneziano model is a very useful theoretical 'toy', which, as we shall find in chapter 9, can readily be extended to multi-particle processes.

So far the model is suitable only for $\pi^+\pi^- \to \pi^+\pi^-$ which has exotic $I_u = 2$. If we assume that the f′ is decoupled from $\pi\pi$ (see section 5.2) the full amplitude will also have just the ρ–f exchange-degenerate trajectory as its leading trajectory (once the P component has been subtracted), but it is necessary to impose the isospin crossing relations (6.7.10), and the Bose statistics requirement that an amplitude of even isospin is even under the spatial parity transformation $z \to -z$, and vice versa. Thus the t-channel isospin amplitudes $A_t^I(s,t)$ might be written

$$\left.\begin{aligned} A_t^0(s,t) &= a(V(s,t)+V(t,u))+bV(s,u) \quad \text{even under } s\leftrightarrow u \\ A_t^1(s,t) &= c(V(s,t)-V(t,u)) \quad \text{odd under } s\leftrightarrow u \\ A_t^2(s,t) &= V(s,u) \quad \text{even } s\leftrightarrow u, \text{ exotic } t \end{aligned}\right\} \tag{7.4.16}$$

(where a, b and c are constants), provided $V(s,t)$ is symmetric under $s\leftrightarrow t$, etc. Then applying the crossing relation (6.7.10)

$$A_s^I = \sum_{I_t} M(I_s, I_t)\, A_t^I$$

to (7.4.16) with the $\pi\pi$ crossing matrix of table 6.3, we find that to ensure that there are no poles in the exotic A_s^2 amplitude, i.e. to eliminate from it $V(s,t)$ and $V(s,u)$ terms, we need $a = \frac{3}{2}c$, and $b = -\frac{1}{2}$, while to make A^0_s symmetric under $t \leftrightarrow u$ demands $c = 1$, so

$$\left.\begin{aligned} A^s_0(s,t) &= \tfrac{3}{2}(V(s,t)+V(t,u)) - \tfrac{1}{2}V(s,u) \\ A^1_t(s,t) &= V(s,t) - V(t,u) \\ A^2_t(s,t) &= V(s,u) \end{aligned}\right\} \tag{7.4.17}$$

(Lovelace 1968). The residues of the t-channel poles in (7.4.17) in the three isospin states $I_t = (0,1,2)$ are obviously in the ratio 3:2:0 which gives an eigenvector of the $\pi\pi$ crossing matrix with eigenvalue 1, i.e.

$$\begin{pmatrix} \frac{1}{3} & 1 & \frac{5}{3} \\ \frac{1}{3} & \frac{1}{2} & -\frac{5}{6} \\ \frac{1}{3} & -\frac{1}{2} & \frac{1}{6} \end{pmatrix} \begin{pmatrix} 3 \\ 2 \\ 0 \end{pmatrix} = \begin{pmatrix} 3 \\ 2 \\ 0 \end{pmatrix} \tag{7.4.18}$$

As $s \to \infty$ ($u \to -\infty$) at fixed t (7.4.7) in (7.4.17) gives

$$A^1_t \xrightarrow[s\to\infty]{} \frac{g\pi(\alpha' s)^{\alpha(t)}}{\Gamma(\alpha(t))\sin\pi\alpha(t)}[\mathrm{e}^{-i\pi\alpha(t)} - 1] \tag{7.4.19}$$

the -1 coming from the $V(t,u)$ term. The square bracket in (7.4.19) is, of course, just the signature factor expected for the odd-signature $I_t = 1$ ρ pole. Similarly for A^0_t, which is even under $s \leftrightarrow u$, the terms $V(s,t) + V(t,u) \sim (\mathrm{e}^{-i\pi\alpha(t)} + 1)\, s^{\alpha(t)}$ for the even-signature f. We need to be careful about $V(s,u)$ however. This contains no poles in t, and hence should not contribute to the asymptotic behaviour in this limit. Now from (7.4.6) we find that

$$V(s,u) \sim \mathrm{e}^{-cs}), \quad s \to \infty, \quad t \text{ fixed} \tag{7.4.20}$$

where c is a constant, provided that $\alpha'_s = \alpha'_u$, i.e. the slopes of the trajectories in the s and u channels are the same. For the crossing-symmetric $\pi\pi$ amplitude clearly this will always be true.

Now $V(s,t)$ in (7.4.7) vanishes when

$$\alpha(s) + \alpha(t) = 1, \quad \text{i.e. } 2\alpha^0 + \alpha' s + \alpha' t = 1 \tag{7.4.21}$$

This zero will coincide with the Adler zero required by current algebra theory (see for example Renner (1968), Adler and Dashen (1968)) which makes the $\pi\pi$ amplitude vanish at the unphysical point $s = t = u = m^2_\pi$, if

$$\alpha^0 = \tfrac{1}{2} - \alpha' m^2_\pi \tag{7.4.22}$$

(Weinberg 1966), and since the trajectory must reach $\alpha = 1$ for $t = m^2_\rho$ we have

$$\alpha' = \frac{1}{2(m^2_\rho - m^2_\pi)} \approx 0.88\,\mathrm{GeV}^{-2}, \quad \alpha^0 = 0.48 \tag{7.4.23}$$

in quite good agreement with (5.3.1) and figs. 5.5 and 6.6. Using these parameters for the trajectory good agreement is found between (7.4.4) and current algebra requirements (see Lovelace 1968) so despite its obvious defects the Veneziano model has many surprising and desirable properties for $\pi\pi$ scattering.

7.5 Duality and SU(3)

The construction of the $\pi\pi$ model (7.4.17) depends on the fact that once the P has been eliminated there is only a single leading trajectory in all the channels of $\pi\pi$ scattering, i.e. the isospin-degenerate ρ–f trajectory (since we assumed that the f′ does not couple to $\pi\pi$). It is thus convenient to refer to $V(s,t)$ in (7.4.17) as $V_{\rho\rho}(s,t)$ since ρ (and f) poles occur in both s and t. Exchange degeneracy was necessary because, using an obvious notation for the factorizable exchange couplings,

$$\left.\begin{aligned} \operatorname{Im}\{A(\pi^+\pi^-)\} &= (f_{\pi\pi})^2+(\rho_{\pi\pi})^2 \\ \operatorname{Im}\{A(\pi^+\pi^+)\} &= (f_{\pi\pi})^2-(\rho_{\pi\pi})^2 \end{aligned}\right\} \tag{7.5.1}$$

and strong exchange degeneracy gives

$$(f_{\pi\pi})^2 = (\rho_{\pi\pi})^2 \tag{7.5.2}$$

and eliminates poles from the exotic $I = 2$, π^+–π^+ amplitude.

If we now consider Kπ scattering, related to $\pi\pi$ by SU(3), there will be the same ρ–f trajectory in the t channel, $\pi\pi \to K\bar{K}$, but the exchange-degenerate K*–K** trajectory appears in both the s and u channels. To achieve the required symmetry we thus write

$$\left.\begin{aligned} A_t^0 &= a(V_{\rho K^*}(t,s)+V_{\rho K^*}(t,u)) \quad \text{even } s\leftrightarrow u \\ A_t^1 &= b(V_{\rho K^*}(t,s)-V_{\rho K^*}(t,u)) \quad \text{odd } s\leftrightarrow u \end{aligned}\right\} \tag{7.5.3}$$

the V_{ij} being like (7.4.4) but with different trajectories in the two channels ($I_t = 2$ is not possible for K$\bar{K}$). However, in view of (7.4.20) we require $\alpha'_\rho = \alpha'_{K^*}$, so only the intercepts of the trajectories can be different. To obtain the s-channel isospin amplitudes we use the πK crossing matrix of table 6.3 in the crossing relation (6.7.10), and to eliminate poles in the exotic $I_s = \frac{3}{2}$ state we need $a = (\sqrt{\frac{3}{2}})b$. This gives

$$\left.\begin{aligned} \operatorname{Im}\{A(K^+\pi^+)\} &= f_{KK}f_{\pi\pi}-\rho_{KK}\rho_{\pi\pi} \\ \operatorname{Im}\{A(K^+\pi^0)\} &= f_{KK}f_{\pi\pi}+\rho_{KK}\rho_{\pi\pi} \end{aligned}\right\} \tag{7.5.4}$$

and $f_{KK}f_{\pi\pi} = \rho_{KK}\rho_{\pi\pi}$, which together with our solution to (7.5.2) requires

$$f_{KK} = \rho_{KK} \tag{7.5.5}$$

Then KK and $K\bar{K}$ elastic scattering are similar, except that the $I = 0$ f and ω exchanges and the $I = 1$ ρ and A_2 exchanges all occur. So we can write

$$\left.\begin{aligned}\operatorname{Im}\{A(K^+K^-)\} &= (f_{KK})^2+(A_{2KK})^2+(\omega_{KK})^2+(\rho_{KK})^2\\ \operatorname{Im}\{A(K^+K^+)\} &= (f_{KK})^2+(A_{2KK})^2-(\omega_{KK})^2-(\rho_{KK})^2\\ \operatorname{Im}\{A(K^+K^0)\} &= (f_{KK})^2-(A_{2KK})^2-(\omega_{KK})^2+(\rho_{KK})^2\\ \operatorname{Im}\{A(K^+\bar{K}^0)\} &= (f_{KK})^2-(A_{2KK})^2+(\omega_{KK})^2-(\rho_{KK})^2\end{aligned}\right\} \tag{7.5.6}$$

the sign changes being those demanded by the signature and charge-conjugation properties of the exchanges. Since both K^+K^+ and K^+K^0 are exotic ($S = 2$) we require

$$(f_{KK})^2 = (\omega_{KK})^2 \quad \text{and} \quad (A_{2KK})^2 = (f_{KK})^2 \tag{7.5.7}$$

with the ω and A_2 trajectories degenerate with f and ρ, which is indeed approximately true in fig. 5.4. However (7.5.7) and (7.5.3) imply

$$\rho_{KK} = \omega_{KK} \tag{7.5.8}$$

while exact SU(3) for the couplings would give (see Gourdin 1967)

$$(\sqrt{3})\,\rho_{KK} = \omega_{KK} \tag{7.5.9}$$

We can satisfy both these requirements by remembering that with broken SU(3) the physical ω particle may be a mixture of octet and singlet states (see (5.2.17)), and then the SU(3) symmetry requirement for the couplings becomes

$$(\sqrt{3})\,\rho_{KK} = \omega_{8KK} \tag{7.5.10}$$

so if we take the ideal mixing angle given by (5.2.18), $\cos\theta = 3^{-\frac{1}{2}}$, both (7.5.8) and (7.5.10) will be satisfied. This means that the exchange-degenerate $\phi + f'$ trajectory will also be exchanged in KK scattering (but not in $\pi\pi$). And this is very desirable since (7.5.7) and (7.5.5) imply that $\operatorname{Im}\{A(K^+\bar{K}^0)\}$ in (7.5.6) vanishes; that is to say without a $\phi + f'$ contribution there would be no resonances in the $K^+\bar{K}^0$ channel despite the fact that it is not exotic.

All these relations can readily be described if the various particles are represented by their quark content, shown in table 5.2 (Harari 1969, Rosner 1969). All the incoming and outgoing mesons can be represented as $q_i\bar{q}_j$ where q_i, q_j = p, n or λ quarks. The condition we have been imposing on (7.5.1), (7.5.4) and (7.6.5) is that there should be no exotic resonances, so all the internal particles must also have the quantum numbers of the $\{1\} \oplus \{8\}$ representations of SU(3) which

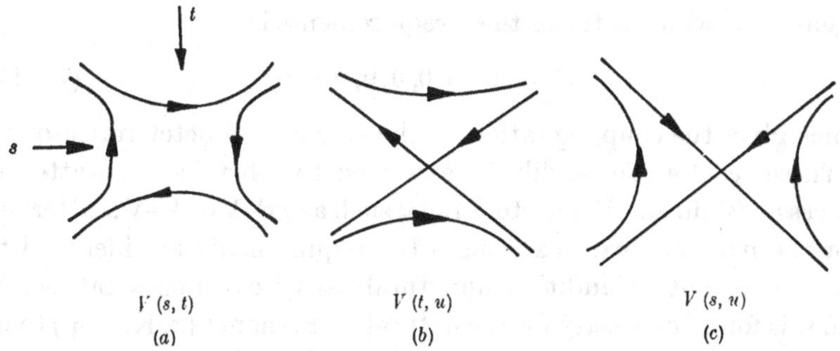

FIG. 7.7 Quark duality diagrams for meson–meson scattering. The arrow represents the direction of the quark; an anti-quark travels in the opposite direction to the arrow.

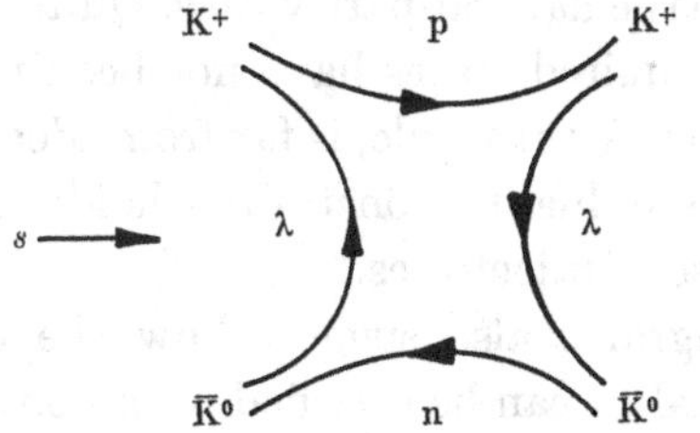

FIG. 7.8 The duality diagram for $K^+\overline{K}^0$ elastic scattering.

are also contained in $q\overline{q}$ (see (5.2.16)). So the duality diagram fig. 7.7 (*a*) can represent $V(s,t)$ for all our PS–PS meson scattering solutions, since it ensures quantum number conservation and only non-exotic $q\overline{q}$ states in both the s and t channels. However, the lines must not cross over each other as in fig. 7.7 (*b*), (*c*) or there would be exotics in one of the channels. But these crossed diagrams are suitable for the $V(s,u)$ and $V(t,u)$ terms respectively. Fig. 7.7 also incorporates our mixing-angle result (7.5.10) since in $K^+\overline{K}^0$ elastic scattering (fig. 7.8) only $\lambda\overline{\lambda}$, and hence with ideal mixing (equation (5.2.19)) only ϕ–f′, can be exchanged in the t channel. The ρ, f, ω and A_2 trajectories do not contribute to this process.

With exact SU(3) symmetry, meson–meson scattering is $\{8\}\otimes\{8\}$ scattering with amplitudes A_t^μ, $\mu = \{1\}, \{8_{ss}\}, \{8_{sa}\}, \{8_{as}\}, \{8_{aa}\}, \{10\}, \{\overline{10}\}, \{27\}$ (see section 6.7). However, since $\{10\}$, $\{\overline{10}\}$ and $\{27\}$ are exotic we need a solution which is an eigenvector of the $\{8\}\oplus\{8\}$ crossing matrix (table 6.4) having eigenvalue 1, and no trajectories in $\{10\}$, $\{\overline{10}\}$ or $\{27\}$ (cf. (7.4.18) for isospin). Because of charge conjugation only symmetric d-type couplings are possible for the tensor $\{8\}$, and only anti-symmetric f-type couplings for the vector $\{8\}$. The eigenvector which satisfies these requirements is

$$A^{\mu} = (16, 5, 0, 0, 9, 0, 0, 0) \tag{7.5.11}$$

which gives the coupling ratios for the singlet and octet trajectories.

These results can readily be extended to other meson scattering processes (Chiu and Finkelstein 1968) such as PS–V or V–V scattering. For the natural-parity exchanges the requirements are identical to the above, but in addition unnatural-parity exchanges can occur, and it is found necessary for the natural-C_n PS nonet (π, K, η, η') to be degenerate with the natural-C_n A^- nonet (B, Q, H?) and for the unnatural-C_n A^+ nonet (A_1, Q, D?) to be exchange degenerate with some axial tensor nonet. For each nonet the symmetry-breaking pattern should be similar to the natural-parity case. Quite apart from the fact that many of the required states have not been identified, we know that the η–η' mixing, for example, is far from ideal, so it would seem that in practice these duality constraints hold only for the leading natural-parity meson trajectories.

The duality diagrams also suggest how the internal symmetry requirements of duality can be satisfied in meson–baryon scattering, since we can represent all the external and internal baryons as $q_i q_j q_k$, i, j, k = p, n or λ quarks, as in fig. 7.9. This ensures that only non-exotic baryons occur in the s channel, and non-exotic mesons in the t channel. The corresponding su diagram has baryons in both channels.

When the SU(3) symmetry is broken, the exchange-degeneracy requirements on the meson exchanges in the $V(s,t)$ and $V(t,u)$ terms in PS–B scatterings are identical to those for PS–PS scattering (see Mandula *et al.* 1969). In fact we have already noted in table 7.1 (p. 220) the exchange-degeneracy requirements for ρ, ω, A_2 and f to prevent exotics in K^+p and pp, which are the same as those for $K^+\pi^+$ and $\pi^+\pi^+$.

Constraints on the baryon spectrum arise from the $V(s,u)$ term which controls backward scattering. The most plausible full solution (see Mandula, Weyers and Zweig 1970) requires the $J^P = \frac{1}{2}^+$ octet to be exchange degenerate with the $\frac{3}{2}^+$ decuplet, $\frac{3}{2}^-$ octet and $\frac{3}{2}^-$ singlet. But evidently this constraint is badly violated since, for example, the Δ trajectory is well separated from that of the N (see figs. 5.6), though the hyperon Λ and Σ trajectories seem to satisfy the constraint quite well (fig. 7.10). A Veneziano model for meson–baryon scattering can be constructed, using $V(s,t)$ etc., like (7.4.4) for the invariant A' and B amplitudes (equation (4.3.11)), with $\alpha \to \alpha - \frac{1}{2}$ for channels containing baryons (see for example White (1971)). A rather thorough discussion

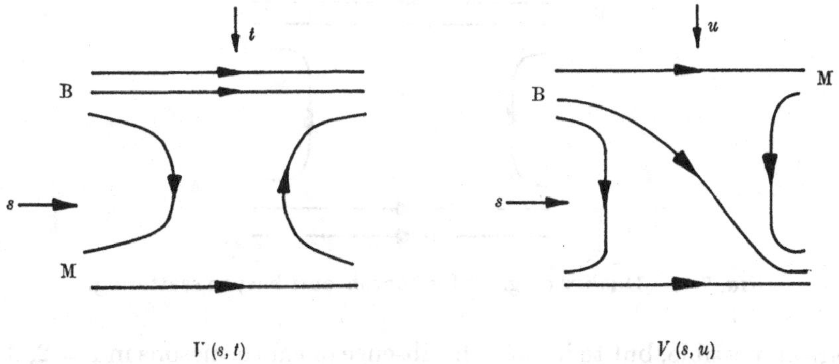

FIG. 7.9 Duality diagrams for meson–baryon scattering.

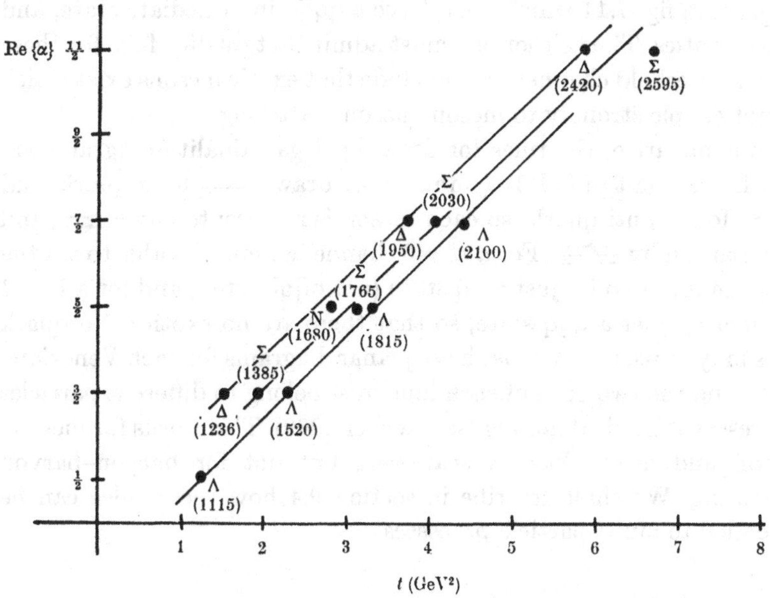

FIG. 7.10 Some examples of exchange-degenerate baryon trajectories. The splitting is much greater in most cases.

of the self-consistent, factorizing solutions for these cases has been given by Rimpault and Salin (1970). It seems probable, however, that to impose factorization constraints is too restrictive since, as we shall discuss below, phenomenologically duality seems to involve sums of cuts and poles rather than just poles.

When we come to examine baryon–anti-baryon scattering there are serious troubles because, for example, in $\Delta\Delta$ scattering $I = 0, 1, 2, 3$

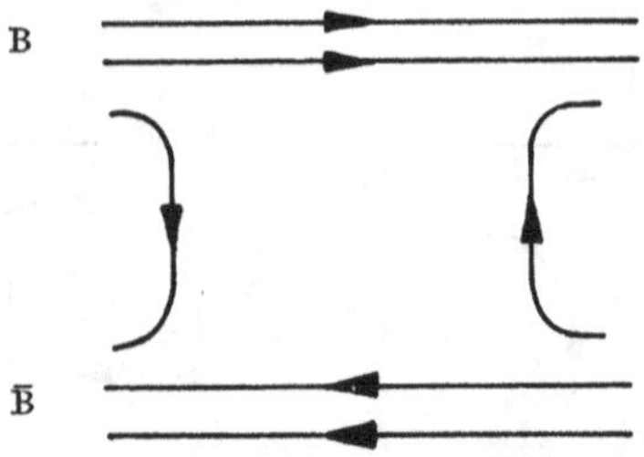

FIG. 7.11 Duality diagram for baryon–anti-baryon scattering.

are all possible, but to impose the absence of exotic mesons in $I = 2, 3$ in both the s and t channels requires that all the isospin amplitudes vanish (Rosner 1968). This is in fact rather obvious from the duality diagram in fig. 7.11 which must have a $qq\bar{q}\bar{q}$ intermediate state, and hence exotics. Thus either one must admit that duality fails for these higher-threshold channels, or conclude that exotic mesons exist which do not couple strongly to meson–meson scattering.

To summarize, the rules for drawing 'legal' duality diagrams are that in the limit of SU(3) symmetry we draw $\rightarrow$ for a quark, and $\leftarrow$ for an anti-quark, so each meson is represented by $\rightleftarrows$, and each baryon by $\Rrightarrow$. For a $B = 0$ channel we must be able to cut the diagram into two by just a $q\bar{q}$ state (not $q\bar{q}q\bar{q}$, etc.), and for a B = 1 channel by just a qqq state, so that there are no exotics. No quark lines may cross, i.e. we must have planar diagrams for each Veneziano term, and the two ends of each line must belong to different particles to preserve the ideal mixing (see Rosner 1969). This works for meson–meson and meson–baryon scattering but not for baryon–baryon scattering. We shall describe in section 9.4 how these rules can be extended to multi-particle processes.

7.6 Phenomenological implications of duality

There are many important consequences of the duality hypothesis which seem to be borne out experimentally. These include the pole dominance of the non-Pomeron part of scattering amplitudes, the absence of exotic resonances (which may help to explain why the quark model works), strong exchange degeneracy and nonsense decoupling, ideal mixing of SU(3) representations, the occurrence of parallel linear trajectories, and the fact that $s_0 = \alpha'^{-1}$. But we have also found that when pressed too hard the self-consistency of the duality scheme breaks down, so it is important to try and discover from experiment the extent to which these duality ideas hold good.

We have noted that although exchange degeneracy and ideal mixing seem to be valid for the vector and tensor mesons this is not the case for other exchanges. However, as these are the dominant exchanges in forward meson–baryon and baryon–baryon scattering, the duality rules work quite well for such processes. For example fig. 6.4 shows that the total cross-sections for exotic pp and K^+p are much flatter than those for $\bar{p}p$ and K^-p, and it seems very plausible that $\mathrm{Im}\{A^{el}(K^+p)\}$ contains just the P, as required by two-component duality. But $\sigma^{tot}(pp)$ does fall at low s, which indicates that the cancellation between the ω and f exchanges is not perfect in this case. These trajectories do of course contribute to $\mathrm{Re}\{A^{el}\}$ (see (6.8.22)). The dips in $d\sigma/dt$ at $|t| \approx 0.55\,\mathrm{GeV}^2$, observed in medium energy $\bar{p}p$ and K^-p elastic scattering, and due to the nonsense zero of the R contribution, are conspicuously absent in pp and K^+p (fig. 6.5). This is a direct verification of the importance of s-channel quantum numbers in controlling the t-channel exchanges, and hence of duality.

Detailed fits of meson–baryon scattering using the Veneziano model for the R term have been attempted. It is first necessary to 'smooth' the amplitude by taking its asymptotic form (7.4.8) even for real positive s. To cope with the baryon spin it has been usual to use the Veneziano model for the invariant amplitudes $A'(s,t)$ and $B(s,t)$ introduced in (4.3.11) rather than helicity amplitudes, because the former have more simple crossing properties. The chief difficulties are that, since no cuts are included, baryon parity doublets automatically appear (see (6.5.13)), the scale factor has to be altered from α'^{-1} to obtain the observed exponential fall of $d\sigma/dt$ with t (note that g in (7.4.4) is a constant), satellite terms have to be introduced, and there is the cross-over zero problem of section 6.8l (see Berger and Fox 1969). So quantitative fits of the data with the Veneziano model are not really possible.

Another interesting consequence of duality (Barger and Cline 1970) is that since with ideal mixing the ϕ is made of $\lambda\bar{\lambda}$ quarks only, it is impossible to exchange a $q\bar{q}$ pair in the quasi-elastic process $\gamma p \to \phi p$, so P alone should be exchanged (fig. 7.12). The very flat energy dependence of this process even at low energies (fig. 7.13) suggests that this is indeed the case.

For inelastic processes, where P cannot contribute, strong exchange degeneracy requires the sort of line-reversal equalities whose (modest)

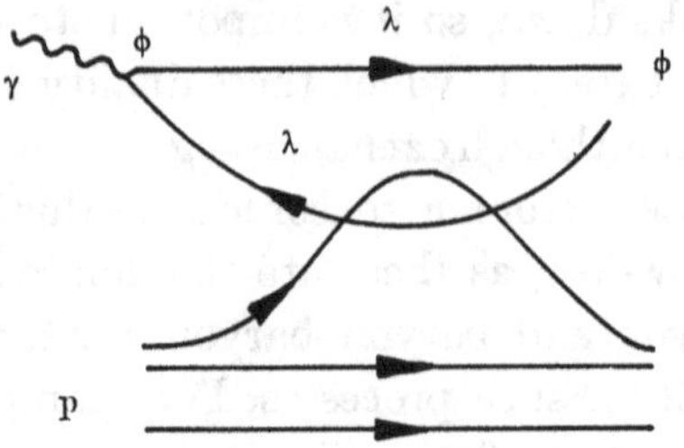

FIG. 7.12 A representation of P exchange in $\gamma p \to \phi p$. As the λ quarks are not exchanged down the diagram this has vaccuum quantum numbers but not $q\bar{q}$ in the t channel.

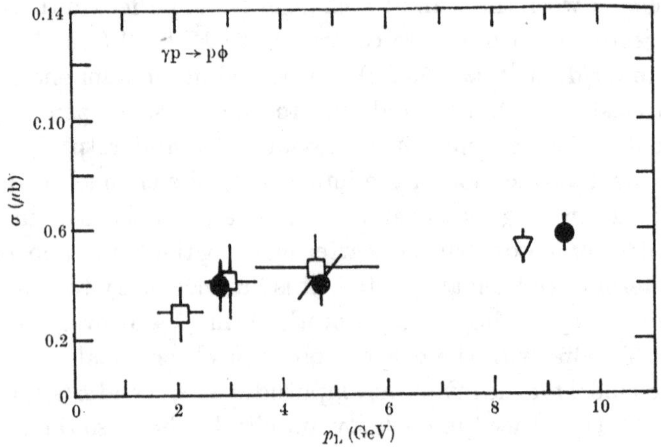

FIG. 7.13 Plot of $\sigma(\gamma p \to \phi p)$ versus laboratory momentum p_L, from Leith (1973).

success was described in section 6.8h. In particular Im$\{A(s, t)\}$ should vanish identically for inelastic processes with exotic s-channel quantum numbers. Examples are $K^+n \to K^0p$ and $Kp \to K\Delta$ for which duality diagrams with $q\bar{q}$ meson exchanges cannot be drawn (fig. 7.14). More interesting are processes like

$$K^-p \to \pi^-\Sigma^+, \quad K^-n \to \pi^-\Lambda, \quad K^-n \to \pi^-\Sigma^0,$$

which are not exotic but for which no legal duality diagram can be drawn, so there must be a cancellation between K** and K* exchanges in Im$\{A\}$. This also means that the resonances which occur in these processes must couple with alternating signs so that $\langle A^r \rangle \approx 0$ when averaged over a few resonances.

Similarly if the t channel is exotic, as in $\pi^-p \to \pi^+\Delta^-$ or $K^-p \to \pi^+\Sigma^-$,

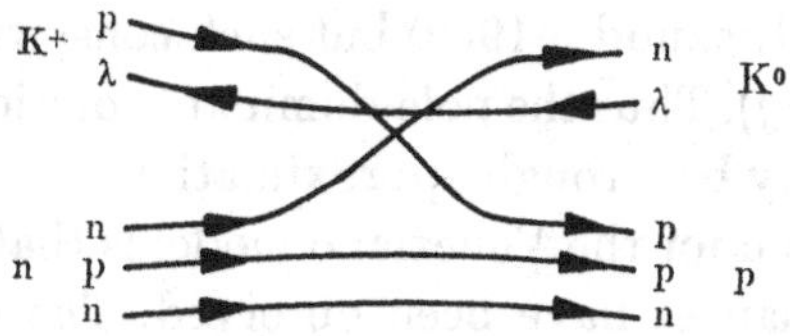

FIG. 7.14 Illegal duality diagram for $K^+n \to K^0p$.

since there are no t-channel exchanges, we must expect the resonances to cancel on average. This seems to work approximately for the former process but not the latter (Kernan and Sheppard 1969, Ferro-Luzzi *et al.* 1971). Duality diagrams make the further prediction that since $\phi = \lambda\bar{\lambda}$ it must decouple from inelastic (non-P exchange) processes involving only non-strange quarks. So processes like $\pi^- p \to \phi n$, $\pi^+ p \to \phi \Delta^{++}$ should not occur. Their cross-sections certainly seem to be very small compared with similar allowed processes such as $\pi^- p \to \omega n$, $\pi^+ p \to \omega \Delta^{++}$.

In general one concludes that the duality, exchange-degeneracy and ideal mixing requirements are moderately well satisfied for V and T exchanges, but certainly not exactly. But for most other exchanges, such as PS, $A^\pm$, or baryon, they are rather badly broken.

We have noted that strong exchange degeneracy demands nonsense decoupling, but found in section 6.8k that the choosing-nonsense hypothesis does not seem to be compatible with factorization, even for V and T exchanges. In fact it seems likely that pole–cut cancellation is needed to account for the dip in $d\sigma/dt\,(\pi N)$ near $\alpha = 0$ (see section 8.7c below). Similarly we have provisionally blamed the cross-over zero in $\operatorname{Im}\{A_{++}\}$ at $|t| \approx 0.15\,\text{GeV}^2$ on pole–cut cancellation (section 6.8l). However, as we mentioned in section 7.3, both of these features are present in the low energy resonance contribution and it therefore seems as though duality works somewhat better than does the hypothesis that Regge pole exchanges dominate, and it might be better to write

$$\langle A^r \rangle \approx A^R + A^c \tag{7.6.1}$$

where A^c is the Regge cut amplitude, rather than (7.3.2).

Further evidence for this comes from π exchange processes like $\gamma p \to \pi^+ n$, $\pi p \to \rho p$, etc., where the resonances produce forward peaks which were explained in section 6.8j (see also section 8.7f below) as due to interference between the π pole and a self-conspiring cut, π_c. So we have $\langle A^r \rangle \approx \pi + \pi_c$. The Veneziano model can only account for such processes by including conspiring trajectories (Armad,

Fayyazuddin and Riazuddin 1969) but such conspiracies are unsatisfactory (section 6.8j). Thus the pole-dominant solutions to the duality constraints can only be a rough approximation.

A further problem for the Veneziano model is that by no means all the required resonances have been observed. The leading ρ, ω, K* trajectories certainly seem to rise linearly to the $J = 3$ or 4 level, and baryon states up to perhaps $J = \frac{19}{2}$ are known, with no indication that higher-spin resonances may not be found eventually. But the daughter trajectories are much less well established. This may be partly because partial-wave analysis of the non-peripheral partial waves (i.e. $J < (\sqrt{s})R$, see section 2.2) is difficult because of contamination by the higher waves. However, there is no evidence for a ρ′(1275) daughter of the ρ, degenerate with the f (see fig. 7.5), and in fact strong evidence that it does not appear in the $\pi\pi$ channel. There is evidence of a heavier broad ρ′(1600), which couples more to 4π than 2π (see Particle Data Group 1974). This could be the daughter of the g(1680), which suggests that perhaps only the odd daughters of the ρ trajectory occur.

Many more baryon resonances are known, but fig. 5.6 shows that it is not a simple matter to fit them into daughter sequences. In any case high-mass, low-spin resonances are expected to be wide because of the large number of decay channels available to them, so the narrow resonance approximation will probably be poor at the daughter level, and it seems more plausible to regard the daughter sequences of the Veneziano model as simply a δ-function approximation to the channel discontinuities. In the next chapter we shall show why absorption is expected to be much stronger for low partial waves than higher ones, and it seems likely that pole dominance works best for the peripheral partial waves, $J \approx (\sqrt{s})R$. Of course with linear trajectories there will be resonances in the super-peripheral partial waves up to $J_{\max} \approx \alpha' s$ so pole dominance may in fact be satisfactory for $\alpha' s \gtrsim J \gtrsim (\sqrt{s})R$, but in the Veneziano model the resonances with $J \gg (\sqrt{s})R$ have rather small widths, and those with $J \approx (\sqrt{s})R$ dominate (see fig. 7.15). (This must be so because the Veneziano model reproduces the observed peripheral forward peak.)

Despite these limitations the Veneziano model has had one additional and rather surprising success, in predicting amplitude zeros. The Γ-function in the denominator of (7.4.15) means that $V(s,t)$ has a zero along the line

$$\alpha(s) + \alpha(t) = p \tag{7.6.2}$$

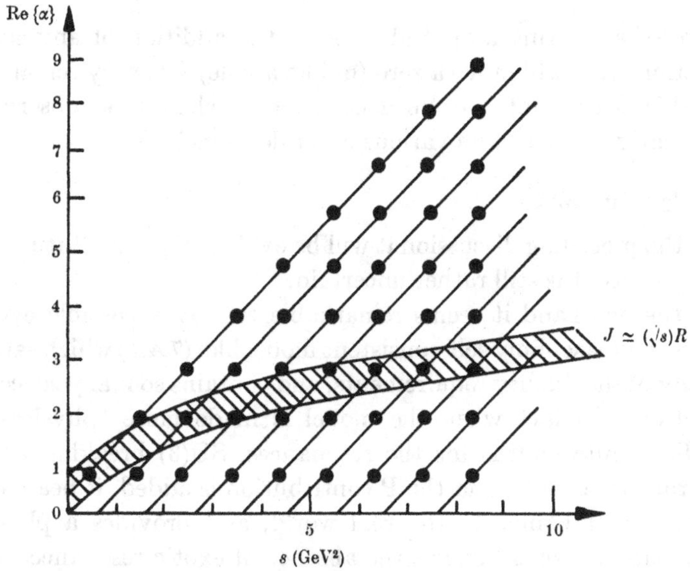

FIG. 7.15 The resonances of the Veneziano model and the peripheral region (shaded).

where p = an integer ⩾ n. With parallel linear trajectories (7.6.2) implies

$$\alpha_s^0 + \alpha_t^0 + \alpha'(s+t) = p \tag{7.6.3}$$

or, from (1.7.21), $$u = \frac{\alpha_s^0 + \alpha_t^0 - p}{\alpha'} + \Sigma = \text{constant} \tag{7.6.4}$$

So zeros of the amplitude are predicted along lines of constant u.

The occurrence of these zeros in the unphysical region $s, t > 0$ is of course necessary to prevent double poles (see fig. 7.4), but the zeros are also predicted to continue into the physical region. Of course if the other terms $V(s,u)$, $V(t,u)$ are added these zeros may be removed, but in a process such as $K^-p \to \bar{K}^0n$, for which the u channel is exotic so only $V(s,t)$ occurs, dips may be expected at fixed u, spaced by $\alpha'^{-1} \approx 1\,\text{GeV}^2$. These dips should occur despite the fact that there are no u-channel poles, because they stem from a cancellation between the s-channel Λ, Σ^0 poles and the t-channel ρ, ω, f, A_2 poles. Fixed zeros are in fact found at $u = -0.1$, -0.7 and $-1.7\,\text{GeV}^2$ (Odorico 1971). This is not exactly where the Veneziano model would predict them, but in view of its approximate nature some displacement is to be expected.

Odorico (1972) has shown that such fixed zeros are quite a general feature of scattering amplitudes. Since the addition of any sort of correction term will move a zero (unlike a pole) it is very remarkable that this feature of the Veneziano model should be observable, particularly in view of its various other deficiencies.

7.7 Conclusions

From the preceding discussion it will be evident that the status of the duality concept is still rather uncertain.

On the one hand it seems remarkable that it is possible even to construct a reasonably self-consistent model like (7.4.4) which satisfies so many of the duality requirements, and contains so many successful predictions. In fact when the model is made more 'physical' by including finite widths for the resonances, SU(3) breaking for the trajectory intercepts, and the P contribution is added, it bears quite a strong resemblance to the real world, and provides a plausible explanation for such facts as the absence of exotic resonances, ideal mixing, parallel linear exchange-degenerate trajectories, and $\alpha'^{-1} \approx s_0$. But unfortunately this physical model is not self-consistent because of the ancestor problem, the occurrence of exotics in $B\bar{B}$ channels, etc., and it does not agree quantitatively with experiment.

This could be because duality is only approximately valid. Alternatively, it might be an exact principle, all our difficulties stemming from the failure to incorporate unitarity, and especially Regge cuts, properly. But there do not seem to be any very compelling arguments in favour of duality as a basic law of strong interactions. All the very tight restrictions of dual models which give them predictive power come from the adoption of meromorphic scattering amplitudes (see for example Oehme 1970) (i.e. amplitudes containing only poles, no cuts), and once cuts are permitted it is not even clear how to formulate the duality idea.

One suggestion has been that one can regard the Veneziano model as a sort of 'Born approximation' for strong interactions, which should be iterated in the unitarity equations (as in section 3.5) to produce the physical S-matrix. In this case loop diagrams like fig. 1.11 (*b*) will occur corresponding to the re-normalization of the masses, and couplings of the resonances. We shall examine some of these ideas briefly in chapters 9 and 11. So far they still seem to suffer from the usual ambiguities concerning the convergence of the Born series and double counting of terms, though such problems may eventually be overcome.

There is, however, one further very important feature of the Veneziano model which we shall look at in chapter 9. It is comparatively easy to generalize to many-particle scattering amplitudes, and provides a parameterization of the amplitudes which exhibits both resonance dominance at low energies and Regge asymptotic behaviour, with factorizable couplings for all the trajectories, in all the different channels. This has greatly facilitated the application of Regge theory to many particle processes. So, even if the duality idea should turn out not to be a fundamental principle of strong interaction dynamics, dual models will still have their uses, both as a mnemonic for many of the basic facts of two-body processes, and as a simplifying model for more complex ones.

8
Regge cuts

8.1 Introduction

In section 4.8 we demonstrated that the occurrence of Gribov–Pomeranchuk fixed poles at wrong-signature nonsense points, generated by the third double spectral function, ρ_{su}, requires that there be cuts in the t-channel angular-momentum plane. Otherwise it is impossible to satisfy t-channel unitarity. We have also found in section 6.8 that, despite the many successes of Regge pole phenomenology, there are some features of the data that poles alone cannot explain. These are mainly failures of factorization, and it seems natural to try an invoke Regge cuts, which correspond to the exchange of two or more Reggeons and so are not expected to factorize, to make good these defects.

Unfortunately we still have a much less complete understanding of the properties of Regge cuts than of the properties of poles. On the phenomenological side, this is mainly because it is difficult to be sure whether cuts or poles are responsible for what is observed, since the main tests, $\log s$ behaviour (see (8.5.12) below) and lack of factorization, are hard to apply. Though cuts do not have to factorize some models suggest that they do, at least approximately. We shall review some of these problems in section 8.7.

Also the various theoretical models which have been used to gain insight into the behaviour of Regge poles (discussed in chapter 3) are harder to apply to cuts. For example in potential scattering, which has only elastic unitarity and no third double spectral function, there are no Regge cuts if the potentials are well behaved. Though if the potential is singular, say

$$U(r) = \frac{V_0}{r^2} + \overline{V}(r) \tag{8.1.1}$$

where $\overline{V}(r)$ is regular as $r \to 0$, the radial Schroedinger equation becomes

$$\frac{d^2\phi_l(r)}{dr^2} + \left[k^2 - \frac{l(l+1)+V_0}{r^2} - \overline{V}(r)\right]\phi_l(r) = 0 \tag{8.1.2}$$

which has the same form as (3.3.3) if l is replaced by L, where

$$L(L+1) \equiv l(l+1) + V_0 \tag{8.1.3}$$

so the solutions will be meromorphic in L. But a pole at $L = \alpha$ gives, on inverting (8.1.3), branch points in the l plane at

$$l = \tfrac{1}{2}\{-1 \pm [1 - 4V_0 + 4\alpha(\alpha+1)]^{\frac{1}{2}}\} \tag{8.1.4}$$

so the singular part of the potential produces cuts in the l plane. But there is no reason to suppose that similar cuts will occur in strong interactions, because they do not seem to be related to multi-Reggeon exchange.

Instead it is necessary to rely mainly on Feynman-diagram models to deduce the properties of Regge cuts. But, as we shall find in the next section, there are difficulties associated with the many-to-one correspondence between Feynman diagrams and unitarity diagrams and the convergence of the perturbation series, which limit the applicability of these models in strong interactions. Gribov (1968) has developed an ingenious scheme for inserting Regge poles themselves into Feynman diagrams, giving a 'Reggeon calculus' from which one can deduce the discontinuities across J-plane cuts, analogous to the unitarity diagram approach to s-plane discontinuities. This calculus, to be discussed in section 8.3, has allowed considerable progress though the theory is still incomplete.

We shall also examine some popular approximation methods for calculating cut contributions, in particular the absorption and eikonal models, before going on to examine the phenomenological application of these ideas in the final section. Much of the discussion is rather technical and the reader is advised to skip the more difficult parts at the first reading. If he is mainly interested in phenomenology he could go straight to section 8.7 and refer back as necessary.

8.2 Regge cuts and Feynman diagrams*

We found in section 3.4 that a single Regge pole exchange corresponds to the set of ladder Feynman diagrams like fig. 8.1, where we sum over all possible numbers of rungs as in (3.4.12). Regge cuts arise from the exchange of two (or more) Reggeons, and so the simplest type of diagram which might be expected to produce a Regge cut is fig. 8.2. This is a planar diagram, and the rules for obtaining the asymptotic behaviour of such diagrams are comparatively simple, because they depend only on the 'end-point' contributions (see section 3.4).

The asymptotic power behaviour of a ladder diagram as

* This section may be omitted at first reading.

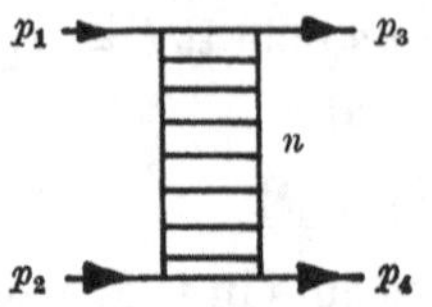

FIG. 8.1 A ladder Feynman diagram which contributes to a t-channel Regge pole.

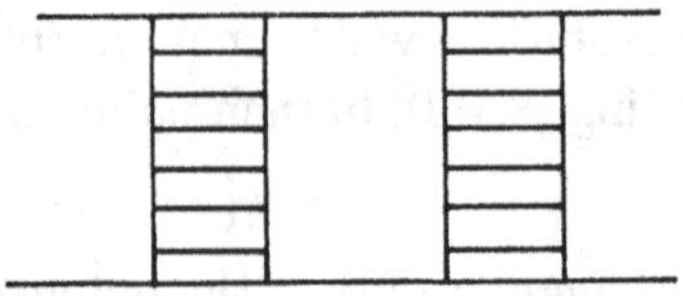

FIG. 8.2 A two-ladder diagram which might be expected to produce a Regge cut.

$s = (p_1+p_2)^2 \to \infty$, $t = (p_1-p_3)^2$ fixed, is s^{-1} (from (3.4.11)), independent of the number of rungs because just one propagator is needed to cross the diagram; and the leading s behaviour is $s^{-1}(\log s)^{n-1}$ because there are n different independent paths by which one might cross the diagram. This result can be generalized for (most) planar diagrams as follows (Eden *et al.* (1966) p. 138).

We look for paths through the graph (i.e. connected sets of internal lines) which if short-circuited split the graph into two parts which have only a single vertex and no lines in common, p_1 and p_3 being coupled to one side, and p_2 and p_4 to the other (assuming we are considering $s \to \infty$, t fixed). The three different ways of doing this for fig. 8.3 (a) are shown in figs. 8.3 (b), (c), (d). We select those paths which are of the minimum length, i.e. those which involve short-circuiting the smallest number of lines. Thus figs. 8.3 (c) and (d) are included because they short-circuit only two lines, but fig. 8.3 (b) which involves three lines is excluded. These paths of minimum length are called 'd-lines'. The rule is that the asymptotic power of s for a diagram whose d-lines are of length m is s^{-m}. So fig. 8.3 (a) with d-lines of length 2 behaves like $\sim s^{-2}$, while the ladder fig. 8.1, with d-lines of length 1, $\sim s^{-1}$.

If there are n such d-lines (all of the same minimum length m) for a given diagram, then the asymptotic behaviour will be

$$\sim s^{-m}(\log s)^{n-1}, \quad m, n \geqslant 1 \tag{8.2.1}$$

Thus since fig. 8.3 (a) has 2 d-lines its behaviour is $\sim s^{-2}\log s$. This rule obviously also works for ladder diagrams to give (3.4.11). Some graphs, involving 'singular configurations', are exceptions to these rules (see Eden *et al.* p. 141) but we shall not need to consider them here.

If we apply (8.2.1) to the two-ladder graph (fig. 8.2) we see that its d-lines are the two paths across the top and bottom of the diagram, each of length 3, and so $m = 3$, $n = 2$. Thus all diagrams like fig. 8.2

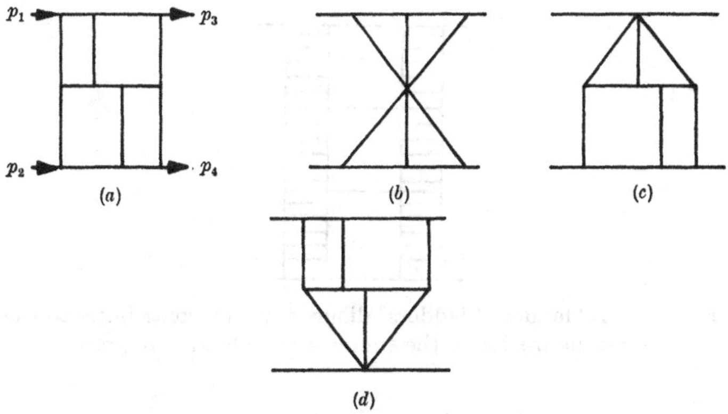

FIG. 8.3 (a) A Feynman diagram; and (b), (c) and (d) the three ways of short-circuiting it, as described in the text.

behave like $s^{-3}\log s$ independent of the number of rungs in the two ladders. So the sum of all such diagrams, with all possible numbers of rungs, may be expected also to have this behaviour (provided the sum converges), and so to give rise to a fixed singularity at $l = -3$ (see (2.7.4)), not a moving Regge cut.

Regge behaviour stems from summing over all powers of $\log s$ in (3.4.12), and it is the fact that only the first power of $\log s$ occurs in the leading asymptotic behaviour of all the diagrams like fig. 8.2 which prevents Reggeization. If we sum sets of such diagrams, like fig. 8.4, the sum would give us a Regge pole like (3.4.12) but with $\alpha(\infty) = -3$. The small ladders simply give re-normalizations of the basic ladder diagram fig. 8.1. This shows why planar diagrams, whose asymptotic behaviour comes just from end-point singularities, contribute only to the Regge poles, not to the cuts.

However, if we take the discontinuity of fig. 8.2 across the two-particle intermediate state, as shown in fig. 8.5(a), the two-body unitarity condition (1.5.7) gives the two particle discontinuity

$$\frac{1}{2i}(A^+ - A^-) \equiv \Delta_2\{A(s,t)\} = \frac{q_s}{32\pi^2\sqrt{s}}\int A^1(s,t_1)\,A^{2*}(s,t_2)\,d\Omega_s \tag{8.2.2}$$

where $d\Omega_s$ is the element of solid angle in the intermediate state (see fig. 2.1) and $A^1(s,t_1)$ and $A^2(s,t_2)$ are the ladder amplitudes shown in the figure. If this equation is decomposed into s-channel partial waves we get (see (2.2.7))

$$\Delta_2\{A_l(s)\} = \frac{2q_s}{\sqrt{s}}A_l^1(s)A_l^{2*}(s) \tag{8.2.3}$$

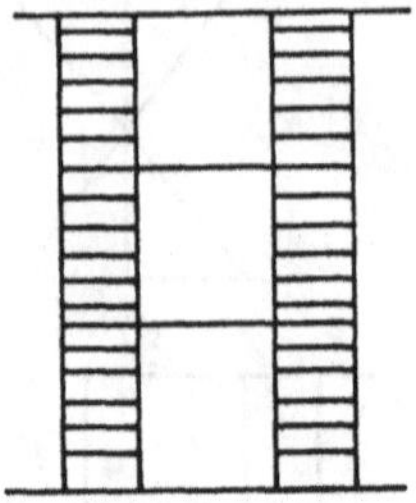

FIG. 8.4 A 'ladder of ladders' diagram which contributes to the re-normalization of the simple 4-rung ladder diagram.

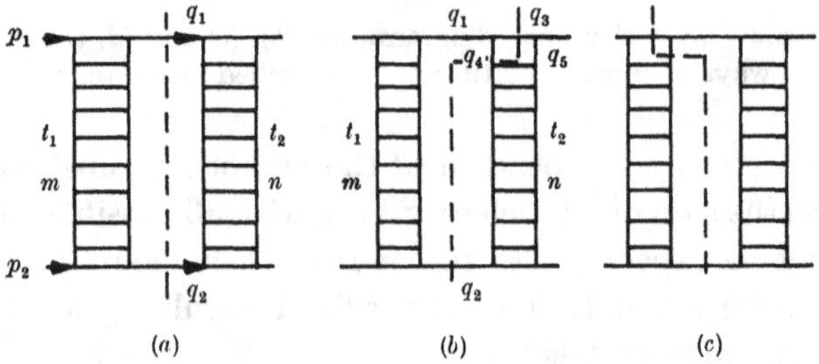

FIG. 8.5 (a) Fig. 8.2 cut across the two-body intermediate state; (b) a cut across a three-body state; (c) a similar cut.

where $$A_l^1(s) = \frac{1}{32\pi}\int_{-1}^{1} A^1(s,t_1)\, P_l(z_1)\, dz_1, \quad z_1 \equiv z_s(s,t_1) \tag{8.2.4}$$

etc., and so on summing the partial-wave series (2.2.2),

$$\begin{aligned}\Delta_2\{A(s,t)\} &= 16\pi \sum_l (2l+1)\, P_l(z_s)\, \Delta_2\{A_l(s)\} \\ &= 16\pi \sum_l (2l+1)\, P_l(z_s) \frac{2q_s}{\sqrt{s}} \frac{1}{32\pi}\int_{-1}^{1} A^1(s,t_1)\, P_l(z_1)\, dz_1 \\ &\quad \times \frac{1}{32\pi}\int_{-1}^{1} A^{2*}(s,t_2)\, P_l(z_2)\, dz_2 \end{aligned} \tag{8.2.5}$$

But (Goldberger and Watson (1964) p. 595; Henyey *et al.* (1969))

$$\sum_l (2l+1)\, P_l(z_s)\, P_l(z_1)\, P_l(z_2) = \frac{2}{\pi}\frac{\theta(\Delta)}{\Delta^{\frac{1}{2}}} \tag{8.2.6}$$

where $$\Delta \equiv 1 - z_s^2 - z_1^2 - z_2^2 + 2z_s z_1 z_2 \tag{8.2.7}$$

and $\theta(\Delta)$ is the step function

$$\theta(\Delta) = 0, \quad \Delta < 0; \quad \theta(\Delta) = 1, \quad \Delta > 0 \tag{8.2.8}$$

and so

$$\Delta_2\{A(s,t)\} = \frac{q_s}{16\pi^2\sqrt{s}}\int_{-1}^{1} dz_1 \int_{-1}^{1} dz_2 A^1(s,t_1)\, A^{2*}(s,t_2)\frac{\theta(\Delta)}{\Delta^{\frac{1}{2}}} \tag{8.2.9}$$

Then for large s and small t, from (1.7.22),

$$z_1 \approx 1+\frac{2t_1}{s}, \quad z_2 \approx 1+\frac{2t_2}{s}, \quad z_s \approx 1+\frac{2t}{s} \tag{8.2.10}$$

so (8.2.9) becomes

$$\Delta_2\{A(s,t)\} \approx \frac{1}{16\pi^2 s}\int_{-\infty}^{0} dt_1 \int_{-\infty}^{0} dt_2 A^1(s,t_1)\, A^{2*}(s,t_2)\, \frac{\theta(-\lambda)}{(-\lambda(t,t_1,t_2))^{\frac{1}{2}}} \tag{8.2.11}$$

where $$\lambda(t,t_1,t_2) \equiv t^2+t_1^2+t_2^2-2(tt_1+tt_2+t_1t_2) \tag{8.2.12}$$

(see (1.7.11)). The result, from (1.5.3)–(1.5.7) and (8.2.2)–(8.2.11), that for $s \to \infty$, t small,

$$\int \frac{d^4q}{(2\pi)^4}\, 2\pi\,\delta((p_1+q)^2-m^2)\, 2\pi\,\delta((p_2-q)^2-m^2)$$
$$\to \int \frac{d^2\mathbf{q}_\perp}{(2\pi)^2} \to \frac{1}{8\pi^2|s|}\int_{-\infty}^{0}\int_{-\infty}^{0} dt_1\, dt_2 \frac{\theta(-\lambda)}{(-\lambda)^{\frac{1}{2}}} \tag{8.2.13}$$

will frequently be of use for phase-space integrations in the high energy limit.

So if, for example, we represent each ladder sum by a linear Regge pole amplitude

$$A^i(s,t) \sim s^{\alpha_i(t)} = e^{(\alpha_i^0+\alpha_i' t)\log s} \tag{8.2.14}$$

since it is found that (O'Donovan 1969)

$$\int_{-\infty}^{0} dt_1 \int_{-\infty}^{0} dt_2\, e^{(b_1t_1+b_2t_2)}\frac{\theta(-\lambda)}{(-\lambda)^{\frac{1}{2}}} = \pi\frac{e^{[b_1b_2/(b_1+b_2)]t}}{b_1+b_2} \tag{8.2.15}$$

(8.2.11) gives, for $\log s \to \infty$,

$$\Delta_2\{A(s,t)\} \sim \frac{s^{(\alpha_1{}^0+\alpha_2{}^0-1)+[\alpha_1'\alpha_2'/(\alpha_1'+\alpha_2')]t}}{(\alpha_1'+\alpha_2')\log s} \tag{8.2.16}$$

which corresponds to a Regge cut at

$$\alpha_c(t) = \alpha_1^0+\alpha_2^0-1+\left(\frac{\alpha_1'\alpha_2'}{\alpha_1'+\alpha_2'}\right)t \tag{8.2.17}$$

with a finite discontinuity at the branch point (see (2.7.4)). As long as the trajectories $\alpha_i(t_i)$ are monotonically increasing functions of t_i in $-\infty < t_i < 0$, the leading behaviour of (8.2.11) will come from the region $\lambda = 0$, which from (8.2.12) implies (since $t_1, t_2, t \leqslant 0$)

$$\sqrt{-t} = \sqrt{-t_1}+\sqrt{-t_2} \tag{8.2.18}$$

so more generally $\alpha_c(t) = \max\{\alpha_1(t_1)+\alpha_2(t_2)-1\}$ (8.2.19)

subject to (8.2.18). The reader can easily check that (8.2.19) gives (8.2.17) for linear trajectories.

This argument mistakenly led Amati, Fubini and Stanghellini (1962) to suppose that fig. 8.2 would give rise to a Regge cut (now called an AFS cut). However, we know that the asymptotic behaviour of the diagram is actually $s^{-3}\log s$ not (8.2.16), so this Regge cut behaviour of the two-particle discontinuity must be cancelled by the other discontinuities of fig. 8.2, such as fig. 8.5 (*b*) (Mandelstam 1963). This cancellation has been demonstrated nicely by Halliday and Sachrajda (1973).

The discontinuity across the two-particle cut fig. 8.5 (*a*) may be written

$$\Delta_2\{A\} = \tfrac{1}{2}\iint \frac{d^4q_1}{(2\pi)^4}\frac{d^4q_2}{(2\pi)^4}(2\pi)\,\delta(q_1^2-m^2)\,(2\pi)\,\delta(q_2^2-m^2) \times (2\pi)^4\,\delta^4(q_1+q_2-p_1-p_2)\,A_m A_n^* \quad (8.2.20)$$

As usual $s = (p_1+p_2)^2, \quad t = (p_1-p_3)^2 = q^2$ (8.2.21)

and we introduce the four-vectors

$$p_1' = p_1 - \frac{m^2}{s}p_2, \quad p_2' = p_2 - \frac{m^2}{s}p_1 \quad (8.2.22)$$

which have the property that (using (1.7.4))

$$p_1'^2 = p_2'^2 = 0 + O\left(\frac{1}{s^2}\right) \quad \text{and} \quad 2p_1'.p_2' = s \quad (8.2.23)$$

Then introducing Sudakov variables α_i, β_i and $\boldsymbol{q}_{i\perp}$ for each four-vector q_i (see Halliday and Saunders 1968)

$$q_i = \alpha_i p_1' + \beta_i p_2' + \boldsymbol{q}_{i\perp}, \quad i = 1, 2 \quad (8.2.24)$$

where $\boldsymbol{q}_{i\perp}$ is a two-vector perpendicular to the plane containing p_1' and p_2', and α_i and β_i give the components of q_i in the directions of p_2' and p_1', respectively. Thus we have

$$d^4q_i\,\delta(q_i^2-m_i^2) = \tfrac{1}{2}|s|\,d\alpha_i\,d\beta_i\,d^2\boldsymbol{q}_{i\perp}\,\delta(\alpha_i\beta_i s-\mu_i^2) \quad (8.2.25)$$

where $\mu_i^2 \equiv m^2 + \boldsymbol{q}_{i\perp}^2$ (8.2.26)

and $\delta^4(p_1+p_2-\Sigma q_i) = \delta(\Sigma\alpha_i-1)\,\delta(\Sigma\beta_i-1)\,\delta(\Sigma\boldsymbol{q}_{i\perp})\dfrac{2}{|s|}$ (8.2.27)

and so

$$\Delta_2\{A\} = \frac{1}{(2\pi)^2}\int \frac{d\alpha_1\,d\beta_1}{2\alpha_1\,2\beta_2}\delta(\alpha_1+\alpha_2-1)\,\delta(\beta_1+\beta_2-1)\,A_m A_n^* \times d^2\boldsymbol{q}_{1\perp}\,d^2\boldsymbol{q}_{2\perp}\,\delta(\boldsymbol{q}_{1\perp}+\boldsymbol{q}_{2\perp}) \quad (8.2.28)$$

The momentum transfer down the left-hand ladder is

$$t_1 = (p_1 - q_1)^2 \to (\alpha_1 - 1)\beta_1 s - \boldsymbol{q}_{\perp}^2 \tag{8.2.29}$$

which must remain finite as $s \to \infty$ if we are to remain in the Regge regime, so as $s \to \infty$ we are interested in the integration region $\beta_1 \sim 1/s$, $\alpha_1 \sim$ constant, and so from the δ functions in (8.2.28) we must have $\beta_2 \sim 1$, $\alpha_2 \sim 1/s$, $\alpha_1 \sim 1$. So as $s \to \infty$

$$\Delta_2\{A\} = \frac{1}{16\pi^2 s}\int A_m A_n^* \,\mathrm{d}^2\boldsymbol{q}_{1\perp} \tag{8.2.30}$$

Then if we insert (3.4.11) for the asymptotic behaviour of the ladder diagrams we get

$$\Delta_2\{A\} = \frac{g^4}{16\pi^2 s^3}\frac{(\log s)^{m+n-2}}{(m-1)!\,(n-1)!}\int K(t_1)^{m-1}K(t_2)^{n-1}\,\mathrm{d}^2\boldsymbol{q}_{1\perp} \tag{8.2.31}$$

which is just the result needed to obtain (8.2.16) after summation over all numbers of rungs.

But if we consider the discontinuity of fig. 8.5 (*b*), the left-hand side has

$$A^{\mathrm{L}} = A_m \frac{g}{q_1^2 - m^2}$$

and the right-hand side has one rung subtracted so

$$A^{\mathrm{R}} = A_{n-1}\frac{g}{q_5^2 - m^2}$$

and it is found after integration over q_2, q_3 and q_4 that

$$\Delta_3\{A\} = -\frac{g^4}{16\pi^2 s^3}\frac{(\log s)^{m+n-2}}{(m+n-2)!\,(m-1)!\,(n-2)!}\int K(t_1)^{m-1}K(t_2)^{n-1}\,\mathrm{d}^2\boldsymbol{q}_{1\perp} \tag{8.2.32}$$

Then adding fig. 8.5 (*c*) which is the same with $m \leftrightarrow n$ we get an exact cancellation of the leading behaviour of fig. 8.5 (*a*), i.e. (8.2.31). Similarly there is a cancellation among the leading behaviours of all the other possible unitary dissections of fig. 8.2, and so no Regge cut actually appears. (In fact the AFS cut occurs on the unphysical sheet reached through the two-body cut in s.)

The above is a very good example of the dangers which lurk in the many-to-one correspondence between Feynman and unitarity diagrams.

To obtain a Regge cut we must look at non-planar diagrams in

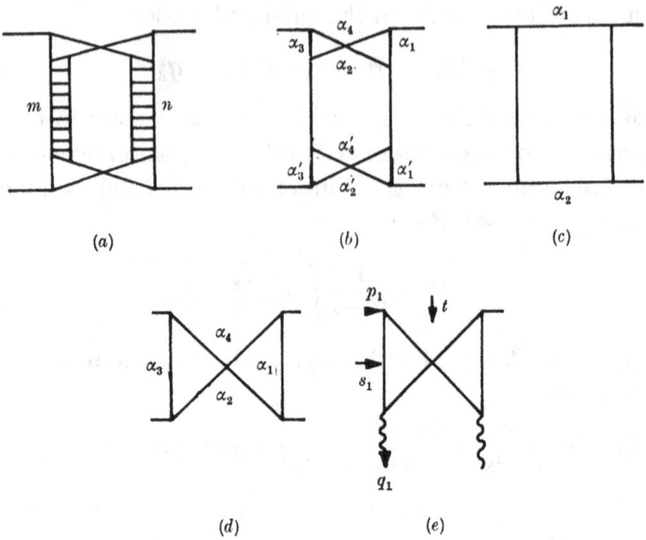

Fig. 8.6 (*a*) Mandelstam's double-cross diagram. (*b*) The most elementary form of (*a*) with the Feynman parameters. (*c*) The box diagram which does not have a 'pinch' asymptotic behaviour. (*d*) The cross diagram. (*e*) The cross diagram in particle–Reggeon scattering.

which the leading behaviour comes from the pinching of singularities (Eden *et al.* p. 158). The simplest such diagram is Mandelstam's 'double-cross' diagram, fig. 8.6(*a*), of which the simplest form is fig. 8.6(*b*). This has 6 d-lines each of length 2 and so the end-point behaviour is $\sim s^{-2}(\log s)^5$. However we have seen that the cross is the simplest diagram which can produce a Gribov–Pomeranchuk fixed pole at $l = -1$ in the t-channel angular-momentum plane (see (2.8.7)), so it should have an $\sim s^{-1}$ behaviour.

The coefficient of s in the Feynman denominator of (3.4.4) is

$$(\alpha_1\alpha_3-\alpha_2\alpha_4)(\alpha_1'\alpha_3'-\alpha_2'\alpha_4') \equiv x_1x_2 \quad \text{(say)} \tag{8.2.33}$$

and when we integrate over the α's $(0\to 1)$ there will be points where both brackets x_1 and x_2 vanish. In this region the integral takes the form

$$\int_{a_1}^{b_1} \mathrm{d}x_1 \int_{a_2}^{b_2} \mathrm{d}x_2 \frac{1}{(x_1x_2s+d)^2} = -\frac{1}{sd}\log\left[\frac{(a_1a_2s+d)(b_1b_2s+d)}{(a_1b_2s+d)(a_2b_1s+d)}\right]+\dots \tag{8.2.34}$$

Now as s tends to ∞ the argument of the log tends to 1, and as $\log 1 = 0$ we get the expected s^{-2} behaviour, but only if all $a_i, b_i > 0$ or all

$a_i, b_i < 0$. If say $a_1, a_2 < 0$ and $b_1, b_2 > 0$ then as s tends to ∞ the numerator in the log tends to $\infty + i\epsilon$ while the denominator tends to $\infty - i\epsilon$, so the log tends to $2\pi i$ giving instead (8.2.34) $\sim -2\pi i(sd)^{-1}$. So the vanishing of the brackets in (8.2.33) gives a pinch asymptotic behaviour which is different from that of the end-point singularities. In the box diagram fig. 8.6 (c) x_1 and x_2 are replaced by α_1 and α_2 respectively which vanish only at the end-points.

If we now return to fig. 8.6 (a) the leading singularity comes from the pinch singularities of the crosses, together with the end-point singularities of the ladders, (3.4.11), and the asymptotic behaviour is found to be

$$\frac{ig^4}{16\pi^2}\int dt_1 \int dt_2 \frac{(N(t,t_1,t_2))^2\,\theta(-\lambda)}{(-\lambda(t,t_1,t_2))^{\frac{1}{2}}}\frac{K(t_1)^{m-1}\,K(t_2)^{n-1}\,(\log s)^{m+n-2}}{(m-1)!\,(n-1)!\,s^3} \tag{8.2.35}$$

where $K(t)$ is the box diagram function (3.4.9), and N is the Feynman integral of the cross diagram fig. 8.6 (d) in the pinch configuration, i.e.

$$N = \int_0^1 d\alpha_1 \ldots d\alpha_4 \frac{\delta(\Sigma\alpha - 1)\,\delta(\alpha_1\alpha_3 - \alpha_2\alpha_4)}{d(t,t_1,t_2,\alpha)} \tag{8.2.36}$$

d being its Feynman denominator. N appears squared because fig. 8.6 (a) contains two identical crosses. Then if we sum (8.2.35) over all possible numbers of rungs in the ladders we get

$$A = \frac{ig^4}{16\pi^2}\int dt_1 \int dt_2 \frac{(N(t,t_1,t_2))^2\,\theta(-\lambda)}{(-\lambda(t,t_1,t_2))^{\frac{1}{2}}}\, s^{\alpha(t_1)+\alpha(t_2)-1} \tag{8.2.37}$$

$$\alpha(t) = -1 + K(t)$$

which agrees with AFS result (8.2.19) for the position of the cut, and gives (8.2.17) if the trajectories are linear.

So the Mandelstam diagram, fig. 8.6 (a), produces a branch point whose location in the l plane is identical to that of the AFS cut, fig. 8.5 (a). But (8.2.37) differs from (8.2.11) not only through the occurrence of N^2, but because (8.2.11) involves A_n^* where as (8.2.35) does not. (Equation (8.2.11) has been divided by 2i because we have taken the discontinuity.) And whereas the AFS cut occurs only on unphysical sheets, the Mandelstam cut is present on the physical sheet and so contributes to the asymptotic behaviour.

Another way of seeing why the non-planar structure is necessary is to note that we can write

$$N(t,t_1,t_2) = \int_{-\infty}^{\infty} ds_1 A_1(s_1,t,t_1,t_2) \tag{8.2.38}$$

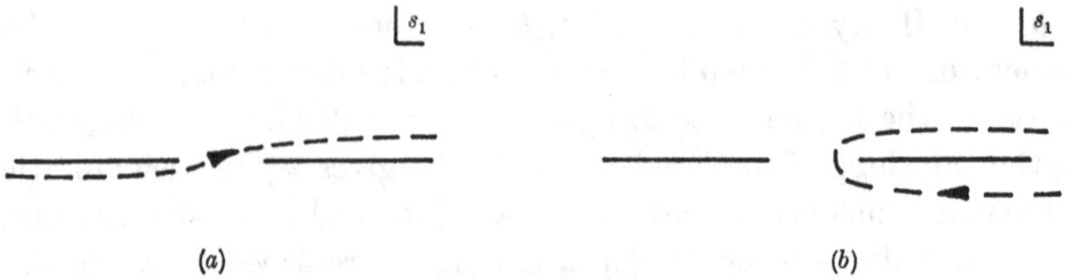

FIG. 8.7 (a) Integration contour along the real s_1 axis. (b) Deformed contour round the positive s_1 cuts.

where $A_1(s_1, t, t_1, t_2)$ is the Reggeon–particle scattering amplitude for the cross (fig. 8.6(e)), and $s_1 = (p_1 - q_1)^2$, the integration contour being as in fig. 8.7(a). Since $A_1 \sim s^{-2}$ we can close the contour at infinity as in fig. 8.7(b) and obtain

$$N(t, t_1, t_2) = 2i \int_{4m^2}^{\infty} \text{Im}\{A_1(s_1, t, t_1, t_2)\} \, ds_1 \tag{8.2.39}$$

which is just the residue of the Gribov–Pomeranchuk fixed pole in A_1, i.e. in (4.8.4) at $J_0 = -1$, $\lambda = \lambda' = 0$. However, if the amplitude did not have the cross-structure, and hence only had singularities for positive s_1, we should be able to close the contour in the upper half-plane, and so find $N = 0$, which is what happens in the AFS case. The contribution of the pole at $q_1^2 = s = m^2$ in fig. 8.5(a) is cancelled by the right-hand cut due to the singularities of the vertex function coupling this particle to the Reggeon, the simplest contributions to which are the lines q_3 and q_4 in fig. 8.5(b) – see Rothe (1967), Landshoff and Polkinghorne (1971).

A rather more physical understanding of this result can be obtained by considering scattering of composite particles, say deuterons. In d–d scattering, in addition to the single exchange diagram fig. 8.8(a), there are various double scattering diagrams, figs. 8.8(b), (c). Of these fig. 8.8(b) becomes very improbable at high energies because it requires a given pair of nucleons to scatter off each other twice, despite the fact that they are passing each other very rapidly. On the other hand fig. 8.8(c), which involves each nucleon in only a single scattering, can perfectly well occur even at very high energies. So the planar diagram (b) dies away at high energies, but (c), which depends in an essential way on the deuterons being composite, remains. Obviously (c) has the same structure as the Mandelstam diagram, fig. 8.6(a). (The connection between Regge cuts and Glauber's multiple scattering theory is complicated, however; see Glauber (1959), Abers *et al.* (1966), Harrington (1970).)

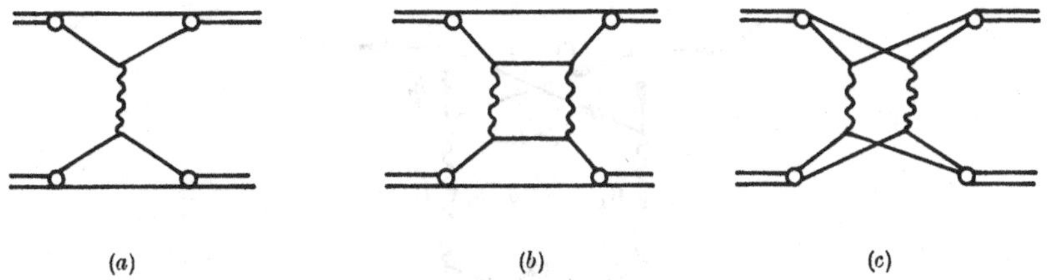

Fig. 8.8 Deuteron–deuteron scattering. (*a*) Single interaction between one pair of nucleons. (*b*) Double interaction between a pair of nucleons. (*c*) Double interaction between different pairs of nucleons.

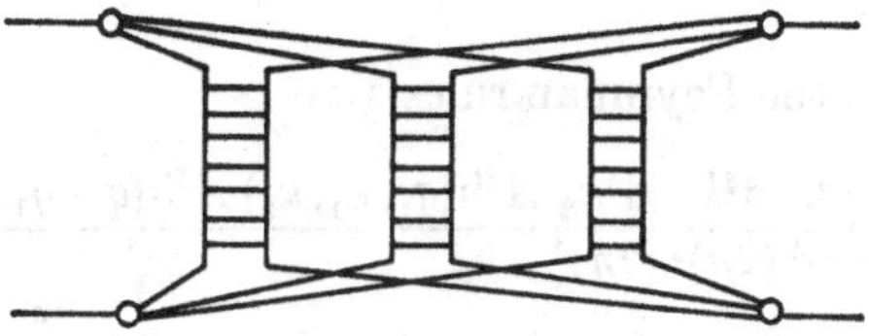

Fig. 8.9 A three-Reggeon cut diagram.

A three-ladder diagram, with non-planar couplings between each as in fig. 8.9, gives rise to a three-Reggeon cut, and so on.

The above discussion should be sufficient to demonstrate that cuts are much more difficult to deal with than poles because, quite apart from the technical difficulties (which we have skated over in this brief account), to calculate the magnitude of the cut contribution one has to make use of the off-mass-shell properties of the Feynman integrals, and not just the discontinuities of the integrals for which the particles are on the mass shell. Hence we are left with the function $N(t, t_1, t_2)$ which is known only as a Feynman integral, not as a physical quantity.

A somewhat more systematic way of analysing these problems has been invented by Gribov: the Reggeon calculus.

8.3 The Reggeon calculus*

The Reggeon calculus (Gribov 1968) uses Feynman integrals for the couplings of the Reggeons, but replaces the ladders directly by Regge poles. The plausibility of doing this depends on results such as (8.2.37).

Thus the Mandelstam diagram fig. 8.5 is replaced directly by fig. 8.10 where R_1 and R_2 are Regge poles. There are then just three closed loops, two corresponding to the crosses, plus the loop including

* This section may be omitted at first reading.

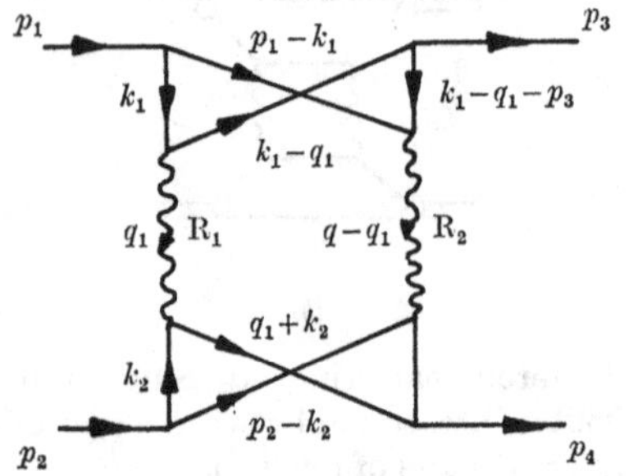

FIG. 8.10 Two-Reggeon cut diagram. $q_2 \equiv q-q_1$, $q \equiv p_1-p_3$.

the Reggeons, so the Feynman rules give

$$A^{\mathrm{c}}(s,t) = \frac{\mathrm{i}g^4}{2}\int \frac{\mathrm{d}^4q_1}{(2\pi)^4}\frac{\mathrm{d}^4k_1}{(2\pi)^4}\frac{\mathrm{d}^4k_2}{(2\pi)^4}\frac{A^{\mathrm{R}_1}(q_1,k_1,k_2)\,A^{\mathrm{R}_2}(q-q_1,p_1-k_1,p_2-k_2)}{\prod_{m=1}^{8} d_m} \tag{8.3.1}$$

where the d's are the Feynman propagators

$$\left.\begin{aligned} d_1 &= k_1^2-m^2+\mathrm{i}\epsilon \\ d_2 &= (p_1-k_1)^2-m^2+\mathrm{i}\epsilon, \quad \text{etc.}\end{aligned}\right\} \tag{8.3.2}$$

If we introduce Sudakov variables like (8.2.24)

$$\left.\begin{aligned} q_1 &= \alpha p_2' + \beta p_1' + \boldsymbol{q}_{1\perp} \\ k_i &= \alpha_i p_2' + \beta_i p_1' + \boldsymbol{k}_{i\perp}\end{aligned}\right\}, \quad i = 1,2 \tag{8.3.3}$$

the denominators become

$$\left.\begin{aligned} d_1 &= \alpha_1\beta_1 s - \mu_1^2 + \mathrm{i}\epsilon \\ d_2 &= (\alpha_1 - m^2/s)(\beta_1-1)s - \mu_2^2 + \mathrm{i}\epsilon, \quad \text{etc.}\end{aligned}\right\} \tag{8.3.4}$$

where $$\mu_i^2 = m^2 + \boldsymbol{k}_{i\perp}^2$$

and the integration volumes become

$$\mathrm{d}^4q_1 = \tfrac{1}{2}|s|\,\mathrm{d}\alpha\,\mathrm{d}\beta\,\mathrm{d}^2\boldsymbol{q}_{1\perp} \quad \text{etc.} \tag{8.3.5}$$

Since $A^{\mathrm{R}_1}(q_1,k_1,k_2)$ is a Regge pole amplitude we require that it should vanish if the momentum transfer becomes large, $q_1^2 \gg m^2$, and the 'masses' coupled to it, k_1^2, k_2^2, should also be $\not\gg m^2$. But its energy variable $s_1 = (k_1+k_2)^2 \approx 2k_1k_2 = \beta_1\alpha_2 s$ is large, so that the dominant region of integration in (8.3.1) is

$$\boldsymbol{q}_{1\perp}^2, \boldsymbol{k}_{1\perp}^2, \boldsymbol{k}_{2\perp}^2 \leqslant m^2; \quad \alpha,\beta,\alpha_1,\beta_2 \sim \frac{m^2}{s}; \quad \beta_1,\alpha_2 \sim 1$$

Then with a factorized form for the Reggeon

$$A^{R_1}(q_1, k_1, k_2) = \gamma(q_1^2, k_1^2, (q_1-k_1)^2) \times \gamma(q_1^2, k_2^2, (q_1+k_2)^2)\, \xi_{\alpha_1(q_1^2)}(2k_1 k_2)^{\alpha_1(q_1^2)} \quad (8.3.6)$$

where $\alpha_1(q_1^2)$ is the Regge trajectory (not to be confused with the Sudakov variable α_1 in (8.3.3) !), and

$$\xi_{\alpha_1(q_1^2)} \equiv \frac{e^{-i\pi\alpha_1(q_1^2)} + \mathscr{S}_1}{\sin[\pi\alpha_1(q_1^2)]} \quad (8.3.7)$$

is the signature factor, and with a similar expression for A^{R_2}, we end up with

$$A^c(s,t) = \frac{ig^4}{2|s|}\int \frac{d^2 \boldsymbol{q}_{1\perp}}{(2\pi)^2} N^2_{\alpha_1\alpha_2}(q, \boldsymbol{q}_{1\perp})\, s^{\alpha_1(q_1^2)\alpha_2(q_2^2)}\, \xi_{\alpha_1(q_1^2)}\, \xi_{\alpha_2(q_2^2)} \quad (8.3.8)$$

where

$$N_{\alpha_1\alpha_2} = \int \frac{d^2 \boldsymbol{k}_{1\perp}}{(2\pi)^2} \frac{d\beta_1\, d\alpha_1\, d\alpha\, s^2\gamma^2(\beta_1)^{\alpha_1(q_1^2)}(1-\beta_1)^{\alpha_2(q_2^2)}}{\prod_{m=1}^{4} d_m} \quad (8.3.9)$$

is the Feynman integral over the upper cross, and is the same as (8.2.36) except for the incorporation of γ^2, and the occurrence of the Sudakov parameters raised to the power $\alpha(q^2)$, due to the spins of the Reggeons.

The result (8.3.8) obviously agrees with (8.2.37) except that we have now included the signature factors of the Reggeons properly (remembering (8.2.13)). The two-dimensional nature of the remaining integration agrees with the results (3.4.9) and (8.2.13), and stems from the fact that after partial-wave projection (over two angles) only two of the four space–time dimensions remain to be integrated over. It is evident that the signature of this cut is just the product of the signatures of the two poles, i.e.

$$\mathscr{S}_c = \mathscr{S}_1 \mathscr{S}_2 \quad (8.3.10)$$

This is because $|s|$ appears in the denominator, so that under $s \to -s$ A^c transforms like the product of the poles. The four terms obtained from multiplying the two signature factors are shown in fig. 8.11.

To examine the J-plane structure we take the Mellin transform (2.10.3),

$$A_J(t) = \frac{1}{\pi}\int_0^\infty D_s(s,t)\, s^{-J-1}\, ds \quad (8.3.11)$$

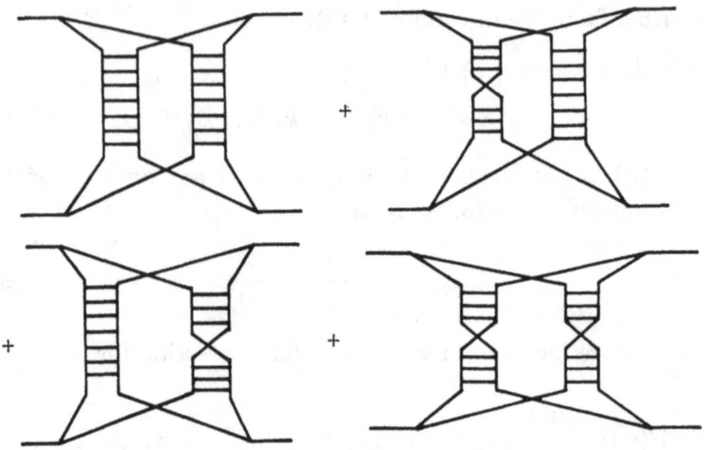

FIG. 8.11 The set of four diagrams, including crossed terms $s \leftrightarrow u$, obtained from the product of the signature factors in (8.3.8).

with the s-discontinuity of (8.3.8),

$$D_s = \frac{g^4}{2|s|} \int \frac{d^2 q_{1\perp}}{(2\pi)^2} N^2_{\alpha_1\alpha_2}(q, q_{1\perp})\, s^{\alpha_1(q_1^2)+\alpha_2(q_2^2)} \operatorname{Re}\{\xi_{\alpha_1} \xi_{\alpha_2}\} \tag{8.3.12}$$

where from (8.3.7)

$$\operatorname{Re}\{\xi_{\alpha_1} \xi_{\alpha_2}\} = \cos\left[\frac{\pi}{2}\left(\alpha_1 + \alpha_2 + \sum_{i=1,2} \frac{1-\mathscr{S}_i}{2}\right)\right] \tag{8.3.13}$$

and obtain
$$A_J(t) = \frac{g^4}{2} \int \frac{d^2 q_{1\perp}}{(2\pi)^2} \frac{N^2_{\alpha_1\alpha_2}(q, q_{1\perp}) \operatorname{Re}\{\xi_{\alpha_1} \xi_{\alpha_2}\}}{J+1-\alpha_1(q_1^2)-\alpha_2(q_2^2)} \tag{8.3.14}$$

which exhibits the cut at

$$\begin{aligned} J &= \max\{\alpha_1(q_1^2) + \alpha_2((q-q_1)^2) - 1\} \\ &= \max\{\alpha_1(-q_{1\perp}^2) + \alpha_2(-(q_\perp - q_{1\perp})^2) - 1\} \end{aligned} \tag{8.3.15}$$

corresponding to (8.2.19).

The discontinuity across this two-Reggeon cut is

$$\begin{aligned} \Delta_2(J,t) \equiv \Delta_{J_2}\{A_J(t)\} = \mathrm{i}g^4 \int \frac{d^2 q_{1\perp}}{(2\pi)^2} N^2_{\alpha_1\alpha_2}(q, q_{1\perp}) \\ \times \operatorname{Re}\{\xi_{\alpha_1} \xi_{\alpha_2}\}\, \delta(J+1-\alpha_1(q_1^2)-\alpha_2(q_2^2)) \end{aligned} \tag{8.3.16}$$

which may be rewritten

$$\begin{aligned} \Delta_{J_2}\{A_J(t)\} = (-1)\sin\left[\frac{\pi}{2}\left(J - \sum_i \frac{1-\mathscr{S}_i}{2}\right)\right] \mathrm{i}g^4 \\ \times \int \frac{d^2 q_{1\perp}}{(2\pi)^2} \int \frac{d^2 q_{2\perp}}{(2\pi)^2} (2\pi)^2 \delta^2(q_{1\perp} + q_{2\perp} - q_\perp) \\ \times \delta(J+1-\alpha_1(-q_{1\perp}^2) - \alpha_2(-q_{2\perp}^2))\, N^2_{\alpha_1\alpha_2} \end{aligned} \tag{8.3.17}$$

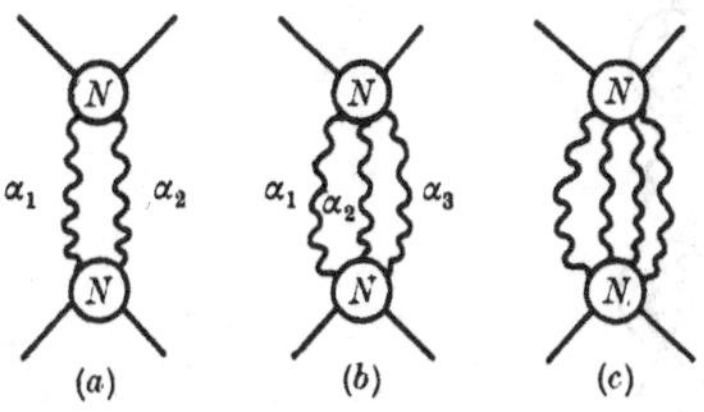

FIG. 8.12 Some multi-Reggeon cuts.

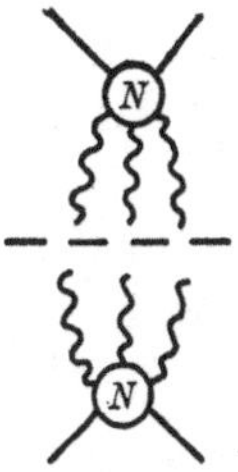

FIG. 8.13 Discontinuity across the three-Reggeon cut. Each Reggeon has momentum $q_{i\perp}$ and 'energy' $E_i = 1-\alpha_i(-q_{i\perp}^2)$.

Similarly for higher order cuts such as fig. 8.12 the discontinuity is

$$\Delta_{J_n}\{A_J(t)\} = (-1)^{n-1}\sin\left[\frac{\pi}{2}\left(J-\sum_i\frac{1-\mathcal{S}_i}{2}\right)\right] \mathrm{i}g^{2n}$$
$$\times\int\frac{\mathrm{d}^2\boldsymbol{q}_{1\perp}}{(2\pi)^2}\dots\int\frac{\mathrm{d}^2\boldsymbol{q}_{n\perp}}{(2\pi)^2}(2\pi)^2\delta^2(\boldsymbol{q}_{1\perp}+\dots+\boldsymbol{q}_{n\perp}-\boldsymbol{q}_{\perp})$$
$$\times\delta(J+n-1-\sum_i\alpha_i(-\boldsymbol{q}_{1\perp}^2))\,N^2_{\alpha_1\dots\alpha_n} \quad (8.3.18)$$

This equation can be regarded as the discontinuity across the Feynman graph like fig. 8.13, in which each Reggeon is regarded as a quasi-particle in a two-dimensional space, with momentum $\boldsymbol{q}_{i\perp}$ and 'energy' $E_i = 1-\alpha_i(-\boldsymbol{q}_{1\perp}^2)$, the 'energy' and momentum being conserved at each vertex, since the δ-functions in (8.3.18) then correspond to those of the Cutkosky rules (1.5.11). The 'phase space' is

$$\prod_i\frac{\mathrm{d}^2\boldsymbol{q}_{i\perp}}{(2\pi)^2}(2\pi)^2\delta(\sum_i\boldsymbol{q}_{i\perp}-\boldsymbol{q}_{\perp})\,2\pi\delta(\sum_i E_i-E)\equiv\mathrm{d}\Phi_n^J \quad (8.3.19)$$

where $E\equiv 1-J$, and (8.3.18) can be rewritten as

$$\Delta_{J_n}\{A_J(t)\} = (-1)^{n-1}\sin\left[\frac{\pi}{2}\left(1-E-\sum_i\frac{1-\mathcal{S}_i}{2}\right)\right]\mathrm{i}g^{2n}\int\mathrm{d}\Phi_n^J N^2_{E_1\dots E_n} \quad (8.3.20)$$

The next step is to try and generalize the above prescription so that the N's instead of being just Feynman integrals, can themselves contain Reggeon amplitudes. Thus N may be expected to contain Regge poles and cuts like fig. 8.14. This is more complicated, however, because it is necessary to be clear about which side of their branch cuts the N's in the above formulae are to be evaluated. To determine this it is necessary to regard the Reggeons as two-body states (at

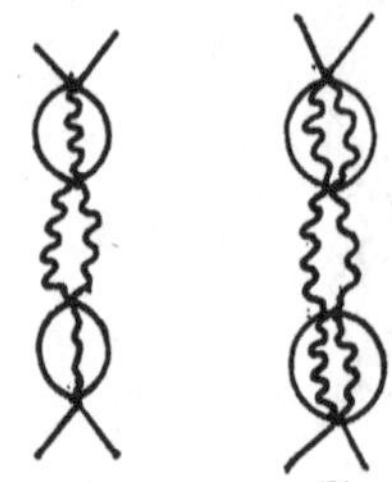

FIG. 8.14 (*a*) Pole and (*b*) cut contributions to the N's in fig. 8.12 (*a*).

least), and so the two-Reggeon cut involves four-body unitarity in the t channel.

It turns out (Gribov, Pomeranchuk and Ter-Martirosyan 1965, White 1972, 1974) that the results are almost exactly analogous to the discontinuity formulae for s-plane singularities, and we can write, for example (cf. (1.3.16)),

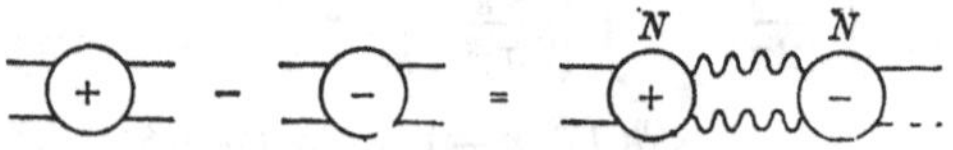

(8.3.21)

or

$$\Delta_{J_2}\{A_J(t)\} = (-1)\sin\left[\frac{\pi}{2}\left(J-\sum_i\frac{1-\mathscr{S}_i}{2}\right)\right]\int \mathrm{d}\Phi_2^J N_{\alpha_1\alpha_2}(J_+)\, N_{\alpha_1\alpha_2}(J_-) \tag{8.3.22}$$

where $J_\pm \equiv J \pm i\epsilon$ are evaluated above/below the cuts in N. This generalization looks rather obvious, but in fact a great deal of care is needed to ensure that the correct discontinuities have been taken, particularly keeping in mind the signature properties of the Reggeons.

This similarity between the unitarity equations in the s plane and the J plane, with the Cutkosky-like rule (8.3.22) has led various authors to try and construct a Reggeon field theory in a space with two spacelike and one timelike dimension (Gribov and Migdal 1968, 1969, Cardy and White 1973, 1974, Migdal, Polyakov and Ter-Martirosyan 1974, Abarbanel and Bronzan 1974*a*, *b*). For linear trajectories

$$\alpha(-\mathbf{K}^2) = \alpha^0 - \alpha'\mathbf{K}^2$$

the Regge pole becomes

$$\frac{1}{J-\alpha} \to \frac{1}{E-\alpha'\mathbf{K}^2+(1-\alpha^0)} \tag{8.3.23}$$

reminiscent of the propagator of a non-relativistic particle of mass $m = (2\alpha')^{-1}$ (cf. (1.13.25)), and with an 'energy gap' $1-\alpha_0$, i.e. the velocity is $v = [4\alpha'(E-1+\alpha^0)]^{\frac{1}{2}}$. So one can produce a field theory in which the Reggeon field obeys the Schroedinger equation. There are the usual problems of re-normalization (see fig. 8.14) and convergence, compounded by a fundamental uncertainty as to whether it makes sense to replace ladders by bare Reggeons and then re-normalize them. For example the presence of a pole or cut above $J = -1$ in N means that $A(s_1, t, t_1, t_2)$ in (8.2.38) $\sim s_1^{\alpha(t)}$ or $\sim s_1^{\alpha_c(t)}$, and so the integral defining N will not converge without re-normalization. What is worse, for the P with $\alpha^0 = 1$ there is no energy gap, so the P is analogous to a massless particle in conventional field theory, and all the singularities pile up at $J = 1$, just like the 'infra-red' problem caused by the massless photon in quantum electrodynamics. The asymptotic behaviour of cross-sections thus depends on the solutions near the critical point $J = 1$. These have been studied using re-normalization group methods. So far only limited progress has been made with this approach, and we shall not pursue it further (see Abarbanel *et al.* (1975) for a review).

To summarize and generalize these results, we have found that the exchange of n Reggeons $R_1, \ldots, R_n$ gives rise to a cut branch point at (from (8.3.18), cf. (8.2.19))

$$\alpha_{c_n}(t) = \max\left\{\sum_{i=1}^{n} \alpha_i(t_i) - n + 1\right\} \tag{8.3.24}$$

where the maximum value is over the allowed region of integration, and for increasing trajectories this is bounded by (cf. (8.2.18))

$$\sum_{i=1}^{n} \sqrt{-t_i} = \sqrt{-t} \tag{8.3.25}$$

We shall often refer to this as the $\mathrm{R}_1 \otimes \mathrm{R}_2 \otimes \ldots \otimes \mathrm{R}_n$ Regge cut, where the $\otimes$ implies the phase-space integration (8.3.8) or (8.3.18). If the trajectories are identical these rules give

$$\alpha_{c_n}(t) = n\alpha(t/n^2) - n + 1 \tag{8.3.26}$$

and if they are linear, $\alpha(t) = \alpha^0 + \alpha' t$,

$$\alpha_{c_n}(t) = \alpha' t/n + n(\alpha^0 - 1) + 1 \tag{8.3.27}$$

The signature of the cut is the product of the signatures of the poles (cf. (8.3.10))

$$\mathscr{S}_c = \prod_i \mathscr{S}_i \tag{8.3.28}$$

We remarked in the introduction that Regge cuts are necessary to

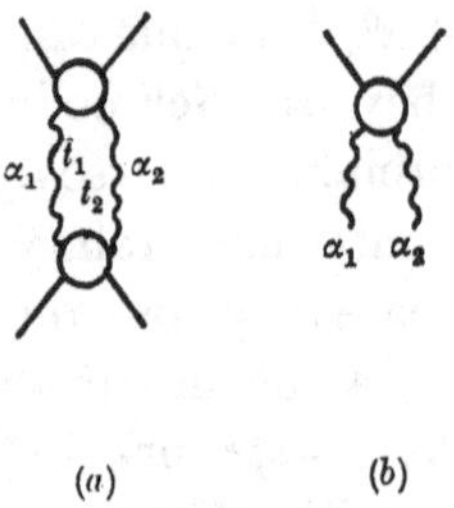

FIG. 8.15 (a) Two-Reggeon intermediate state in the t channel. (b) Reggeon–particle scattering amplitude.

ensure consistency of the Gribov–Pomeranchuk fixed poles with t-channel unitarity, and we should check that the above branch points can do this (see Jones and Teplitz 1967, Bronzan and Jones 1967, Schwarz 1967, Hwa 1967).

For scalar external particles the highest Gribov–Pomeranchuk singularity is at $J = -1 \equiv J_0$, say, in an even-signature amplitude. If the Regge cut is to overlie the t-channel unitarity cut beginning at the threshold $t = t_T = 4m^2$, we obviously require that $\alpha_c(t_T) = J_0$. If the particles constituting this threshold (fig. 8.15) lie on trajectories $\alpha(t)$, we must have $\alpha(m^2) = 0$ for scalar particles, and substituting this in (8.3.26) for $n = 2$ we get

$$\alpha_{c_2}(4m^2) = 2\alpha(m^2) - 1 = -1 \tag{8.3.29}$$

so the cut branch point coincides with the fixed pole at threshold. Then if we continue in t_1 and t_2 up the trajectories to α_1, α_2 = integers > 0, the highest Gribov–Pomeranchuk pole in the amplitude fig. 8.15(a) will be at $J = \alpha_1 + \alpha_2 - 1$, since α_1 is the largest possible helicity for a particle of spin α_1, and the Reggeon branch point evidently remains in the correct place to prevent conflict with unitarity. The full cut structure is a good deal more complicated than this brief account suggests, however, particularly for unequal mass particles–see Schwarz (1967) and Olive and Polkinghorne (1968).

In the t plane the branch point occurs at $t = t_c(J)$ where $t_c(J)$ is defined by

$$\alpha_c(t_c(J)) = J \tag{8.3.30}$$

So, from (8.3.29), $t_c(-1) = 4m^2$. As J is increased from -1, t_c moves along the elastic branch cut until the first inelastic threshold t_I is reached, whereupon it passes through the inelastic branch cut on to the unphysical sheet. So $\alpha_c(t)$ has a branch point at J_I where $t_c(J_I) = t_I$, and the cut discontinuity $\Delta_2(J, t)$ has a branch point here.

This connection between the Gribov–Pomeranchuk poles and the cuts is of course not accidental but arises because the cuts are generated by the fixed poles in the cross diagram N (fig. 8.6(e)) through two-Reggeon unitarity. The way in which this works in perturbation theory was demonstrated by Olive and Polkinghorne (1968), and Landshoff and Polkinghorne (1969). If we write the cross as

$$= \frac{G_1(t, J)}{J+1}$$

then the next-order diagram is

$$= \frac{G_2(t, J)}{J+1}$$

and the discontinuity of G_2 across the cut must contain the fixed pole, so

$$\text{Disc}_2\{G_2\} = \frac{\rho G_1}{J+1}$$

where ρ is a phase-space factor. So from these two diagrams we have, above the cut,

$$A_J^{\text{I}} = \frac{G_1 + G_2}{J+1} \tag{8.3.31}$$

but below the cut,

$$A_J^{\text{II}} = \frac{G_1 + G_2}{J+1} + \frac{\mathrm{i}\rho G_1^2}{(J+1)^2} \tag{8.3.32}$$

This appears to have generated a double pole, but using the t-channel unitarity equation (i.e. like (2.2.7) with $A_J^{\text{I}} \equiv A_J(t_+)$, $A_J^{\text{II}} = A(t_-)$) we get

$$A_J^{\text{II}} = \frac{A_J^{\text{I}}}{1 - \mathrm{i}\rho A_J^{\text{I}}} = A_J^{\text{I}} + \mathrm{i}\rho (A_J^{\text{I}})^2 + (\mathrm{i}\rho)^2 (A_J^{\text{I}})^3 + \dots \tag{8.3.33}$$

so if to all orders

$$A_J^{\text{I}} = \frac{G}{J+1}, \quad G = G_1 + G_2 + \dots \tag{8.3.34}$$

then

$$A_J^{\text{II}} = \frac{G}{J+1-\mathrm{i}\rho G} = \frac{G}{J+1} + \frac{\mathrm{i}\rho G^2}{(J+1)^2} + \dots \tag{8.3.35}$$

i.e. A_J^{II} contains a sequence of multiple poles which sum to give a finite value for A_J^{II} as $J \to -1$. So the cut sequence, fig. 8.16, permits there to be a fixed pole on the physical sheet (I) and nothing worse. Such cuts are clearly essential if continuation in angular momentum is to be compatible with t-channel unitarity in any theory which includes a third double spectral function ρ_{su}.

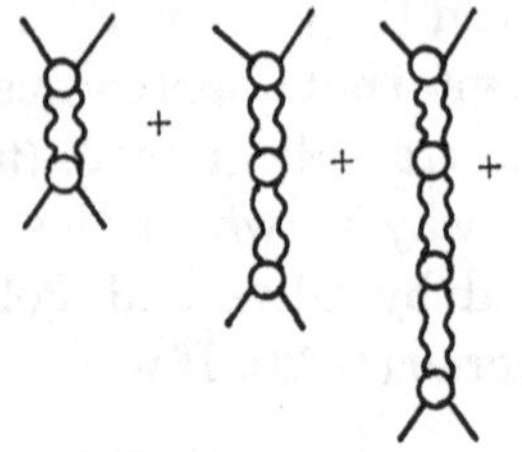

FIG. 8.16 The t-channel iteration of two-Reggeon cut.

However, if the presence of this cut discontinuity is to be compatible with the t-channel elastic unitarity condition (8.3.33) we need (Bronzan and Jones 1967), for $t > t_{\mathrm{T}}$,

$$\Delta_t\{A_J(t)\} \to 0 \quad \text{as} \quad t \to t_{\mathrm{c}}(J) \tag{8.3.36}$$

and so

$$\Delta_J\{A_J(t)\} \to 0 \quad \text{as} \quad J \to \alpha_{\mathrm{c}}(t) \tag{8.3.37}$$

i.e. the discontinuity across the Regge cut must vanish at the branch point. This is not true of the cut (8.3.8) which, as we have seen in (8.2.16), gives

$$A(s,t) \sim s^{\alpha_{\mathrm{c}}}(\log s)^{-1} \quad \text{and so} \quad A_J(t) \sim \log(J-\alpha_{\mathrm{c}}) \tag{8.3.38}$$

(from (2.7.4)). And of course the logarithm has a finite discontinuity ($=\pi$) between one sheet and the next at $J = \alpha_{\mathrm{c}}$. However, t-channel unitarity requires that we include the full sequence of cuts, fig. 8.16, and so $N_{\alpha_1\alpha_2}$ in (8.3.8) will contain the two-Reggeon cut (fig. 8.14 (b)) and so satisfies the unitarity condition (White 1972)

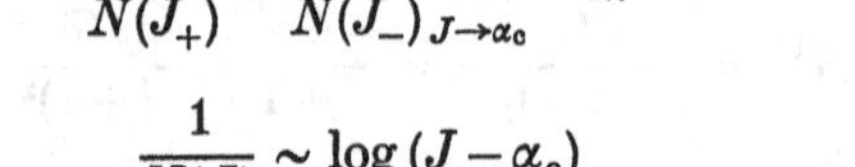

or

$$N(J_+) - N(J_-) \underset{J\to\alpha_{\mathrm{c}}}{\sim} \mathrm{i}\pi N(J_+)\,N(J_-) \tag{8.3.39}$$

so

$$\frac{1}{N(J_+)} - \frac{1}{N(J_-)} \underset{J\to\alpha_{\mathrm{c}}}{\sim} -\mathrm{i}\pi$$

i.e.

$$\frac{1}{N(J)} \sim \log(J-\alpha_{\mathrm{c}})$$

which in (8.3.16) gives

$$\Delta\{A_J(t)\} \underset{J\to\alpha_{\mathrm{c}}}{\sim} (\log(J-\alpha_{\mathrm{c}}))^{-2} \tag{8.3.40}$$

so the singularity is softened to an inverse logarithmic cut, not a logarithm.

Substituted in (2.7.8) this gives

$$A^{c}(s,t) \underset{s\to\infty}{\sim} \int^{\alpha_c} \mathrm{d}J \frac{1}{(\log(J-\alpha_c))^2} s^J \sim s^{\alpha_c} \int^{0} \mathrm{d}x \frac{\mathrm{e}^{x\log s}}{(\log x)^2}$$

$$\sim \frac{s^{\alpha_c}}{\log s} \int^{1} \frac{\mathrm{d}y}{(\log(\log y) - \log(\log s))^2} \sim \frac{s^{\alpha_c}}{\log s(\log(\log s))^2} \quad (8.3.41)$$

(where we have made the successive substitutions $x = J - \alpha_c$, $\log y = x \log s$). Since $\log(\log s)$ varies so slowly with s, this correction is of very little practical importance, but it is necessary to keep in mind that compatibility with t-channel unitarity involves not just fig. 8.15, but the infinite sum fig. 8.16. Unitarization makes even less difference to three-and-more-Reggeon cuts since (8.3.20) gives $A^{c}_{J}(t) \sim (J-\alpha_c)^{n-2}\log(J-\alpha_c)$ which for $n > 2$ already has a vanishing discontinuity at the branch point, and through (2.7.4) produces the asymptotic behaviour $\sim s^{\alpha_c}(\log s)^{-(n-1)}$.

8.4 The absorption and eikonal models

Although the Feynman-diagram models and the Reggeon calculus have told us a good deal about the properties to be expected of Regge cuts, they do not give the strength of the cuts relative to the poles, and so give very little idea of how important cuts are likely to be in practice.

Thus (8.2.37) and (8.3.8) suggest that for a two-Reggeon cut amplitude, $R_1 \otimes R_2$, we should write

$$A^{c}(s,t) = \frac{\mathrm{i}}{16\pi^2|s|} \iint_{-\infty}^{0} \mathrm{d}t_1\, \mathrm{d}t_2 \frac{\theta(-\lambda)}{(-\lambda(t,t_1,t_2))^{\frac{1}{2}}}$$
$$\times (N(t,t_1,t_2))^2 A^{R_1}(s,t_1) A^{R_2}(s,t_2) \quad (8.4.1)$$

where the A^{R_i} are physical Regge pole amplitudes. We have absorbed g^2 into the definition of the vertices N which should include all contributions to ρ_{su}, and is of course unknown (but see section 10.9). However, various models have been suggested for calculating cuts which reproduce a structure like (8.4.1) but with a specific prescription for N, and we review two of them here. Though neither is particularly compelling both do at least have the merit of providing a simple way of including spin-kinematics etc.

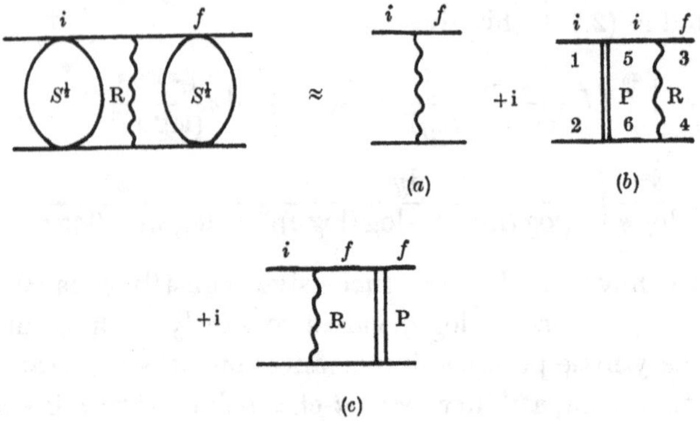

FIG. 8.17 Reggeized absorption model.

a. The Reggeized absorption model

This is used for inelastic reactions where quantum numbers are exchanged. The basic idea is to use a Regge pole, R, to carry the exchanged quantum numbers, but to include modifications caused by elastic scattering in the initial and final states, as in fig. 8.17. Since the elastic amplitude is predominantly imaginary the effect is to reduce the lower partial waves, which corresponds physically to absorption of the incoming flux into channels other than those being considered. It is possible to use the full elastic scattering amplitude, but it is more illuminating to represent it by its dominant Regge pole, the Pomeron, P.

Specifically the hypothesis is that the s-channel partial waves for the processes $1+2$ (channel i) $\rightarrow 3+4$ (channel f) may be written in the form (Henyey *et al.* 1969)

$$A_J^{if}(s) = (S_J^{ii}(s))^{\frac{1}{2}}\, A_J^{if}(s)\, (S_J^{ff}(s))^{\frac{1}{2}} \tag{8.4.2}$$

where $A_J^{if\mathrm{R}}(s)$ is the s-channel partial-wave projection of the t-channel Reggeon, and $S_J^{ii}(s)$ is the partial-wave S-matrix for elastic scattering in the initial state, etc. Since we shall want to sum over all the helicities of the particles it is convenient to regard i and f as helicity labels, and (8.4.2) as a matrix product relation. Then we write the elastic S-matrix as

$$S_J^{ii}(s) \approx 1 + 2\mathrm{i}\rho(s)\, A_J^{ii\mathrm{P}}(s) \tag{8.4.3}$$

where $A_J^{ii\mathrm{P}}(s)$ is the partial-wave projection of the P exchange amplitude and $\rho^i(s) = 2q_{s12}s^{-\frac{1}{2}}$ is the kinematical factor (2.2.9). On substituting (8.4.3), and a similar expression for S_J^{ff}, into (8.4.2), and expanding the square-roots, we get

$$A_J^{if}(s) \approx A_J^{if\mathrm{R}}(s) + \mathrm{i}\rho^i(s)\, A_J^{ii\mathrm{P}}(s)\, A_J^{if\mathrm{R}}(s) + \mathrm{i}\rho^f(s)\, A_J^{if\mathrm{R}}(s)\, A_J^{ff\mathrm{P}}(s) + \dots \quad (8.4.4)$$

The first term is just the Reggeon R, while the second and third terms give cuts due to the exchange of R and P together, i.e. R ⊗ P cuts. The full amplitude is obtained by summing the partial-wave series (4.4.9). So for example from the second term in (8.4.4) we get

$$A_{H_s}^{\mathrm{R}\otimes\mathrm{P}}(s,t) = \mathrm{i}\rho^i(s)\, 16\pi \sum_J (2J+1) \sum_{\mu_5\mu_6} A^{ii\mathrm{P}}_{\mu_1\mu_2\mu_5\mu_6 J}(s)\, A^{if\mathrm{R}}_{\mu_5\mu_6\mu_3\mu_4 J}(s)\, d^J_{\mu\mu'}(z_s)$$

$$\mu = \mu_1 - \mu_2, \quad \mu' = \mu_3 - \mu_4 \quad (8.4.5)$$

Then if we make a partial-wave projection of the pole amplitudes (dropping the channel labels for simplicity), we get

$$A_{H_s}^{\mathrm{R}\otimes\mathrm{P}}(s,t) = \sum_{\mu_5\mu_6} \mathrm{i}\rho(s)\, 16\pi \sum_J (2J+1)\, d^J_{\mu\mu'}(z_s) \frac{1}{32\pi}\int_{-1}^{1} A^{\mathrm{P}}_{\mu_1\mu_2\mu_5\mu_6}(s,z_1) \times d^J_{\mu\mu''}(z_1)\,\mathrm{d}z_1 \frac{1}{32\pi}\int_{-1}^{1} A^{\mathrm{R}}_{\mu_5\mu_6\mu_3\mu_4}(s,z_2)\, d^J_{\mu''\mu'}(z_2)\,\mathrm{d}z_2, \quad \mu'' = \mu_5 - \mu_6 \quad (8.4.6)$$

where z_1 and z_2 are the cosines of the scattering angles between the initial and intermediate, and intermediate and final states, respectively (see fig. 2.1), which satisfy (2.2.4), viz.

$$z_1 = z z_2 + (1-z^2)^{\frac{1}{2}}(1-z_2^2)^{\frac{1}{2}} \cos\phi \quad (8.4.7)$$

But (Henyey *et al.* 1969)

$$\sum_J (2J+1)\, d^J_{\mu\mu'}(z_s)\, d^J_{\mu\mu''}(z_1)\, d^J_{\mu''\mu'}(z_2) = \frac{2}{\pi}\frac{\theta(\Delta)}{\Delta^{\frac{1}{2}}} \cos(\mu\phi_1 + \mu'\phi_2 + \mu''\phi_3) \quad (8.4.8)$$

(cf. (8.2.6)) and in the high-energy limit (8.2.10) the ϕ dependence may be neglected, and $\rho(s) \to 1$, and so we obtain

$$A_{H_s}^{\mathrm{R}\otimes\mathrm{P}}(s,t) = \sum_{\mu_5\mu_6} \frac{\mathrm{i}}{16\pi^2|s|} \iint_{-\infty}^{0} \mathrm{d}t_1\,\mathrm{d}t_2\, A^{\mathrm{P}}_{\mu_1\mu_2\mu_5\mu_6}(s,t_1)\, A^{\mathrm{R}}_{\mu_5\mu_6\mu_3\mu_4}(s,t_2) \times \frac{\theta(-\lambda)}{(-\lambda(t,t_1,t_2))^{\frac{1}{2}}} \quad (8.4.9)$$

which is identical with (8.4.1) for spinless scattering, with $N(t,t_1,t_2) = 1$. It also agrees with the AFS result (8.2.11) except for the complex conjugation of the Reggeon amplitude. In fact (8.2.11) corresponds to taking

$$\mathrm{Im}\{\mathrm{A}^{\mathrm{c}}_J(s)\} = \rho^i(s)\,\mathrm{Re}\{A_J^{ii\mathrm{P}}(s)\,(A_J^{if\mathrm{R}}(s))^*\} \quad (8.4.10)$$

instead of (8.4.4). However, since the P is almost pure imaginary this complex conjugation would give essentially the opposite sign for the cut. The absorptive sign in (8.4.9) agrees with the Mandelstam result (8.2.37), and the Reggeon calculus (8.3.8), rather than the AFS sign of (8.2.11).

There are some fairly obvious defects in this approach. First, fig. 8.17 (*b*) is a planar diagram, and we found in section 8.2 that planar diagrams should not give rise to cuts. The reason why we get a similar answer is that the particle propagators across the diagram $\sim 1/s$, and so have the same power behaviour as the crosses of fig. 8.6, but really fig. 8.17 (*b*) looks more like a re-normalization of the box-diagram contribution to the Regge pole in fig. 8.1. Secondly, if the Reggeon is regarded as a ladder it already includes inelastic intermediate states in the s channel, and so to absorb it again may involve double counting. This is clearly related to the re-normalization problem. However, we shall find in section 8.6 that one of the main defects of Regge poles is that they give too large a contribution to the low partial waves, so phenomenologically some extra absorption is certainly necessary, and probably should be provided by cuts. Also the elastic intermediate state $|5, 6\rangle$ is only one of a large number of diffractively produced states which can arise through P exchange, and we should probably consider the sum of all diagrams like fig. 8.18. They are sometimes included rather crudely by multiplying (8.4.9) by an enhancement factor $\lambda > 1$. Note that λ must be independent of s, otherwise the position of the cut will be moved, despite the fact that more diffractive states open as s is increased.

We thus conclude that though the absorption idea is useful in confirming the basic form of (8.4.1), it cannot be taken too seriously as a quantitative model for Regge cuts.

b. The eikonal model

This is directly related to the eikonal method for high energy potential scattering discussed in section 1.14, and it gives a way of computing the high energy limit of sums of diagrams like fig. 8.29 corresponding to many-Reggeon exchange (Arnold 1967). In fact, however, the nature of the exchange is not very important so we begin by considering the exchange of scalar particles rather than Reggeons (see Levy and Sucher 1969, Abarbanel and Itzykson 1969, Chang and Yan 1970, Tiktopoulos and Trieman 1971, Cardy 1971).

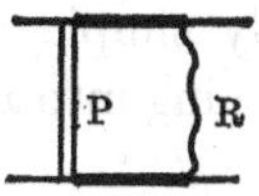

FIG. 8.18 Diffractively produced intermediate states in the absorption model giving additional terms in (8.8.4).

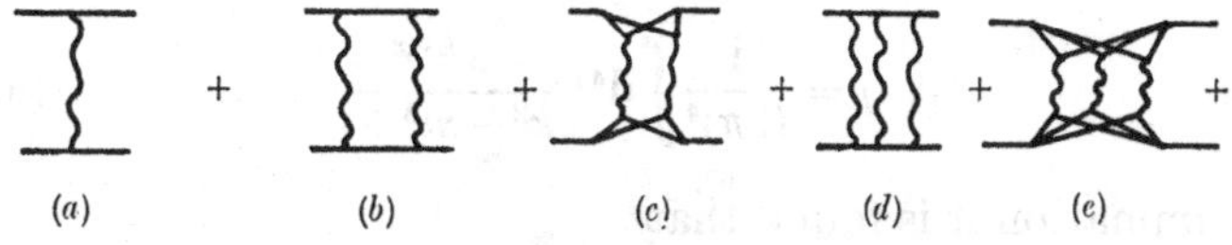

FIG. 8.19 A sequence of multi-Reggeon exchange diagrams.

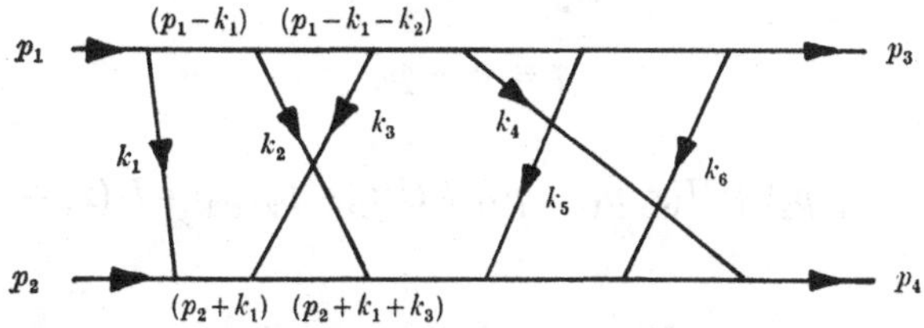

FIG. 8.20 A crossed-rung ladder.

A typical n-rung diagram is fig. 8.20, a generalized ladder in which some of the rungs cross over each other. The Feynman rules give

$$A^n(s,t) = g^{2n} \prod_{i=1}^{n} \int \frac{d^4k_i}{(2\pi^4)} \frac{1}{k_i^2-m^2} (2\pi)^4 \delta^4(p_1-p_1'-\Sigma k_i)$$
$$\times \{[(p_1-k_1)^2-m^2][(p_1-k_1-k_2)^2-m^2]\dots[(p_1-k_1-\dots-k_n)^2-m^2]$$
$$\times [(p_2+k_1)^2-m^2]\dots[(p_2+k_1+\dots+k_n)^2-m^2]\}^{-1} \quad (8.4.11)$$

We work in the high energy small-angle scattering approximation in which very little momentum is given up to each of the rungs, so the recoil of particles 1 and 2 at each successive scattering is small, in which case we can make the replacement

$$(p_1 \pm k)^2 - m^2 = \pm 2p_1 . k + k^2 \approx \pm 2p_1 . k \quad (8.4.12)$$

throughout. This clearly corresponds to the eikonal assumptions of section 1.14. It is then necessary to sum over all permutations of the ordering of the rungs arriving at particle 2 for the given ordering $k_1 \dots k_n$ of rungs leaving particle 1. With the approximations (8.4.12) the symmetry of the integrand makes it possible with some effort (see Levy and Sucher (1969) for details) to rearrange the sum over permutations to a remarkably simple form. The integrations can then be performed by transforming into x space using

$$\frac{1}{k_i^2 - m^2} = \int \mathrm{d}^4x \Delta_{\mathrm{F}}(x)\, \mathrm{e}^{-\mathrm{i}k_i \cdot x} \tag{8.4.13}$$

where the Feynman propagator is

$$\Delta_{\mathrm{F}}(x) = \frac{\mathrm{i}}{(2\pi)^4} \int \mathrm{d}^4k \frac{\mathrm{e}^{\mathrm{i}k \cdot x}}{k^2 - m^2 + \mathrm{i}\epsilon} \tag{8.4.14}$$

and on summation it is found that

$$A^n(s,t) = \frac{g^2}{n!} \int \mathrm{d}^4x\, \mathrm{e}^{-\mathrm{i}q \cdot x} \Delta_{\mathrm{F}}(x)\, (\mathrm{i}\chi)^{n-1} \tag{8.4.15}$$

where $$q \equiv p_1 - p_3$$
and

$$\chi \equiv -(U(x, p_1, p_2) + U(x, p_1, -p_4) + U(x, -p_3, p_2) + U(x, -p_2, -p_4)) \tag{8.4.16}$$

where

$$U(x, p_i, p_j) \equiv g^2 \int \frac{\mathrm{d}^4k}{(2\pi)^4} \frac{\mathrm{e}^{\mathrm{i}k \cdot x}}{(k^2 - m^2 + \mathrm{i}\epsilon)(-2p_i . k + \mathrm{i}\epsilon)(2p_j . k + \mathrm{i}\epsilon)} \tag{8.4.17}$$

Clearly in the high energy, small-angle limit $p_3 \simeq p_1$, $p_2 \simeq p_4$ so χ depends only on $2p_1 \cdot p_2$ (i.e. s), $p_1 \cdot k$ and $p_2 . k$. On performing the contour integration in k contributions appear just from the vanishing of the denominators, so putting $(2p_i . k + \mathrm{i}\epsilon)^{-1} \to 2\pi\mathrm{i}\delta(2p_i . k)$ the four terms in (8.4.16) give

$$\chi(x, p_1, p_2) = \frac{g^2}{(2\pi)^2} \int \mathrm{d}^4k \frac{\mathrm{e}^{\mathrm{i}k \cdot x}}{k^2 - m^2 + \mathrm{i}\epsilon} \delta(2p_1 . k)\, \delta(2p_2 . k) \tag{8.4.18}$$

and integrating k in the plane of p_1 and p_2 (see fig. 1.12) this becomes

$$\chi(s, \mathbf{b}) = \frac{g^2}{8\pi^2 s} \int \mathrm{d}^2\mathbf{k} \frac{\mathrm{e}^{-\mathrm{i}\mathbf{k} \cdot \mathbf{b}}}{t - m^2 + \mathrm{i}\epsilon} \tag{8.4.19}$$

We can then perform the ϕ integration as in (1.14.10) to obtain

$$\chi(s, b) = \frac{1}{8\pi s} \int_{-\infty}^{0} \mathrm{d}t\, J_0(b\sqrt{-t}) \frac{g^2}{t - m^2} \tag{8.4.20}$$

of which the inverse is (from (1.14.14))

$$\frac{g^2}{t - m^2} = 4\pi s \int_0^{\infty} b\, \mathrm{d}b\, J_0(b\sqrt{-t})\, \chi(s, b) = A^{\mathrm{B}}(s, t) \tag{8.4.21}$$

which from (8.4.13) $$= g^2 \int \mathrm{d}^4x\, \Delta_{\mathrm{F}}(x)\, \mathrm{e}^{-\mathrm{i}q\cdot x} \tag{8.4.22}$$

in the high energy, small-angle approximation. So (8.4.15) gives

$$A^n(s,t) = \frac{4\pi s}{n!}\int_0^\infty b\,\mathrm{d}b\, J_0(b\sqrt{-t})\,\chi(s,b)\,(\mathrm{i}\chi(s,b))^{n-1} \tag{8.4.23}$$

and on summing over all possible numbers of rungs we get

$$A(s,t) = \sum_{n=1}^{\infty} A^n(s,t) = 8\pi s\int_0^\infty b\,\mathrm{d}b\, J_0(b\sqrt{-t})\,\frac{\mathrm{e}^{\mathrm{i}\chi(s,b)}-1}{2\mathrm{i}} \tag{8.4.24}$$

The first term of this series ($n = 1$) is just (8.4.21), the single-particle-exchange Born approximation. The second term is the sum of all the two-particle exchange graphs

$$A^2(s,t) = 4\pi s\int_0^\infty b\,\mathrm{d}b\, J_0(b\sqrt{-t})\,\frac{\mathrm{i}\chi^2}{2} \tag{8.4.25}$$

which when we substitute (8.4.20) gives

$$A^2(s,t) = 2\pi s\mathrm{i}\int_0^\infty b\,\mathrm{d}b\, J_0(b\sqrt{-t})\,\frac{1}{(8\pi s)^2}$$
$$\times \int_{-\infty}^0 \mathrm{d}t_1 J_0(b\sqrt{-t_1})\,A^{\mathrm{B}}(s,t_1)\int_{-\infty}^0 \mathrm{d}t_2 J_0(b\sqrt{-t_2})\,A^{\mathrm{B}}(s,t_2) \tag{8.4.26}$$

But (Heneyey *et al.* 1969, Erdelyi *et al.* (1953) vol. 2)

$$\int_0^\infty b\,\mathrm{d}b\, J_0(b\sqrt{-t_1})\,J_0(b\sqrt{-t_2})\,J_0(b\sqrt{-t}) = \frac{2}{\pi}\frac{\theta(-\lambda)}{(-\lambda(t,t_1,t_2))^{\frac{1}{2}}} \tag{8.4.27}$$

(cf. (8.2.6) and (8.4.8)) so

$$A^2(s,t) = \frac{\mathrm{i}}{16\pi^2 s}\iint_{-\infty}^0 \mathrm{d}t_1\,\mathrm{d}t_2\,A^{\mathrm{B}}(s,t_1)\,A^{\mathrm{B}}(s,t_2)\,\frac{\theta(-\lambda)}{(-\lambda(t,t_1,t_2))^{\frac{1}{2}}} \tag{8.4.28}$$

This would agree with (8.4.1), with $N = 1$, if we were to take Regge poles instead of (8.4.21) as the Born approximation, which shows that the precise form of the exchange does not really matter provided the approximation (8.4.12) remains valid. In fact it can be shown (Tiktopoulos and Trieman 1970) that if the particle exchanges in fig. 8.20 are replaced by ladders, the leading diagrams are those in which the couplings at the ends of the ladders cross as in fig. 8.19 (*c*), (*e*), rather

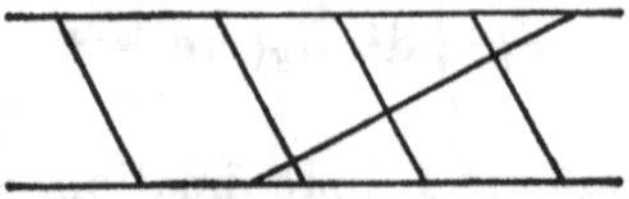

FIG. 8.21 A diagram which violates the eikonal approximation in ϕ^3 theory. Only three propagators are needed to get across the diagram, but there are four along each side.

than the planar fig. 8.19 (*b*), (*d*) as one would expect from section 8.2. So the eikonal series (8.2.24) can be regarded as the sum of the Regge cuts due to any number of Reggeons with their couplings 'nested'.

There must, however, be doubt about the applicability of these results to hadronic physics. First they are not actually true in ϕ^3 field theory because the approximation (8.4.12) is invalid. For example, fig. 8.21 has a *d*-line of length 3 and hence $\sim s^{-3}$. But in the eikonal approximation we suppress the possibility of large momenta travelling across the diagram, because the momentum should mainly travel along the sides, which involves 4 propagators, and hence in this approximation fig. 8.21 $\sim s^{-4}$. So this diagram would violate the eikonal approximation in ϕ^3 theory. However, we have seen in chapter 6 that experimentally momentum transfers are cut off exponentially, so in this respect the approximate version of the field theory seems more realistic than the theory itself. Models with elementary vector meson exchanges have also been examined (Cheng and Wu 1969, 1970). In this case the s-dependence of the exchanged propagators (see (2.6.10)) ensures the validity of the eikonal approximation without a cut-off, but it also means that the Reggeons lie above 1 for all t. We shall discuss this further in the next section.

Fig. 8.19 includes only one set of relevant graphs. In the previous section we mentioned the necessary for iterating the ladders in t as well as s (as in fig. 8.16) to be compatible with t-channel unitarity. There are also more complicated diagrams like fig. 8.22 (called 'checkerboard' diagrams), in which the Reggeons interact during the exchange, which seem to violate the eikonal result (Blankenbecler and Fried 1973, Swift 1975), and diagrams (of which the diffraction diagram fig. 8.18 is an example) in which the leading particle fragments and recombines. Quite apart from the difficulties of including these effects, there is also the usual worry as to whether they are not already partially included (implicitly) as re-normalization corrections to simpler diagrams. This is a fundamental problem with any field-theoretic model. (For a review see Blankenbecler, Fulco and Sugar 1974).

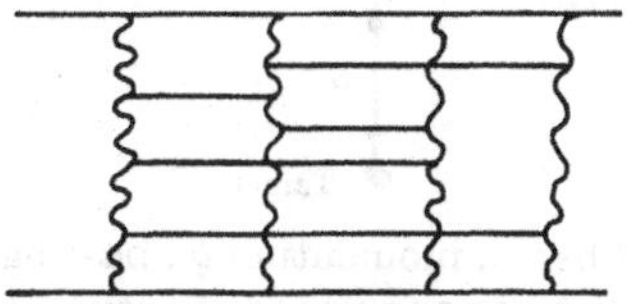

FIG. 8.22 An example of a 'checkerboard' diagram in which the Reggeons interact during the exchange.

In spite of these reservations, the eikonal model offers several advantages. First it ensures satisfaction of s-channel unitarity bounds (in analogy with section 1.14). Secondly it is easily generalized to include different types of Reggeons, and different helicity amplitudes. And thirdly the model is comparatively easy to evaluate. To demonstrate this it is convenient to start from the s-channel partial-wave series for an elastic-scattering helicity amplitude, (4.4.9),

$$A_{Hs}(s,t) = 16\pi \sum_{J=M}^{\infty} (2J+1)\, A_{HJ}(s)\, d^J_{\mu\mu'}(z_s) \tag{8.4.29}$$

At high energies and small angles, $s \gg t$, and large J (Durand and Chiu 1965),

$$d^J_{\mu\mu'}(z_s) \approx J_n((J+\tfrac{1}{2})\,\theta_s), \quad n \equiv |\mu-\mu'| \tag{8.4.30}$$

and

$$\cos\theta_s \equiv z_s \approx 1+\frac{t}{2q_s^2}, \quad \text{so} \quad \theta_s \approx \left(\frac{-t}{q_s^2}\right)^{\frac{1}{2}} \tag{8.4.31}$$

The classical impact parameter b (fig. 8.23) for a particle passing the target with angular momentum J is given by

$$J = q_s b - \tfrac{1}{2} \tag{8.4.32}$$

(the $\frac{1}{2}$ is arbitrary since we are working with large J) so we can replace $\sum_J$ by $\int_0^\infty q_s\,\mathrm{d}b$ and hence (8.4.29) becomes

$$A_{H_s}(s,t) = 16\pi \int_0^\infty q_s\,\mathrm{d}b\, 2q_s b\, A_{HJ}(s)\, J_n(b\sqrt{-t}) \tag{8.4.33}$$

We then express the elastic partial-wave amplitude in unitary form in terms of the phase shift (2.2.10)

$$A_{HJ}(s) = \frac{e^{2i\delta_{HJ}(s)}-1}{2i\rho(s)} \tag{8.4.34}$$

and define the eikonal phase $\chi_H(s,b)$ in terms of this phase shift by

$$\chi_H(s,b) = 2\delta_{HJ}(s) \tag{8.4.35}$$

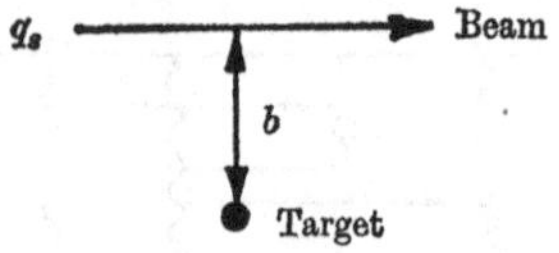

FIG. 8.23 Classical beam, momentum q_s, passing target at impact parameter b has angular momentum $J = q_s b$.

using (8.4.32). This gives the impact parameter amplitude

$$A_H(s,b) = \frac{e^{i\chi_H(s,b)}-1}{2i\rho(s)} \underset{s\to\infty}{\to} \frac{e^{i\chi_H(s,b)}-1}{2i} \tag{8.4.36}$$

Physically this replacement means that we are supposing that each part of the wave front of the incident beam passes through the target with its impact parameter unchanged, only its phase being altered. So at high energies conservation of J is replaced by conservation of b through (8.4.23). This corresponds to the derivation in section 1.14. Then (putting $q_s \to \frac{1}{2}\sqrt{s}$)

$$\begin{aligned} A_{H_s}(s,t) &= 8\pi s \int_0^\infty b\,\mathrm{d}b A_H(s,b) J_n(b\sqrt{-t}) \\ &= 8\pi s \int_0^\infty b\,\mathrm{d}b \frac{e^{i\chi_H(s,b)}-1}{2i} J_n(b\sqrt{-t}) \end{aligned} \tag{8.4.37}$$

which agrees with (8.4.21) for the non-flip amplitude, $n = 0$, if we define the eikonal function in terms of the Regge pole exchange Born approximation to the helicity amplitude (like (8.4.20))

$$\chi_H(s,b) = \chi_H^R(s,b) \equiv \frac{1}{8\pi s}\int_{-\infty}^0 \mathrm{d}t J_n(b\sqrt{-t})\, A_{H_s}^R(s,t) \tag{8.4.38}$$

Expanding the exponential in (8.4.37) gives us the series of cuts produced by R exchange, i.e. $R + R \otimes R + R \otimes R \otimes R + \ldots$. Since we want to sum over intermediate-state helicities a matrix product of the χ's in helicity space is implied.

For an inelastic process we can invoke the so-called 'distorted wave Born approximation' (see for example Newton (1966)), and replace (8.4.36) by

$$A_H(s,b) = \chi_H^R(s,b)\, e^{i\chi_H^{el}(s,b)} = \chi_H^R + i\chi_H^R\,\chi_H^{el} + \ldots \tag{8.4.39}$$

which obviously corresponds, up to second order, to the absorptive prescription (8.4.4) if we use P for the elastic amplitude. So combining (8.4.39) and (8.3.36) the eikonal/absorption prescription for a Regge cut involving the exchange of n_1 Reggeons R_1, n_2 of R_2, etc. is

$$A^{\mathrm{c}}_{H_s}(s,t) = -\mathrm{i}4\pi s\int_0^\infty b\,\mathrm{d}b\,J_n(b\sqrt{-t})\,\frac{(\mathrm{i}\chi_H^{\mathrm{R}_1})^{n_1}}{n_1!}\frac{(\mathrm{i}\chi_H^{\mathrm{R}_2})^{n_2}}{n_2!}\cdots \tag{8.4.40}$$

where each χ is calculated according to (8.4.38), and we must sum over intermediate-state helicities.

It may seem surprising that we have chosen to evaluate the cut contributions in s-channel helicity amplitudes, but in fact this is easiest because the cuts involve s-channel unitarization.

We explore some of the properties of (8.4.40) in the next section.

8.5 Evaluation of Regge cut amplitudes

The expression (8.4.40) is readily evaluated provided the Regge poles are expressed sufficiently simply. If we take a linear trajectory, $\alpha(t) = \alpha^0+\alpha' t$, and an exponential residue, $\gamma(t) = G\mathrm{e}^{at}$, with the phase (6.8.15), we have for the Regge pole amplitude

$$A^{\mathrm{R}}_{H_s}(s,t) = -x(-t)^{n/2}\left(\frac{s}{s_0}\right)^{\alpha^0}\mathrm{e}^{-\mathrm{i}\pi\alpha^0/2}\,G\mathrm{e}^{ct} \tag{8.5.1}$$

where $x \equiv 1/-\mathrm{i}$ for $\mathscr{S} = \pm 1$, and

$$c \equiv a+\alpha'\left(\log\left(\frac{s}{s_0}\right)-\mathrm{i}\frac{\pi}{2}\right) \tag{8.5.2}$$

When substituted in (8.4.38) this gives

$$\chi^{\mathrm{R}}_{H_s}(s,b) = \frac{-xG(s/s_0)^{\alpha^0}\,\mathrm{e}^{-\mathrm{i}\pi\alpha^0/2}}{8\pi s}\int_{-\infty}^{0}\mathrm{d}t\,J_n(b\sqrt{-t})\,(-t)^{n/2}\,\mathrm{e}^{ct} \tag{8.5.3}$$

which is evaluated using the result

$$\int_{-\infty}^{0}\mathrm{e}^{ct}(-t)^{(n/2)+m}J_n(b\sqrt{-t})\,\mathrm{d}t = \left(\frac{b}{2}\right)^n\left(-\frac{\partial}{\partial c}\right)^m\left(\frac{\mathrm{e}^{-b^2/4c}}{c^{n+1}}\right) \tag{8.5.4}$$

This may be obtained from Magnus and Oberhettinger (1949, p. 131) when it is realized that multiplying the integrand by t is equivalent to differentiating it with respect to $-c$. So

$$\chi^{\mathrm{R}}_{H_s}(s,b) = \frac{-xG(s/s_0)^{\alpha^0}\mathrm{e}^{\mathrm{i}\pi\alpha^0/2}}{8\pi s}\left(\frac{b}{2c}\right)^n\frac{\mathrm{e}^{-b^2/4c}}{c} \tag{8.5.5}$$

This expression gives the impact parameter profile of a Regge pole. Except for non-flip amplitudes ($n = 0$) it must vanish at $b = 0$, and all the amplitudes are Gaussian in b for large b (because of the assumed exponential t dependence), the width of the profile, given by c, increasing with $\log s$. This accords with our discussion in section 6.8*e*.

Such expressions are substituted in (8.4.40) for each Reggeon, and the integral may be evaluated using (8.5.4) by interchanging b and $\sqrt{-t}$, i.e.

$$\int_0^\infty e^{-b^2/4c}(b^2)^{(n/2)+m} J_n(b\sqrt{-t})\, b\, db = (-t)^{n/2}\left(-4c^2\frac{\partial}{\partial c}\right)^m [(2c)^{n+1} e^{ct}] \tag{8.5.6}$$

Thus specializing to $n = 0$ we obtain for the $R_1 \otimes R_2$ cut

$$A^{c2}_{H_s}(s,t) = \frac{i}{8\pi s}\frac{\mathscr{x}_1 G_1 \mathscr{x}_2 G_2}{c_1+c_2}\left(\frac{s}{s_0}\right)^{\alpha_1{}^0+\alpha_2{}^0} e^{-(i\pi/2)(\alpha_1{}^0+\alpha_2{}^0)} e^{[c_1c_2/(c_1+c_2)]t} \tag{8.5.7}$$

so we see that the cut will have a flatter t dependence than the pole, and its impact parameter profile will be

$$\sim e^{-(b^2/4)[(c_1+c_2)/c_1c_2]} \tag{8.5.8}$$

which has a shorter range than the pole (8.5.5).

Now for $\log(s/s_0) \gg a/\alpha'$ (~ 4 typically)

$$\frac{c_1c_2}{c_1+c_2} \to \frac{\alpha_1'\alpha_2'}{\alpha_1'+\alpha_2'}\left(\log\left(\frac{s}{s_0}\right) - \frac{i\pi}{2}\right)$$

and so

$$A^{c2}_{H_s}(s,t) \to -\frac{G_1G_2}{8\pi s_0}\frac{(s/s_0)^{\alpha_c(t)}\mathscr{x}_1\mathscr{x}_2 e^{-(i\pi/2)\alpha_c(t)}}{a_1+a_2+(\alpha_1'+\alpha_2')(\log s/s_0 - i\pi/2)} \sim \frac{s^{\alpha_c(t)}}{\log s} \tag{8.5.9}$$

where
$$\alpha_c(t) \equiv \alpha_1^0 + \alpha_2^0 - 1 + \left(\frac{\alpha_1'\alpha_2'}{\alpha_1'+\alpha_2'}\right)t \tag{8.5.10}$$

which agrees with (8.2.17) and (8.3.24) for the position of the cut. It also gives the signature factor $\mathscr{x}_1\mathscr{x}_2 e^{-i\pi\alpha_c/2}$ which corresponds to the product of the signatures of the two poles as in (8.3.28). But the presence of $(\log s)^{-1}$ in the asymptotic behaviour indicates that the cut is hard, i.e. has a finite discontinuity at the branch point. As $\log s \to \infty$ the phase of the cut corresponds to its power behaviour (as in section 6.8*f*) but for finite $\log s$ the denominator modifies this phase.

Similarly for n identical Reggeons (8.4.40) with (8.5.5) gives

$$\left.\begin{aligned} A^{cn}_{H_s}(s,t) &= -i\frac{[-i\mathscr{x}G(s/s_0)^{\alpha^0} e^{-i\pi\alpha^0/2}]^n}{(8\pi c s)^{n-1}}\frac{e^{ct/n}}{n n!} \\ &\to \frac{(\mathscr{x}G)^n}{(8\pi s_0 c)^{n-1}}\left(\frac{s}{s_0}\right)^{\alpha_c(t)} e^{(-i\pi/2)\alpha_c(t)} \sim \frac{s^{\alpha_c(t)}}{(\log s)^n} \\ \alpha_c(t) &\equiv n\alpha^0 - (n-1) + \frac{\alpha'}{n}t \end{aligned}\right\} \tag{8.5.11}$$

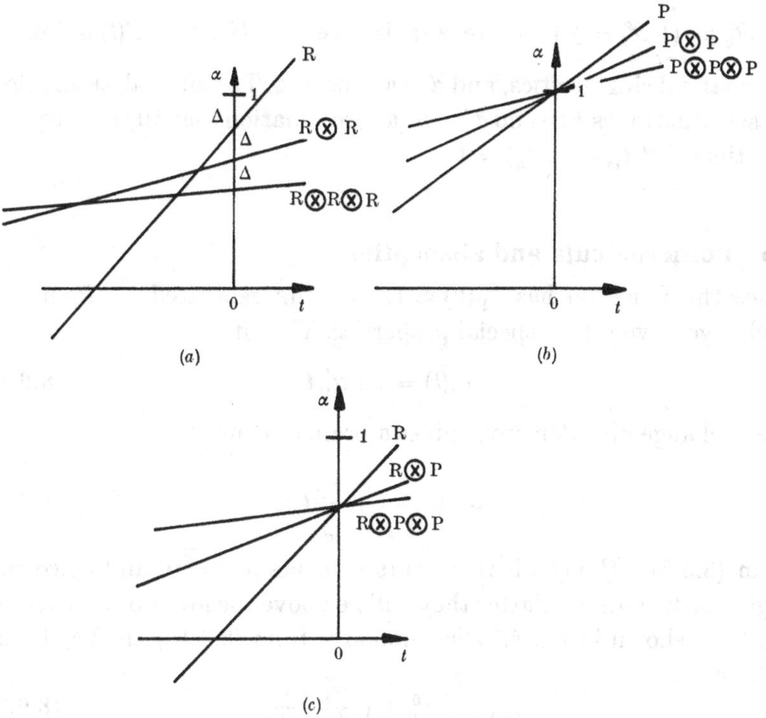

FIG. 8.24 (a) Regge trajectory R, and R ⊗ R, R ⊗R⊗R cuts. $\Delta \equiv (1-\alpha^0)$ gives the spacing of successive cuts at $t = 0$, but the higher order cuts are flatter. (b) Pomeron cuts converging on $\alpha = 1$ at $t = 0$. (c) Reggeon R and the sequence of R ⊗ P, R ⊗ P⊗P cuts which converge on α^0_{R} at $t = 0$.

which again agrees with (8.2.17) and (8.3.24), and gives

$$\Delta(J,t) \sim (J-\alpha_c(t))^{n-1}$$

as $J \to \alpha_c$. The positions of such cuts are shown in fig. 8.24.

The factor $(-t)^{\frac{1}{2}|\mu-\mu'|}$ in (8.5.6) ensures that cuts of all orders have the correct helicity-flip factor, as long as it is present in the input poles. However, if the poles also have evasive t factors (see section 6.5) cuts generated by these poles will not usually contain these additional factors. This is because the cuts do not have a definite t-channel parity, and so it is natural for them to conspire, not evade.

Combining these results we can write a general expression for an n-Reggeon cut

$$A^{c}_{H_s}(s,t) = (-t)^{\frac{1}{2}|\mu-\mu'|} F(t) \left(\frac{s}{s_0}\right)^{\alpha_c(t)} e^{(i\pi/2)\alpha_c(t)} \left(\log\left(\frac{s}{s_0}\right)+d\right)^{-n+1} \begin{Bmatrix} 1 \\ -i \end{Bmatrix} \tag{8.5.12}$$

for $\mathscr{S}_c \equiv \prod_{i=1}^{n} \mathscr{S}_i = \pm 1$, where $\alpha_c(t)$ is given by (8.3.24), $F(t)$ is free of kinematical singularities, and d is a constant. The eikonal/absorption prescription gives $F(t)$ and d in the approximation that all the coupling functions $N(t, t_1, t_2, \ldots, t_n) = 1$.

8.6 Pomeron cuts and absorption

Since the Pomeron has $\alpha_P(0) \approx 1$, the cuts generated by Pomeron exchange have rather special properties. Thus if

$$\alpha_P(t) = 1 + \alpha'_P t \tag{8.6.1}$$

the exchange of n Pomerons gives a branch point at

$$\alpha_{c_n}(t) = 1 + \frac{\alpha'_P}{n} t \tag{8.6.2}$$

from (8.3.27). Hence all these cuts coincide at $t = 0$, and since the higher order cuts are flatter they will be above the lower order cuts for $t < 0$, as shown in fig. 8.24 (b). Similarly from (8.5.10) an R $\otimes$ P cut will be at

$$\alpha_{c_2}(t) = \alpha_R^0 + \left(\frac{\alpha'_R \alpha'_P}{\alpha'_R + \alpha'_P}\right) t \tag{8.6.3}$$

and an R $\otimes$ (P)^n cut will be at

$$\alpha_c(t) = \alpha_R^0 + \left(\frac{\alpha'_R (\alpha'_P)^n}{\alpha'_R + n\alpha'_P}\right) t \tag{8.6.4}$$

so all the cuts coincide with $\alpha_R(t)$ at $t = 0$ and lie above it for $t < 0$ (fig. 8.24 (c)).

This coincidence of the P pole and its cuts at $t = 0$ means that successive terms in the sum over all numbers of P exchanges differ only by powers of $(\log s)^{-1}$, not powers of s. Hence the re-normalization and unitarity problems mentioned in section 8.3 seem much more severe for Pomerons than for other trajectories. Indeed we shall find in sections 10.8 and 11.7 that naïve iteration of P exchange in t would give a leading behaviour which violates the Froissart bound (that is $A(s, t = 0) \leqslant s \log^2 s$) so that pole dominance, even at $t = 0$, does not seem to be self-consistent. So it is clear that iteration in s, giving 'absorption' of P exchange, must also be important, but unfortunately a proper unitarization in both s and t is beyond our competence. The Reggeon field theory mentioned in section 8.3 suggests that eventually $A(s, t = 0) \sim s(\log s)^{\nu}$ (where $\nu \approx \frac{1}{6}$ to the first approximation: see

Abarbanel *et al.* (1975)), but it only explicitly takes into account t-channel unitarity, and compatibility with s-channel unitarity is not obvious.

If the leading J-plane singularity is to be self-consistent, it should reproduce itself when inserted in (8.3.24). Obviously linear forms cannot do this, but if one takes (Schwarz 1967)

$$\alpha(t) = 1 \pm \mathrm{i}\alpha'\sqrt{-t}\,, \quad \alpha' \text{ constant} \tag{8.6.5}$$

then $$\alpha_{c_n}(t) \equiv \max\left\{\sum_{i=1}^{n} \alpha(t_i) - n + 1\right\} = \alpha(t) \quad \text{for any } n, \tag{8.6.6}$$

with $$\sum_{i=1}^{n} \sqrt{-t_i} = \sqrt{t}\,. \tag{8.6.7}$$

So the poles are complex for $t < 0$, but there is no violation of Mandelstam analyticity because the two poles have equal and opposite imaginary parts (see section 3.2). Indeed it generally happens that when poles and cuts collide unitarity requires the trajectories to be complex (see Zachariasen 1971). The fact that $\mathrm{Im}\{\alpha\} \neq 0$ for $t < 0$ gives $s^{\alpha(t)} = s^{\alpha_{\mathrm{R}}(t)}\,\mathrm{e}^{\mathrm{i}\alpha_{\mathrm{I}}(t)\log s}$ (where $\alpha_{\mathrm{R,I}}$ are the real and imaginary parts of α respectively) and so the phase–energy relation (8.6.14) will not hold, and the Regge power behaviour will be modulated by oscillations in $\log s$. There has so far been no sign of these effects, and the effective trajectories of fig. 6.6 all seem to be linear in t. It may be of course that we need $\log s \gg a/\alpha'$ to observe such effects, in which case it will be some time before they can be verified.

With $\alpha_{\mathrm{P}}(0) = 1$, $x = 1$, we obtain for the sum of all P exchanges, from (8.5.11),

$$A_{H_s}(s,t) = \sum_{n=1}^{\infty} A^{c_n}(s,t) = \sum_{n=1}^{\infty} -\mathrm{i}\,\frac{[-G(s/s_0)]^n}{(8\pi c s)^{n-1}}\,\frac{\mathrm{e}^{ct/n}}{n n!} \tag{8.6.8}$$

which, setting $s_0 = 1$, gives

$$\sigma^{\mathrm{tot}}_{12\to\mathrm{all}}(s) = \frac{1}{s}\,\mathrm{Im}\{A^{12\to12}(s,0)\} = G - \frac{G^2}{32\pi c} + \frac{G^3}{1152\pi^2c^2} - \dots$$

$$\to G\left(1 - \frac{G}{32\pi\alpha'\log s} + \dots\right) \tag{8.6.9}$$

so if we assume that the series converges (which may be false) we predict that $\sigma^{\mathrm{tot}}(s) \to$ constant logarithmically from below. This rise of σ^{tot} depends crucially on the sign of the cut being that of the eikonal/absorption model and the Reggeon calculus, not the AFS sign, which with our pure imaginary P amplitude (at $t = 0$) would give a positive second term. Unfortunately, the magnitude of the cut term in (8.6.9) is insufficient to account for the rise of $\sigma^{\text{tot}}(\text{pp})$ (fig. 6.4) as it stands. However, if we use the freedom suggested by (8.3.8) and (8.2.37) to multiply the cut strength by an arbitrary factor N^2, we can choose N to fit the data. But this makes the cuts very strong, and the convergence of the series even more doubtful (see section 8.7*a* below).

Of course if $\alpha_{\text{P}}(0) < 1$ the pole and cuts are separated by a finite amount $\Delta \equiv (1-\alpha_{\text{P}}(0))$ at $t = 0$ (fig. 8.24*a*), but this makes the observed rise of σ^{tot} hard to understand. In fact fig. 6.6 suggests that $\alpha_{\text{P}}(0) > 1$, but this can only be compatible with the Froissart bound (2.4.9) if unitarization produces strong cancelling cuts. We can see how this works as follows. If $\alpha_{\text{P}}(0) > 1$, then from (8.5.5) with $n = 0$,

$$\chi_H^{\text{P}}(s,b) = -\frac{G_{\text{P}}(s/s_0)^{(\alpha_{\text{P}}{}^0-1)}\,\mathrm{e}^{-\mathrm{i}\pi\alpha_{\text{P}}{}^0/2}\,\mathrm{e}^{-b^2/4c_{\text{P}}}}{8\pi s_0 c_{\text{P}}}$$

$$= -\frac{G_{\text{P}}}{8\pi s_0 c_{\text{P}}}\,\mathrm{e}^{-\mathrm{i}\pi\alpha_{\text{P}}{}^0/2}\,\mathrm{e}^{[-b^2/4c_{\text{P}}+(\alpha_{\text{P}}{}^0-1)\log(s/s_0)]} \tag{8.6.10}$$

so if $$b^2 < b_0^2 \equiv 4(\alpha_{\text{P}}^0-1)\,\alpha_{\text{P}}'\log^2\left(\frac{s}{s_0}\right) \tag{8.6.11}$$

then $$\chi_H^{\text{P}}(s,b)\to 0 \qquad \text{as } s\to\infty,$$

but for $b^2 > b_0^2$ $$\chi_H^{\text{P}}(s,b)\to 0$$

So from (8.4.36) $$A_H(s,b)\to\tfrac{1}{2}\mathrm{i},\quad b^2 < b_0^2$$

$$\to 0,\quad b^2 > b_0^2 \tag{8.6.12}$$

This is like complete absorption on a black disk of radius b_0. Now $J_0(b\sqrt{-t})\to 1$ for $t\to 0$, so from (8.4.37)

$$\operatorname{Im}\{A_{H_s}(s,0)\} = 8\pi s\int_0^\infty b\,\mathrm{d}b\,\operatorname{Im}\{A_H(s,b)\}\to 4\pi s\int_0^{b_0} b\,\mathrm{d}b = 2\pi s b_0^2 \tag{8.6.13}$$

and so $$\sigma_{12}^{\text{tot}}(s)\to 8\pi\alpha_{\text{P}}'(\alpha_{\text{P}}^0-1)\log^2\left(\frac{s}{s_0}\right) \tag{8.6.14}$$

in accord with the Froissart bound. However, with the numbers of fig. 6.6, $\alpha_{\text{P}}' \approx 0.25\,\text{GeV}^{-2}$, $\alpha_{\text{P}}^0 \approx 1.07$, this gives

$$\sigma_{12}^{\text{tot}}(s)\to 3.6\log^2\left(\frac{s}{s_0}\right)\text{mb}$$

which with $s_0 = 1$ is much in excess of current measurements ($\sigma^{\text{tot}}(\text{pp}) \approx 44$ mb for $\log s \approx 8$ at ISR). So if this model is correct very much higher energies will have to be achieved before the observed pole-like behaviour $\sigma^{\text{tot}}(\text{pp}) \sim 27 s^{0.07}$ mb turns into the $\log^2 s$ asymptotic behaviour. If more complex diagrams than fig. 8.19, like fig. 8.22, are permitted, then $\alpha_{\text{P}}(0) > 1$ gives a grey disk instead ($\text{Im}\{A(s,b)\} < \frac{1}{2}$) but the main conclusions are unaltered; see Bronzan (1974), Cardy (1974*a*).

It is evident from (8.4.40) that for inelastic scattering, where P cannot be exchanged, in addition to the dominant R exchange there will be a sequence of $\text{R} \otimes (\text{P})^n$ cuts which should dominate for $t < 0$, $s \to \infty$. Thus if we use (8.5.1) for R and take $\alpha_{\text{P}}(0) = 1$, the $\text{R} \otimes \text{P}$ cut from (8.5.7) will be

$$A_{H_s}^{\text{R}\otimes\text{P}}(s,t) = \frac{\lambda x_{\text{R}} G_{\text{R}} G_{\text{P}}}{8\pi a_0} \left(\frac{s}{s_0}\right)^{\alpha_{\text{R}}{}^0} \frac{\text{e}^{-\text{i}\pi\alpha_{\text{R}}{}^0/2}}{c_{\text{R}}+c_{\text{P}}} \text{e}^{c_{\text{R}}c_{\text{P}}/(c_{\text{R}}+c_{\text{P}})t} \tag{8.6.15}$$

where λ is the enhancement factor.

This has the same asymptotic phase as the pole (8.5.1) but the opposite sign at $t = 0$, and for t near zero where $\alpha_c \approx \alpha_{\text{R}}$. Also the t dependence of the cut is shallower than that of the pole, so even if the pole dominates the cut at $t = 0$, there will be a cancellation (destructive interference) between cut and pole for some $t < 0$, as shown in fig. 8.25. Approximately (neglecting higher order cuts),

$$A_{H_s}(s,t) \approx A^{\text{R}} + A^{\text{R}\otimes\text{P}} = -x_{\text{R}} G_{\text{R}} \left(\frac{s}{s_0}\right)^{\alpha_{\text{R}}{}^0} \text{e}^{-\text{i}\pi\alpha_{\text{R}}{}^0/2}$$
$$\times \left(\text{e}^{c_{\text{R}}t} - \frac{\lambda G_{\text{P}}}{8\pi s_0} \text{e}^{[c_{\text{R}}c_{\text{P}}/(c_{\text{R}}+c_{\text{P}})]t}\right) \tag{8.6.16}$$

and for small $|t|$, where the phase of the pole and cut are similar, there will be an almost simultaneous zero in $\text{Re}\{A_{H_s}\}$ and $\text{Im}\{A_{H_s}\}$ and so there may be a dip in $\text{d}\sigma/\text{d}t$. As we shall see in the next section, this may provide an explanation for some of the dips discussed in section 6.8*k*.

It is interesting to examine the impact parameter structure of this amplitude. From (8.4.39) we can write

$$A_H(s,b) \approx \chi_H^{\text{R}}(s,b) + \text{i}\lambda\chi_H^{\text{R}}(s,b)\,\chi_H^{\text{P}}(s,b)$$
$$= \chi_H^{\text{R}}(s,b)\,(1 + \text{i}\lambda\chi_H^{\text{P}}(s,b))$$

which since P is almost pure imaginary becomes

$$\approx \chi_H^{\text{R}}(s,b)\,(1 - \lambda|\chi_H^{\text{P}}(s,b)|) \tag{8.6.17}$$

which is exhibited in fig. 8.26, from which it is seen that the effect of

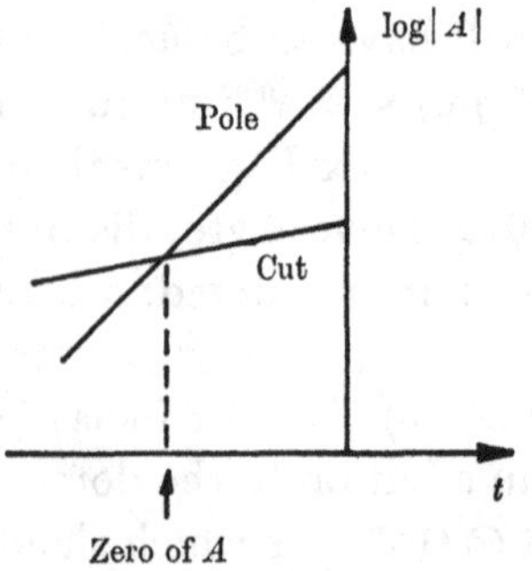

FIG. 8.25 Pole and cut magnitudes as a function of t at fixed s. The different exponential t dependences in (8.6.16) result in an approximate zero where the pole and cut magnitudes are the same (neglecting the phase difference which is small for small $|t|$).

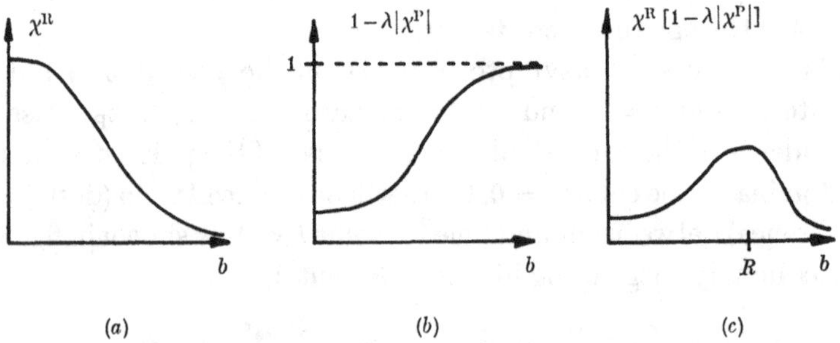

FIG. 8.26 (*a*) The Gaussian impact-parameter profile of a Regge pole from (8.5.5) with $n = 0$. (*b*) The absorption *S*-matrix. (*c*) The impact profile of the absorbed amplitude (8.6.17) showing the reduction of the amplitude at small b, i.e. low J. The resulting peak is at $b = R \approx 1\,\text{fm}$.

absorption is to reduce the Reggeon amplitude at small b through the destructive effect of the shorter range cut. By a suitable choice of λ we can eliminate small-b scattering (complete absorption) so that the predominant part of the scattering amplitude is at $b \approx 1\,\text{fm}$, the periphery of the target. For helicity-flip amplitudes the kinematical b^n in (8.5.5) means that the Regge pole amplitude is already fairly peripheral, and the effect of absorption is much smaller in this case.

One can roughly approximate this peripheral profile by a δ-function in b-space at radius R

$$A_H(s,b) \approx \frac{-xG_H}{8\pi s} e^{-i\pi\alpha/2}\left(\frac{s}{s_0}\right)^{\alpha}\frac{1}{R}\delta(b-R) \qquad (8.6.18)$$

where $R = 1\,\text{fm}$, which when substituted in (8.4.37) gives

$$A_H(s,t) = -xG_H e^{(-i\pi/2)\alpha(t)}\left(\frac{s}{s_0}\right)^{\alpha(t)} J_n(R\sqrt{-t}) \qquad (8.6.19)$$

for the sum of pole and cut. In this approximation we have completely ignored the difference between $\alpha_R(t)$ and $\alpha_c(t)$ in both the power behaviour and phase, and have dropped the $\log s$ factor from the cut. With $R = 1$ fm the first zeros of the Bessel function appear at $-t = 0.2$, 0.55 and 1.2 GeV^2 for $n = 0, 1, 2$, respectively (Henyey *et al.* 1969), positions which correspond to some of the amplitude zeros noted in section 6.8. We shall examine this result further in the next section.

8.7 Regge cut phenomenology

Having established the general features of Regge cut contributions we can now try and discover whether they can make good the deficiencies of Regge poles found in section 6.8.

Once we know the Regge pole trajectories, the positions of the various branch points, and hence the power behaviour and asymptotic phase of the cut contributions, are fixed by (8.3.24). Also the kinematical restrictions on the cuts are very simple for s-channel helicity amplitudes, which we used in (8.5.12). This leaves us with two main problems. First, we need to find $F(t)$ and d in (8.5.12). They are predicted by the eikonal/absorption model, but when we compare (8.4.28) with the Reggeon calculus result (8.3.8) it seems doubtful if the model will be reliable in this respect. If N in (8.3.8) is regarded as an unknown, then so are F and d. Secondly, we cannot be sure at what energy an expression like (8.5.12) becomes applicable. In the various derivations of sections 8.2, 8.3 and 8.4 we have taken only the leading $\log(s/s_0)$ behaviour of each diagram, which suggests that $\log(s/s_0) \gg 1$ is needed, which is seldom achieved in practice. This problem is worst for cuts involving the P where successive terms in the eikonal expansion differ only by $(\log(s/s_0))^{-1}$.

Keeping these uncertainties in mind we can now review some of the difficulties left over from section 6.8.

a. Total cross-sections and elastic scattering

The rising $\sigma^{\text{tot}}(s)$ shown in fig. 6.4 require either $\alpha_P(0) > 1$, which will eventually violate the Froissart bound unless there are cut corrections, or $\alpha_P(0) = 1$ and very strong cuts (see (8.6.9) *et seq.*). Thus for pp scattering we can write

$$\sigma^{\text{tot}}_{\text{pp}}(s) = \mathrm{i}G_P\left(1 - \frac{\lambda G_P}{32\pi\alpha'_P \log s} + \dots\right) \tag{8.7.1}$$

where λ is the enhancement factor, and one needs $G_P \approx 85$ mb, $\lambda \approx 2$ to fit the data. But the shallower t dependence of the cut means that cut and pole cancel for $|t| \approx 0.5\,\text{GeV}^2$, predicting a dip in $d\sigma/dt$ which is not observed. There does not seem to be any simple way out of this dilemma (see Collins *et al.* 1974). But $\alpha_P(0) \approx 1.07$ fits all the σ^{tot} data perfectly and there does not seem to be any sign at small $|t|$ of the cuts which will eventually be needed to ensure satisfaction of the Froissart bound. But since $\sigma^{\text{tot}}(s) = 27s^{0.07}$ mb does not violate the Froissart bound $\sigma^{\text{tot}}(s) \leqslant 60 \log^2 s$ mb until $s \simeq 10^{75}\,\text{GeV}^2$, this is hardly surprising.

It is possible to explain the dip in $d\sigma/dt(\text{pp})$ at $|t| \approx 1.4\,\text{GeV}^2$ (fig. 8.27 (*a*)) as interference between the P and the $\text{P} \otimes \text{P}$ cut (as in (8.6.16) with $\text{R} \rightarrow \text{P}$) and the effective trajectory plot (fig. 8.27 (*b*)) supports this idea. But since the dip occurs at such large $|t|$ we need a very small λ, $\approx \frac{1}{15}$. Also since the forward peak has $d\sigma/dt \sim e^{12t}$, while for $|t| > 1.4$ $d\sigma/dt \sim e^{2t}$, (8.6.16) which gives $A^c \sim e^{c_P t/2}$ will not do. We need to put some t dependence into N in (8.4.1), say $N^2 = \lambda e^{bt}$ with $b < 0$. So if fig. 8.27 is an example of Regge cuts in elastic scattering their properties must be very different from those of the simple eikonal model (Collins *et al.* 1974).

b. The cross-over zero

This zero in the ω and ρ non-flip coupling to nucleons is discussed in section 6.8 *l*. The fact that it appears at $|t| \approx 0.15\,\text{GeV}^2$ is just what one might have expected from the destructive interference of the R pole and $\text{R} \otimes \text{P}$ cut, from (8.6.19) with $n = 0$. Since this zero does not factorize it seems almost inevitable that cuts should provide the explanation, and it certainly seems to vindicate the absorption idea (Henyey *et al.* 1969).

c. Nonsense dips

Table 6.7 shows that the explanation of dips like that in

$$d\sigma/dt(\pi^- p \rightarrow \pi^0 n)$$

at $|t| \approx 0.55\,\text{GeV}^2$ as being due to a nonsense zero at $\alpha_\rho(t) = 0$ in A_{+-} is incompatible with factorization. Again the t value is just where (8.6.19) predicts a zero in the $n = 1$ amplitude, so cut–pole interference seems to provide a preferable explanation. There is a difficulty,

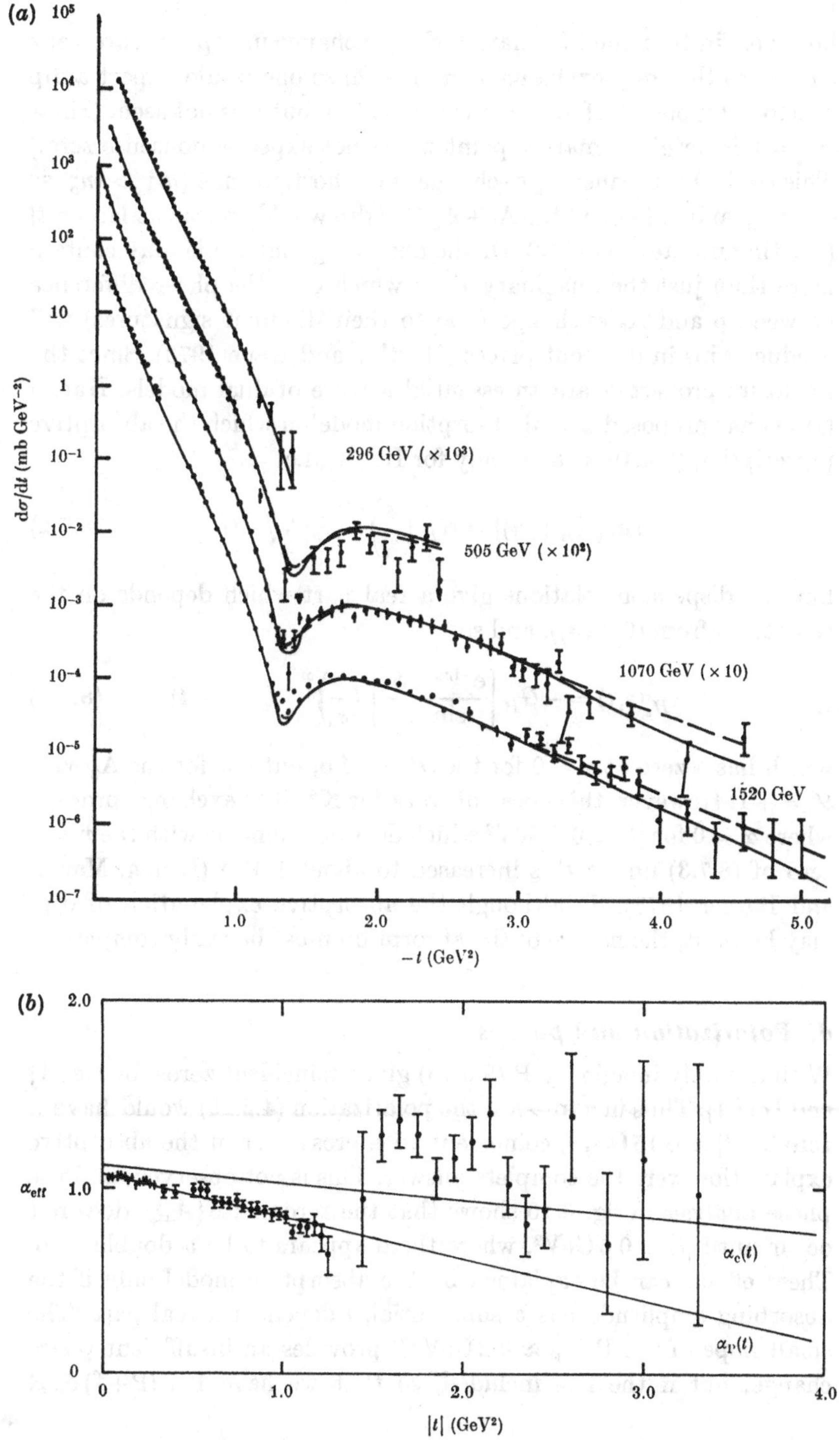

FIG. 8.27 (a) Fit $d\sigma/dt$ (pp) data with P and P$\otimes$P cut (from Collins *et al.* 1974). (b) The effective trajectory of the data compared with α_P and α_c. The pole dominates for $|t| < 1.4\,\text{GeV}^2$ and the cut for $|t| > 1.4\,\text{GeV}^2$.

however, in that the kinematics of A_2 exchange in $\pi^-p \to \eta n$ are very similar to that of ρ exchange in $\pi^-p \to \pi^0 n$ so one would expect a dip due to cut–pole interference in this case too, but it is not seen. (Since $\alpha_{A_2} = 0$ is a right-signature point we do not expect a nonsense zero.) This could be because A_2 exchange is of shorter range ($m^2_{A_2} > m^2_\rho$) so $c_{A_2} < c_\rho$, and in (8.6.16) the $A_2 + A_2 \otimes P$ dip would appear at larger $|t|$ (Martin and Stevens 1972). Or the absorbing amplitude may contain more than just the imaginary P, in which case the phase difference between ρ and A_2 exchange (due to their different signatures) will produce dips in different places (Hartley and Kane 1973). Since the signature properties are an essential feature of dual models, Harari (1971) has proposed a dual absorption model in which the absorptive prescription (8.6.19) is used only for $\text{Im}\{A\}$, i.e.

$$\text{Im}\{A_{H_s}(s,t)\} = G_H \left(\frac{s}{s_0}\right)^{\alpha(t)} J_n(R\sqrt{-t}) \tag{8.7.2}$$

but the dispersion relations give a real part which depends on the signature (from (6.8.18)), and so

$$A_{H_s}(s,t) = -G_H \left[\frac{e^{-i\pi\alpha} + \mathcal{S}}{\sin \pi\alpha}\right] \left(\frac{s}{s_0}\right)^{\alpha(t)} J_n(R\sqrt{-t}) \tag{8.7.3}$$

which has a zero at $\alpha = 0$ for the $\mathcal{S} = -1$ ρ, but not for the A_2 with $\mathcal{S} = +1$. However, this does not work for K*, K** exchange process, where $\alpha = 0$ for $|t| \approx 0.2\,\text{GeV}^2$ which does not coincide with the $n = 1$ zero of (8.7.3) unless R is increased to about 1.6 fm (Irving, Martin and Barger 1973). So although the absorptive explanation of dips may be right, the nature of the absorption must be fairly complex.

d. Polarization and phases

With a purely imaginary P (8.6.16) gives coincident zeros for $\text{Re}\{A\}$ and $\text{Im}\{A\}$. Thus in $\pi^-p \to \pi^0 n$ the polarization (4.2.22) would have a zero at $|t| \approx 0.15\,\text{GeV}^2$, coincident with cross-over, if the absorptive explanation were the complete answer. This is not observed, and the phase analysis in fig. 6.10 shows that the zero of $\text{Re}\{A_{++}\}$ does not occur until $|t| \approx 0.5\,\text{GeV}^2$, where there appears to be a double zero. These effects can be explained by the absorption model only if the absorbing amplitude has a substantial, t-dependent, real part. The small slope of the P ($\alpha'_P \approx 0.2\,\text{GeV}^{-2}$) provides an insufficient phase change, but if the f is included, so that we have $R + (P + f) \otimes R$

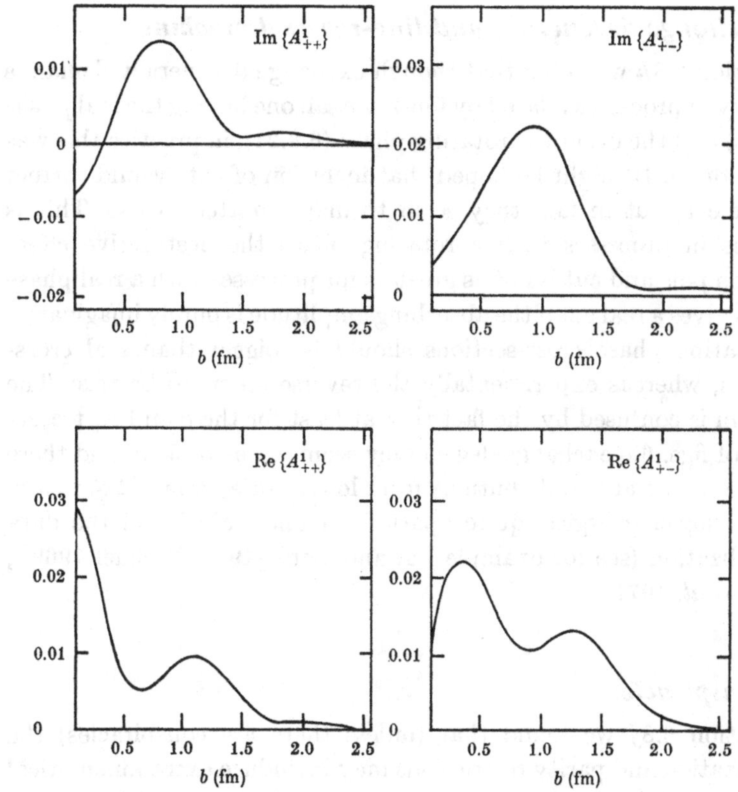

FIG. 8.28 The impact parameter amplitudes for $I_t = 1$ πN elastic scattering at 6 GeV, corresponding to fig. 6.10, from Halzen and Michael (1971).

instead, fairly satisfactory fits can be obtained (Collins and Swetman 1972). Other, more *ad hoc*, modifications of the phase have also been proposed (Hartley and Kane 1973). But whether this sort of approach is correct given that, as discussed in section 6.8*m*, only Im $\{A_{++}\}$ is different from what we expect from a nonsense-choosing ρ pole, is not certain. The impact parameter decomposition (8.4.38) of the amplitude in fig. 6.10 gives fig. 8.28. Since a pole without nonsense factors gives fig. 8.26 (*a*), it is clear that, for small b, Im $\{A_{++}(s, b)\}$ is not just being absorbed but over-absorbed (i.e. its sign is reversed) which conflicts with any simple physical interpretation of what absorption is supposed to mean.

e. Exchange degeneracy and line-reversal breaking

In section 6.8*h* we noted that though exchange degeneracy demands equality of processes related by line reversal, one having the real phase (6.8.22) and the other the rotating phase (6.8.21), in practice this was often not so. It might be hoped that inclusion of cuts would correct this defect, but in fact they seem to make matters worse. This is because in processes with a rotating phase the destructive effect between pole and cut is not as great as for processes with a real phase (which gives a real cut if the absorbing amplitude is purely imaginary). So rotating phase cross-sections should be bigger than real cross-sections, whereas experimentally the reverse seems to be true. The problem is confused by the fact that at least for the ρ and A_2 trajectories of figs. 6.6 exchange degeneracy seems to be broken, and there may be important contributions from lower trajectories, R ⊗ f cuts, etc., at lower energies, quite apart from uncertainties in the data normalization (see for example Lai and Louie 1970, Michael 1969*a*, Irving *et al.* 1971).

f. Conspiracies

In section 6.8*j* we found that (unless there are conspiracies) the factorization and parity restrictions may introduce extra kinematical factors into Regge pole amplitudes, causing them to vanish at $t = 0$. This is particularly important for π exchange in processes such as $\gamma p \to \pi^+ n$, $\pi p \to \rho p$, $\bar{p}p \to \bar{n}n$ which would have amplitudes like

$$A(s,t) \sim \frac{t}{t-m_\pi^2}, \quad t \to 0 \qquad (8.7.4)$$

However, we saw in table 6.6 that in practice spikes often occur. Since cuts are self-conspiring they do not have to vanish at $t = 0$ in non-flip amplitudes, but of course they will not contain the pion pole. But if we take $\pi + \pi \otimes P$, where $\pi \otimes P$ is slowly varying near $t = 0$, we get

$$A(s,t) \sim \frac{t}{t-m^2} - 1 = \frac{m_\pi^2}{t-m_\pi^2} \qquad (8.7.5)$$

which has the pion pole but no evasive t factor. The effect of the cut is to absorb away the S-wave contribution of the pion pole (S-wave because it is independent of t and hence z_s). This is sometimes called the Williams model (Williams 1970) or 'poor man's absorption'. This procedure can account for the forward structure in the processes listed in table 6.6. At $t = 0$ the amplitude is purely cut (no pole), so the magnitude of the cut is unambiguously determined, and it is very large. In $\gamma p \to \pi^+ n$, for example, a model like (8.6.17) needs $\lambda \approx 3$ (Kane *et al.* 1970).

g. *Shrinkage and pole-dominated cuts*

Since cuts are flatter in t than poles (fig. 8.24) they should become dominant for large s and $|t|$, so one would expect that the amount of shrinkage would decrease, and α_{eff} would become flatter as $|t|$ increases. This does indeed happen in a few cases such as pp elastic scattering (fig. 8.27 (*b*)) and photo-production. But these processes are quite atypical since most hadronic inelastic channels like fig. 6.6 show linear α_{eff}, $\alpha' \approx 0.9\,\text{GeV}^{-2}$, out to the largest measured $|t|$. Thus $\pi^- p \to \pi^0 n$ has $\alpha_{\text{eff}} = \alpha_\rho \approx 0.55 + 0.9t$ despite the fact that the $\rho \otimes \text{P}$ cut with

$$\alpha_c(t) \approx \alpha_\rho^0 + \frac{\alpha'_\rho \alpha'_{\text{P}}}{\alpha'_\rho + \alpha'_{\text{P}}} t \approx 0.55 + 0.2t \tag{8.7.6}$$

(see (8.6.3)) is supposed to dominate A_{++} for $|t| > 0.2\,\text{GeV}^2$, and A_{+-} for $|t| > 0.55\,\text{GeV}^2$, if the arguments of sections 8.7*b* and *c* are correct. This persistence of a pole-like α_{eff} is extremely puzzling. It may indicate that it is quite wrong to blame the failures of factorization on cuts. But perhaps a more likely explanation is that at current energies the cut contribution does not come mainly from the region of the discontinuity near the branch point as (8.4.1) assumes. One reason for this is probably the necessity for the cut discontinuity to vanish at the branch point (see section 8.3), a feature which is not built into the eikonal/absorption calculation. Another more controversial possibility (Cardy 1974*b*) is that the N's in fig. 8.12 are in fact dominated by poles, so that the leading contribution to the cut is given by fig. 8.14 (*a*), and

$$\left.\begin{aligned} A_J(t) &\sim \frac{(J-\alpha_c)^2}{(J-\alpha_{\text{R}})^2} \log(J-\alpha_c) \\ \Delta_J\{A_J(t)\} &\sim \frac{(J-\alpha_c)^2}{(J-\alpha_{\text{R}})^2} \end{aligned}\right\} \tag{8.7.7}$$

where $(J-\alpha_{\text{R}})^{-1}$ is the Reggeon propagator, $\log(J-\alpha_c)$ arises from the cut loop-integration, and $J-\alpha_c$ occurs at each triple-Reggeon vertex to make the discontinuity vanish at $J = \alpha_c$. When (8.7.7) is substituted in (4.6.2) we find that $A(s,t) \sim s^{\alpha_c}(\log s)^{-3}$ as $\log s \to \infty$, but $A(s,t) \sim s^{\alpha_R}$ for finite log s because the pole provides the dominant region of the discontinuity. Such a model can certainly be made to fit the data (e.g. Collins and Fitton 1975) but as we have no prescription for calculating the magnitude of the cut discontinuity in terms of the pole parameters there is a good deal of arbitrariness. Also fig. 8.14 (*a*) suggests that the sum of pole and cut should factorize, which is clearly no good.

h. *Exotic exchanges*

Because of this uncertainty about the importance of cut contributions it would be very useful to be able to examine amplitudes where no Regge pole can be exchanged, so cuts alone should appear. Clearly $R \otimes P$ cuts are no good because they have the same quantum numbers as R itself, so we must look for $R \otimes R$ exchanges. If $\alpha_R(0) \approx 0.5$ then (8.63) gives $\alpha_{RR}(0) \approx 0$, so we expect a rapid decrease of these cross-sections with energy, $\sim s^{-2} (\log s)^{-2}$.

For example $\pi^- p \to \pi^+ \Delta^-$ involves the exchange of 2 units of charge, $I_t = 2$, so the leading exchange should be a $\rho \otimes \rho$ cut. Unfortunately the forward differential cross-section for this process, and many of the other exotic exchange processes listed in table 6.5, have proved too small to measure except close to threshold. Some processes which have been observed are $\pi^- p \to K^+ \Sigma^-$ and $K^- p \to \pi^+ \Sigma^-$ ($\rho \otimes K^*$ exchange) and $K^- p \to K^+ \Xi^-$ ($K^* \otimes K^*$ exchange). There is some evidence that the $\sim s^{-2}$ behaviour is setting in for $s > 5\,\text{GeV}^2$, and that the magnitude of the cut is compatible with estimates using (8.4.1) with $N^2 = \lambda = 1-1.5$ (see Phillips 1967, Michael 1969*b*, Quigg 1971). Another measured process is $K^- p \to p K^-$ which requires charge = 2, strange, baryon exchange, so one would expect the leading singularity to be the $K^* \otimes \Delta$ cut, $\sim s^{-3}$, but up to 6 GeV a s^{-10} decrease of $d\sigma/dt$ is found.

If better data on this class of processes can be obtained, it should help to clarify our ideas about cuts considerably.

i. *Regge cuts and duality*

In section 7.6 we remarked that since amplitude structures such as the cross-over zero in $\text{Im}\{A_{++}(\pi^- p \to \pi^0 n)\}$ and the forward peak in $\gamma p \to \pi^+ n$, which may be due to cuts, are also present in the FESR average of the s-channel resonances, these resonances must be dual to the sum $R + R \otimes P$ not just R.

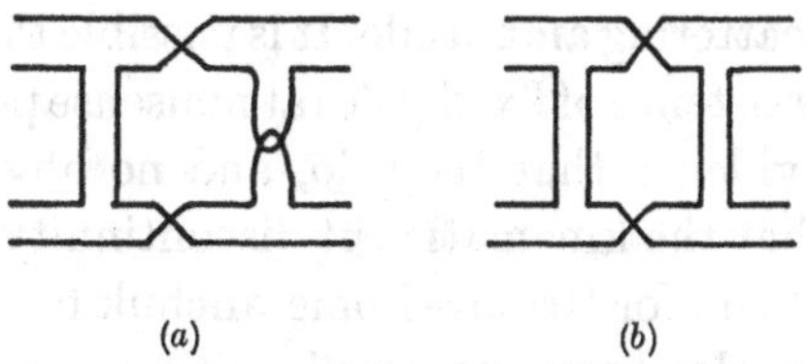

Fig. 8.29 Duality diagrams for (*a*) R⊗P cut, (*b*) R⊗R cut in meson–meson scattering.

Duality diagrams for these R ⊗ P cuts can be drawn as in fig. 8.29 (*a*), where we have been careful to include the double-cross structure, so that each quark scatters only once (cf. fig. 8.8 (*c*)). The diagram for an R ⊗ R cut in meson–meson scattering (fig. 8.29 (*b*)) shows that it is dual to the P in the *s* or *u* channels. For meson–baryon scattering there is only one R ⊗ R diagram, because the baryon quarks must all travel in the same direction, and this diagram can only be drawn if both the *s* and *u* channels are non-exotic, so we can expect R ⊗ R to contribute to the resonances in these channels (in the sense of duality). It also means that there should be no R ⊗ R cuts in a process like $K^-p \rightarrow \bar{K}^0n$ since the λ quark must travel straight across the diagram.

Worden (1973) has shown that the R ⊗ R cuts should cancel in some processes such as $\pi^-p \rightarrow \pi^0n$ because of exchange degeneracy. Briefly his argument may be interpreted as follows. Because of the crosses, and the fact that each signatured Regge pole is the sum of two parts (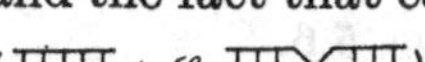), the f ⊗ ρ and ω ⊗ A_2 cuts will cancel if f, ω, ρ and A_2 are exchange degenerate in both their trajectories and their couplings. Although the duality diagrams apply only to Im $\{A\}$, the phase–energy relation ensures that the cancellation works for Re $\{A\}$ too. This is rather a disturbing result because, as we mentioned in section 8.7*d*, many of the phase problems of the R + R ⊗ P absorption model can be solved by the inclusion of R ⊗ f cuts as well. However, since exchange degeneracy is not exact it is not clear how compelling this argument is.

j. *Fixed cuts*

In addition to the moving Regge cuts there are also fixed cuts whose positions are independent of *t*. These are the fixed square-root branch points at sense–nonsense points (see section 4.8) with branch cuts running from $J = M-1$ to $-M$. However, since $d^J_{\lambda\lambda'}(z_t)$ has compensating branch points these cuts do not contribute to the asymptotic behaviour of the scattering amplitude. It is possible that their presence might permit the existence of fixed poles at nonsense points $J_0 < M-1$, but there is no evidence that they do, and no obvious mechanism exists to ensure that the kinematic cut discontinuity can contain the pole (as Regge cuts do for Gribov–Pomeranchuk fixed poles).

Fixed cuts have also been suggested as a way of coping with the generalized MacDowell symmetry for baryon Regge poles and the absence of parity doublets (see (6.5.13)). Carlitz and Kisslinger (1970) have suggested that scattering amplitudes may have a fixed cut at $J = \alpha^0$ (where $\alpha^0 \equiv \alpha(t=0)$) and that the negative-parity trajectory (say) will move through it on to an unphysical sheet for positive $\sqrt{t}$ so that it will not give any physical poles. For example

$$A^{\eta}_{HJ}(t) = \beta(t)\,\frac{(\alpha')^{\frac{1}{2}}\sqrt{t}+\eta(J-\alpha^0)^{\frac{1}{2}}}{J-\alpha^0-\alpha' t}\,\frac{1}{(J-\alpha^0)^{\frac{1}{2}}} \tag{8.7.8}$$

has the pole at $J = \alpha^0 + \alpha' t$ and the cut at $J = \alpha^0$, and the constraint (6.5.13) is automatically satisfied. But in the $\eta = -1$ amplitude there are no poles for positive t. However, such models have not proved very satisfactory phenomenologically (Halzen *et al.* 1970). More recently it has been shown by Savit and Bartels (1975) that similar cuts occur in Reggeon field theory due to the interaction of the fermion with Pomerons. These cuts not only swallow up the unwanted wrong-parity states, but also turn a bare trajectory $\sim \sqrt{t}$ into a re-normalized trajectory approximately $\sim t$. This may explain figs. 5.6.

The rather sad conclusion to be drawn from this whole section is that despite the development of various models which have improved our understanding of Regge cuts and unitarity in the J plane, and despite the partial success of absorption ideas in correcting some of the worst phenomenological defects of Regge poles, we still do not really know how important cuts are. This is probably because we can expect Regge poles to be useful for all $s/s_0 \gg 1$, but cut theories are only really applicable for $\log(s/s_0) \gg 1$ and even at CERN-ISR the maximum value of $\log s$ is only 8.

9
Multi-Regge theory

9.1 Introduction

So far we have limited our attention to four-particle scattering amplitudes (i.e. to processes of the form $1+2 \to 3+4$). These have the advantage of being kinematically rather similar to the potential-scattering amplitudes, for which the basic ideas of Regge theory were originally developed. In particular they depend on only two independent variables, s and t, and so it is a fairly straightforward matter to make analytic continuations in J and t. Also there is a wealth of two-body-final-state data with which to compare the predictions of the theory.

Though the initial state of any physical scattering process will always in practice be a two-particle state (counting bound states such as deuterons as single particles), except at very low energies particle production is always likely to occur. And as the energy increases two-body and quasi-two-body final states make up a diminishing fraction of all the events. So it is very desirable to be able to extend our understanding of Regge theory so as to obtain predictions for many-body final states. Theoretically, this is even more necessary, since models like fig. 3.3 for Regge poles or fig. 8.6 for Regge cuts demonstrate how even in $2 \to 2$ amplitudes Regge theory makes essential use of many-body unitarity. So if we are to have any hope of making Regge theory self-consistent (in the bootstrap sense, for example) we must be able to describe such intermediate states in terms of Regge singularities.

In principle this is a fairly simple matter since if we consider for example the amplitude fig. 9.1 (a) with $s_{12}, s_{34}, s_{45} \to \infty$ we may expect from fig. 9.1 (b) that

$$A \sim (s_{34})^{\alpha_1(t_1)}(s_{45})^{\alpha_2(t_2)}\beta(t_1, t_2, s_{12}, s_{34}, s_{45}) \tag{9.1.1}$$

and indeed this is so. However, there are several problems to be solved before we can be sure that this result is right. It is necessary to understand how to define scattering angles, and thence partial-wave amplitudes, for many-body processes, and how to continue them analytically both in J and in the channel invariants. We must also be clear about

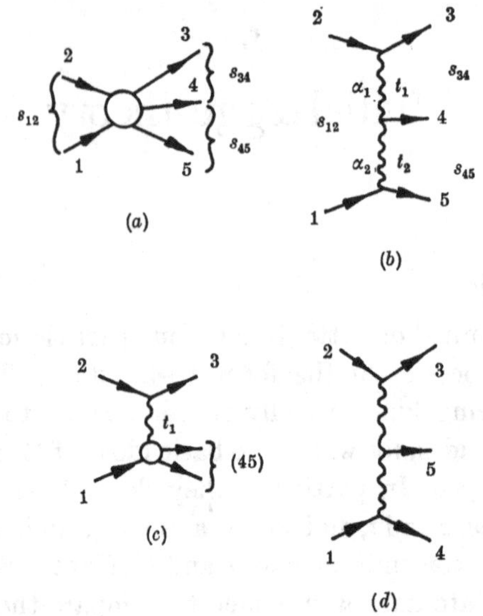

FIG. 9.1 (*a*) The amplitude for $1+2\to 3+4+5$. (*b*) A double-Regge model for this process. (*c*) $1+2\to 3+(45)$. (*d*) Another double-Regge coupling.

which variables are being kept fixed and which tend to infinity when we take a given Regge limit, the singularity structure of the amplitude in these variables, and the order in which the limits are to be taken. And the central vertex in fig. 9.1 (*b*) involves Reggeons whose spin and helicity depend on α so we must check on the resulting kinematical factors.

In fact most of these questions cannot yet be tackled rigorously because to do so would require a more detailed understanding of the singularity structure of many-particle amplitudes than has so far been achieved. Hence we shall adopt a rather simple-minded approach, and assume that the methods which we adopted in chapters 1 and 2 can be extended in the most obvious way without mishap. A more thorough account of present theoretical knowledge can be found in Brower, de Tar and Weis (1974).

In the next section we review the kinematics of many-body processes, and we then go on to consider the different Regge asymptotic limits which may be taken. This is followed by a more detailed discussion of the $2\to 3$ amplitude, on the basis of which we postulate some general rules for any multi-Regge amplitudes. It is rather remarkable that the dual models of chapter 7 can readily be extended to many-body amplitudes, and as they provide a good deal of insight into the nature of multi-Regge couplings we outline the main results. The chapter concludes with a very short discussion of some phenomenological applications of the theory.

9.2 Many-particle kinematics

We consider first the process $1+2 \to 3+4+5$ shown in fig. 9.1. For simplicity we suppose that all the external particles are spinless.

The square of the centre-of-mass energy is (cf. (1.7.5))

$$s \equiv s_{12} = (p_1+p_2)^2 = (p_3+p_4+p_5)^2 \equiv s_{345} \tag{9.2.1}$$

Similarly, for the outgoing two-body channels we have the sub-energies

$$s_{34} = (p_3+p_4)^2, \quad s_{45} = (p_4+p_5)^2 \quad \text{and} \quad s_{35} = (p_3+p_5)^2 \tag{9.2.2}$$

The 6 crossed-channel invariants, involving both incoming and outgoing particles are,

$$\left.\begin{aligned} t_1 \equiv t_{23} = (p_2-p_3)^2, \quad t_{24} = (p_2-p_4)^2, \quad t_{25} = (p_2-p_5)^2 \\ t_2 \equiv t_{15} = (p_1-p_5)^2, \quad t_{14} = (p_1-p_4)^2, \quad t_{13} = (p_1-p_3)^2 \end{aligned}\right\} \tag{9.2.3}$$

Clearly any three-particle invariant will be equal to some two-particle invariant (as in (9.2.1)) because of four-momentum conservation, so the 10 variables defined in (9.2.1), (9.2.2) and (9.2.3) include all the independent invariants. But evidently they cannot all be independent because we showed in section 1.4 than an n-line amplitude has only $3n-10$ independent variables, so with $n = 5$ only 5 can be regarded as independent variables. In the centre-of-mass frame of particles 4 and 5, i.e. where $\mathbf{q}_4 + \mathbf{q}_5 = 0$, s_{45} is the square of the total energy of these particles, i.e.

$$s_{45} \equiv (p_4+p_5)^2 = (E_4+E_5, \mathbf{0})^2 = (E_4+E_5)^2 = m_{45}^2 \tag{9.2.4}$$

and m_{45} is called the 'invariant mass' of the 'quasi-particle' (45). So if we regard the reaction of fig. 9.1 (*a*) as the process $1+2 \to 3+(45)$ shown in fig. 9.1 (*c*) we have, like (1.7.21),

$$s_{12}+t_{23}+t_{13} = m_1^2+m_2^2+m_3^2+s_{45} \equiv \Sigma_{45} \tag{9.2.5}$$

with similar relations for other pairings of particles.

A convenient choice of independent invariants suggested by fig. 9.1 (*b*) is

$$s_{12}, s_{34}, s_{45}, t_1 \text{ and } t_2, \tag{9.2.6}$$

but this depends on how we choose to couple the particles together, and fig. 9.1 (*d*) for example, suggests a quite different choice.

In the centre-of-mass frame $\mathbf{q}_1+\mathbf{q}_2=0$ the energies and momenta of particles 1 and 2 are given by (1.7.8), (1.7.9) and (1.7.10), i.e.

$$E_1=\frac{1}{2\sqrt{s}}(s+m_1^2-m_2^2),\quad q_{s12}^2=\frac{1}{4s}\lambda(s,m_1^2,m_2^2) \tag{9.2.7}$$

etc. Similarly if we regard (45) as a single particle of mass $m_{45}=\sqrt{s_{45}}$ as above, it is clear that in this frame

$$E_3=\frac{1}{2\sqrt{s}}(s+m_3^2-s_{45}),\quad q_{s3}^2=\frac{1}{4s}\lambda(s,m_3^2,s_{45}) \tag{9.2.8}$$

with similar expressions for particles 4 and 5.

Also the scattering angle between the direction of motion of particle 3 and that of particle 2 is given by (1.7.17) with $\sqrt{s_{45}}$ instead of m_4, i.e.

$$z_{23}=\cos\theta_{s23}=\frac{s^2+s(2t_1-\Sigma_{45})+(m_1^2-m_2^2)(m_3^2-s_{45})}{\lambda^{\frac{1}{2}}(s,m_1^2,m_2^2)\,\lambda^{\frac{1}{2}}(s,m_3^2,s_{45})} \tag{9.2.9}$$

and the physical region for this scattering process is given by (1.7.24) with the obvious substitutions.

The four-momentum conservation relation (9.2.1)

$$s_{12}=(p_3+p_4+p_5)^2$$

with (9.2.2) and (1.7.4) gives

$$s_{12}=s_{34}+s_{45}+s_{35}-m_3^2-m_4^2-m_5^2 \tag{9.2.10}$$

so for a given fixed s_{12} only two of the three sub-energies are independent, and the boundary of the physical region, determined by (1.7.24) with the substitutions described above, is as shown in fig. 9.2. This is known as a Dalitz plot (Dalitz 1953). If there is a resonance, r, which decays into particles (4+5), as in fig. 9.3, we can expect that for a given fixed s_{12} there will be a peak in the cross-section as a function of s_{45} along the line $s_{45}=M_r^2$. Likewise if 3 and 4 resonate there will be a peak at fixed s_{34}, while if 3 and 5 resonate there will be a diagonal line across the plot at fixed s_{35}. So a plot like fig. 9.2 is very useful for deciding which pairs of particles, if any, are resonating.

But our main interest lies in examining Regge exchanges like fig. 9.1 (*b*), and for this purpose we need to be able to define angular momenta for the various *t* channels. Thus one of the crossed processes to fig. 9.1 is fig. 9.4 (*a*), i.e.

$$2+\bar{3}\to(\bar{1}5)+4 \tag{9.2.11}$$

where we treat $\bar{1}5$ as a quasi-particle of mass $(p_1-p_5)^2=t_2$. The centre-of-mass energies and momenta can all be obtained from

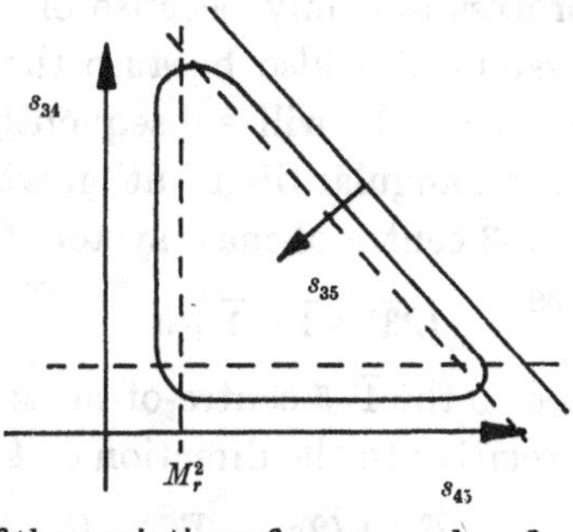

FIG. 9.2 Dalitz plot of the variation of s_{34}, s_{45} and s_{35} for a given s_{12} constrained by (9.2.10). The boundary of the physical region determined by (1.7.24) with the obvious substitutions is shown. The dotted lines mark positions where resonance peaks may occur.

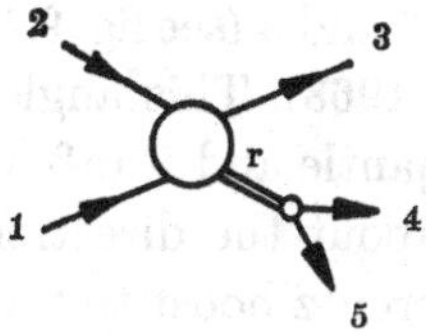

FIG. 9.3 The amplitude for $1+2 \to 3+\mathrm{r}$, $\mathrm{r} \to 4+5$.

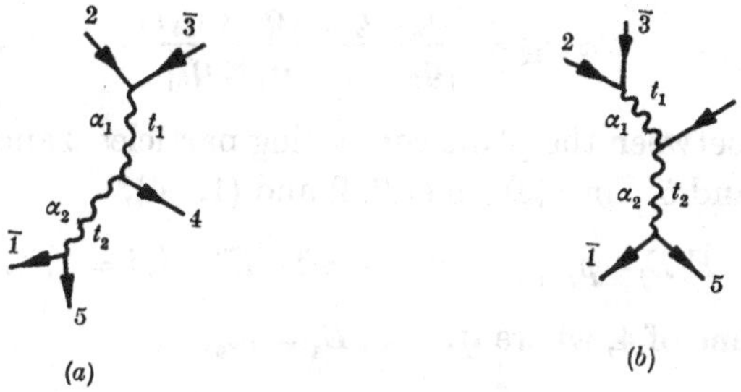

FIG. 9.4 (a) The crossed-channel process $2+\bar{3} \to (\bar{1}5)+4$.
(b) The crossed-channel process $(2\bar{3})+\bar{4} \to \bar{1}+5$.

(1.7.15) with the obvious substitutions, where now $t \to t_{23} \equiv t_1$, and the centre-of-mass scattering angle of particle 4 with respect to the direction of $\bar{3}$ is given by (1.7.19), viz.

$$\cos\theta_{34} \equiv z_{t34} \equiv z_1 = \frac{t_1^2 + t_1(2s_{34} - \Sigma_{15}) + (m_2^2 - m_3^2)(t_2 - m_4^2)}{\lambda^{\frac{1}{2}}(t_1, m_2^2, m_3^2)\,\lambda^{\frac{1}{2}}(t_1, t_2, m_4^2)}$$

where $$\Sigma_{15} \equiv m_2^2 + m_3^2 + m_4^2 + t_2 \qquad (9.2.12)$$

This is the scattering angle in the centre-of-mass system of $2+\bar{3}$, i.e $\mathbf{q}_2 + \mathbf{q}_3 = 0$. But the process (9.2.11) differs from a $2 \to 2$ spinless-particle scattering process not only because of the variation of the 'mass' of the $(\bar{1}5)$ system, but also because the $(\bar{1}5)$ quasi-particle carries angular momentum. It will subsequently 'decay' into the particles $\bar{1}$ and $\bar{5}$ with an angular distribution which depends on the helicity of $(\bar{1}5)$ in the 2–$\bar{3}$ centre-of-mass system (like (4.2.13)).

Then for the process

$$(2\bar{3})+\bar{4}\rightarrow\bar{1}+5 \tag{9.2.13}$$

(fig. 9.4(*b*)) we proceed to the $\bar{1}$–5 centre-of-mass frame, in which the scattering angle of 5 relative to the direction of $\bar{4}$ is

$$\cos\theta_{45}\equiv z_{t45}\equiv z_2=\frac{t_2^2+t_2(2s_{45}-\Sigma_{23})+(t_1-m_4^2)(m_1^2-m_5^2)}{\lambda^{\frac{1}{2}}(t_2,t_1,m_4^2)\,\lambda^{\frac{1}{2}}(t_2,m_1^2,m_5^2)} \tag{9.2.14}$$

The azimuthal angle ω_{12} between the plane containing particles 4 and 5 and that containing 3 and 4 (see fig. 9.5) is called the Toller angle (or helicity angle) (Toller 1968). This angle may be evaluated with some effort (see Chan, Kajantie and Ranft 1967) as follows.

Since ω_{12} is the angle about the direction of particle 4 it will be unaltered if we make a Lorentz boost to the rest frame of particle 4. This makes the kinematics much easier to cope with. In this rest frame the Toller angle is defined by

$$\cos\omega_{12}=\frac{(\mathbf{q}_2\times\mathbf{q}_3)\cdot(\mathbf{q}_1\times\mathbf{q}_5)}{|\mathbf{q}_2\times\mathbf{q}_3|\,|\mathbf{q}_1\times\mathbf{q}_5|} \tag{9.2.15}$$

i.e. the angle between the plane containing particles 2 and 3 and that containing 1 and 5. Since, from (1.7.2) and (1.7.4),

$$\mathbf{q}_i\cdot\mathbf{q}_j=E_iE_j-p_i\cdot p_j,\quad \mathbf{q}_i^2=-m_i^2+E_i^2,\quad i,j=1,\dots,5 \tag{9.2.16}$$

in the rest frame of 4, where $\mathbf{q}_4=0$, $E_4=m_4$,

$$E_i=\frac{p_i\cdot p_4}{m_4} \tag{9.2.17}$$

But $$s_{ij}\equiv(p_i+p_j)^2=p_i^2+p_j^2+2p_i\cdot p_j=m_i^2+m_j^2+2p_i\cdot p_j \tag{9.2.18}$$

so
$$\left.\begin{aligned}E_i&=\frac{1}{2m_4}(s_{i4}-m_i^2-m_4^2),\quad i=3,5\\ &=\frac{1}{2m_4}(t_{i4}-m_i^2-m_4^2),\quad i=1,2\end{aligned}\right\} \tag{9.2.19}$$

Now

$$\begin{aligned}|\mathbf{q}_2\times\mathbf{q}_3|&=|\mathbf{q}_2|\,|\mathbf{q}_3|\sin\theta_{23}\\&=|\mathbf{q}_2|\,|\mathbf{q}_3|(1-\cos^2\theta_{23})^{\frac{1}{2}}=[\mathbf{q}_2^2\,\mathbf{q}_3^2-(\mathbf{q}_2\cdot\mathbf{q}_3)^2]^{\frac{1}{2}}\end{aligned} \tag{9.2.20}$$

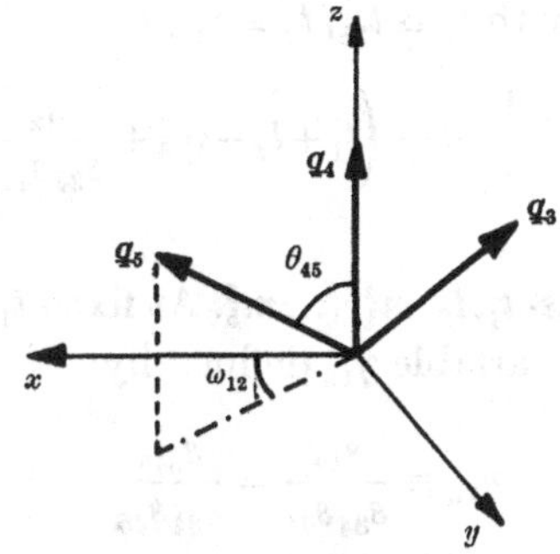

FIG. 9.5 Angles in the $\bar{1}$–5 centre-of-mass system. q_4 is along the z axis, q_3 is in the x–z plane, and ω_{12} is the angle between the plane containing q_3 and q_4 and that containing q_4 and q_5, i.e. between q_5 and the x–z plane.

and Lagrange's identity gives

$$(\mathbf{q}_2\times\mathbf{q}_3).(\mathbf{q}_1\times\mathbf{q}_5) = (\mathbf{q}_2.\mathbf{q}_1)(\mathbf{q}_3.\mathbf{q}_5)-(\mathbf{q}_2.\mathbf{q}_5)(\mathbf{q}_3.\mathbf{q}_1) \quad (9.2.21)$$

and all these scalar products can be evaluated using (9.2.16). Thus

$$\mathbf{q}_2.\mathbf{q}_3 = E_2E_3-\mathbf{p}_2.\mathbf{p}_3 = \frac{1}{2m_4}(t_{24}-m_2^2-m_4^2)\times\frac{1}{2m_4}(s_{34}-m_3^2-m_4^2)-\frac{t_{23}-m_2^2-m_3^2}{2} \quad (9.2.22)$$

so
$$(\mathbf{q}_2.\mathbf{q}_3)^2\to\left(\frac{t_{24}s_{34}}{4m_4^2}\right)^2-\frac{t_{24}s_{34}t_{23}}{4m_4^2} \quad \text{for} \quad t_{24}, s_{34}, t_{23} \gg m_i^2 \quad (9.2.23)$$

And
$$q_2^2\,q_3^2 = (E_2^2-m_2^2)(E_3^2-m_3^2)\to\frac{s_{34}^2t_{24}^2}{(2m_4)^4} \quad (9.2.24)$$

in the same limit, giving

$$|\mathbf{q}_2\times\mathbf{q}_3|\to\left(\frac{t_{23}t_{24}s_{34}}{4m_4^2}\right)^{\frac{1}{2}} \quad (9.2.25)$$

But, like (9.2.5), we have

$$t_{24}+t_{23}+s_{34} = m_2^2+m_3^2+m_4^2+t_{15} \quad (9.2.26)$$

so that
$$t_{24}\to -s_{34}, \quad s_{34}\to\infty \quad \text{at fixed} \quad t_{23}, t_{15}$$

and hence
$$|\mathbf{q}_2\times\mathbf{q}_3|\to\frac{s_{34}}{2m_4}\sqrt{-t_{23}} \quad (9.2.27)$$

Similarly
$$|\mathbf{q}_4\times\mathbf{q}_5|\to\frac{s_{45}}{2m_4}\sqrt{-t_{15}} \quad (9.2.28)$$

and with more effort we find

$$(\mathbf{q}_2\times\mathbf{q}_3).(\mathbf{q}_1\times\mathbf{q}_5)\to\frac{1}{8m_4^2}\{s_{12}[(t_{23}+t_{15}-m_4^2)^2-4t_{23}t_{15}] + s_{34}\,s_{45}(t_{23}+t_{15}-m_4^2)\} \quad (9.2.29)$$

so that from (9.2.15) with $t_1 \equiv t_{23}$, $t_2 \equiv t_{15}$,

$$\cos\omega_{12} \approx \frac{1}{2\sqrt{-t_1}\sqrt{-t_2}}\left(t_1+t_2-m_4^2+\frac{s_{12}}{s_{34}s_{45}}\lambda(t_1,t_2,m_4^2)\right) \tag{9.2.30}$$

in the limit $s_{12}, s_{34}, s_{45} \gg t_1, t_2, m_1^2, \ldots, m_5^2$. At fixed t_1, t_2 it is often more convenient to use the variable η_{12} defined by

$$\eta_{12} \equiv \frac{s_{12}}{s_{34}s_{45}} = \frac{s_{345}}{s_{34}s_{45}} \tag{9.2.31}$$

rather than ω_{12}.

The set of variables $\quad t_1, t_2, z_1, z_2$ and η_{12} (9.2.32)

provide an alternative to (9.2.6), and one which is more useful for Reggeization.

To extend this approach to the six-particle amplitude, fig. 9.6(*a*), we simply note that it becomes similar to the five-particle amplitude, fig. 9.1 if we regard $(1\bar{6})$ as a particle, and replace s_{12} by $s_{345} \equiv (p_3+p_4+p_5)^2$, but in addition to the scattering angles z_1 and z_2 and the Toller variable $\eta_{12} = s_{345}/s_{34}s_{45}$ we also have z_3, the centre-of-mass scattering angle for $(\overline{234})+5 \to \bar{1}+6$, and the Toller angle ω_{23}, the angle between the plane containing particles 5 and 6 and that containing 4 and 5 in the $\bar{1}$–6 rest frame. Or instead we can use $\eta_{23} \equiv s_{456}/s_{45}s_{56}$. The sets of variables

$$t_1, t_2, t_3, s_{34}, s_{45}, s_{56}, s_{345}, s_{456}, \quad \text{or} \quad t_1, t_2, t_3, z_1, z_2, z_3, \eta_{12}, \eta_{23} \tag{9.2.33}$$

give the required 8 independent variables for a 6-line amplitude. Of course these sets are convenient only if we choose to couple the particles as in fig. 9.6(*a*), rather than, say, fig. 9.6(*b*) for which a different set of angular variables is appropriate (see below).

As the number of external lines increases so does the number of different ways of coupling together the particles. But for any given configuration a complete set of variables is provided by the momentum transfers, t_i, the cosines of the scattering angles, z_i, and the Toller variables, η_{ij}, associated with each adjacent pair of t's (t_i and t_j say). And for given fixed values of the t's these angle variables can all be expressed in terms of the s's.

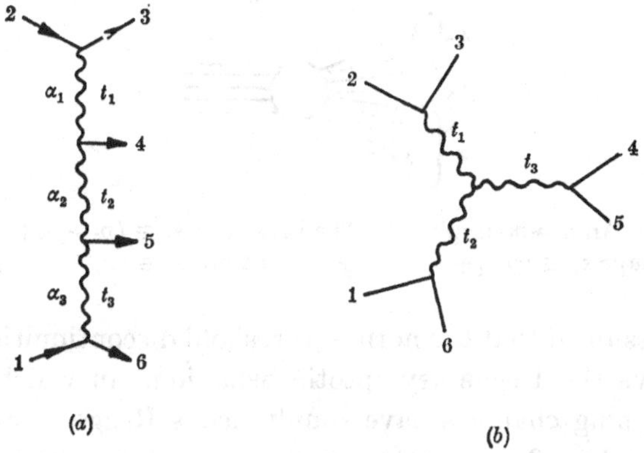

FIG. 9.6 (a) Multi-Regge amplitude for $1+2 \to 3+4+5+6$.
(b) Another multi-Regge coupling.

9.3 Multi-Regge scattering amplitudes

The Froissart–Gribov partial-wave projection (2.5.3), in terms of which Regge poles were defined in $2 \to 2$ scattering, involves integration over the s-discontinuity of the scattering amplitude (2.7.2). The pole appears in the power behaviour of this discontinuity. So when generalizing to a multi-Regge limit of a many-particle scattering process we shall have to concern ourselves with simultaneous discontinuities in several variables.

It is obviously essential that these discontinuities should be independent in the asymptotic limit. For normal threshold discontinuities it is easy to decide when they are independent. In an $n \to m$ scattering amplitude, fig. 9.7, we can distinguish between overlapping channels such as x and y, for which the invariants

$$s_x \equiv s_{1,\dots,i} \equiv (p_1+p_2+ \dots +p_i)^2$$

and
$$s_y \equiv s_{i-1,i,\dots,j} \equiv (p_{i-1}+p_i+ \dots +p_j)^2$$

have the particles i and $i-1$ in common, and non-overlapping channels like s_x and s_z which have no particles in common and are therefore independent. The normal-threshold discontinuity of a given channel is a singularity just in that channel's invariant (e.g. the 12 threshold branch point is at $s_{12} \equiv (p_1+p_2)^2 = (m_1+m_2)^2$) and so normal-threshold discontinuities in non-overlapping channels are independent of each other. But more complicated Landau curves do not have this independence. For example the box diagram, fig. 1.10(b), gives the s–t curve (1.12.10) for the position of the double discontinuity. It is

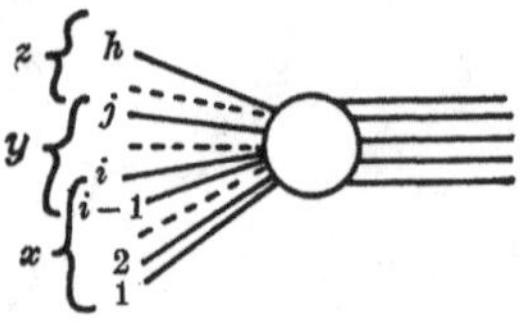

FIG. 9.7 An $n \to m$ amplitude. The invariant $s_x \equiv (p_1+p_2+\ldots+p_i)^2$ overlaps $s_y \equiv (p_{i-1}+p_i+\ldots+p_j)^2$ but not $s_z \equiv (p_j+\ldots+p_n)^2$.

generally assumed that the normal-threshold discontinuities are sufficient to give the Regge asymptotic behaviour, in which case only non-overlapping channels have simultaneous Regge discontinuities. This is trivial in $2 \to 2$ scattering since we obviously do not have simultaneous Regge behaviour in the overlapping $s \equiv (p_1+p_2)^2$ and $t \equiv (p_1-p_3)^2$ channels, but it has not been established for certain in more complex amplitudes. It is, however, true in all the simple models such as ladder diagrams or dual models and we adopt it here (see Brower *et al.* 1974).

There are generally several different asymptotic limits which can be taken for a given amplitude and for a given configuration of the particles, depending on which variables are taken to infinity, and which are held fixed. Thus in the five-particle amplitude, fig. 9.1, we have the following possibilities.

(*a*) *The single-Regge limit.* In this case $z_1 \to \infty$ but t_1 and the other angles and invariants in (9.2.32) are held fixed. This means $s_{34} \to \infty$ from (9.2.12), and hence $s_{12} \to \infty$ from (9.2.10) but s_{45}, t_1 and t_2 are fixed. Also to keep ω_{12} fixed in (9.2.30) (or η_{12} in (9.2.31)) we must keep the ratio s_{12}/s_{34} fixed as both $\to \infty$.

This corresponds to the single-Regge graph fig. 9.1 (*c*). There are obviously three possible single-Regge limits of the amplitude depending on whether we take s_{34}, s_{45} or $s_{35} \to \infty$.

(*b*) *The double-Regge limit.* Here $z_1, z_2 \to \infty$, the other angle and the invariants in (9.2.32) being fixed. This means s_{12}, s_{34} and $s_{45} \to \infty$, t_1, t_2 fixed, and with the ratio $s_{12}/s_{34}s_{45}$ fixed to keep ω_{12} and η_{12} fixed.

This corresponds to the double-Regge graph fig. 9.1 (*b*), but other double-Regge limits like fig. 9.1 (*d*) can be obtained by permuting the final-state particles.

(*c*) *The helicity limit.* This has ω_{12} (and η_{12}) $\to \infty$, with z_1, z_2, t_1 and t_2 all fixed, so $s_{12} \to \infty$ with s_{34}, s_{45}, t_1, t_2 fixed. Since this involves $\cos \omega_{12} \to \infty$ it is clearly not a physical limit.

Obviously (*a*) is just the same as the single-Regge limit in $2 \to 2$

scattering except that one of the final-state 'particles' is actually a two-particle state with fixed invariant mass. It is thus similar to resonance production in quasi-two-body processes and requires little further discussion. But (*b*) and (*c*) are quite new, and depend in an essential way on there being three particles available in the final state. They will be considered below.

This discussion can readily be generalized to any multi-particle final state. In the single-Regge limit those invariants which overlap the given Reggeon line (e.g. s_{12} and s_{34} in fig. 9.1 (*c*)) all tend to infinity, with fixed ratios, and all the other independent invariants (t_1, t_2, s_{45}) are held fixed. In the multi-Regge limit those invariants which overlap any Reggeon line (e.g. s_{12}, s_{34}, s_{45} in fig. 9.1 (*b*)) tend to infinity, and the others are held fixed. The ratios of those invariants which overlap a given Reggeon are held fixed, while those invariants which overlap several Reggeons (for example s_{12} overlaps α_1 and α_2 in fig. 9.1 (*b*)) tend to infinity like the products of the invariants of the individual lines (for example $s_{12} \sim s_{34}s_{45}$). In the helicity limit only those invariants which overlap two Reggeons tend to infinity, with a fixed ratio so that the Toller angle between those two Reggeons tends to infinity.

We shall now examine in more detail the Reggeization of the $2 \to 3$ amplitude, fig. 9.1. Since we are interested in using the results in the *s*-channel physical region, some authors have preferred to use the $O(2, 1)$ group-theory method (whose application in $2 \to 2$ scattering was mentioned in section 6.6); see Bali, Chew and Pignotti (1967), Toller (1969), Jones, Low and Young (1971). However we shall use the Sommerfeld–Watson transform of the *t*-channel partial-wave series, and assume that this can be continued in the *t*'s without difficulty.

In the single-Regge limit (*a*) we are concerned with the *t*-channel process $2 + \bar{3} \to (\overline{15}) + 4$ where $(\overline{15})$ is a quasi-particle (see fig. 9.4 (*a*)). So following section 4.6 we begin with the *t*-channel partial-wave series (4.5.10)

$$\left.\begin{aligned} A(t_1, z_1; \omega_{12}; t_2, z_2) &= \sum_{J_1=0}^{\infty} \sum_{\lambda=-J_1}^{J_1} (2J_1+1)\, A_{J_1}(t_1; t_2, z_2)\, d^{J_1}_{0\lambda}(z_1)\, e^{i\lambda\omega_{12}} \\ &= \sum_{\lambda=-\infty}^{\infty} \sum_{J_1 \geqslant |\lambda|} (2J_1+1)\, A_{J_1}(t_1; t_2, z_2)\, d^{J_1}_{0\lambda}(z_1)\, e^{i\lambda\omega_{12}} \end{aligned}\right\} \tag{9.3.1}$$

where J_1 is the angular momentum of $\bar{3}$ with respect to 2, and in addition to summing over all partial waves we have also summed over all the possible helicities λ for the quasi-particle $(\bar{1}5)$. By angular-momentum conservation $|\lambda|$ cannot be greater than J_1. (Remember that for simplicity we are assuming that all the particles 1,..., 5 are spinless.) The second expression in (9.3.1) seems more appropriate for continuing in J_1 (though in fact it may be better to continue in λ first: see Goddard and White (1971), White (1971, 1973*b*)). The factor $e^{i\lambda\omega_{12}}$ appears because (see (4.4.7) and (4.2.14)) ω_{12} gives the azimuthal angle in the 'decay' $(\bar{1}5)\to\bar{1}+5$, and by definition λ is measured in the direction of motion of $(\bar{1}5)$.

We then replace the sum in (9.3.1) by the Sommerfeld–Watson integral (4.6.1) in the complex J_1 plane, and draw back the integration contour to expose the leading Regge pole $\alpha_1(t_1)$ whose contribution can be written

$$A^{\mathrm{R}}(t_1, z_1; \omega_{12}; t_2, z_2) = \Gamma(-\alpha_1(t_1))\,\gamma_1(t_1)\,(z_1)^{\alpha_1(t_1)} \sum_{\lambda=-\infty}^{\infty} e^{i\lambda\omega_{12}}\gamma_\lambda(t_1; t_2, z_2) \tag{9.3.2}$$

where we have factorized the residue into a part $\gamma_1(t_1)$ for the 2–$\bar{3}$ vertex, and $\gamma_\lambda(t_1; t_2, z_2)$ for the $(\bar{1}5)$–4 vertex, and have included the nonsense factor $\Gamma(-\alpha)$. If we define

$$\beta(t_1, \omega_{12}; t_2, z_2) = \sum_{\lambda=-\infty}^{\infty} e^{i\lambda\omega_{12}}\gamma_\lambda(t_1; t_2, z_2) \tag{9.3.3}$$

the Fourier transform of γ_λ, and take the asymptotic form

$$(z_1)^{\alpha_1} \sim (s_{34})^{\alpha_1},$$

we can rewrite this more conveniently as

$$A^{\mathrm{R}}(s_{12}, s_{34}, s_{45}, t_1, t_2) = \Gamma(-\alpha_1(t_1))\,\gamma_1(t_1)\,\beta(t_1, \omega_{12}; t_2, z_2)\,(s_{34})^{\alpha_1} \tag{9.3.4}$$

just like the $2\to 2$ case (6.8.1).

For the double-Regge limit we start from a double partial-wave decomposition in z_1 and z_2 (Ter-Martirosyan 1965, Kibble 1963), i.e.

$$A(t_1, z_1; \omega_{12}; t_2, z_2) = \sum_{J_1, J_2=0}^{\infty} \sum_{\lambda} (2J_1+1)(2J_2+1) \times A_{J_1J_2\lambda}(t_1, t_2)\, d^{J_1}_{0\lambda}(z_1)\, d^{J_2}_{\lambda 0}(z_2)\, e^{i\lambda\omega_{12}} \tag{9.3.5}$$

where $|\lambda| \leqslant J_1, J_2$. Then if we make the Sommerfeld–Watson transform in both J's and expose the leading Reggeon in each channel, we get in the double-Regge asymptotic limit

$$A^{\mathrm{R}}(s_{12}, s_{34}, s_{45}, t_1, t_2) = \Gamma(-\alpha_1(t_1))\,\gamma_1(t_1)\,(s_{34})^{\alpha_1(t_1)} \times \beta(t_1, \eta_{12}, t_2)\,\Gamma(-\alpha_2(t_2))\,\gamma_2(t_2)\,(s_{45})^{\alpha_2(t_2)} \tag{9.3.6}$$

where $\beta(t_1, \eta_{12}, t_2)$ is the coupling at the central vertex $(\alpha_1 \alpha_2 4)$ and depends on the Toller angle as well as the t's.

Apart, perhaps, from the inclusion of the η_{12} dependence these results are just what one would naïvely expect from drawing diagrams like fig. 9.1 (*b*). However, we have certainly not done full justice to the problem because we have not bothered much about the discontinuities in the different invariants, and in particular we have completely ignored the fact that Reggeons have signature and hence have discontinuities for both positive and negative s, which give the amplitude its phase. We must now remedy this.

The assumption that there are no simultaneous Regge discontinuities in overlapping-channel invariants means that for example the discontinuity in s_{34} must not itself have a discontinuity in s_{45}, though it may have one in s_{12}. So we expect that the s_{34} discontinuity may involve terms like

$$(-s_{34})^{\alpha_1-\alpha_2}(-s_{12})^{\alpha_2} V_2(\eta_{12}) + (-s_{34})^{\alpha_1-\alpha_2}(s_{12})^{\alpha_2} V_2'(\eta_{12}) \qquad (9.3.7)$$

where the V's are real functions of the η's (for negative t_1, t_2). Both terms $\sim |s_{34}|^{\alpha_1} |s_{45}|^{\alpha_2}$ since $s_{12} \sim s_{34}s_{45}$ in the double-Regge limit, but the first term is cut for positive s_{12} as well as s_{34}, while the second is not. We also want the Reggeons to have a definite signature, so that for example the Reggeon α_1 gives a discontinuity for positive s_{34} and an equal one for negative s_{34} (up to a $\pm$ sign depending on its signature $\mathscr{S}_1$) and so we have equal amplitudes under the interchange $2 \leftrightarrow 3$. There are thus four different terms, from fig. 9.8, and combining them, in the physical region where all the Regge functions are real, gives (Drummond, Landshoff and Zakrzewski 1969*b*)

$$A^{\mathrm{R}}(s_{12}, s_{34}, s_{45}, t_1, t_2) = \Gamma(-\alpha_1(t_1))\,\gamma_1(t_1)\,(s_{34})^{\alpha_1(t_1)}\,\Gamma(-\alpha_2(t_2))$$
$$\times\,\gamma_2(t_2)\,(s_{45})^{\alpha_2(t_2)}\,[\xi_1\xi_{21}(\eta_{12})^{\alpha_1(t_1)} V_1(t_1, t_2, \eta_{12}) + \xi_2\xi_{12}(\eta_{12})^{\alpha_2(t_2)} V_2(t_1, t_2, \eta_{12})] \qquad (9.3.8)$$

where $$\xi_i = \mathrm{e}^{-\mathrm{i}\pi\alpha_i} + \mathscr{S}_i, \quad \xi_{ij} = \mathrm{e}^{-\mathrm{i}\pi(\alpha_i-\alpha_j)} + \mathscr{S}_i\mathscr{S}_j \qquad (9.3.9)$$

This may be re-expressed more conveniently as

$$A^{\mathrm{R}}(s_{12}, s_{34}, s_{45}, t_1, t_2) = \gamma_1(t_1)\,R_1(t_1, s_{34})$$
$$\times\,G^4_{12}(t_1, t_2, \eta_{12})\,R_2(t_2, s_{45})\,\gamma_2(t_2) \qquad (9.3.10)$$

where $$R_i(t_i, s) \equiv \xi_i(t_i)\,\Gamma(-\alpha_i(t_i))\,s^{\alpha_i(t_i)} \qquad (9.3.11)$$

and $$G^4_{ij}(t_i, t_j, \eta_{ij}) \equiv \xi_i^{-1}\xi_{ji}(\eta_{ij})^{\alpha_i(t_i)}$$
$$\times\,V_1(t_i, t_j, \eta_{ij}) + \xi_j^{-1}\xi_{ij}(\eta_{ij})^{\alpha_j(t_j)} V_2(t_i, t_j, \eta_{ij}) \qquad (9.3.12)$$

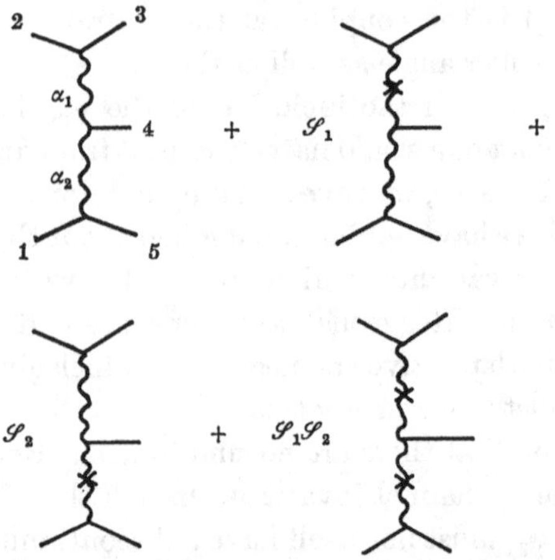

FIG. 9.8 The four different terms in the double-Regge amplitude stemming from the signature properties of the Reggeons. The $\times$ implies that the Reggeon is twisted ($s \to -s$) like the twisted ladders of fig. 8.11.

We can regard (9.3.11) as the Reggeon propagator, and all the phase complexity has been put into G^4_{12}, the coupling of particle 4 to the two Reggeons.

For more complicated amplitudes with extended chains of Reggeons like fig. 9.6 (*a*) we simply increase the number of propagators and G's in the obvious manner. However, with six lines a new type of configuration with a triple-Reggeon coupling, fig. 9.6 (*b*) becomes possible. In this case we can write (Landshoff and Zakrzewski 1969)

$$A^{\mathrm{R}} = \gamma(t_1)\,R_1(t_1, s_{345})\,\gamma(t_2)\,R_2(t_2, s_{456})\,\gamma(t_3) \times R_3(t_3, s_{234})\,G_{123}(t_1, t_2, t_3, \eta_{12}, \eta_{23}, \eta_{31}) \quad (9.3.13)$$

where again all the phase problems are contained in G_{123}. A careful analysis (de Tar and Weis 1971) finds

$$G_{123}(t_1, t_2, t_3, \eta_{12}, \eta_{23}, \eta_{31}) = \xi_3^{-1}\xi_{312}\overline{V}_{12} + \xi_1^{-1}\xi_{123}\overline{V}_{23} + \xi_2^{-1}\xi_{231}\overline{V}_{31} + \xi_1^{-1}\xi_2^{-1}\xi_3^{-1}\,\mathrm{e}^{-\mathrm{i}\pi(\alpha_1+\alpha_2+\alpha_3)}(1 + \mathscr{S}_1\mathrm{e}^{\mathrm{i}\pi\alpha_1} + \mathscr{S}_2\mathrm{e}^{\mathrm{i}\pi\alpha_2} + \mathscr{S}_3\mathrm{e}^{\mathrm{i}\pi\alpha_3})\overline{V}_{123}$$

where

$$\left.\begin{aligned} \overline{V}_{ij} &= (\eta_{ki})^{\alpha_i}(\eta_{jk})^{\alpha_j}V_{ij} \\ \overline{V}_{ijk} &= (\eta_{ij})^{\frac{1}{2}(\alpha_i+\alpha_j-\alpha_k)}(\eta_{jk})^{\frac{1}{2}(\alpha_j+\alpha_k-\alpha_i)}(\eta_{ki})^{\frac{1}{2}(\alpha_k+\alpha_i-\alpha_j)}V_{ijk} \\ \xi_{ijk} &\equiv \mathrm{e}^{-\mathrm{i}\pi(\alpha_i-\alpha_j-\alpha_k)} + \mathscr{S}_i\mathscr{S}_j\mathscr{S}_k \end{aligned}\right\} \quad (9.3.14)$$

and the V's are real functions. Any multi-Regge diagram can be expressed in terms of γ_i, R_i, G_{ij} and G_{ijk} as functions of the appropriate invariants (Weis 1973, 1974).

The other limit to be discussed is the helicity limit (*c*) (see Brower *et al.* 1973*b*). Starting from the double partial-wave series (9.3.5)

$$A(t_1, z_1; \omega_{12}; t_2, z_2) = \sum_{\lambda=-\infty}^{\infty} \sum_{J_1=|\lambda|}^{\infty} \sum_{J_2=|\lambda|}^{\infty} (2J_1+1)(2J_2+1) \times A_{J_1 J_2 \lambda}(t_1, t_2)\, d^{J_1}_{0\lambda}(z_1)\, d^{J_2}_{\lambda 0}(z_2)\, e^{i\lambda\omega_{12}} \quad (9.3.15)$$

we express all three summations as contour integrals like (4.6.1)

$$A(t_1, z_1; \omega_{12}; t_2, z_2) = \left(-\frac{1}{2i}\right)^3 \int d\lambda \int dJ_1 \int dJ_2 \frac{(2J_1+1)(2J_2+1) A_{J_1 J_2 \lambda}(t_1, t_2)}{\sin(\pi\lambda)\sin(\pi(J_1-\lambda))\sin(\pi(J_2-\lambda))} \times d^{J_1}_{0\lambda}(-z_1)\, d^{J_2}_{\lambda 0}(-z_2)\, e^{i\lambda\omega_{12}} \quad (9.3.16)$$

which gives, from the Regge poles in J_1 and J_2, taking the asymptotic form of the $d^{\alpha_i}(-z_i)$ (even though we shall not in fact be making the z_i large),

$$A^R(t_1, z_1; \omega_{12}; t_2, z_2) = -\frac{1}{2i} \int d\lambda (-s_{34})^{\alpha_1(t_1)} (-s_{45})^{\alpha_2(t_2)} \times \frac{e^{i\lambda\omega_{12}}}{\sin \pi\lambda} \Gamma(\lambda-\alpha_1)\Gamma(\lambda-\alpha_2)\, \beta_\lambda(t_1, t_2)\, \gamma_1(t_1)\, \gamma_2(t_2) \quad (9.3.17)$$

where β_λ is the central coupling. Then using the fact that

$$\cos \omega_{12} = \tfrac{1}{2}(e^{i\omega_{12}} + e^{-i\omega_{12}}) \sim \eta_{12}$$

we can rewrite this as

$$\begin{aligned} A^R(t_1, z_1; \omega_{12}; t_2, z_2) &= \frac{1}{2\pi i} \int d\lambda (-s_{34})^{\alpha_1} (-s_{45})^{\alpha_2} (-\eta_{12})^{\lambda}\, \Gamma(\lambda-\alpha_1)\Gamma(\lambda-\alpha_2)\,\Gamma(-\lambda) \\ &\qquad \times \beta_\lambda(t_1, t_2)\, \gamma_1(t_1)\, \gamma_2(t_2) \\ &= \frac{1}{2\pi i} \int d\lambda (-s_{34})^{\alpha_1-\lambda} (-s_{45})^{\alpha_2-\lambda} (-s_{12})^{\lambda}\, \Gamma(\lambda-\alpha_1)\,\Gamma(\lambda-\alpha_2) \\ &\qquad \times \Gamma(-\lambda)\, \beta_\lambda(t_1, t_2)\, \gamma_1(t_1)\, \gamma_2(t_2) \end{aligned} \quad (9.3.18)$$

(see White (1972*a*), Brower *et al.* (1974) for details). Then for $s_{12} \to \infty$, s_{34}, s_{45}, t_1, t_2 fixed we find, on opening the λ contour, that the leading asymptotic behaviour stems from the 'helicity poles' of the Γ-functions at $\lambda = \alpha_i$, and gives terms

$$A^R \sim (s_{12})^{\alpha_1} \quad \text{and} \quad \sim (s_{12})^{\alpha_2}$$

So in this helicity limit the Regge behaviour arises from the nonsense Γ-factors which relate the coupling of each Reggeon to the helicity of the other Reggeon.

We shall find that this limit is useful in the next chapter, but for multi-Regge analysis it is of course the various multi-Regge limits which concern us.

9.4 Multi-particle dual models*

In chapter 7 we introduced the idea of duality: that the Regge poles in the t channel already include the resonance poles in the s channel, at least in some average sense, and so it is a mistake to try to add these two types of contributions. The Veneziano model like (7.4.4), which we shall here take to be

$$V(s,t) = g\frac{\Gamma(-\alpha(s))\,\Gamma(-\alpha(t))}{\Gamma(-\alpha(s)-\alpha(t))} \tag{9.4.1}$$

gives a specific, though not unique, realization of this property, with Regge behaviour both in s at fixed t, and in t at fixed s. We now want to discuss the generalization of this result for many-particle amplitudes (see Veneziano 1974*a*, Schwarz 1973, Mandelstam 1974). It seems clear that this must be possible because for example in fig. 9.4 (*a*) we treated $(\overline{1}5)$ like a single particle, and if we choose a positive value of t_2 such that $\alpha_2(t_2) = n$, a right-signature integer, we have a physical, and presumably dual, $2 \to 2$ process.

First it should be noted that in $2 \to 2$ scattering there is a different dual amplitude for each planar ordering of the particles (see fig. 7.7) so that the $V(s,t)$ term is represented by fig. 9.9 (*a*) for which $s \leftrightarrow t$ involves just a cyclic permutation of 1, 2, 3, 4. But since $s \leftrightarrow u$ requires a non-cyclic permutation there is also a $V(s,u)$ term, fig. 9.9 (*b*), which must be added separately, as must $V(t,u)$. So generalizing this idea of planar duality we can expect that the set of diagrams, fig. 9.10, which all have the same cyclic ordering of particles $1, \ldots, 5$ will be dual to each other, but that for example the diagrams of fig. 9. 11 will comprise a separate dual term. In all there are 12 inequivalent orderings of the particles and hence 12 dual terms. Secondly the two Reggeons $\alpha_1(t_1)$ and $\alpha_2(t_2)$ in fig. 9.10 (*a*) depend on completely unrelated variables t_{23} and t_{15}, so it is rather obvious that they cannot be dual to each other. It is Reggeons in overlapping channels, like t_{23} and s_{34} which have particle 3 in common (see fig. 9.10 (*a*), (*b*)), which will be dual to each other.

* This section may be omitted at first reading.

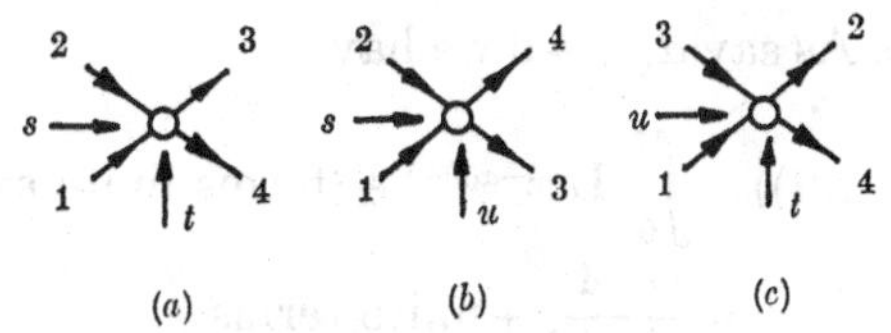

Fig. 9.9 The three inequivalent planar orderings of the particles which give the three terms in a $2 \to 2$ Veneziano amplitude like (7.4.17).

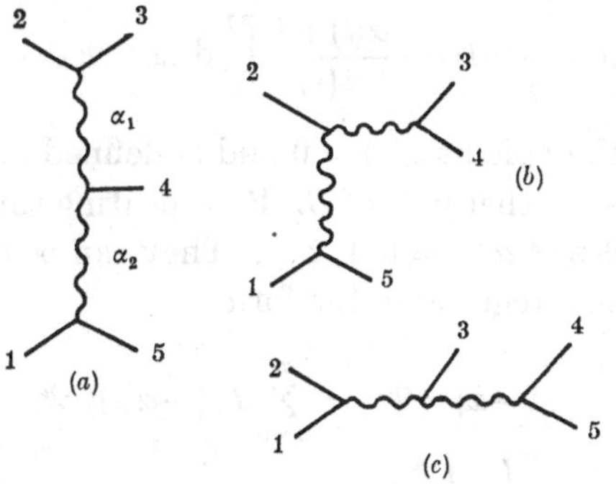

Fig. 9.10 Three different Reggeon amplitudes which involve the same planar cyclic ordering of particles 1, ..., 5 and so should be represented by a single dual amplitude.

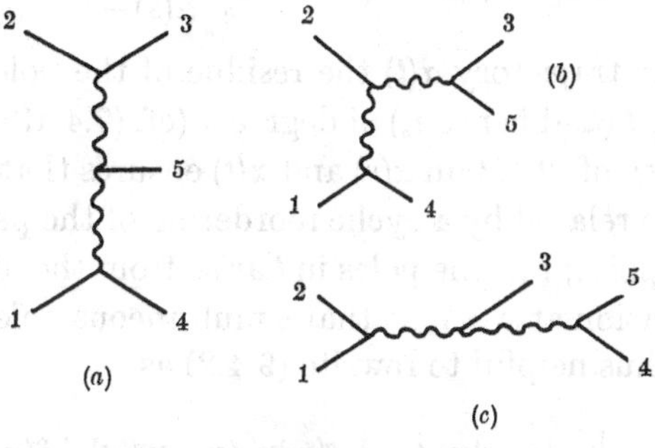

Fig. 9.11 Some Reggeon amplitudes which are dual to each other, but not to those in fig. 9.10.

To extend (9.4.1) we begin by rewriting it as

$$V(s,t) = gB_4(-\alpha(s), -\alpha(t)) \equiv g\int_0^1 \mathrm{d}x\, x^{-\alpha(s)-1}(1-x)^{-\alpha(t)-1} \tag{9.4.2}$$

where B_4 is known as the Euler β-function (see Veneziano (1968), Magnus and Oberhettinger (1949) p. 4). This integral is only defined for $\alpha(s)$, $\alpha(t) < 0$. As say $\alpha(s) \to 0$ we have

$$B_4(-\alpha(s), -\alpha(t)) \to \int_0^1 \mathrm{d}x\, x^{-\alpha(s)-1} + (\text{terms finite at } \alpha(s) = 0)$$
$$= -\frac{1}{\alpha(s)} + \text{finite terms} \qquad (9.4.3)$$

so the pole at $\alpha(s) = 0$ arises from the divergence of the integrand at $x = 0$. We can continue past this singularity by integrating by parts, giving

$$B_4(-\alpha(s), -\alpha(t)) = \frac{\alpha(t)+1}{\alpha(s)} \int_0^1 \mathrm{d}x\, x^{-\alpha(s)}(1-x)^{-\alpha(t)-2} \qquad (9.4.4)$$

which exhibits the pole at $\alpha(s) = 0$ and is defined for $\alpha(s) < 1$ where of course there is another pole of B_4. By repeating this process we find a sequence of poles at $\alpha(s) = 0, 1, 2, \ldots$. They can be obtained directly by expanding the integrand in the form

$$(1-x)^{-\alpha(t)-1} = \sum_{n=0}^{\infty} P_n(-\alpha(t))\, x^n \qquad (9.4.5)$$

where $$P_n(-\alpha) \equiv \frac{(-1)^n}{n!}(-\alpha-1)(-\alpha-2)\ldots(-\alpha-n)$$

and integrating each term to give

$$B_4(-\alpha(s), -\alpha(t)) = \sum_{n=0}^{\infty} \frac{P_n(-\alpha(t))}{\alpha(s)-n} \qquad (9.4.6)$$

So with a linear trajectory $\alpha(t)$ the residue of the pole at $\alpha(s) = n$ is a polynomial in t (and hence z_s) of degree n (cf. (7.4.13)).

The symmetry of (9.4.2) in $\alpha(s)$ and $\alpha(t)$ ensures that the channels s and t, which are related by a cyclic reordering of the particles $1, \ldots, 4$, have identical poles; but the poles in t arise from the other end of the range of integration at $x \to 1$, so that simultaneous poles in s and t are avoided. It is thus helpful to rewrite (9.4.2) as

$$V(s_{12}, t_{23}) = g \int_0^1 \mathrm{d}x_{12}\, \mathrm{d}x_{23} (x_{12})^{-\alpha(s_{12})-1} (x_{23})^{-\alpha(t_{23})-1}\, \delta(x_{12}+x_{23}-1) \qquad (9.4.7)$$

where we have associated an x variable with each channel which contains a pole (which arises for $x \to 0$), but by including the δ function have ensured that the overlapping s_{12} and t_{23} channels do not have simultaneous poles. It is also possible to insert an arbitrary function $f(x_{12}, x_{23})$ into the integrand of (9.4.7), analytic in $0 \leqslant x \leqslant 1$, in which case expanding f in a power series in the x's would give a sequence of Veneziano satellite terms like (7.4.15).

For the five-particle amplitude, fig. 9.10, we write similarly (Bardakci and Ruegg 1968, Virasoro 1969)

$$V(s_{12}, s_{34}, s_{45}, t_{23}, t_{15}) = gB_5(-\alpha(s_{12}), -\alpha(s_{34}), -\alpha(s_{45}), -\alpha(t_{23}), -\alpha(t_{15}))$$

$$\equiv g\int_0^1 dx_{12}\,dx_{34}\,dx_{45}\,dx_{23}\,dx_{15}(x_{12})^{-\alpha(s_{12})-1}\ldots(x_{15})^{-\alpha(t_{15})-1}f(x_{12},\ldots,x_{15}) \tag{9.4.8}$$

which has poles for each of the possible pairings of external particles (in this planar configuration). The function f must be chosen so as to prevent simultaneous poles in overlapping channels like, for example, s_{34}, t_{23} and s_{45}, so it must not be possible for x_{34} and x_{23} or x_{45} to vanish simultaneously. So we require f to vanish unless

$$\left.\begin{aligned} x_{34} &= 1-x_{23}x_{45} && a\\ x_{45} &= 1-x_{34}x_{15} && b\\ x_{15} &= 1-x_{45}x_{12} && c\\ x_{12} &= 1-x_{15}x_{23} && d\\ x_{23} &= 1-x_{12}x_{34} && e \end{aligned}\right\} \tag{9.4.9}$$

This gives five equations for five unknowns but they are not all independent equations, and in fact two of the variables remain free. These can conveniently be taken to be x_{23} and x_{15}. Then d gives x_{12} in terms of these, and e and a give

$$x_{34} = \frac{1-x_{23}}{1-x_{15}x_{23}}, \quad x_{45} = \frac{1-x_{15}}{1-x_{15}x_{23}}$$

respectively; equations b and c are consistent with these results. So we can write from a, b and e

$$f(x_{12},\ldots,x_{15}) = \delta(1-x_{34}-x_{23}x_{45})\,\delta(1-x_{45}-x_{34}x_{15})\,\delta(1-x_{23}-x_{12}x_{34}) \tag{9.4.10}$$

We could also multiply by any analytic function of the x's to give satellite terms. These δ-functions can be used to perform the integrations over x_{34}, x_{45} and x_{23} giving

$$B_5(-\alpha(s_{12}), -\alpha(s_{34}), -\alpha(s_{45}), -\alpha(t_{23}), -\alpha(t_{15}))$$

$$= \int_0^1 dx_{23}\,dx_{15}(1-x_{15}x_{23})^{-\alpha(s_{12})-1}\left(\frac{1-x_{23}}{1-x_{15}x_{23}}\right)^{-\alpha(s_{34})-1}$$

$$\times\left(\frac{1-x_{15}}{1-x_{15}x_{23}}\right)^{-\alpha(s_{45})-1}(x_{23})^{-\alpha(t_{23})-1}(x_{15})^{-\alpha(t_{15})-1}(1-x_{15}x_{23})^{-1} \tag{9.4.11}$$

or

$$B_5(-\alpha(s_{12})-\alpha(s_{34}),\,-\alpha(s_{45}),\,-\alpha(t_{23}),\,-\alpha(t_{15})) = \int_0^1 dx_{23}\,dx_{15}$$
$$\times\,(x_{23})^{-\alpha(t_{23})-1}\,(x_{15})^{-\alpha(t_{15})-1}\,(1-x_{23})^{-\alpha(s_{34})-1}(1-x_{15})^{-\alpha(s_{45})-1}$$
$$\times\,(1-x_{15}x_{23})^{-\alpha(s_{12})+\alpha(s_{45})+\alpha(s_{34})} \quad (9.4.12)$$

The complete five-particle dual amplitude is the sum of 12 terms like (9.4.12) involving different planar orderings of the five external particles. These are necessary to give the Reggeons signature since, for example, the signature properties of $\alpha(t_{23})$ and $\alpha(t_{15})$ require the four diagrams of fig. 9.8.

To examine the poles of this amplitude we put

$$-\alpha(s_{12})+\alpha(s_{45})+\alpha(s_{34}) \equiv -\beta \quad (9.4.13)$$

and expand $$(1-x_{15}x_{23})^{-\beta} = \sum_{n=0}^{\infty}(x_{15}x_{23})^n P_n(-\beta) \quad (9.4.14)$$

and integrate term-by-term to obtain (Hopkinson and Plahte 1968)

$$B_5 = \sum_{n=0}^{\infty} P_n(-\beta)\int_0^1 dx_{23}\,dx_{15}(x_{23})^{-\alpha(t_{23})-1+n}(x_{15})^{-\alpha(t_{15})-1+n}$$
$$\times\,(1-x_{23})^{-\alpha(s_{34})-1}(1-x_{15})^{-\alpha(s_{45})-1}$$
$$= \sum_{n=0}^{\infty} P_n(-\beta)\,B_4(-\alpha(t_{23})+n,\,-\alpha(s_{34}))$$
$$\times\,B_4(-\alpha(t_{15})+n,\,-\alpha(s_{45})) \quad (9.4.15)$$

Then if we expand the first B_4 as in (9.4.6)

$$B_5 = \sum_{m=0}^{\infty}\frac{1}{-\alpha(t_{23})+m}\sum_{n=0}^{m}P_n(-\beta)\,P_{m-n}(-\alpha(_{34}))$$
$$\times\,B_4(-\alpha(t_{15})+n,\,-\alpha(s_{45}))$$

giving a residue of the pole at $\alpha(t_{23}) = m$ of degree m in s_{34}, the angular variable for the t_{23} channel, so we have a daughter sequence of spins $k = 0, \ldots, m$. The residue contains the four-point Veneziano formula for $(2\overline{3})+\overline{4}\to\overline{1}+5$ as one would expect from factorization in fig. 9.10(*a*). However, while the highest trajectory contains just single resonances at $\alpha(t_{23}) = m$, all the daughter trajectories are multiply degenerate (Fubini and Veneziano 1969, Fubini, Gordon and Veneziano 1969), so simple amplitude factorization does not hold except on the leading trajectory. By excluding Veneziano satellites we have kept the daughter spectrum as simple as possible (Gross 1969), but none the less there are a very large number of particles. In fact for a given m the number of levels is given by the number of ways of choosing non-negative integers n_i which satisfy

$$n_1+2n_2+3n_3+\ldots = m$$

For large m this increases as $e^{(2\pi/\sqrt{6})m}$. It is of course a moot point whether one should take this seriously as a prediction of the model or whether it simply stems from the fact that we are unrealistically trying to represent a continuous branch cut by a sequence of poles.

To obtain the double-Regge limit of (9.4.12) we make the replacements

$$\alpha(s) = \alpha^0+\alpha' s \underset{s\to\infty}{\longrightarrow} \alpha' s, \quad x_{23} \equiv \frac{y_{23}}{-s_{34}}, \quad x_{15} \equiv \frac{y_{15}}{-s_{45}}$$

so

$$(1-x_{23})^{-\alpha(s_{34})-1} \to \left(1+\frac{y_{23}}{s_{34}}\right)^{-\alpha' s_{34}} \to e^{-y_{23}\alpha'}$$

$$(1-x_{15})^{-\alpha(s_{45})-1} \to e^{-y_{25}\alpha'}$$

$$(1-x_{23}x_{15})^{-\alpha(s_{12})+\alpha(s_{34})+\alpha(s_{45})} \to e^{(-y_{23}y_{15}s_{12}/s_{35}s_{45})\alpha'}$$

and hence

$$B_5 \to (-s_{34})^{\alpha(t_{23})}(-s_{45})^{\alpha(t_{15})} \int_0^\infty dy_{23}\, dy_{15} (y_{23})^{-\alpha(t_{23})-1} (y_{15})^{-\alpha(t_{15})-1} \times e^{-(y_{23}+y_{15}+(y_{23}y_{15}s_{12}/s_{34}s_{45}))\alpha'} \quad (9.4.16)$$

This gives the double-Regge form (9.3.10) with an explicit form for the dependence on the Toller angle in V which can be shown to be (Drummond *et al.* 1969*a*)

$$V_1(t_1,t_2,\eta_{12}) = \frac{1}{\Gamma(-\alpha_1)\Gamma(-\alpha_2)} \sum_{n=0}^{\infty} \frac{\Gamma(-\alpha_1-n)\Gamma(-\alpha_2+\alpha_1-n)}{n!(\eta_{12})^n} \quad (9.4.17)$$

and similarly for V_2 (where $t_1 = t_{23}$, $t_2 = t_{25}$, $\alpha_1 = \alpha(t_{23})$, $\alpha_2 = \alpha(t_{25})$).

To generalize (9.4.8) to an N-particle amplitude we write for a given cyclic labelling of the particles (Chan 1968, Koba and Nielson 1969)

$$V_N = gB_N = g\int_0^1 f(x) \prod_{m,n} (x_{mn})^{-\alpha_{mn}-1} dx_{mn} \quad (9.4.18)$$

and the full amplitude will be the sum of $\frac{1}{2}(N-1)!$ terms for all the inequivalent non-cyclic permutations of the particles. A given $\alpha_{mn} \equiv \alpha(s_{mn})$ is specified by the channel invariant

$$s_{mn} = (p_m+p_{m+1}+\cdots+p_n)^2 \quad (9.4.19)$$

as shown in fig. 9.12(*b*), and to prevent simultaneous poles occurring in overlapping channels we must insert into $f(x)$

$$\delta(x_{mn}+\prod_{k,l} x_{kl}-1) \quad (9.4.20)$$

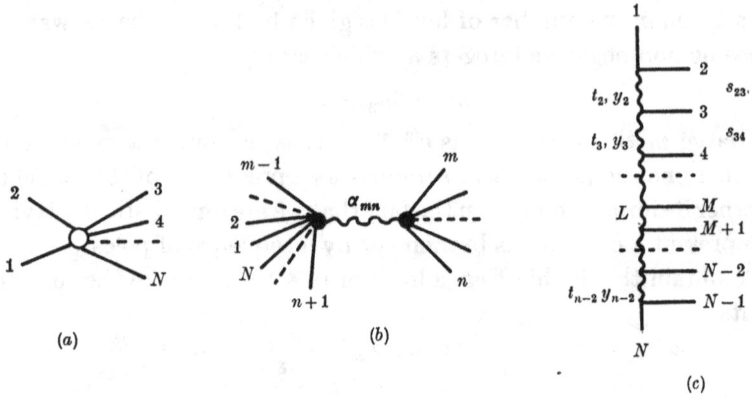

FIG. 9.12 (*a*) $1+2\to 3+\ldots+N$ amplitude with a cyclic ordering of the particles. (*b*) The α_{mn} trajectory exchange. (*c*) Labelling for $y_n = x_{1n}$.

where the kl are all the channels which overlap mn. To exhibit these we define $N-3$ variables

$$y_n \equiv x_{1n}, \quad n = 2, 3, \ldots, N-2 \tag{9.4.21}$$

as shown in fig. 9.12 (*c*). Then all the other x's are related to these by (Chan and Tsou 1969)

$$x_{mn} = \frac{a_{m,n-1}\, a_{m-1,n}}{a_{m,n}\, a_{m-1,n-1}}, \quad 2 \leqslant m < n \leqslant N-1 \tag{9.4.22}$$

where

$$a_{mn} \equiv 1 - \prod_{k=m}^{n} y_k, \quad y_1 = y_{N-1} \equiv 0 \tag{9.4.23}$$

and it is found that the constraint (9.4.20) is incorporated by writing

$$B_N = \int_0^1 \mathrm{d}y_2 \ldots \mathrm{d}y_{N-2} \prod_{i=2}^{N-3} (1-y_i y_{i+1})^{-1} \prod_{m,n} (x_{mn}(y))^{-\alpha_{mn}-1} \tag{9.4.24}$$

This agrees with the result (9.4.12) for $N = 5$, and the resulting multi-Regge behaviour corresponding to fig. 9.12 (*c*) is

$$B_N \to \Gamma(-\alpha(t_2))\,(-s_{23})^{\alpha(t_2)}\, V(t_2, t_3, \eta_{23})\, \Gamma(-\alpha(t_3))\,(-s_{34})^{\alpha(t_3)}$$
$$\times V(t_3, t_4, \eta_{34}) \ldots \Gamma(-\alpha(t_{N-2}))\,(-s_{N-2,N-1})^{\alpha(t_{N-2})} \tag{9.4.25}$$

where the V's are given by (9.4.17). This accords with (9.3.10) except that of course our single planar amplitude lacks the signature factors.

It is also possible to include internal symmetry in these multi-particle dual models. This is achieved by incorporating the quark ($q\bar{q}$) structure of the mesons, just as we did in section 7.5 (Chan and Paton 1969).

Each meson is represented by a matrix, the rows corresponding to the quark index, and the columns to the anti-quark index. Thus if we consider just the isospin symmetry the quarks are the $I = \frac{1}{2}$ isodoublets (5.2.2), and a meson will be represented by a 2×2 matrix: a Kronecker $\delta_{\alpha\beta}$ if it is an isoscalar $I = 0$ (equation (5.2.7)), and the isospin Pauli matrices (5.2.5) $(\tau_i)_{\alpha\beta}$, $i = 1, 2, 3$ if it is the ith component of an isotriplet $I = 1$ (equation (5.2.8)). $I = 0$ and 1 are the only values which can be made from two $I = \frac{1}{2}$ quarks so there are no exotic states. It is convenient tointroduce the notation $(\tau_0)_{\alpha\beta} = \delta_{\alpha\beta}$ so that the set τ_i, $i = 0, 1, 2, 3$, includes all four possible isospin states which a particle may have.

The Chan–Paton rule is that to include isospin in a B_N corresponding to a given cyclic ordering of the particles $1, \ldots, N$ we multiply it by a factor $\frac{1}{2}\operatorname{tr}(\tau_{i_1}, \tau_{i_2}, \tau_{i_3}, \ldots, \tau_{i_N})$ (where tr = trace). This factor has the same cyclic symmetry as that of B_N, and gives the correct $q\bar{q}$ structure with no exotics in any intermediate state. This can be seen by writing for the L exchange particle in fig. 9.12 (c).

$$\tfrac{1}{2}\operatorname{tr}(\tau_{i_1} \ldots \tau_{i_N}) = \sum_{i_L=0}^{3} [\tfrac{1}{2}\operatorname{tr}(\tau_{i_1} \ldots \tau_{i_M}\tau_{i_L})]\,[\tfrac{1}{2}\operatorname{tr}(\tau_{i_L}\tau_{iM+1} \ldots \tau_{i_N})] \tag{9.4.26}$$

which obviously has the desired factorization and isospin content for the residue of particle L, with exchange degeneracy between $I = 0$ and $I = 1$ particles. This can be extended from SU(2) to SU(3) simply by replacing the τ's by the λ matrices of table 5.1. But of course the method is only applicable in the limit of exact SU(3) degeneracy, which is far from the actual experimental situation.

In the last few years this dual formalism has undergone many developments which we shall not attempt to cover in any detail. The reader desiring to follow them can consult such excellent reviews as those of Veneziano (1974a), Schwarz (1973), Mandelstam (1974) and Scherk (1975).

We mentioned in section 3.3 that straight trajectories like those of the dual model are produced by a relativistic harmonic oscillator potential, and it has proved possible to re-express the dual model in an operator formalism in which particle states are created by an infinite set of harmonic oscillator creation operators a_μ^n, $n = 0, 1, \ldots, \infty$, operating on the basic vacuum state (Fubini *et al.* 1969, Fubini and Veneziano 1970, 1971). This makes it much easier to discuss such features as the resonance spectrum, and in particular the degeneracy of the daughters. But there is a fundamental problem that to ensure the Lorentz covariance of the theory the creation operators must be four-dimensional ($\mu = 0, 1, 2, 3$) and the inclusion of the time dimension produces so-called 'ghost' states, with negative residues, which would violate causality (see section 1.4). The same problem occurs in quantum electrodynamics where the creation of time-like photons would cause difficulties were it not for the fact that the Lorentz gauge condition ensures that such states are eliminated (Bjorken and Drell 1965). This is possible because the massless nature of the photon means that there can be no longitudinal photons either (the helicity $\lambda = \pm 1$ only, not 0), so the longitudinal and time-like components can be arranged to cancel.

It has been found that likewise in dual models, if $\alpha(0) = 1$ for the leading trajectory, then an infinite set of gauge conditions can be imposed which eliminates all the ghosts. In fact this is true for up to 26 space–time dimensions. But of course such a restriction is very unphysical and makes it quite impossible to regard the model as a prototype for real physics even in the meromorphic limit. It does mean, however, that the resulting dual field theory is closely related to other field theories with massless particles, in particular to quantum electrodynamics with massless photons and electrons, to the Yang–Mills field theory, and to quantum gravity with a massless spin $= 2$ graviton. In fact these field theores can be obtained as limits of dual field theory when the trajectory slope $\alpha' \to 0$ (see Veneziano 1974).

A further development has been to visualize this operator formalism as describing the motion of a quantized massless relativistic string (Goddard *et al.* 1973, Mandelstam 1973, Scherk 1975). A meson may be thought of as a string with free ends moving under internal tension counter-balanced by the centrifugal force due to its rotation (fig. 9.13). The maximum angular momentum for a given energy ($\equiv$ mass) occurs when the string is rigid, as in fig. 9.13 (a), and simply rotates, while lower-angular-momentum states of the same energy occur if there are also vibrational modes (like those of a violin string) whose frequencies will be multiples of the fundamental rotation frequency. This produces the daughter spectrum at a given mass. Internal symmetry can be incorporated by imagining the string to have quarks tied to its ends.

The motion of the string will in time trace out a world sheet like a twisted ribbon (fig. 9.13 (c)) and the gauge conditions correspond to the requirement that only vibrations perpendicular to this world sheet

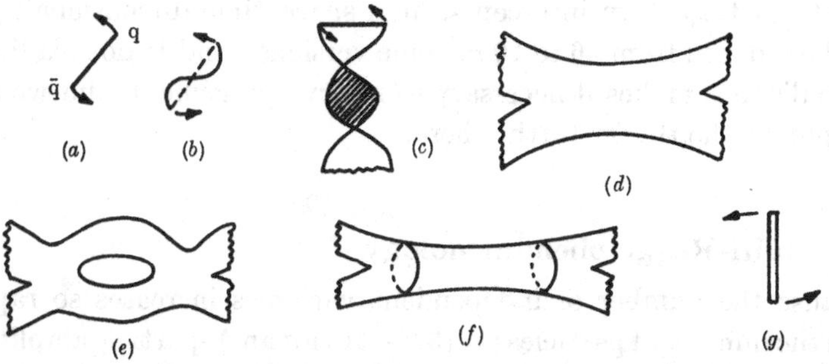

FIG. 9.13 (*a*) A rotating string with quarks at its ends. (*b*) A vibrational mode of the string. (*c*) World sheet of a rotating string. (*d*) String–string scattering. (*e*) Re-normalization loop in string–string scattering. (*f*) A tube corresponding to the Pomeron. (*g*) Highest angular-momentum state for a closed string.

occur. A consistent unitary quantum theory of such a string is possible only if $\alpha(0) = 1$ and the dimensionality of space–time is $D = 26$.

One can picture the interactions of such strings as in fig. 9.13(*d*), which looks very like the duality diagram of fig. 7.7 (*a*) (see Olive 1974). To unitarize the theory one must of course be able to include loops like fig. 9.13 (*e*), but such loops give infinite contributions which are not susceptible to the usual renormalization techniques of standard field theory because of the infinite number of intermediate states available. However, there is also another type of loop, namely a tube (fig. 9.13(*f*)), which is the world sheet of a closed string. The maximum angular momentum of such a closed string, for a given energy, occurs when it is pulled rigid as in fig. 9.13 (*g*), and it has twice the angular momentum of the corresponding open string, so $\alpha(0) = 2$. In fact it can be shown that

$$\alpha_{\text{tube}} = 2 + \frac{\alpha'}{2} t$$

where α' is the slope of the open-string trajectory. Since the closed string has no ends it can carry no quarks, and so has vacuum quantum numbers, and it has therefore been identified with the Pomeron. The fact that the intercept is at 2 rather than 1 is another embarrassment, but perhaps if the intercept of the ordinary Reggeons could be brought down to $\alpha(0) = \frac{1}{2}$ then the Pomeron would come down to 1 as well. In the zero-slope limit the Pomeron field theory reduces to that of a graviton.

This dual field theory could be the first hint of a fundamental theory of strong interactions in which dual Reggeons play the central role. However, the fact that at present the theory seems to be restricted to integer trajectory intercepts, high space–time dimensionality (D can be reduced from 26 to 10 in some versions), and is not readily renormalizable, makes it necessary to reserve judgement, and we shall not pursue the theory further here.

9.5 Multi-Regge phenomenology

Because the number of independent variables increases so rapidly with the number of particles ($=(3N-10)$ for an N-particle amplitude) many-particle processes have been much less well explored than those with two particles in the final state ($N=4$). Thus to examine thoroughly the $2\rightarrow 3$ amplitude we need, ideally, sufficient events to map the probability distribution in five different variables, or four at a given incident energy. Further, since the double-Regge region requires s_{12}, s_{34}, $s_{45}\rightarrow\infty$ with $s_{12}/s_{34}s_{45}$ fixed, to get both s_{34} and s_{45} large enough we need a very large s_{12}. But at such large s_{12} the given three-body final state will be found in only a small fraction of the events. For this reason it has become more usual to try and analyse many-body reactions 'inclusively' as we shall describe in the next chapter, rather than concentrating on a particular final state exclusively. Nevertheless, it is important to discover what Regge theory has to say about individual many-body processes.

We shall concentrate on $2\rightarrow 3$ scattering as in fig. 9.1. From (1.8.5) the double differential cross-section, integrated over t_{23}, t_{15} at fixed s_{12}, will be (see (1.8.17))

$$\frac{\mathrm{d}^2\sigma}{\mathrm{d}s_{34}\,\mathrm{d}s_{45}}(s_{12},s_{34},s_{45}) = \frac{1}{2\lambda^{\frac{1}{2}}(s,m_1^2,m_2^2)}\int\prod_{i=3}^{5}\left(\frac{\mathrm{d}^3p_i}{2p_{i0}(2\pi)^3}\right)$$
$$\times(2\pi)^4\,\delta^4(p_1+p_2-p_3-p_4-p_5)\,\delta(s_{34}-(p_3+p_4)^2)$$
$$\times\,\delta(s_{45}-(p_4+p_5)^2)\,|A(1+2\rightarrow 3+4+5)|^2 \qquad (9.5.1)$$

which gives the distribution of events in the Dalitz plot, fig. 9.2, as a function of s_{34} and s_{45} for a given s_{12}. (If the particles have spin a sum over the helicities of A_H is implied as usual – see (4.2.5).)

The single-Regge limits like fig. 9.1 (c) are characterized by a fixed small value of one of these invariants, say s_{45}, with $s_{34}\sim s_{12}\rightarrow\infty$, and so there are three single-Regge regions as shown in fig. 9.14(a). For example in $\pi^+p\rightarrow\pi^+\pi^0p$ we may have $\pi^+p\rightarrow(\pi^0\pi^+)\,p$, $\pi^+p\rightarrow\pi^0(\pi^+p)$ and $\pi^+p\rightarrow\pi^+(\pi^0p)$. Particular examples where two of the final-state particles are correlated as resonances, such as $(\pi^0\pi^+)=\rho^+$ or

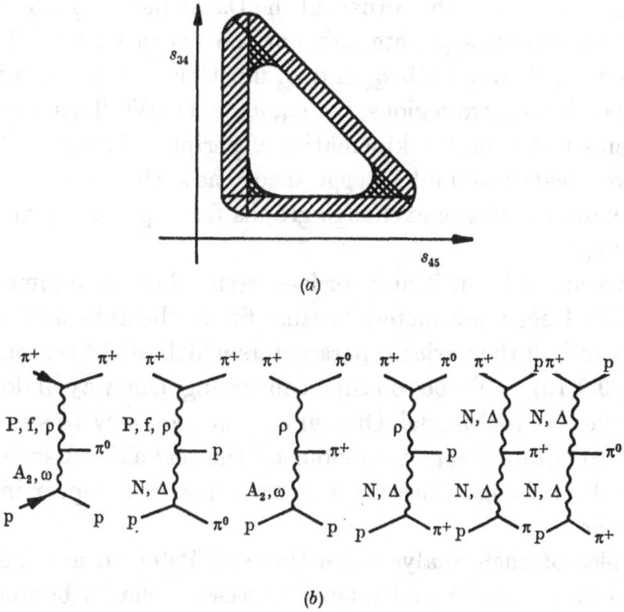

FIG. 9.14 (*a*) Dalitz plot for large s_{12} showing the three single-Regge regions (hatched) and the three double-Regge regions (cross-hatched). (*b*) Double-Regge exchange diagrams for $\pi^+p \to \pi^+\pi^0p$.

$(\pi^+p) = \Delta^{++}$, give quasi-two-body reactions of the type already discussed in chapter 6, and in fact single-Regge analysis is identical to that for two-body final states except for the dependence on $s_{45} = m_{45}^2$, the invariant mass, and the (45) 'decay' angular distribution.

Of greater interest are the various double-Regge limits, like fig. 9.1(*b*) which requires s_{12}, s_{34}, $s_{45} \to \infty$, $\eta_{12} = s_{12}/s_{34}s_{45}$ fixed. Now from (9.2.30) η_{12} is related to ω_{12}, and since ω_{12} is a physical angle it is restricted to $\cos\omega_{12} \geqslant -1$ which gives (after some manipulation, see Chan *et al.* (1967))

$$(\sqrt{-t_{23}}+\sqrt{-t_{15}})^2+m_4^2 \geqslant \frac{s_{34}s_{45}}{s_{12}} = \frac{1}{\eta_{12}} \qquad (9.5.2)$$

Now Regge theory is applicable only when the interaction is peripheral, and we expect that the amplitudes will be negligible for large values of t. Empirically this stems partly from the exponential t dependence of Regge couplings and partly from Regge shrinkage, but it is also necessary on theoretical grounds that $s \gg t$ for each Reggeon. Hence we must have $|t_{23}|$, $|t_{15}|$ small (i.e. $\not> 1\,\text{GeV}^2$), which means that $1/\eta_{12}$ in (9.5.2) is restricted to similar small values. So the three double-

Regge regions are near the corners of the Dalitz plot (as in fig. 9.14 (*a*)) where the products $s_{34}s_{45}$ etc. are not too big in view of the given fixed large s_{12}, though both s_{34} and s_{45} must be large enough to be in their respective Regge regions, i.e. $s_{34}, s_{45} \gg 1\,\text{GeV}^2$. This 'cornering' effect stems just from the kinematics of peripheral interactions, and is not a verification of multi-Regge theory as such.

The six double-Regge exchange graphs for $\pi^+ p \to \pi^+ \pi^0 p$ are shown in fig. 9.14 (*b*).

To proceed further it is more or less essential to place some restrictions on the Regge parameters because fits to the data with all these diagrams and all the variable parameters which might reasonably be put into (9.3.10) would be too time-consuming. One way of doing this is to invoke the dual model. Of course, it is necessary to smooth out the poles to obtain Regge behaviour on the real axis. Also one must eliminate P exchange since the Pomeron does not appear in simple dual models.

Examples of such analyses are those of Peterson and Tornqvist (1969) on $K^- p \to \pi^0 \pi^+ \Lambda$ and related processes, chosen because no P exchange can occur, and those of Chan *et al.* (1970) who examined $K^+ p \to K^0 \pi^+ p$, $K^- p \to \bar{K}_0 \pi^- p$, and $\pi^- p \to K^0 K^- p$. The allowed planar diagrams are shown in fig. 9.15, and using them good agreement with the data was obtained. On inserting the known trajectory functions there remains just one free parameter, the overall normalization. See Berger (1971 *a*) for a more complete survey.

A more simple version with many of the same features is the Chan–Loskiewicz–Allison (1968) model in which one writes, labelling the particles as in fig. 9.12 (*c*) for convenience,

$$A_N = \prod_{i=2}^{N-2} (G_i s_i + F_i)(s_i + 1)^{\alpha_i^0 - 1} (e^a s_i + 1)^{\alpha' t_i} \tag{9.5.3}$$

where

$$s_i \equiv s_{i\,i+1} \equiv (p_i + p_{i+1})^2, \quad t_i = [p_1 - (p_2 + p_3 + \dots + p_i)]^2 \tag{9.5.4}$$

This has the property that for all $s_i \gg 1$ it gives the multi-Regge form

$$A_N \sim \prod_{i=2}^{N-2} G_i (s_i)^{\alpha_i^0} e^{(a + \log s_i)\alpha' t_i} \tag{9.5.5}$$

like (9.3.6), but it neglects all the Toller-angle and spin effects at the vertices. For $s_i \to 0$ the *i*th term $\to F_i$ a constant, which provides a very crude parameterization of low sub-energy effects (which in fact provide the bulk of the events) but without the resonance structure which is

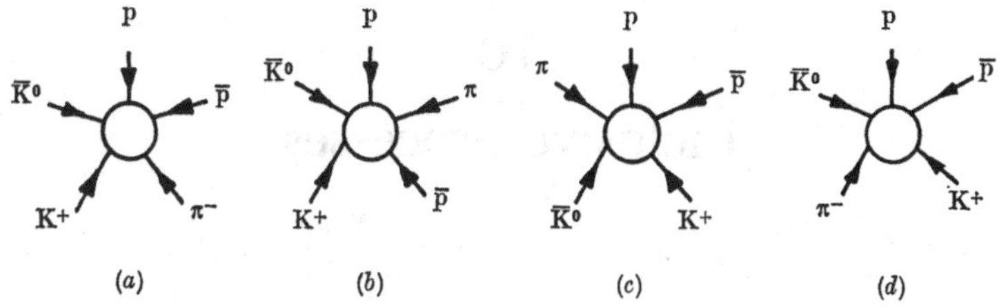

FIG. 9.15 Different orderings for the process $K^+p \to K^0\pi^+p$ (all particles drawn ingoing) with no exotic pairings. These are all the planar diagrams allowed by duality, but (*d*) is an illegal duality diagram because the λ quark would have to cross from $\overline{K}^0$ to K^+.

necessary for a really good description of the data. The full amplitude is a sum of terms like A_N for all inequivalent permutations of the particles. Though not good enough for detailed quantitative work this parameterization provides a manageable approximation with many of the desired qualitative features. Plahte and Roberts (1969) have produced an improved version.

The conclusions of this chapter may be summarized as follows. A consistent multi-Regge theory seems to be possible, though at present to derive it one has to make unproven if plausible assumptions about the singularity structure which determines the Regge asymptotic behaviour. A dual model with such a multi-Regge structure can be constructed, though the internally self-consistent factorizing version of the model bears at most a rather limited resemblance to nature. However, it might eventually lead to a fundamental theory of strong interactions. Phenomenologically multi-Regge theory can be tested only on that rather small fraction of the events for a given process which occur in the multi-Regge region of phase space. It appears to be satisfactory, and, despite their obvious limitations, dual models have enjoyed some phenomenological success. But many-particle amplitudes depend on too many variables for a really detailed comparison of theory and experiment to be made. Hence for example it has so far been possible to more or less ignore the Regge-cut corrections to the dominant pole exchanges.

It will be evident that a better way of analysing inelastic scattering processes is necessary, and this is provided by the Mueller–Regge approach to inclusive cross-sections, which is the subject of the next chapter.

10
Inclusive processes

10.1 Introduction

Though many-body final states provide the bulk of the high energy scattering cross-section, individual final states are hard to analyse. They are hard to extract experimentally because it is essential to test (using energy, momentum, and quantum-number arguments) that the final-state particles observed in the detecting apparatus were the only particles produced, and to exclude all the many other different types of events which could have occurred. In particular the production of neutral particles is especially hard to detect. And, as we have found in the previous chapter, final states are also hard to analyse theoretically both because the number of independent variables increases rapidly with the number of particles, and because only a fraction of the events occur in regions of phase space which are easy to parametrize, such as the low sub-energy resonance region, or the high sub-energy Regge region.

Because of these problems it has been found more useful to concentrate attention on so-called 'inclusive processes', that is, processes in which a given particle or set of particles is found to occur in the final state, but no questions are asked about all the other particles which may also be present in this final state. Thus we have the single-particle inclusive cross-section for the process

$$1+2 \to 3+X \tag{10.1.1}$$

(fig. 10.1 (*a*)) where 3 is a specified type of particle (for example it may be specifically a π^-, or more generally any negatively charged particle), and X includes all the particles which may be produced with 3, given the need to conserve energy, momentum and quantum numbers. Obviously we must have, to conserve four-momentum and charge,

$$p_X = p_1+p_2-p_3, \quad Q_X = Q_1+Q_2-Q_3 \tag{10.1.2}$$

etc. Similarly, the two-particle inclusive process is

$$1+2 \to 3+4+X \tag{10.1.3}$$

where 3 and 4 are specified types of particles, and X is anything (fig. 10.1 (*b*)).

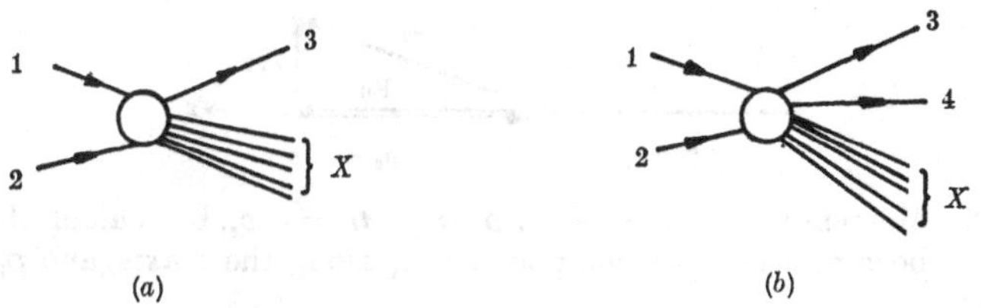

FIG. 10.1 (*a*) The single-particle inclusive process $1+2 \rightarrow 3+X$. (*b*) The two-particle inclusive process $1+2 \rightarrow 3+4+X$.

Such processes are fairly easy to identify experimentally since all one has to do is to verify that a particle (or particles) of the specified type(s) has been detected. It is necessary to measure the momentum only of the detected particle(s) (in addition to the beam momentum) to determine the event completely, because, for the process (10.1.1) for example, there are only three independent variables (s_{12}, t_{13} and M_X), as we shall see in the next section.

Also, through a rather ingenious generalization of the optical theorem, due to Mueller, it is surprisingly simple to obtain Regge predictions about the high energy behaviour of such processes. So in recent years a great deal more progress has been made in understanding many-body processes through this inclusive approach than by analysing particular exclusive final states such as $1+2 \rightarrow 3+4+5$.

This chapter is devoted to the Regge analysis of inclusive processes. We begin by discussing their kinematics, and the definition of an inclusive cross-section, before introducing Mueller's theorem which is then used to make a variety of Regge predictions. Useful reviews of this subject have been made by Horn (1972), Frazer *et al.* (1972) and Morrison (1972).

10.2 The kinematics of inclusive processes

We consider the process (10.1.1) shown in fig. 10.1 (*a*). As usual we work in the *s*-channel centre-of-mass system in which the four-momenta are

$$\left.\begin{aligned} p_1 &= (E_1, 0, 0, p_z), \quad p_1^2 = E_1^2 - p_z^2 = m_1^2 \\ p_2 &= (E_2, 0, 0, -p_z), \quad p_2^2 = E_2^2 - p_z^2 = m_2^2 \\ p_3 &= (E_3, \mathbf{p}_{3\mathrm{T}}, p_{3\mathrm{L}}), \quad p_3^2 = E_3^2 - \mathbf{p}_{3\mathrm{T}}^2 - p_{3\mathrm{L}}^2 = m_3^2 \end{aligned}\right\} \quad (10.2.1)$$

The *z* axis is defined as the direction of motion of particle 1, and (as in fig. 10.2) we have resolved the momentum of 3 into its longi-

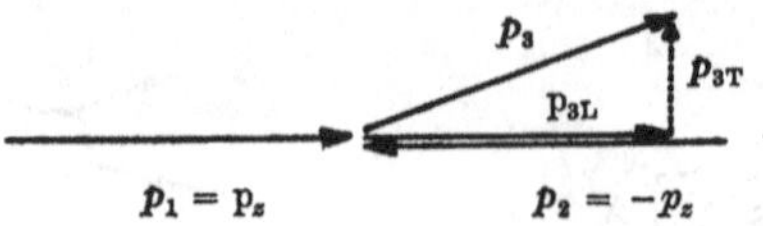

FIG. 10.2 Momenta in $1+2 \rightarrow 3+X$: $\boldsymbol{p}_1 = p_z$, $\boldsymbol{p}_2 = -p_z$, both along the z axis, and $\boldsymbol{p}_3$ has been resolved into components p_{3L} along the z axis, and $\boldsymbol{p}_{3T}$ transverse to it.

tudinal component, p_{3L}, along this z axis, and its components transverse to this axis which are represented by the two-component vector $\boldsymbol{p}_{3T}$. This distinction is very useful because it is found experimentally that though at high energy p_{3L} may take on almost any kinematically allowed value, from $p_{3L} \approx p_z$ if 3 is produced as a fragment of particle 1, to $p_{3L} \approx -p_z$ if it is produced from 2, the transverse component is usually rather small, very few events having $|\boldsymbol{p}_{3T}| > 0.5\,\text{GeV}/c$. In fact $\langle \boldsymbol{p}_{3T} \rangle \approx 0.3$–$0.4\,\text{GeV}/c$ whatever the beam energy.

Usually the majority of the particles in the final state are pions, presumably because the pion is the lightest hadron, with much smaller numbers of kaons, baryons etc., so typically $m_3 \lesssim 1\,\text{GeV}/c^2$. It is convenient to introduce the 'longitudinal mass' μ_3 defined by

$$\mu_3 \equiv (m_3^2 + \boldsymbol{p}_{3T}^2)^{\frac{1}{2}} \tag{10.2.2}$$

which is also generally $\leqslant 1\,\text{GeV}/c^2$, so that, from (10.2.1), μ_3 gives the effective mass associated with the longitudinal momentum, i.e.

$$E_3^2 = \mu_3^2 + p_{3L}^2 \tag{10.2.3}$$

As usual $s = s_{12} \equiv (p_1+p_2)^2$, so that E_1 and E_2 are given by (1.7.8) and (1.7.9), and $p_z = q_{s12}$ is given by (1.7.10), and so

$$p_z^2 \xrightarrow[s\to\infty]{} \frac{s}{4}, \quad E_1, E_2 \xrightarrow[s\to\infty]{} \frac{\sqrt{s}}{2} \quad \text{for} \quad s \gg m_1^2, m_2^2 \tag{10.2.4}$$

For the final state

$$s = (p_3 + p_X)^2 \tag{10.2.5}$$

and we define the 'missing mass' by

$$M^2 \equiv M_X^2 \equiv (p_1+p_2-p_3)^2 = s + m_3^2 - 2E_3\sqrt{s} \tag{10.2.6}$$

from (10.2.1) with (1.7.5). Obviously M takes the place of m_4 in the expressions (1.7.9) and (1.7.12) for the final-state energy and momentum, so

$$\boldsymbol{p}_3^2 = \boldsymbol{p}_{3T}^2 + p_{3L}^2 = \frac{1}{4s}[s-(m_3+M)^2]\,[s-(m_3-M)^2]$$

$$\xrightarrow[s,\,M^2\to\infty]{} \frac{(s-M^2)^2}{4s} \xrightarrow[s\gg M^2]{} \frac{s}{4} \tag{10.2.7}$$

$$E_3 = \frac{1}{2\sqrt{s}}(s+m_3^2-M^2) \xrightarrow[s,\,M^2\to\infty]{} \frac{s-M^2}{2\sqrt{s}} \xrightarrow[s\gg M^2]{} \frac{\sqrt{s}}{2} \tag{10.2.8}$$

Since p_{3T}^2 is small $$p_{3L}^2 \approx p_3^2 \to \frac{(s-M^2)^2}{4s}$$

and so $$\frac{M^2}{s} \approx 1 - \frac{2p_{3L}}{\sqrt{s}} \qquad (10.2.9)$$

Another independent variable is

$$t_{13} = t \equiv (p_1 - p_3)^2 = m_1^2 + m_3^2 - 2p_1 \cdot p_3 = m_1^2 + m_3^2 - 2E_1 E_3 + 2p_z p_{3L}$$

$$\xrightarrow[s\to\infty]{} -\sqrt{s}(E_3 - p_{3L}) = -\sqrt{s}\frac{E_3^2 - p_{3L}^2}{E_3 + p_{3L}} = -\frac{s\mu_3^2}{s - M^2} \qquad (10.2.10)$$

using (10.2.4), followed by (10.2.3), (10.2.7) and (10.2.8). Similarly

$$u \equiv (p_2 - p_3)^2 \to -s(E_3 + p_{3L}) \qquad (10.2.11)$$

and like (1.7.18) $$s + t + u = m_1^2 + m_2^2 + m_3^2 + M^2 \qquad (10.2.12)$$

So s, t and M^2 form a complete set of variables from which all the other kinematical quantities can readily be obtained.

However, two other variables are also frequently used. One of these is the Feynman variable, or 'reduced longitudinal momentum' x, defined by (Feynman 1969)

$$x_3 \equiv \frac{p_{3L}}{p_{3L\,\max}} \qquad (10.2.13)$$

Now from (10.2.9) the maximum value of p_{3L} occurs when $M^2 \to 0$ so

$$x_3 \approx \frac{2p_{3L}}{\sqrt{s}} \quad \text{or} \quad x_3 \approx 1 - \frac{M^2}{s} \qquad (10.2.14)$$

(though in fact $M_{\min}^2$ is the mass of the lightest particle which can be produced, and is > 0).

Sometimes (10.2.14) is used to define x instead of (10.2.13), but the equations are equivalent only to the extent that $m_{1,2,3}$ and $|p_{3T}|$ can be neglected compared with s and M^2. Clearly $x_1 = 1$ and $x_2 = -1$, and if $x_3 \approx 1$ it means that 3 has acquired most of the momentum of 1 and we can say that 3 is a 'fragment' of 1, or if $x \approx -1$, 3 is a fragment of 2 (see fig. 10.3). The 'central region' $x_3 \approx 0$ implies that 3 is approximately stationary in the centre-of-mass system and so is not directly connected with 1 or 2. These ideas will be made a bit more precise below. From (10.2.10) and (10.2.14) we have

$$t \to -\frac{\mu_3^2}{1 - x_3} \qquad (10.2.15)$$

so that s, x_3 and p_{3T}^2 provide a complete set of variables.

The other commonly employed variable is the rapidity y, defined by (de Tar 1971)

$$y_3 \equiv \tfrac{1}{2}\log\left(\frac{E_3+p_{3\mathrm{L}}}{E_3-p_{3\mathrm{L}}}\right) \tag{10.2.16}$$

from which we obtain, using (10.2.3),

$$\sinh y_3 = \frac{p_{3\mathrm{L}}}{\mu_3}, \quad \cosh y_3 = \frac{E_3}{\mu_3} \tag{10.2.17}$$

and so the components of p_3 are

$$p_3 = (\mu_3 \cosh y_3, \mathbf{p}_{3\mathrm{T}}, \mu_3 \sinh y_3) \tag{10.2.18}$$

This variable has the advantage that under a Lorentz boost by velocity v along the z axis (we use $c \equiv 1$ so $\beta \equiv v$, $\gamma \equiv (1-v^2)^{-\frac{1}{2}}$ in the usual notation)

$$p_3 = (E_3, \mathbf{p}_{3\mathrm{T}}, p_{3\mathrm{L}}) \xrightarrow[]{\text{Lorentz transformation}} (\gamma(E_3+vp_{3\mathrm{L}}), \mathbf{p}_{3\mathrm{T}}, \gamma(p_{3\mathrm{L}}+vE_3)) \tag{10.2.19}$$

and if these transformed values are substituted into (10.2.16)

$$y_3 \xrightarrow[]{\text{Lorentz transformation}} y_3 + \tfrac{1}{2}\log\left(\frac{1+v}{1-v}\right) \tag{10.2.20}$$

So the rapidity has very simple transformation properties along the beam axis. In fact a particle of rest mass m moving along the z axis with velocity v has $E = \gamma m$, $p_{\mathrm{L}} = \gamma m v$ and hence

$$y = \tfrac{1}{2}\log\left(\frac{1+v}{1-v}\right) \xrightarrow[v \ll 1]{} v$$

so in the non-relativistic limit, $v \ll c \equiv 1$, rapidity → velocity (which accounts for the name). But, unlike velocities, rapidities simply add like (10.2.20), even relativistically.

In the centre-of-mass system

$$y_1 = \tfrac{1}{2}\log\left(\frac{E_1+p_z}{E_1-p_z}\right) = \tfrac{1}{2}\log\left(\frac{(E_1+p_z)^2}{E_1^2-p_z^2}\right)$$

$$= \tfrac{1}{2}\log\left(\frac{(E_1+p_z)^2}{m_1^2}\right) \xrightarrow[s\to\infty]{} \tfrac{1}{2}\log\left(\frac{s}{m_1^2}\right) \tag{10.2.21}$$

using (10.2.1) and (10.2.4), and likewise $y_2 \to \frac{1}{2}\log(m_2^2/s)$

so

$$y_1 - y_2 \to \log\frac{s}{m_1 m_2}, \quad \text{or} \quad s \to m_1 m_2 \,\mathrm{e}^{(y_1-y_2)} \tag{10.2.22}$$

Also, from (10.2.15) and (10.2.2), in the centre-of-mass system,

$$y_3 = \tfrac{1}{2}\log\left(\frac{(E_3+p_{3L})^2}{\mu_3^2}\right) \tag{10.2.23}$$

and since, from (10.2.8) and (10.2.7), the extreme values (which occur when $M^2 \to 0$) are $E_3 \approx \sqrt{s}/2$, $p_{3L} \approx \pm\sqrt{s}/2$, we find

$$y_{3\,\mathrm{max}} = \tfrac{1}{2}\log\left(\frac{s}{\mu_3^2}\right), \quad y_{3\,\mathrm{min}} = -\tfrac{1}{2}\log\left(\frac{s}{\mu_3^2}\right) \tag{10.2.24}$$

so the range of y_3 is

$$Y_3 \equiv y_{3\,\mathrm{max}} - y_{3\,\mathrm{min}} = \log\left(\frac{s}{\mu_3^2}\right) \tag{10.2.25}$$

The maximum occurs when 3 takes on the longitudinal momentum of 1, and the minimum when it takes on that of 2, as in figs. 10.3 (*a*), (*b*), while $y_3 = 0$ corresponds to 3 being at rest in the centre-of-mass system. It is sometimes convenient to introduce the reduced rapidity

$$\tilde{y}_3 \equiv \frac{2y_3}{Y_3} \tag{10.2.26}$$

which like x_3 has the range $-1 \leqslant \tilde{y}_3 \leqslant 1$. However, $\tilde{y}_3$ and x_3 are not identical except at the three points -1, 0, $+1$, since as $s \to \infty$ all particles whose $|p_{3L}| \not\to \infty$ move towards $x = 0$. A boost to the laboratory frame (particle 2 at rest) is just, from (10.2.20),

$$y_3 \to y_3 + \tfrac{1}{2}\log\left(\frac{s}{\mu_3^2}\right) \tag{10.2.27}$$

as shown in fig. 10.4 (*a*). From (10.2.10)) and (10.2.11) y_3 is related to s, t, u and M^2 by

$$y_3 \to \tfrac{1}{2}\log\left(\frac{u}{t}\right) \to \log\left(\frac{M^2-s-t}{t}\right) \tag{10.2.28}$$

The quantities s, y_3, p_{3T}^2 thus provide another complete set of variables for the single-particle inclusive process.

10.3 Inclusive cross-sections

In (1.8.5) we wrote down an expression for the cross-section $\sigma_{12\to n}$, giving the probability per unit incident flux of n particles being produced in the final state; and in (1.8.7) we summed these to obtain the total cross-section $\sigma_{12}^{\mathrm{tot}} \equiv \sigma_{12\to\mathrm{all}}$. Correspondingly the cross-section

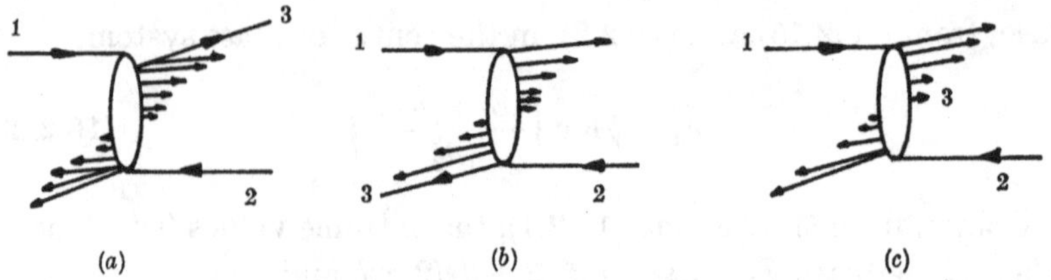

FIG. 10.3 Particle 3 produced (*a*) as a fragment of 1, (*b*) as a fragment of 2, and (*c*) in the central region where it is not associated directly with either incoming particle.

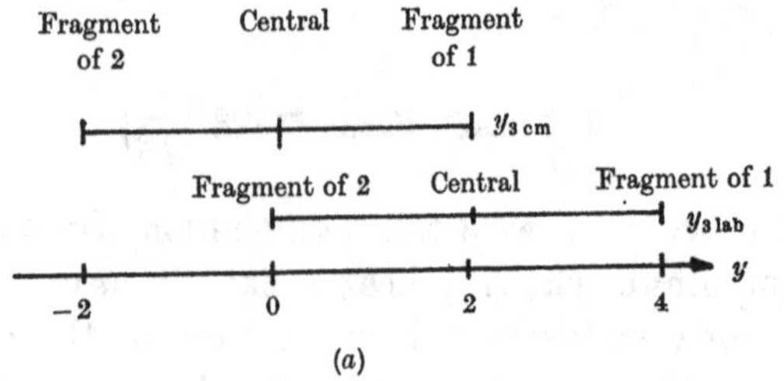

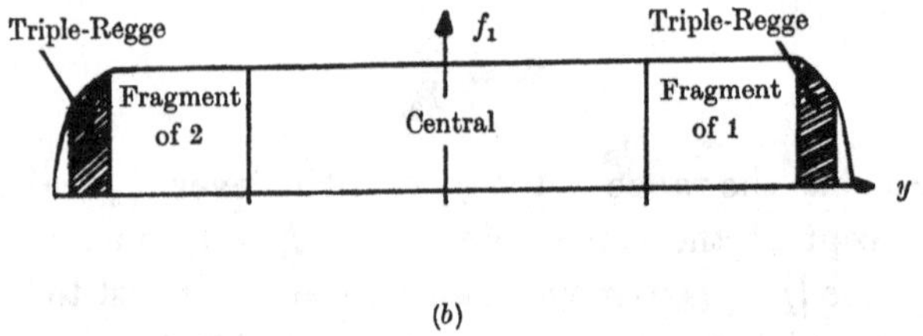

FIG. 10.4 (*a*) Transformation from laboratory-frame (2 at rest) to centre-of-mass frames rapidities for $Y = 4$; y_{cm} is simply displaced from y_{lab} by 2 units. (*b*) The different regions of the rapidity plot to be discussed below.

for producing at least one particle of type 3 plus anything is given by

$$\sigma_{12\to3X} = \frac{1}{4q_s\sqrt{s}}\sum_{n=0}^{\infty}\sum_{n_3=1}^{\infty}\int \mathrm{d}\Phi_{n+n_3}|\langle p_1'\ldots p_n', p_3^1\ldots p_3^{n_3}|\,A\,|p_1p_2\rangle|^2 \tag{10.3.1}$$

where the p_3^l, $l = 1, \ldots, n_3$, are the momenta of the n_3 particles of type 3 in the final state, and $p_1', \ldots, p_n'$ are the momenta of the n other particles which also appear ($n+n_3 \geqslant 2$). So the probability per unit incident flux of detecting a particle of type 3 within the phase-space volume element $\mathrm{d}^3\mathbf{p}_3$ (i.e. within the element of solid angle $\mathrm{d}\Omega$, with momentum between p_3 and $p_3+\mathrm{d}p_3$) is given by (cf. (1.8.17))

$$\frac{\mathrm{d}^3\sigma}{\mathrm{d}^3\mathbf{p}_3} = \frac{1}{4q_s\sqrt{s}}\sum_{n=0}^{\infty}\sum_{n_3=1}^{\infty}\int \mathrm{d}\Phi_{n+n_3}\sum_{l=1}^{n_3}\delta^3(\mathbf{p}_3-\mathbf{p}_3^l)\,|\langle|\,A\,|\rangle|^2 \tag{10.3.2}$$

where we have summed over all the n_3 particles of type 3 in the final state. However, this cross-section is frame-dependent, and instead it is preferable to use the invariance of $\mathrm{d}^3\boldsymbol{p}/2E(2\pi)^3$ (shown in (1.2.7)) to define the invariant single-particle distribution by

$$f_1^{12\to 3X}(\boldsymbol{p}_3, s) \equiv 16\pi^3 E_3 \frac{\mathrm{d}^3\sigma}{\mathrm{d}^3\boldsymbol{p}_3} \tag{10.3.3}$$

This may also be expressed in terms of our other variables. For example, using $\mathrm{d}^3\boldsymbol{p}_3 = \pi|\boldsymbol{p}_3|\,\mathrm{d}|\boldsymbol{p}_3|^2\,\mathrm{d}(\cos\theta)$ with (10.2.10) and (10.2.7) we obtain

$$f_1 = 16\pi^3 \frac{2p_z\sqrt{s}}{\pi}\frac{\mathrm{d}^2\sigma}{\mathrm{d}t\,\mathrm{d}M^2} \underset{s\to\infty}{\longrightarrow} 16\pi^2 s \frac{\mathrm{d}^2\sigma}{\mathrm{d}t\,\mathrm{d}M^2} \tag{10.3.4}$$

or, writing $\mathrm{d}^3\boldsymbol{p}_3 = \pi\,\mathrm{d}\boldsymbol{p}_{3\mathrm{T}}^2\,\mathrm{d}p_{3\mathrm{L}}$ and noting that, from (10.2.18),

$$\frac{\mathrm{d}p_{3\mathrm{L}}}{\mathrm{d}y_3} = \mu_3 \frac{\mathrm{d}\sinh y_3}{\mathrm{d}y_3} = \mu_3 \cosh y_3 = E_3$$

we get

$$f_1 = 16\pi^2 \frac{\mathrm{d}^2\sigma}{\mathrm{d}(\boldsymbol{p}_{3\mathrm{T}}^2)\,\mathrm{d}y} \tag{10.3.5}$$

Or since from (10.2.14), (10.2.17) and (10.2.3)

$$\frac{\mathrm{d}x_3}{\mathrm{d}y_3} = \frac{2\mu_3}{\sqrt{s}}\cosh y_3 = \frac{2E_3}{\sqrt{s}} = \left(x_3^2 + \frac{4\mu_3^2}{s}\right)^{\frac{1}{2}}$$

we find

$$f_1 = 16\pi^2\left(x^2 + \frac{4\mu_3^2}{s}\right)^{\frac{1}{2}} \frac{\mathrm{d}^2\sigma}{\mathrm{d}x\,\mathrm{d}(\boldsymbol{p}_{3\mathrm{T}}^2)} \underset{s\to\infty}{\longrightarrow} 16\pi^2 x_3 \frac{\mathrm{d}^2\sigma}{\mathrm{d}x_3\,\mathrm{d}(\boldsymbol{p}_{3\mathrm{T}}^2)} \tag{10.3.6}$$

All of the expressions (10.3.3)–(10.3.6) are used in the literature.

The total single-particle inclusive cross-section is

$$\int f_1(\boldsymbol{p}_3, s)\frac{\mathrm{d}^3\boldsymbol{p}_3}{16\pi^3 E_3} = \frac{1}{4q_s\sqrt{s}}\sum_{n+n_3=2}^{\infty}\int \mathrm{d}\Phi_{n+n_3}\sum_{l=1}^{n_3}$$

$$\times \int \mathrm{d}^3\boldsymbol{p}_3\,\delta^3(\boldsymbol{p}_3 - \boldsymbol{p}_3{}^l)|\langle|\,A\,|\rangle|^2 = \sum_{n_3=1}^{\infty} n_3\sigma(1+2\to n_3+X') \tag{10.3.7}$$

where $\sigma(1+2\to n_3+X')$ is the total cross-section for producing n_3 particles of type 3, plus X', which represents everything else produced but includes no particles of type 3. (So σ is given by (10.3.1) summed over n but not over n_3.) The weighting by n_3 occurs because of the extra summation over l in (10.3.2). So if we define the average multiplicity of particles of type 3 by

$$\langle n_3\rangle = \frac{\sum_{n_3=0}^{\infty} n_3\sigma(1+2\to n_3+X')}{\sum_{n_3=0}^{\infty}\sigma(1+2\to n_3+X')} = \frac{\sum_{n_3=0}^{\infty} n_3\sigma(1+2\to n_3+X')}{\sigma_{12\to\mathrm{all}}^{\mathrm{tot}}} \tag{10.3.8}$$

then $$\int f_1(p_3, s)\frac{\mathrm{d}^3p_3}{16\pi^3 E_3} = \langle n_3\rangle\,\sigma_{12}^{\text{tot}} \tag{10.3.9}$$

so the total inclusive cross-section is the total cross-section weighted by the average multiplicity. The physical reason for this weighting is, of course, that if the detecting apparatus is set up to register an event every time a particle of type 3 enters then those events in which two particles of type 3 occur will be counted twice, and so on. This multiple counting gives inclusive cross-sections many of their special properties.

It is sometimes convenient to introduce

$$\rho_1(p_3, s) \equiv \frac{f_1(p_3, s)}{\sigma_{12}^{\text{tot}}(s)} \tag{10.3.10}$$

so that $$F_1(s) \equiv \int \rho_1(p_3, s)\frac{\mathrm{d}^3p_3}{16\pi^3 E_3} = \langle n_3(s)\rangle \tag{10.3.11}$$

Empirically it is found (fig. 10.5) that for large $\langle n_3\rangle$ and s

$$\langle n_3(s)\rangle \approx A + B\log s$$

which, since $\sigma_{12}^{\text{tot}} \approx$ constant, means that $\int f_1\,\mathrm{d}^3p_3/16\pi^3E_3$ is increasing like $\log s$. So as the collision energy increases only a decreasing fraction of it is used to produce new particles, the rest being taken up by the kinetic energy of the final-state particles. We shall see below how this can be explained.

Likewise, we can define the two-particle inclusive distribution, giving the probability per unit flux of producing, in the process $1+2\to 3+4+X$, a particle of type 3 in d^3p_3 and a particle of type 4 in d^3p_4, by

$$\begin{aligned} f_2(p_3, p_4, s) &\equiv 4(2\pi)^6 E_3 E_4 \frac{\mathrm{d}^3\sigma}{\mathrm{d}^3p_3\,\mathrm{d}^3p_4} \\ &= \frac{1}{4q_s\sqrt{s}}\sum_{n=0}^{\infty}\sum_{n_3=1}^{\infty}\sum_{n_4=1}^{\infty}\mathrm{d}\Phi_{n+n_3+n_4}\sum_{l=1}^{n_3}2E_3(2\pi)^3\delta^3(p_3-p_3^l)\sum_{m=1}^{n_4}2E_4(2\pi)^3 \\ &\quad\times\delta^3(p_4-p_4^m)\,|\langle p_1'\ldots p_n'; p_3^1\ldots p_3^{n_3}; p_4^1\ldots p_4^{n_4}|A|\mathrm{p}_1\mathrm{p}_2\rangle|^2 \end{aligned} \tag{10.3.12}$$

Then like (10.3.7)

$$\begin{aligned} \int f_2(p_3, p_4, s)\frac{\mathrm{d}^3\mathrm{p}_3}{16\pi^3E_3}\frac{\mathrm{d}^3\mathrm{p}_4}{16\pi^3E_4} &= \sigma(1+2\to 3+4+X') \\ &\quad + 2\sigma(1+2\to 3+3+4+X') + 2\sigma(1+2\to 3+4+4+X') \\ &\quad + 4\sigma(1+2\to 3+3+4+4+X') + \ldots \\ &\equiv \langle n_3 n_4\rangle\,\sigma_{12}^{\text{tot}}(s) \end{aligned} \tag{10.3.13}$$

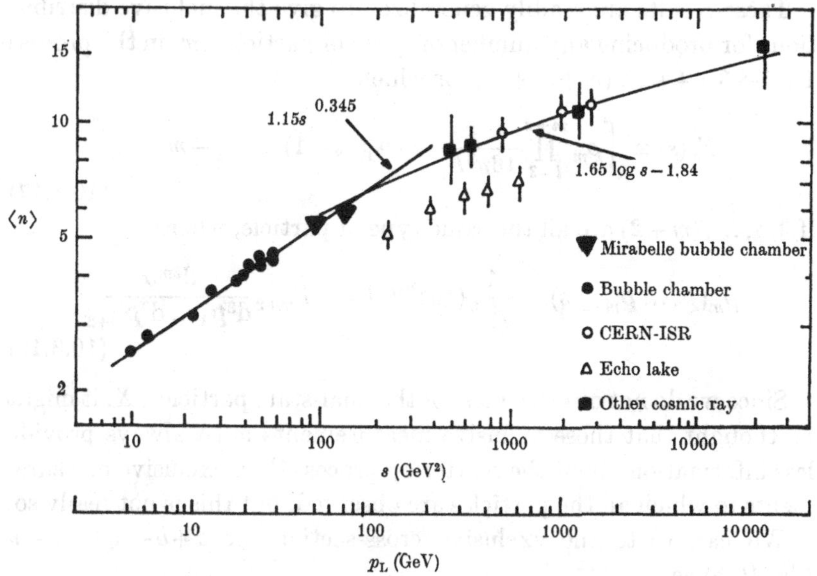

FIG. 10.5 The average charged multiplicity $\langle n \rangle$ in pp scattering versus s, showing the logarithmic increase, from Morrison (1972).

where X' includes no particle of type 3 or 4, and where $\langle n_3 n_4 \rangle$ is the average of the product of the multiplicities of 3 and 4. This assumes that 3 and 4 are distinct types of particles (for example 3 might be pions and 4 protons, or 3 might be negatively charged particles and 4 positively charged ones). If 3 and 4 are the same type of particle then

$$\int f_2(p_3, p_4, s) \frac{\mathrm{d}^3 p_3}{16\pi^3 E_3} \frac{\mathrm{d}^3 p_4}{16\pi^3 E_4} = 2\sigma(1+2 \to 3+3+X') + 6\sigma(1+2 \to 3+3+3+X') + \ldots \equiv \langle n_3(n_3-1)\rangle \, \sigma_{12}^{\mathrm{tot}}(s) \quad (10.3.14)$$

since in a given event producing n_3 particles of type 3 there are n_3 different ways of choosing the first particle to be detected, and $n_3 - 1$ ways of choosing the second particle.

Similar to (10.3.10) we can define

$$\rho_2(p_3, p_4, s) \equiv \frac{f_2(p_3, p_4, s)}{\sigma_{12}^{\mathrm{tot}}(s)} \quad (10.3.15)$$

and combining (10.3.13) and (10.3.14) we find

$$F_2(s) \equiv \int \rho_2 \frac{\mathrm{d}^3 p_3}{16\pi^3 E_3} \frac{\mathrm{d}^3 p_4}{16\pi^3 E_4} = \langle n_3 n_4 - \delta_{34} n_3 \rangle \quad (10.3.16)$$

These results are readily generalized to give the inclusive distributions for producing any number of types of particles, m, in the process $1+2 \to 3+4+\ldots(m+2)+X$, for which

$$F_m(s) \equiv \int \rho_m \prod_{i=3}^{m+2} \frac{d^3 p_i}{16\pi^3 E_i} = \langle n_3(n_3-1)\ldots(n_3-m+1)\rangle \tag{10.3.17}$$

if $3, 4, \ldots, (m+2)$ are all the same type of particle, where

$$\rho_m(p_3 \ldots p_{m+2}, s) \equiv \frac{1}{\sigma_{12}^{\text{tot}}} (16\pi^3)^m E_3 \ldots E_{m+2} \frac{d^{3m}\sigma}{d^3 p_3 \ldots d^3 p_{m+2}} \tag{10.3.18}$$

Since we do not observe most of the final-state particles, X, it might be thought that these inclusive measurements must always provide less information about the scattering process than exclusive measurements in which all the particles are observed, but this is not really so.

We can write the exclusive cross-section for $a+b \to 1+\ldots+n$ (fig. 10.6) as

$$(16\pi^3)^n E_1 \ldots E_n \frac{d^{3n}\sigma^{\text{ex}}}{d^3 p_1 \ldots d^3 p_n}$$

but if we observe, say, only l of these, the inclusive cross-section for $a+b \to l+X$ is

$$(16\pi^3)^l E_1 \ldots E_l \frac{d^{3l}\sigma^{\text{in}}}{d^3 p_1 \ldots d^3 p_l} = \sum_{n=l}^{\infty} \frac{1}{(n-l)!} \int (16\pi^3)^l \times E_1 \ldots E_l \frac{d^{3n}\sigma^{\text{ex}}}{d^3 p_1 \ldots d^3 p_n} d^3 p_{l+1} \ldots d^3 p_n \tag{10.3.19}$$

if we treat all the n particles as identical. So, as expected, the inclusive cross-sections can be obtained from the exclusive ones. But conversely a given n-particle exclusive cross-section can be obtained from all the $n+l$ inclusive ones, since

$$(16\pi^3)^n E_1 \ldots E_n \frac{d^{3n}\sigma^{\text{ex}}}{d^3 p_1 \ldots d^3 p_n} = \sum_{l=0}^{\infty} \frac{(-l)^l}{l!} \int (16\pi^3)^n \times E_1 \ldots E_n \frac{d^{3(n+l)}\sigma^{\text{in}}}{d^3 p_1 \ldots d^3 p_{n+l}} d^3 p_{n+1} \ldots d^3 p_{n+l} \tag{10.3.20}$$

The counting is explained for $n = 3$ in fig. 10.7: we take the three-body inclusive process, but subtract all those processes where at least four bodies are produced, remembering that because of the identity of the particles the five-body exclusive cross-section contributes 2! times to the three-body inclusive cross-section; and so on.

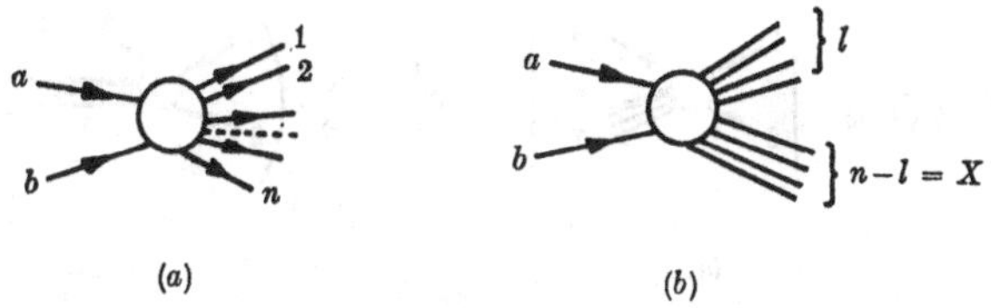

FIG. 10.6 (a) The n-body exclusive cross-section. (b) Contribution of the n-body final state to the l-particle inclusive cross-section.

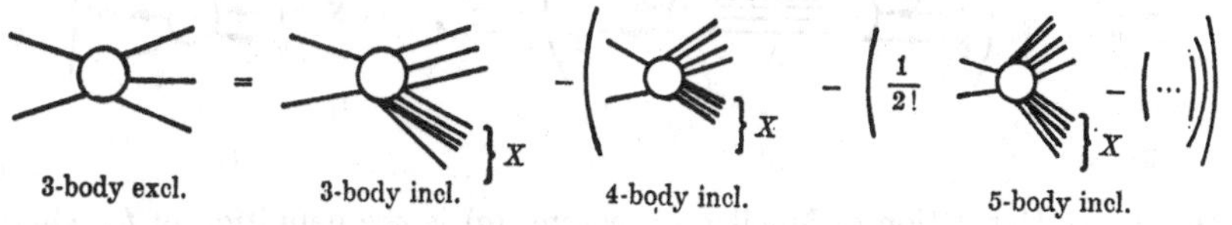

FIG. 10.7 The three-body exclusive cross-section in terms of three- and more-body inclusive cross-sections, as in (10.3.20).

Hence the complete set of inclusive cross-sections contains exactly the same information as the complete set of exclusive ones. Of course many-body inclusive cross-sections are too hard to measure and analyse, as are many-body exclusive cross-sections, and so in practice few-body inclusive cross-sections give complementary information to few-body exclusive ones.

The next step is to derive Mueller's theorem which allows us to make Regge predictions for these inclusive distributions.

10.4 Mueller's generalized optical theorem

In section 1.9, and graphically in fig. 1.6, we gave a derivation of the optical theorem relating the total cross-section $\sigma(12 \to X)$ to the imaginary part of the forward elastic amplitude $A^{el}(12 \to 12)$. Mueller (1970) has obtained a generalization of this result which provides the basis for Regge predictions of inclusive distributions. This is shown in fig. 10.8 and gives

$$f_1(p_3, s) = \frac{1}{2q_s\sqrt{s}} \operatorname{Disc}_X\{A(12\overline{3})\} \to \frac{1}{s} \operatorname{Disc}_X\{A(12\overline{3})\} \quad (10.4.1)$$

where $A(12\overline{3})$ is the amplitude for the process $1+2+\overline{3} \to 1'+2'+\overline{3}'$.

In the first step we use the completeness relation for $\Sigma|A(12 \to 3X)|^2$. The second step uses the crossing property of section 1.6 to analytically continue the amplitude from an outgoing 3 to an incoming $\overline{3}$; and then the unitarity relation (1.9.3) is used to relate this to the discontinuity

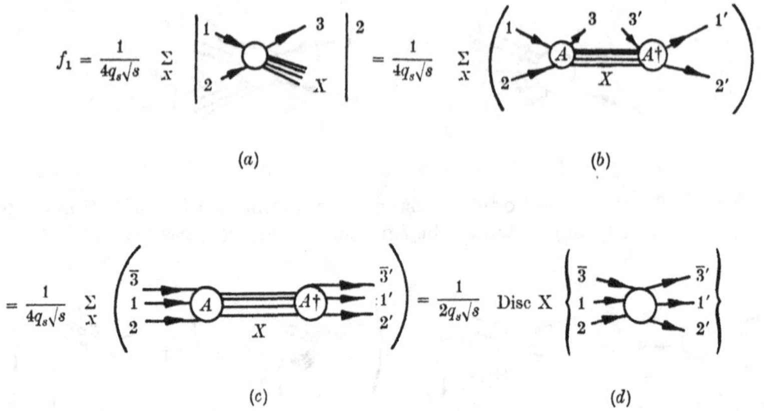

FIG. 10.8 Derivation of Mueller's theorem. (*a*) is the definition of f_1, where $4q_s\sqrt{s}$ is the flux factor (1.8.4). To get (*b*) we use the completeness relation, then (*c*) is obtained by crossing 3 and 3′, and (*d*) is the unitarity relation for the $3 \to 3$ amplitude. (The factor 2 arises from the definition (10.4.2).)

of the forward elastic scattering amplitude for $12\overline{3} \to 12\overline{3}$ in the variable

$$M_X^2 = M^2 = (p_1+p_2-p_3)^2 = s_{123}$$

Here

$$\mathrm{Disc}_X\{A(12\overline{3};\, s_{12\overline{3}}, s, t)\} \equiv \frac{1}{2\mathrm{i}}(A(12\overline{3}; s_{12\overline{3}}+\mathrm{i}\epsilon, s, t) - A(12\overline{3}; s_{12\overline{3}}-\mathrm{i}\epsilon, s, t)) \tag{10.4.2}$$

i.e. the discontinuity is taken across the $s_{12\overline{3}}$ branch cut but keeping on the same side of cuts in s and t. Since the initial state has to be identical to the final state we must have $t_{11'} = t_{22'} = t_{33'} = 0$ (where $t_{11'} \equiv (p_1 - p_1')^2$, etc.) just as we needed $t = 0$ in (1.9.6).

The obvious problem associated with this derivation, which is not present with fig. 1.6, is that we have had to make an analytic continuation in p_3 to the unphysical scattering amplitude $A(12\overline{3})$, and we cannot be sure whether the discontinuity will be affected by so doing. The discontinuity in (10.4.2) is across M^2 keeping on the same side of the cuts in $s \equiv s_{12}$, whereas clearly in fig. 10.8 (*b*) we are above the threshold cut in this variable in A but below it at $A^\dagger$. The independence of normal-threshold discontinuities mentioned in section 9.3 guarantees that the discontinuity in the one variable is unaffected by taking the discontinuity across the other, but anomalous thresholds etc. could spoil the result. However, the general consensus of informed opinion seems to be that this is unlikely (see Cahill and Stapp 1972, 1973, Polkinghorne 1972).

Even so this generalization is clearly more difficult to use than the ordinary optical theorem because in (1.9.6) the total cross-section for a given p_1 and p_2 is related to the elastic amplitude for the same physical values of p_1 and p_2, but (10.4.1) relates the inclusive distribution for $1+2\to 3+X$ to the (in any case unmeasurable) process $1+2+\overline{3}\to 1+2+\overline{3}$ in an unphysical region of $p_{\overline{3}}$. However, even if we cannot measure $A(12\overline{3})$ we can certainly write down a Regge parameterization for it, just as we used the Regge parameters of $A^{\text{el}}(12)$ to predict the behaviour of $\sigma_{12}^{\text{tot}}(s)$ in (6.8.4). It is this which makes inclusive reactions such a valuable testing ground for Regge theory, as we shall see in the following sections.

So far we have neglected the spins of the particles. More strictly we should average over the possible helicities of 1 and 2, and sum over those of 3, so (10.3.2) gives

$$f_1(p_3, s) = \frac{1}{4q_s\sqrt{s}(2\sigma_1+1)(2\sigma_2+1)} \sum_X \sum_{\mu_1\mu_2\mu_3} |A_{\mu_1\mu_2\mu_3}(12\to 3X)|^2$$

$$= \frac{1}{2q_s\sqrt{s}(2\sigma_1+1)(2\sigma_2+1)}$$

$$\times \sum_{\mu_1\mu_2\mu_3} \text{Disc}_X[\langle\mu_1\mu_2\mu_3|A(12\overline{3}\to 12\overline{3})|\mu_1\mu_2\mu_3\rangle] \quad (10.4.3)$$

through the optical theorem (10.4.1). So far, rather few polarization or density matrix measurements have been made, so we shall simply neglect spin below, which means strictly that at each Reggeon vertex we are averaging over the different possible helicities. But if for example 3 has spin $=\frac{1}{2}$, its polarization P_{3y} is given by (cf. (4.2.22))

$$P_{3y}f_1(p_3, s) = \frac{1}{4q_s\sqrt{s}(2\sigma_1+1)(2\sigma_2+1)}$$

$$\times \sum_{\mu_1\mu_2} \text{Im}\{\text{Disc}_X[\langle\mu_1\mu_2-|A(12\overline{3}\to 12\overline{3})|\mu_1\mu_2+\rangle]\} \quad (10.4.4)$$

where $\pm \equiv \pm\frac{1}{2}$. Alternatively inclusive density matrices can be defined like (4.2.10) and clearly they will tell us about the helicity dependence of the Reggeons' couplings to the particles (see Phillips, Ringland and Worden 1972, Goldstein and Owens 1975).

10.5 Fragmentation and the single-Regge limit

In the region where x_3 or $\tilde{y}_3 \approx 1$, i.e. particle 3 is almost at rest in the Lorentz frame of particle 1, we can regard 3 as a fragment of 1, as in fig. 10.3 (*a*). This is called the 'fragmentation region' of 1, and the inclusive distribution in this region is sometimes written as $f_1(1 \overset{2}{\to} 3)$, i.e. $1 \to 3$ under the impact of 2. Indeed 3 may well be the same particle as 1, since then no quantum numbers need be exchanged. The frequent occurrence of the beam particle in the final state, with high p_{L} but small p_{T}, and hence close to the forward direction, is called the 'leading particle effect'.

So in this region we are concerned with high energies, $s = s_{12} \to \infty$, but $t = t_{13}$ fixed and small. And from (10.2.14) fixed x_3 implies $M^2 \to \infty$ with fixed M^2/s. Now M^2 is the total energy for the $ab\bar{c}$ elastic scattering process in fig. 10.8 (*d*), and large M^2, small t suggests a single Regge pole exchange picture as in fig. 10.9 (*a*), so we write

$$f_1(1 \overset{2}{\to} 3; p_3, s) = \sum_i \gamma_i \left(t, \frac{M^2}{s}\right) \left(\frac{M^2}{s_0}\right)^{\alpha_i(0)-1} \tag{10.5.1}$$

where we have summed over all the Reggeons which can be exchanged. The argument of α_i is 0 because always $t_{22'} = 0$ for this forward three-body process. It should not be confused with $t \equiv t_{13}$ which gives the (fixed) invariant mass of the quasi-particle $(1\bar{3})$. From the similarity of fig. 9.1 (*a*) to fig. 9.1 (*b*) it is evident from (9.2.30) and (9.2.31) that the value of M^2/s determines the angle between the planes containing $1\bar{3}$ and $2\bar{3}$. In (10.5.1) s_0 is the usual scale factor, which experience with $2 \to 2$ scattering suggests should be $\approx 1\,\mathrm{GeV}^2$. We neglect the possibility of Regge cuts which would modify (10.5.1) by $\log(M^2)$ factors.

The validity of this formula depends on

$$s,\ M^2 \text{ and } u = (p_2 - p_3)^2 \gg m_3^2,\ t \text{ and } s_0.$$

So we need s large as usual, and $M^2/s = 1 - x$ finite; so M^2 must be large also, but not too large since $M^2 \to s$ implies $x \to 0$ (and from (10.2.12) u becomes small) so we would leave the fragmentation region. Obviously for $x_3, \tilde{y}_3 \approx -1$ we have the process $2 \overset{1}{\to} 3$, i.e. 3 is a fragment of 2, and the Regge picture is fig. 10.9 (*b*), so we can account for both fragmentation regions. But clearly it is necessary for these two regions to be well separated, which, as we shall show below (section 10.10) needs $Y = y_{\max} - y_{\min} > 4$, or $s > 60\,\mathrm{GeV}^2$, from (10.2.25).

In an elastic scattering process the dominant exchange should be the Pomeron, P, and if $\alpha_{\mathrm{P}}(0) \approx 1$ we have

$$f_1(p_3, s) \xrightarrow[s \to \infty]{} \gamma_{\mathrm{P}} \left(t, \frac{M^2}{s}\right) \tag{10.5.2}$$

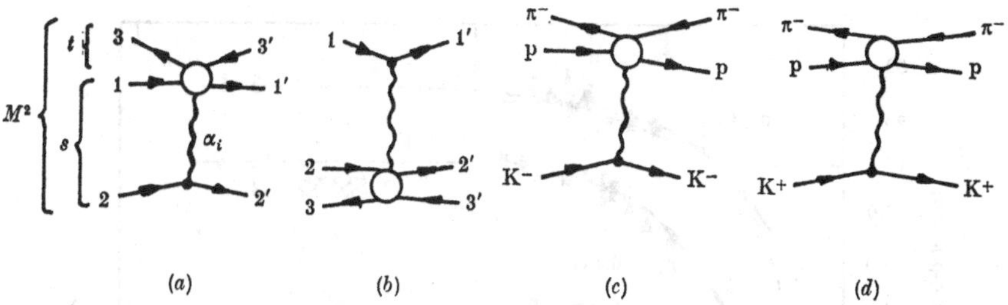

FIG. 10.9 (*a*) The Regge exchange diagram for $\text{Disc}_X\{A(12\bar{3} \to 12\bar{3})\}$ when 3 is in the fragmentation region of 1, i.e. $t_{13} = (p_1 - p_3)^2$ is small. (*b*) The corresponding diagram for the 2-fragmentation region. (*c*), (*d*) The Mueller–Regge diagrams for $p \overset{K^\pm}{\to} \pi^+$.

and so, like $\sigma_{12}^{\text{tot}}(s)$, $f_1(\mathbf{p}_3, s)$ should be approximately independent of s for $s \to \infty$, t, M^2/s fixed, i.e. f_1 should 'scale'.

A cross-section is said to 'scale' if its numerical value is independent of the energy units which are used. Thus $\sigma_{12}^{\text{tot}}(s)$ has values which when expressed as a function of s are independent of the units in which s is measured only if σ_{12}^{tot} is independent of s, which is approximately true at high energies. Likewise in (10.5.2) $f_1 = f_1(t, M^2/s)$ only, so though it depends on s at fixed M^2 (and vice versa) any change of the units in which they are both measured will not affect the ratio M^2/s, so f_1 scales. This is not true generally of (10.5.1) of course.

This scaling result agrees with earlier predictions of Amati *et al.* (1962*a*, *b*), Yang and co-workers (Benecke *et al.* 1969) and Feynman (1969). Yang's prediction was based on the hypothesis of limiting fragmentation, i.e. that the distribution of 3 in the rest frame of 1 should become independent of s for large s. This is because he viewed the scattering particles, 1 and 2, as two Lorentz-contracted disks passing through and exciting each other, followed by a break-up of each disk. Since σ^{el}, $\sigma^{\text{tot}} \to$ constants, the forces between the disks are obviously not changing as $s \to \infty$, and so the break-up of each disk should reach a limiting distribution (in its own rest frame) with no multiple scattering. Feynman's view, like that of Amati and co-workers, was based on the observation that in multi-peripheral and similar models (to be discussed in the next chapter) the distribution of 3 in x_3 and $\mathbf{p}_{3\text{T}}$ becomes independent of s as $s \to \infty$. This agrees with Yang's hypothesis and with the single-Regge limit (10.5.2) for $x_3^2 \gg 4\mu_3^2/s$, but extends the result down to $x = 0$ too, which we shall not deal with until the next section.

This scaling hypothesis works well in many processes. For example

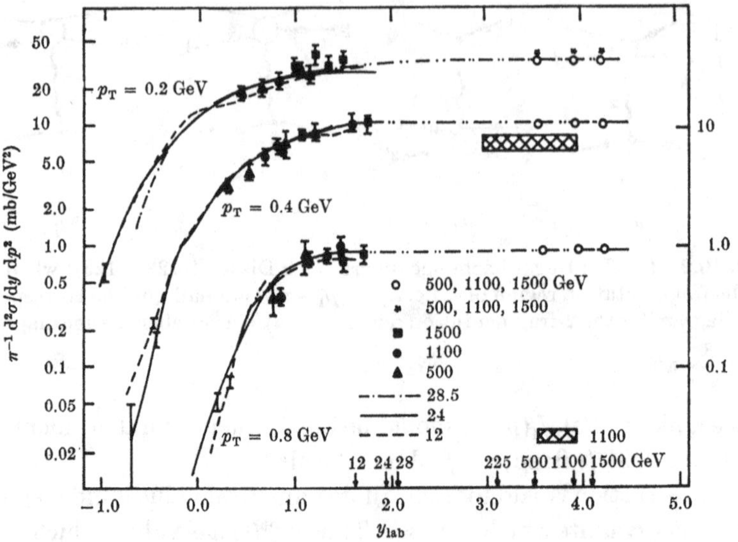

FIG. 10.10 Data for pp → π^+X in the fragmentation region, from Morrison (1972).

in pp → π^+X, shown in fig. 10.10, we see that f_1 is independent of s in the fragmentation region for $s = 50 \to 3000\,\text{GeV}^2$. Of course $\sigma_{pp}^{tot}(s)$ is not constant at high s, so that effectively $\alpha_P(0) > 1$, and it might be expected that ρ_1 (defined in (10.3.10)), rather than f_1, would be the better distribution in which to observe scaling, but generally the data are not sufficiently accurate to distinguish these possibilities.

The great advantage of this Regge view of scaling is that it also predicts how fast the scaling behaviour will be reached (Brower *et al.* 1973*a*, Chan *et al.* 1972*b*) provided we neglect cuts. The next term in the series (10.5.1) will be the normal Reggeons $R = f, \omega, \rho, A_2$ all with $\alpha_R(0) \approx 0.5$, and approximately equal couplings because of exchange degeneracy, so if they all add (as in $p \xrightarrow{K^-} \pi^-$, fig. 10.9 (*c*)) we get

$$f_1(p_3, s) = \gamma_P\left(t, \frac{M^2}{s}\right) + 4\gamma_R\left(t, \frac{M^2}{s}\right)\left(\frac{M^2}{s_0}\right)^{-\frac{1}{2}} \tag{10.5.3}$$

If now we replace 2 by $\overline{2}$ (i.e. K^- is replaced by K^+ as in fig. 10.9 (*d*)) the ω and ρ contributions change sign because they are odd under charge conjugation, giving

$$f_1(1 \xrightarrow{2} 3) - f_1(1 \xrightarrow{\overline{2}} 3) = 4\gamma_R\left(t, \frac{M^2}{s}\right)\left(\frac{M^2}{s_0}\right)^{-\frac{1}{2}} \tag{10.5.4}$$

and comparing for example $p \overset{\mathrm{K}^-}{\rightarrow} \pi^-$ and $p \overset{\mathrm{K}^+}{\rightarrow} \pi^-$ gives $\gamma_{\mathrm{R}}/\gamma_{\mathrm{P}} \approx \frac{1}{3}$, so we need $s \approx 2000\,\mathrm{GeV}^2$ for scaling to hold to within 10 per cent. However, we have found in two-body scattering that, because of duality, exchange degeneracy may result in a mutual cancellation of these secondary terms in exotic processes (see section 7.5), i.e. if $1+2$ have exotic quantum numbers, like $\mathrm{K}^+\mathrm{p}$, then scaling occurs precociously in $\sigma_{12}^{\mathrm{tot}}(s)$, at very low values of s. We can expect this also to be true in inclusive reactions, i.e. that scaling will occur if $(12\overline{3})$ has exotic quantum numbers so that no resonances occur in M^2. However, this is only really analogous to $2 \rightarrow 2$ scattering if $(1\overline{3})$ is not exotic as well, so that we can treat it as a quasi-particle. A more systematic investigation is therefore needed, which we postpone to section 10.6.

As long as poles rather than cuts dominate we can get extra constraints on the inclusive distributions from factorization. Thus we can express fig. 10.9 (a) in the form

$$f_1(1 \overset{2}{\rightarrow} 3; p_3, s) = \sum_i \gamma_{22}^i G_{1\overline{3}}^i \left(t, \frac{M^2}{s}\right) \left(\frac{M^2}{s_0}\right)^{\alpha_i(0)-1} \tag{10.5.5}$$

where $\gamma_{22}^i = \gamma_{22}^i(t_{22'} = 0)$ is the Reggeon coupling to $2\overline{2}$ and $G_{1\overline{3}}^i$ represents the upper vertex. For $s \rightarrow \infty$ this becomes, with $\alpha_{\mathrm{P}}(0) = 1$,

$$f_1(1 \overset{2}{\rightarrow} 3) \rightarrow \gamma_{22}^{\mathrm{P}} G_{1\overline{3}}^{\mathrm{P}} \left(t, \frac{M^2}{s}\right) \tag{10.5.6}$$

but we also have from (6.8.4)

$$\sigma_{12}^{\mathrm{tot}}(s) = \sum_i \gamma_{11}^i \gamma_{22}^i s^{\alpha_i(0)-1} \rightarrow \gamma_{11}^{\mathrm{P}} \gamma_{22}^{\mathrm{P}} \tag{10.5.7}$$

so from (10.3.10)

$$\rho_1(1 \overset{2}{\rightarrow} 3) \rightarrow \frac{G_{1\overline{3}}^{\mathrm{P}}(t, M^2/s)}{\gamma_{11}^{\mathrm{P}}} \tag{10.5.8}$$

which is independent of particle 2, and so $\rho_1(1 \overset{a}{\rightarrow} 3)$ should be independent of a for $s \rightarrow \infty$. This can be tested at finite energies only for exotic $(12\overline{3})$ processes which scale early, such as $\mathrm{p} \overset{\mathrm{K}^+}{\rightarrow} \pi^-$, $\mathrm{p} \overset{\mathrm{p}}{\rightarrow} \pi^-$, $\mathrm{p} \overset{\pi^+}{\rightarrow} \pi^-$, and it is found (see fig. 10.11) that ρ_1 is the same for all three.

The secondary contributions are also related by the exchange degeneracy of the couplings (Miettinen 1972, Chan *et al.* 1972*a*). Thus

$$f_1(\mathrm{p} \overset{\pi^+}{\rightarrow} \pi^-) = \gamma_{\pi^+\pi^+}^{\mathrm{P}} G_{\mathrm{p}\pi^+}^{\mathrm{P}} + \gamma_{\pi^+\pi^+}^{\mathrm{f}} G_{\mathrm{p}\pi^+}^{\mathrm{f}} \left(\frac{M^2}{s_0}\right)^{-\frac{1}{2}} - \gamma_{\pi^+\pi^+}^{\rho} G_{\mathrm{p}\pi^+}^{\rho} \left(\frac{M^2}{s_0}\right)^{-\frac{1}{2}} \tag{10.5.9}$$

where the negative sign of the last term is due to the fact that $\pi^+\pi^+\mathrm{p}$ is

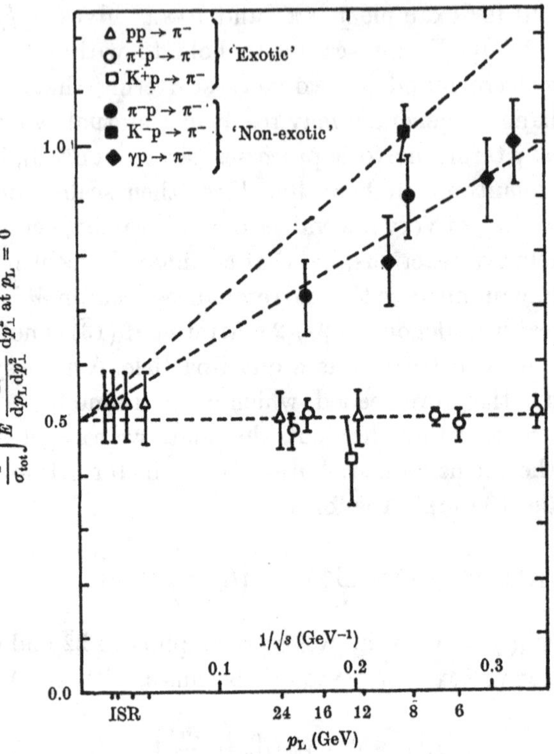

Fig. 10.11 The energy dependence of ρ_1 (equation (10.5.8)) integrated over p^2_{3T} for a variety of processes, showing that it is independent of particle 2, at least for exotic channels, from Miettinen (1973).

exotic (and π^+p is not) so that these secondary f and ρ terms should cancel. But from $\pi^+\pi^+ \to \pi^+\pi^+$ we know that $\gamma^{f}_{\pi^+\pi^+} = \gamma^{\rho}_{\pi^+\pi^+}$ (see (7.5.2)) so we must also have

$$G^{\rho}_{p\pi^+} = G^{f}_{p\pi^+} \tag{10.5.10}$$

Similarly on considering $p \xrightarrow{K^+} \pi^-$ and $p \xrightarrow{K^-} \pi^-$ we deduce that

$$G^{f}_{p\pi^+} = G^{\omega}_{p\pi^+} = G^{A_2}_{p\pi^+}$$

and that all γ^{R}_{KK} are equal and hence

$$f_1(p \xrightarrow{K^-} \pi^-) - f(p \xrightarrow{K^+} \pi^-) = 4\gamma^{R}_{KK}\, G^{R}_{p\pi^+}\left(\frac{M^2}{s_0}\right)^{-\frac{1}{2}} \tag{10.5.11}$$

And for any similar fragmentation we can write

$$f_1(p \xrightarrow{a} \pi^-) = \gamma^{P}_{aa} G^{P}_{p\pi^+} + \sum_{R} \gamma^{R}_{aa}\, G^{R}_{p\pi^+}\left(\frac{M^2}{s_0}\right)^{-\frac{1}{2}} \tag{10.5.12}$$

so, since the behaviour of σ_{ap}^{tot} allows us to deduce γ_{aa}^{R}, we can predict all $f_1(p \xrightarrow{a} \pi^-)$. This is found to work well for $a = \gamma, K^-$ or π^- for example.

Factorization is much more useful in inclusive reactions than in two-body processes because the target is effectively $(1\bar{3})$. Thus even if the actual target (particle 1) is restricted to p or (n) we can still change both vertices in fig. 10.9 (*a*) by changing the beam particle (2) and particle 3.

It is rather remarkable that these factorization tests should work so well, though of course the data are not very accurate in general. It may partly be explained by the fact that we are restricted to $t_{22'} = 0$ where the poles are more important, or it may be the result of pole-enhancement of the cuts (see section 8.7 *g*).

10.6 The central region and the double-Regge limit

We consider next the region $x \approx 0$ where p_{3L} is small. As $s \to \infty$ we have, from (10.2.10) and (10.2.11),

$$t \to -(\sqrt{s})(E_3 - p_{3L}), \quad u \to -(\sqrt{s})(E_3 + p_{3L}) \tag{10.6.1}$$

so that $|t|, |u| \to \infty$ as $s \to \infty$, but

$$\frac{ut}{s} \to (E_3 - p_{3L})(E_3 + p_{3L}) = \mu_3^2 \tag{10.6.2}$$

is fixed. So like η_{12} in (9.2.31), μ_3^2 represents the angle between the plane containing 1 and 3 and the plane containing 2 and 3. Since μ_3^2 is generally small, $\leqslant 1\,\text{GeV}^2$, it requires a very large s to get large $|t|$ and $|u|$, particularly if m_3 is small.

The double-Regge exchange model for this region is shown in fig. 10.21 and gives

$$f_1(p_3, s) = \sum_{i,j} \frac{1}{s} \gamma_{ij}(p_{3T}) \left|\frac{t}{s_0}\right|^{\alpha_i(0)} \left|\frac{u}{s_0}\right|^{\alpha_j(0)} \left(\frac{s_0}{\mu_3^2}\right) \tag{10.6.3}$$

where γ_{ij} represents the product of the three vertices, and the extra factor (s_0/μ_3^2) is arbitrary but convenient, because using (10.6.2) we then get for $s \to \infty$

$$f_1(p_3, s) \to \sum_{i,j} \gamma_{ij}(\mu_3^2) \left|\frac{t}{s_0}\right|^{\alpha_i(0)-1} \left|\frac{u}{s_0}\right|^{\alpha_j(0)-1} \tag{10.6.4}$$

If P dominates asymptotically this gives the Feynman scaling result (fig. 10.13 (*a*))

$$f_1(p_3, s) \to \gamma_{PP}(\mu_3^2) \tag{10.6.5}$$

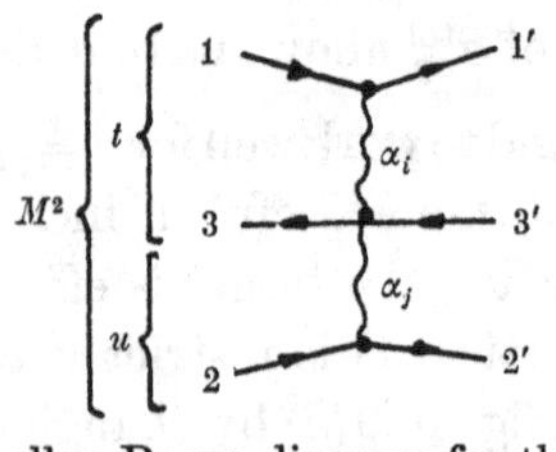

FIG. 10.12 Mueller–Regge diagram for the central region (equation (10.6.3)).

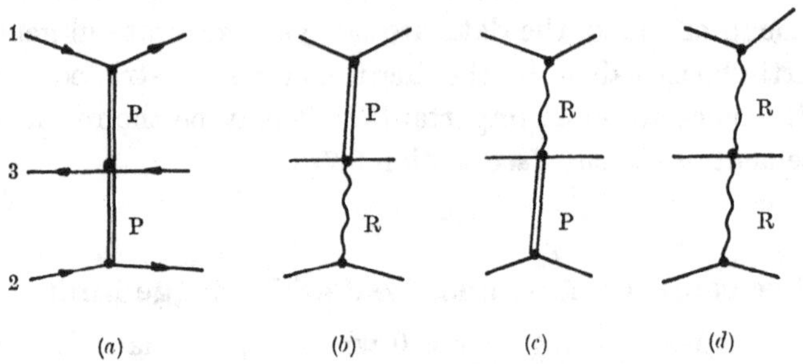

FIG. 10.13 Central region Mueller–Regge diagrams; (a) gives scaling asymptotically while the others give corrections to the scaling behaviour from R exchange.

independent of s, t and u (see fig. 10.14). Using factorization this can be rewritten as

$$f_1(p_3, s) \to \gamma_{11}^{\mathrm{P}} \gamma_{33}^{\mathrm{PP}}(\mu_3^2)\, \gamma_{22}^{\mathrm{P}} \tag{10.6.6}$$

or, using (10.3.10) and (10.5.7),

$$\rho_1(p_3, s) \to \gamma_{33}^{\mathrm{PP}}(\mu_3^2) \tag{10.6.7}$$

which is independent of particles 1 and 2. Also, since from (10.2.28) $y_3 \to \frac{1}{2}\log(u/t)$, this result means that $f_1(p_{3\mathrm{T}}, y_3, s)$ is independent of y_3 and s for small y_3, i.e. $\mathrm{d}\sigma/\mathrm{d}y_3$ at fixed $p_{3\mathrm{T}}$ will have a central plateau, as shown in fig. 10.4 (b). But for this to emerge from between the two fragmentation regions (each of width $\Delta y \approx 2$ – see section 10.10) we need $Y_3 \equiv y_{3\,\max} - y_{3\,\min} > 4$, so with $\mu_3^2 \approx 1\,\mathrm{GeV}^2$ this means $s > 60\,\mathrm{GeV}^2$.

The secondary Reggeons $\mathrm{R}(= \mathrm{f}, \omega, \rho, \mathrm{A}_2)$ with $\alpha_{\mathrm{R}}(0) \approx 0.5$ give corrections to scaling

$$\begin{aligned} f_1(p_3, s) &= \gamma_{\mathrm{PP}}(\mu_3^2) + \gamma_{\mathrm{PR}}(\mu_3^2)\left|\frac{t}{s_0}\right|^{-\frac{1}{2}} \\ &\quad + \gamma_{\mathrm{RP}}(\mu_3^2)\left|\frac{u}{s_0}\right|^{-\frac{1}{2}} + \gamma_{\mathrm{RR}}(\mu_3^2)\left|\frac{t}{s_0}\right|^{-\frac{1}{2}}\left|\frac{u}{s_0}\right|^{-\frac{1}{2}} \\ &\sim \gamma_{\mathrm{PP}}(\mu_3^2) + \gamma_{\mathrm{PR}}(\mu_3^2)\left(\frac{s}{s_0}\right)^{-\frac{1}{4}} + O(s^{-\frac{1}{2}}) \end{aligned} \tag{10.6.8}$$

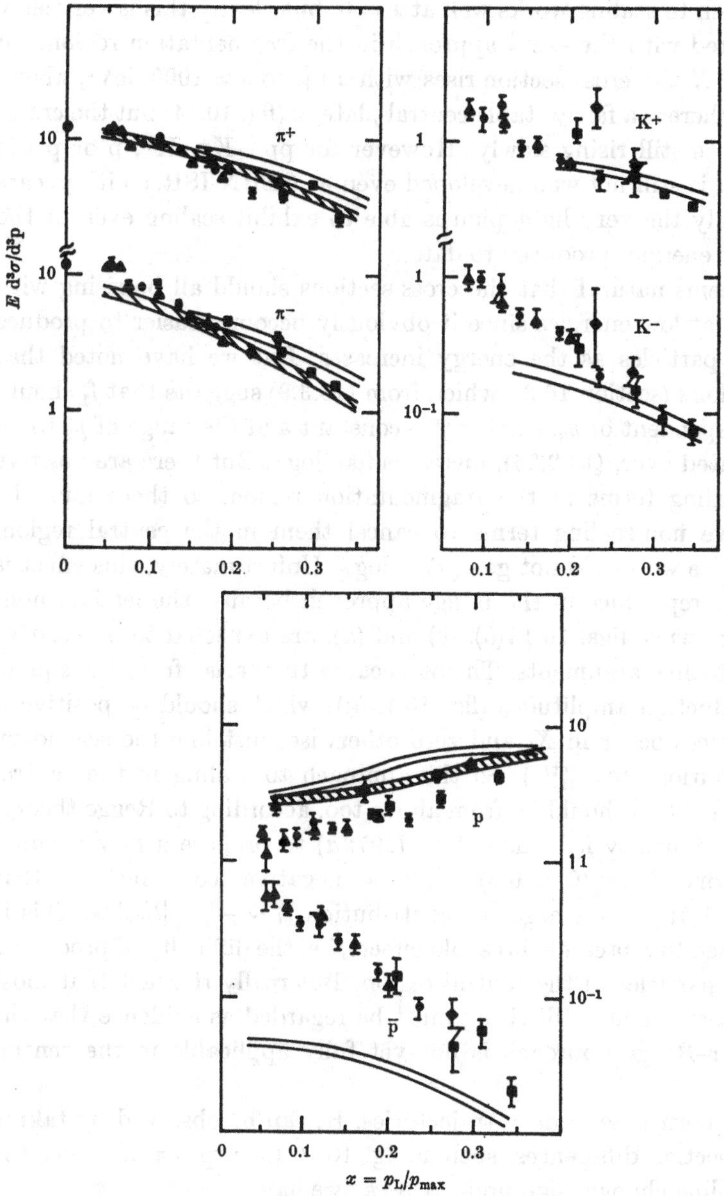

Fig. 10.14 Data for $pp \to \pi^+X$, $K^{\pm}X$, pX and $\bar{p}X$ in the central region showing the approximate scaling behaviour for $\pi^{\pm}$ for high energies, from Jacob (1972).

since from (10.6.1) $t, u \sim \sqrt{s}$. According to Ferbel (1972) this $\sim s^{-\frac{1}{4}}$ approach to scaling works well at $x = 0$, but clearly this is very slow compared with the $\rightarrow s^{-\frac{1}{2}}$ approach in the fragmentation region. In $\text{pp} \rightarrow \pi^{\pm} X$ the cross-section rises with s up to $s \approx 1000\,\text{GeV}^2$, above which there is a fairly stable central plateau (fig. 10.14) but the cross-section is still rising slowly. However for $\text{pp} \rightarrow \text{K}^+$, K^-, p or $\bar{\text{p}}$ the plateau is still not well developed even at CERN-ISR, so it appears that only the very light pion is able to exhibit scaling even at the highest energies produced to date.

It seems natural that the cross-sections should all be rising with energy at low energies since it obviously becomes easier to produce heavy particles as the energy increases. But we have noted that $\langle n \rangle \sim \log s$ (section 10.3), which from (10.3.9) suggests that f_1 should be independent of y_3, since $\sigma_{12}^{\text{tot}} \rightarrow$ constant and the range of y_3 to be integrated over, (10.2.25), increases like $\log s$. But there are positive non-scaling terms in the fragmentation region, so there must be negative non-scaling terms to cancel them in the central region, otherwise we would not get $\langle n \rangle \sim \log s$. Unfortunately, this effect is hard to reproduce in the Regge approach because the leading non-scaling terms, figs. 10.13 (*b*), (*c*) and (*d*), are expected to be positive from duality arguments. This is because they arise from the square of production amplitudes (fig. 10.15 (*a*)) which should be positive if resonances occur in X, and zero otherwise, just like the secondary contributions to $\sigma_{12}^{\text{tot}}(s)$. So the approach to scaling in the central region (10.6.8) should be from above too, according to Regge theory.

This difficulty led Chan *et al.* (1972*a*) to propose a new vacuum trajectory Q ($\alpha_{\text{Q}}(0) \approx 0.5$) with a negative coupling, so that fig. 10.5 (*b*) gives a negative contribution $f_1 \sim -\gamma_{\text{QP}} |t/s_0|^{-\frac{1}{2}}$. This is supposed to represent threshold effects, i.e. the difficulty of producing heavy particles in the central region. But really the fact that most cross-sections are still rising must be regarded as evidence that the Mueller–Regge approach is not yet fully applicable in the central region.

The normal secondary trajectories, R, can be observed by taking cross-section differences, such as fig. 10.16 for $\pi^+\text{p} \rightarrow \pi^+ X$. Since the ρ coupling changes sign under $\pi^+ \leftrightarrow \pi^-$ we have

$$f(\pi^+\text{p} \rightarrow \pi^+ X) - f(\pi^+\text{p} \rightarrow \pi^- X) = 2\gamma_{\text{RP}} \left|\frac{s}{s_0}\right|^{-\frac{1}{2}} \equiv \Delta(\pi^+\text{p} \rightarrow \pi^+ X) \tag{10.6.9}$$

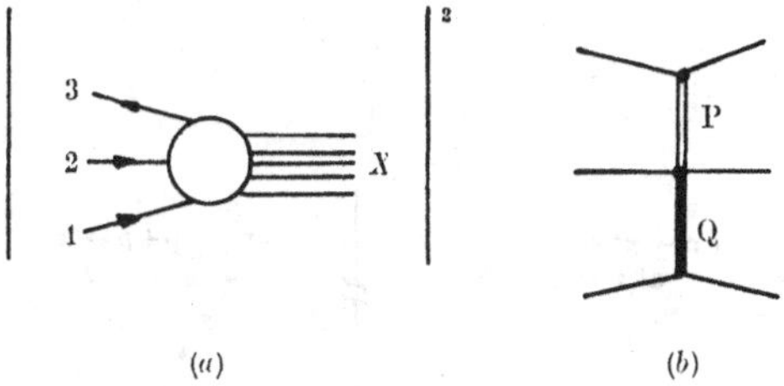

FIG. 10.15 (*a*) The (unphysical) production amplitude whose square contributes to the inclusive distribution. (*b*) The Q exchange which has been invented to parameterize threshold effects.

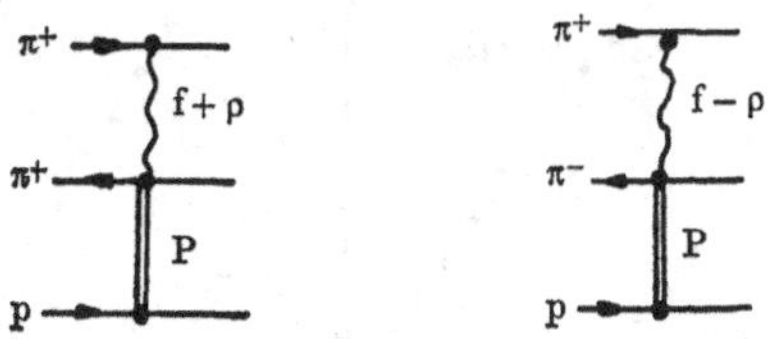

FIG. 10.16 Signs of the R contributions to fig. 10.13(*c*) for $\pi^+p \to \pi^\pm X$.

Then using factorization to write (in a notation obvious from fig. 10.16)

$$\gamma_{RP}(\pi^+p \to \pi^-X) = \gamma^{R}_{\pi^+\pi^+}\gamma^{RP}_{\pi^-\pi^-}\gamma^{P}_{pp}$$

etc., we must have $\gamma^{\rho P}_{\pi^-\pi^-} = -\gamma^{fP}_{\pi^-\pi^-}$ from duality. A generalization allows one to deduce, from SU(3) and exchange degeneracy for the couplings, relations such as

$$\frac{\Delta(\pi^\pm p \to \pi^+X)}{\Delta(pp \to \pi^+X)} = \frac{1}{2}\left(\frac{\gamma^{P}_{\pi\pi}}{\gamma^{P}_{pp}} \pm \frac{\gamma^{\rho}_{\pi\pi}}{\gamma^{\rho}_{pp}}\right), \quad \frac{\Delta(K^\pm p \to \pi^+X)}{\Delta(pp \to \pi^+X)} = \frac{1}{2}\left(\frac{\gamma^{P}_{KK}}{\gamma^{P}_{pp}} \pm \frac{\gamma^{\rho}_{KK}}{\gamma^{\rho}_{pp}}\right) \tag{10.6.10}$$

where we have defined $\Delta(12 \to 3X) \equiv f(12 \to 3X) - f(12 \to \bar{3}X)$. These work well even at quite low energies (Inami 1974) which suggests that extracting the kinematic Q effect in $I = 0$ makes sense, even if one cannot take it seriously as a Regge pole. So it must be the $I = 0$ exchange part which has not yet developed its asymptotic behaviour.

Since in the central region f_1 depends on $\gamma_{ij}(\mu_3^2)$ in (10.6.4) (where μ_3 is defined in (10.2.2)) and since experimentally it is found that $f_1 \sim e^{-4p_{3T}^2}$ for small p_T (see fig. 10.17), we can expect

$$\gamma_{ij} \sim e^{-4\mu_3^2} \tag{10.6.11}$$

So the coupling should be strongly dependent on the mass of the particle which is produced. Substituting m_3^2 for μ_3^2 gives the ratio of π:K:p($\bar{p}$) production as 80:15:5 per cent which is at least qualitatively correct.

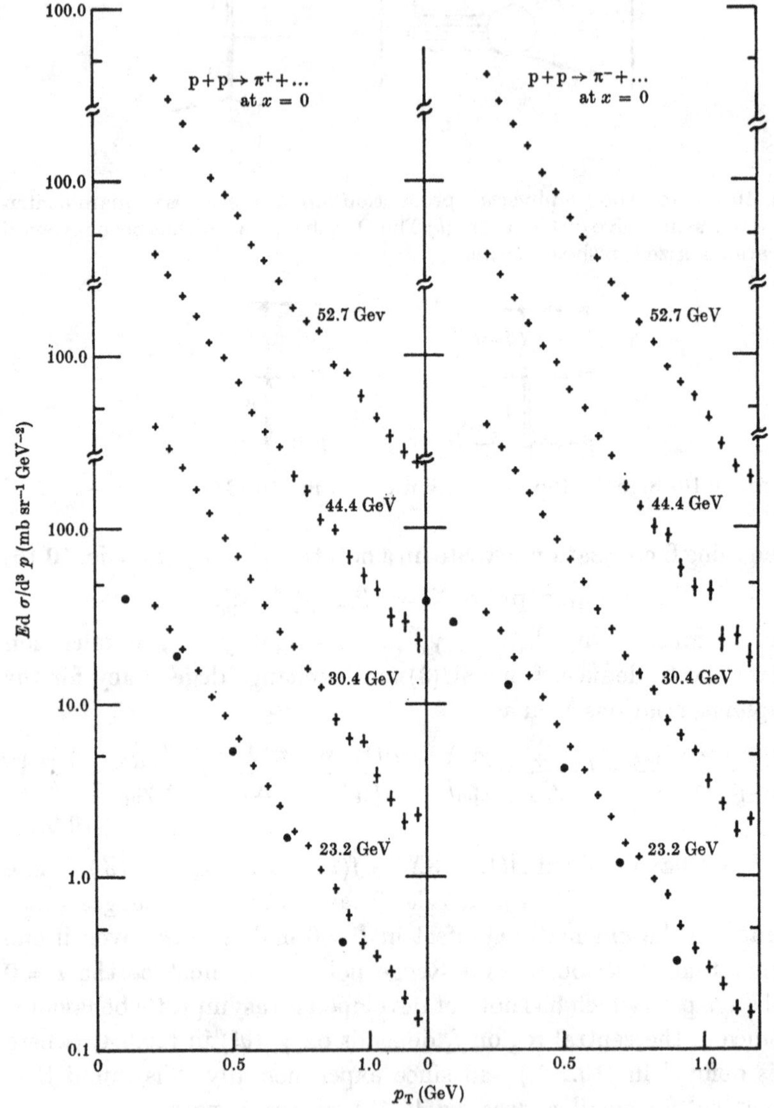

FIG. 10.17 The p_T dependence of f_1 for $pp \to \pi^{\pm} X$, showing the sharp cut-off in p_T, from Jacob (1972).

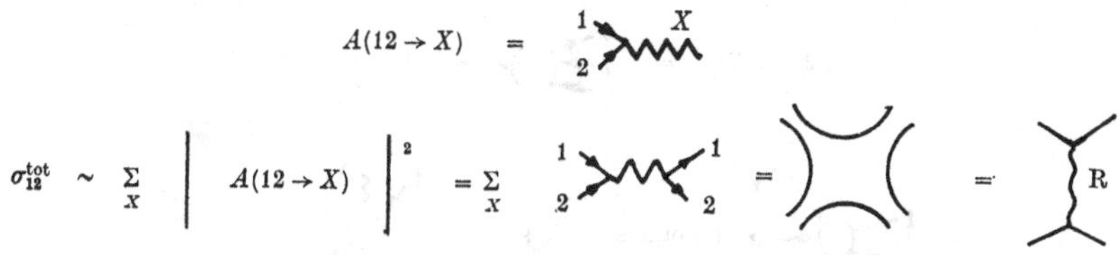

FIG. 10.18 Duality diagram for the R contribution to σ_{12}^{tot} using the optical theorem.

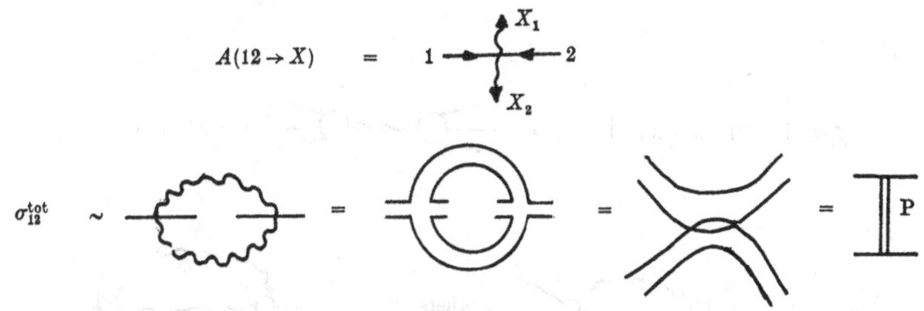

FIG. 10.19 Duality diagram for the P contribution to σ_{12}^{tot} using the optical theorem. Note that no quarks pass down the diagram so the t channel has vacuum quantum numbers; cf. fig. 7.12.

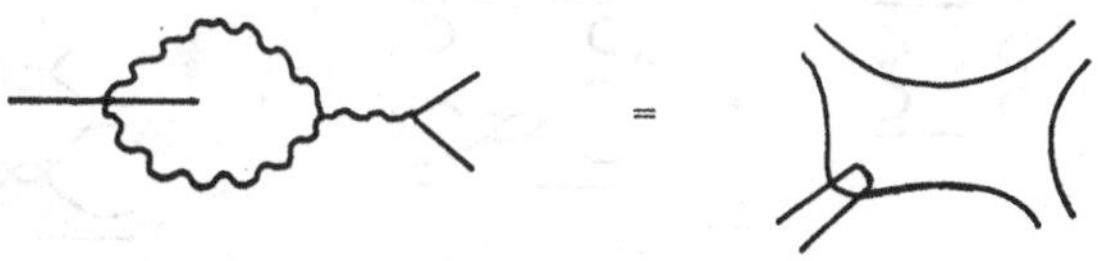

FIG. 10.20 A cross term between figs. 10.18 and 10.19 which is excluded by the rules for duality diagrams.

10.7 Scaling and duality

Total cross-sections such as those for K^+p and pp scale precociously, i.e. are essentially independent of s for rather low s, because these are exotic channels, while the non-exotic K^-p, $\bar{p}p$ fall rapidly at low energies (fig. 6.4). This can readily be explained in terms of duality diagrams as in fig. 10.18 in which the total cross-section for $12 \to X$ is related to the imaginary part of the Regge exchange in the elastic scattering amplitude through the optical theorem. This diagram can be drawn with X as a sum of resonances only if 12 is not exotic, and it gives the R corrections to the scaling P term. Another possible diagram is fig. 10.19 which produces the P as shown, and occurs whether or not 12 is an exotic channel. Note, however, that cross terms like fig. 10.20, which might also be expected, are forbidden by the

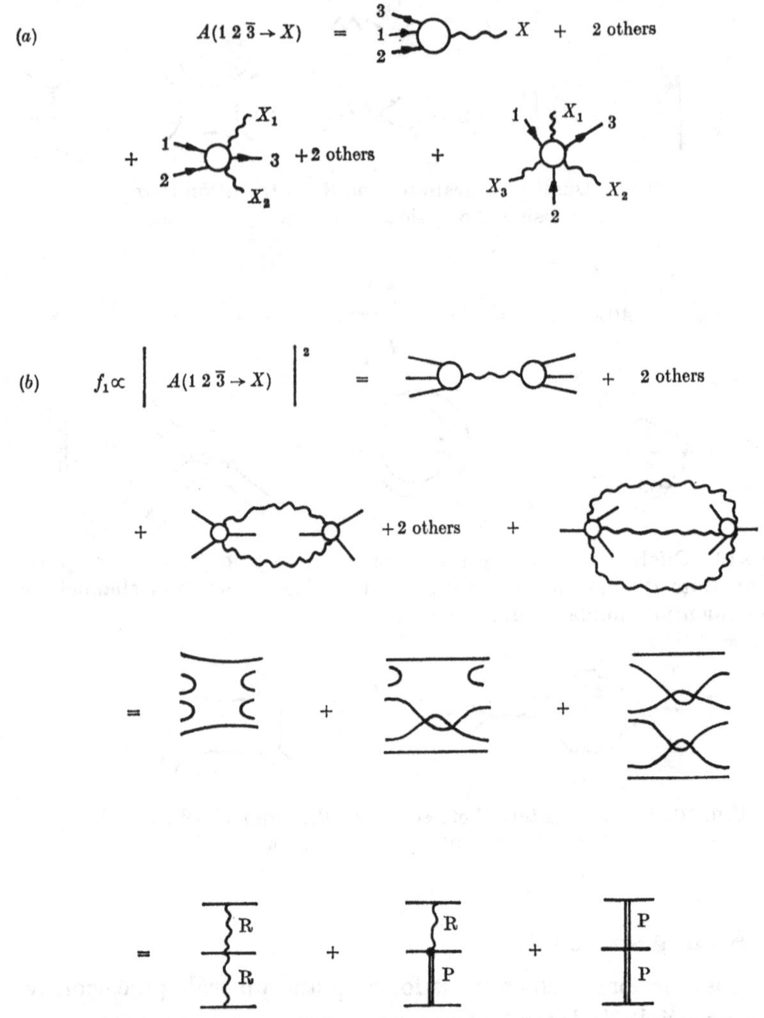

FIG. 10.21 (*a*) The seven terms for $A(12\bar{3} \to X)$. In each case the 'others' are just cyclically inequivalent permutations of the particles. (*b*) The seven corresponding contributions to the inclusive distribution f_1, again excluding cross terms. They are redrawn below as duality diagrams, and as Reggeon and Pomeron exchanges.

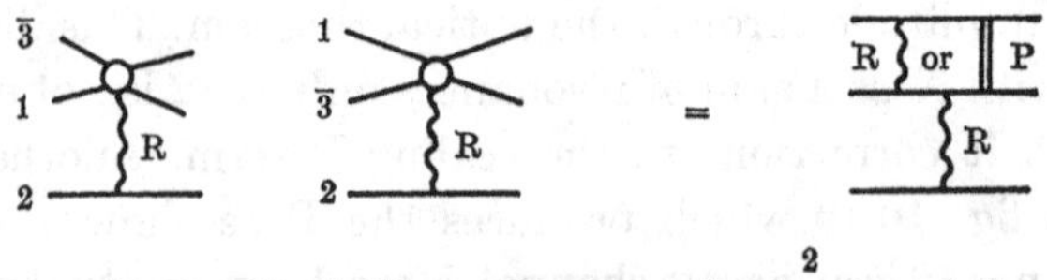

FIG. 10.22 Single Regge diagrams for $1 \to 3$, and the corresponding duality diagrams.

rules for drawing duality diagrams (section (7.5)). A quark loop cannot begin and end on the same particle. So in accord with the Harari–Freund conjecture there are just two terms in σ_{12}^{tot} (see (7.3.1)).

Correspondingly, according to Veneziano (1972) there are seven terms in $12\overline{3} \to X$, shown in fig. 10.21 (*a*), and so if we neglect all cross terms the contributions to f_1 through the generalized optical theorem (10.4.1) are as shown in fig. 10.21 (*b*).

Strictly we get precocious scaling if the last term only is present, which requires that 12, $2\overline{3}$ and $1\overline{3}$ are all exotic. But in the fragmentation region of particle 1 only figs. 10.22 matter. These cannot occur if 12 and $\overline{3}2$ are exotic giving early scaling in this region. A more complete discussion has been given by Einhorn *et al.* (1972*b*) and Tye and Veneziano (1973). Table 10.1 shows a comparison of exoticity and scaling in current data, from which it will be seen that if $\overline{3}$ is a $\pi^{\pm}$ the criterion $12\overline{3}$ exotic seems to work, even if $1\overline{3}$ is exotic and so cannot form a quasi-particle, but on the other hand $pp \to \bar{p}X$ seems to violate all the rules, presumably because for such a heavy particle very high energies will be needed before there is sufficiently copious $\bar{p}p$ production for scaling to develop. It is the lightness of the pion which makes precocious scaling possible.

The fact that duality exchange-degeneracy relations between the Reggeon couplings seem to hold at quite low s in both the fragmentation and central regions suggests that it is the incomplete development of the P term which causes the difficulty.

10.8 Triple-Regge behaviour

In the fragmentation region $1 \to 3$, with a fixed M^2 and $s \to \infty$ we would expect Regge behaviour as shown in fig. 10.23 (*a*)

$$A(12 \to 3X) \underset{s \to \infty}{\longrightarrow} \sum_i \gamma_{13}^i(t)\, \gamma_{2M}^i(t)\, \xi_i(t)\, P_{\alpha_i(t)}(\cos\theta_t) \qquad (10.8.1)$$

where
$$\xi_i(t) = \frac{e^{-i\pi\alpha_i(t)} + \mathscr{S}_i}{\sin \pi\alpha_i(t)} \qquad (10.8.2)$$

is the signature factor and $\gamma_{2M}^i(t)$ is the lower vertex of fig. 23 (*a*). If we insert (10.8.1) into the optical theorem (10.4.1), as in fig. 10.23 (*b*), we get

$$\begin{aligned} f_1(p_3, s) = \frac{1}{2q_s\sqrt{s}} \operatorname{Disc}_{M^2}\{A(12\overline{3} \to 12\overline{3})\} \to \frac{1}{s} \sum_{i,j} \gamma_{13}^i(t)\, \gamma_{13}^{j*}(t) \\ \times \xi_i(t) \xi_j^*(t) (\cos\theta_t)^{\alpha_i(t)+\alpha_j(t)} \\ \times \operatorname{Disc}_{M^2}\{A(i2 \to j2; t, M^2, t_{22'} = 0)\} \end{aligned} \qquad (10.8.3)$$

Table 10.1 *Scaling behaviour and exoticity*

		Exotic?				Scale?		
1	3	$12\bar{3}$	12	$2\bar{3}$	$1\bar{3}$	p fragmentation region	Central	1 fragmentation region
π^+	π^+	No	No	No	No	↓	↓	↓
	π^-	Yes	No	No	Yes	—	↑	↓
	K^0	Yes	No	No	No	—	↑	↓
	p	No	No	No	No	↓	↓	↓
π^-	π^+	Yes	No	No	Yes		↑	
	π^-	No	No	No	No	↓	—	↓
	K^0	No	No	No	No	↑	↑	↑
K^+	π^+	No	Yes	No	No	—	↑	
	π^-	Yes	Yes	No	Yes	—	↑	
	K^0	Yes	Yes	No	Yes	↓	—	↓
K^-	π^+	No	No	No	Yes	↓	↑	↓
	π^-	No	No	No	No	—	↑	↑
	K^0	No	No	No	No	↓	↑	↓
p	π^+	Yes	Yes	No	No	↓	↑	↓
	π^-	Yes	Yes	No	No	↓	↑	↓
	K^0	Yes	Yes	No	No	↑	↑	↑
	p	No	Yes	No	No	↓	↓	↓
	$\bar{p}$	Yes	Yes	Yes	Yes		↑	

For processes of the form $1+p \to 3+X$ we show the tendency of the inclusive distribution in the fragmentation region of the target p, the central region, and the fragmentation region of the beam (particle 1); ↑ means that the cross-section is increasing with energy, ↓ that it is decreasing, and—that an approximately constant scaling behaviour is found. A blank means that suitable data is not available. (Based on Zalewsky 1974.)

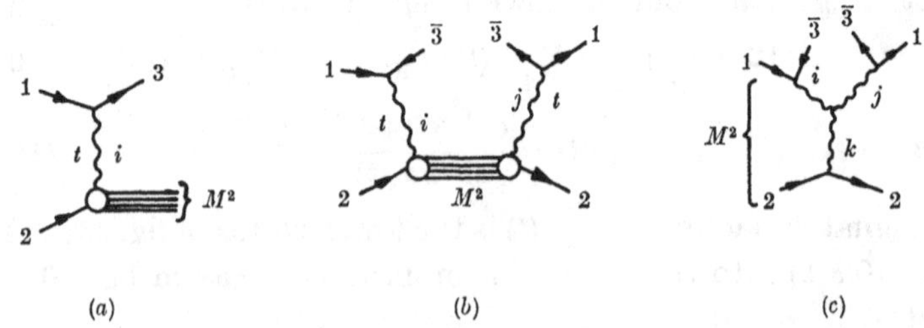

FIG. 10.23 (*a*) Single Reggeon i exchanged in $1+2 \to 3+X$ when 3 is in the fragmentation region of 1, for large s. (*b*) The result of inserting (*a*) into the optical theorem, fig. 10.8. (*c*) The triple-Regge approximation to (*b*) appropriate at large M^2. In (10.8.1) *et seq.* the Reggeon-particle couplings are denoted by γ^i_{13} etc. and the triple-Reggeon coupling in (c) is denoted by $\gamma^{ij,k}$.

where $A(i2 \to j2)$ is the Reggeon-particle scattering amplitude in the bottom half of the figure. Now if $s \gg M^2 \gg t \gg m^2_{1,2,3}$, from (1.7.19)

$$\cos\theta_t \to \frac{s - M^2/2}{q_{t13}\,q_{t2M}} \xrightarrow[s \gg M^2]{} \frac{s}{2q_{t13}\,q_{t2M}} \xrightarrow[M^2 \gg t]{} \frac{s}{M^2} \tag{10.8.4}$$

And for $M^2 \to \infty$ we can put (see fig. 10.23 (c))

$$\mathrm{Disc}_{M^2}\{A(i2 \to j2)\} = \sum_k \gamma^k_{22}(0)\,\gamma^{ij,k}(t,0)\left(\frac{M^2}{s_0}\right)^{\alpha_k(0)} \tag{10.8.5}$$

giving (from (10.3.4))

$$f_1(p_3, s) = 16\pi^2 s \frac{\mathrm{d}^2\sigma}{\mathrm{d}t\,\mathrm{d}M^2} = \frac{1}{s}\sum_{i,j,k} \gamma^i_{13}(t)\,\gamma^{j*}_{13}(t)$$
$$\times\, \xi_i(t)\,\xi^*_j(t)\left(\frac{s}{M^2}\right)^{\alpha_i(t)+\alpha_j(t)} \gamma^k_{22}(0)\,\gamma^{ij,k}(t,0)\left(\frac{M^2}{s_0}\right)^{\alpha_k(0)}$$
$$\equiv \frac{1}{s}\sum_{i,j,k} G^{ij,k}_{13,2}(t)\left(\frac{s}{s_0}\right)^{\alpha_i(t)+\alpha_j(t)}\left(\frac{M^2}{s_0}\right)^{\alpha_k(0)-\alpha_i(t)-\alpha_j(t)} \tag{10.8.6}$$

Note that the Reggeons i, j have mass $t \equiv (p_1 - p_3)^2$, but k has mass $t_{22'} = 0$ since the optical theorem is for forward scattering. All the couplings and signature factors have been incorporated into $G^{ij,k}_{13,2}(t)$.

This expression is valid in the so-called 'triple-Regge' limit when M^2 and $s/M^2 \to \infty$. However, this is really a misnomer because, as we noted in section 10.5, s/M^2 gives the angle between the planes containing $1\bar{3}$ and $2\bar{3}$, and letting this angle tend to infinity is really a helicity limit in the language of section 9.3. However, the leading helicity pole occurs at $\lambda = \alpha$ (see (9.3.18)), so the fact that we are taking a mixed Regge–helicity pole limit in (10.8.6) does not make any difference to the formula to leading order in M^2 (see de Tar and Weis 1971).

From (10.2.14) we see that $s/M^2 \to \infty$ implies that $x_3 \to 1$, $y_3 \to y_{3\,\max}$, so this triple-Regge region is only a small part of the x_3 or y_3 plot near the kinematical limit. Clearly (10.8.6) can only be applied for large s since if we suppose that we need $M^2/s_0 > 10$, and $s/M^2 > 10$ for the Regge expansion to be valid, with $s_0 = 1\,\mathrm{GeV}^2$ this means $s > 100\,\mathrm{GeV}^2$.

Using (10.2.14), (10.8.6) can be rewritten

$$f_1(p_3, s) = \frac{1}{s}\sum_{i,j,k} G^{ij,k}_{13,2}(t)\,(1-x)^{\alpha_k(0)-\alpha_i(t)-\alpha_j(t)}\left(\frac{s}{s_0}\right)^{\alpha_k(0)} \tag{10.8.7}$$

and if M^2 is sufficiently large that only P is needed in the sum over k, and if the leading i and j trajectory with the quantum numbers of $1\bar{3}$ is denoted by i, then

$$f_1(p_3, s) \to \frac{1}{s} |\gamma^i_{1\bar{3}}(t)|^2 |\xi_i(t)|^2 \gamma^{\mathrm{P}}_{2\bar{2}}(0) \gamma^{ii,\mathrm{P}}(t,0) \left(\frac{s}{M^2}\right)^{2\alpha_i(t)-1} \left(\frac{s}{s_0}\right)$$

$$\sim \left(\frac{s}{M^2}\right)^{2\alpha_i(t)-1} = (1-x)^{1-2\alpha_i(t)} \tag{10.8.8}$$

so f_1 is a function of x, or M^2/s, only, which again corresponds to Feynman scaling. And by looking at the s variation at fixed M^2, or the M^2 variation at fixed s, for different values of t, one can determine $\alpha_i(t)$ directly.

Rather comprehensive sets of fits of (10.8.6) to the high energy data have been made by Roy and Roberts (1974) and Field and Fox (1974). In $\mathrm{pp} \to \mathrm{p}X$, since $1\bar{3} = \bar{\mathrm{p}}\mathrm{p}$ has the quantum numbers of the vacuum the leading term will be the triple-Pomeron term

$$f_1^{\mathrm{PP,P}}(p_3, s) = \frac{1}{s} G^{\mathrm{PP,P}}_{\mathrm{pp,p}}(t) \left(\frac{s}{s_0}\right)^{2\alpha_{\mathrm{P}}(t)} \left(\frac{M^2}{s_0}\right)^{\alpha_{\mathrm{P}}(0)-2\alpha_{\mathrm{P}}(t)} \tag{10.8.9}$$

which with $\alpha_{\mathrm{P}}(t) \approx 1 + \alpha'_{\mathrm{P}} t$ gives

$$f_1^{\mathrm{PP,P}} \approx \frac{1}{s_0} G^{\mathrm{PP,P}}_{\mathrm{pp,p}}(t) \left(\frac{s}{M^2}\right)^{1+2\alpha'_{\mathrm{P}} t} \tag{10.8.10}$$

or, also from (10.8.6),

$$\frac{\mathrm{d}^2\sigma}{\mathrm{d}t\,\mathrm{d}M^2} \approx \frac{G^{\mathrm{PP,P}}_{\mathrm{pp,p}}(t)}{16\pi^2 s_0} \frac{s^{2\alpha'_{\mathrm{P}} t}}{(M^2)^{1+2\alpha'_{\mathrm{P}} t}} \tag{10.8.11}$$

The secondary terms come from replacing i, j, k by R, where $\alpha_{\mathrm{R}}(t) \approx 0.5 + \alpha'_{\mathrm{R}} t$ so we can write

$$f_1 = f_1^{\mathrm{PP,P}} + f_1^{\mathrm{RR,P}} + f_1^{\mathrm{PP,R}} + f_1^{\mathrm{RR,R}} \tag{10.8.12}$$

where for example

$$f_1^{\mathrm{RR,P}} = \frac{1}{s} G^{\mathrm{RR,P}}_{\mathrm{pp,p}}(t) \left(\frac{s}{s_0}\right)^{2\alpha_{\mathrm{R}}(t)} \left(\frac{M^2}{s_0}\right)^{\alpha_{\mathrm{P}}(0)-2\alpha_{\mathrm{R}}(t)}$$

$$\approx \frac{1}{s_0} G^{\mathrm{RR,P}}_{\mathrm{pp,p}}(t) \left(\frac{s}{M^2}\right)^{2\alpha'_{\mathrm{R}} t} \tag{10.8.13}$$

The terms in (10.8.12) all have $i = j$. There could also be cross terms like $f^{\mathrm{PR,P}}$ which are usually neglected.

Clearly, by taking different types of particle for 3 one can examine a wide range of quantum numbers for $i = 1\bar{3}$: charge exchange, strangeness exchange, baryon exchange, etc. So far, only a limited amount of data is available but some fits have been made (e.g. Hoyer, Roberts and Roy 1973, Hoyer 1974).

Though the method is only directly applicable for $s > 100\,\mathrm{GeV}^2$ we can extend it to lower values using duality arguments. Thus at low M^2 we can expect resonances (r) to be produced which will be dual to α_k $(k = \mathrm{R})$ in the $i2 \to j2$ amplitude (fig. 10.23(c)). So we expect for $i = j$ in (10.8.6)

$$\left\langle \frac{\mathrm{d}\sigma}{\mathrm{d}t} \right\rangle^{\mathrm{r}} \sim \left(\frac{M^2}{s_0}\right)^{\alpha_{\mathrm{R}}(0)-2\alpha_i(t)} \sim (M^2)^{(\alpha_{\mathrm{R}}{}^0-2\alpha_i{}^0)}\,\mathrm{e}^{-2\alpha_i' t \log(M^2/s_0)} \tag{10.8.14}$$

for linear trajectories. This tells us how the differential cross-section in the two-body process $1+2 \to \mathbf{3}+X$ should vary with M_X^2 at fixed s: it should broaden in t as M^2 increases. An example of how this occurs is shown in fig. 10.24. So the triple-Regge behaviour constrains quasi-two-body scattering as well.

In the triple-Regge fits to $\mathrm{pp} \to \mathrm{p}X$ it is always found that, for small t, $G^{\mathrm{PP,P}}(t) \ll G^{\mathrm{RR,P}}(t)$ but both are non-zero for $t = 0$ (see for example fig. 10.25). The precise value depends on the assumptions made about the secondary terms, but there is now fairly general agreement about this result (cf. Field and Fox 1974, Roy and Roberts 1974, Capella 1973, Lee-Franzini 1973). Since $\gamma_{\mathrm{pp}}^{\mathrm{P}}(t)$ is known from fits to the pp differential cross-section this gives $\gamma^{\mathrm{PP,P}}(t, 0)$ directly (see (10.8.6)). Then if at a given fixed value of t we take out the factors $\gamma_{\mathrm{pp}}^{\mathrm{P}}(t)$, $\xi_{\mathrm{P}}(t)$ and $(s/M^2)^{\alpha_{\mathrm{P}}(t)}$, corresponding to the couplings and propagators of the Reggeons i, j in fig. 10.23(b), the remainder gives (from (10.8.5) and the optical theorem (1.9.6))

$$\sigma_{\mathrm{Pp}}^{\mathrm{tot}}(M^2, t) \to \sum_k \gamma_{22}^k(0)\,\gamma^{\mathrm{PP},k}(t, 0)\left(\frac{M^2}{s_0}\right)^{\alpha_k(0)-1}, \quad k = \mathrm{P}, \mathrm{R}, \ldots \tag{10.8.15}$$

(where we have taken s_0/M^2 as the flux factor) which is the total cross-section for Pomeron–proton scattering as a function of the 'energy', M, and the (mass)^2 of the Pomeron, t. This is plotted in fig. 10.26 from which we see that at large M^2 $\sigma_{\mathrm{Pp}}^{\mathrm{tot}} \to 1\,\mathrm{mb}$ for $t \to 0$. Compared with $\sigma_{\mathrm{pp}}^{\mathrm{tot}} \simeq 40\,\mathrm{mb}$ this shows that the triple-Pomeron coupling $\gamma^{\mathrm{PP,P}}(0, 0) \approx \frac{1}{40}\gamma_{\mathrm{pp}}^{\mathrm{P}}(0)$, so Pomerons couple much more weakly to themselves than they do to other particles. But the coupling is not zero.

This raises a rather difficult point about the self-consistency of P exchange. The diffractive cross-section for $1+2 \to 3+X$ (fig. 10.23(a)

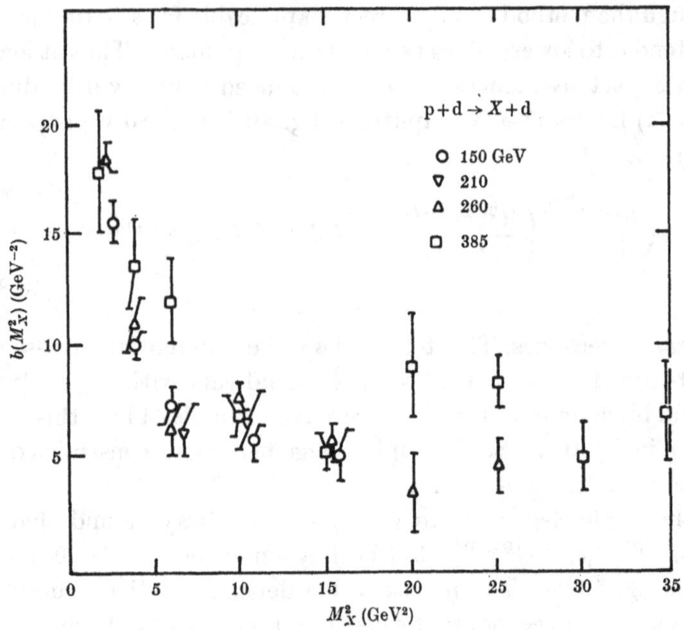

FIG. 10.24 The slope parameter b in $\mathrm{d}^2\sigma/\mathrm{d}t\,\mathrm{d}M^2 \propto \mathrm{e}^{b(M^2)t}$ as a function of M^2 in $\mathrm{p}+\mathrm{d} \to X+\mathrm{d}$, from Loebinger (1974).

with $i = \mathrm{P}$) is, from (10.8.6),

$$\frac{\mathrm{d}^2\sigma}{\mathrm{d}t\,\mathrm{d}M^2} = \frac{G_{13,2}^{\mathrm{PP,P}}(t)}{16\pi^2 s_0^2}\left(\frac{s}{s_0}\right)^{2\alpha_{\mathrm{P}}(t)-2}\left(\frac{M^2}{s_0}\right)^{\alpha_{\mathrm{P}}(0)-2\alpha_{\mathrm{P}}(t)} \tag{10.8.16}$$

So if we put $\alpha_{\mathrm{P}}(t) = \alpha_{\mathrm{P}}^0 + \alpha'_{\mathrm{P}} t$ the total diffractive contribution is given by

$$\sigma_{12}^{\mathrm{D}}(s) = \frac{s^{2\alpha_{\mathrm{P}}{}^0-2}}{16\pi^2 (s_0)^{\alpha_{\mathrm{P}}{}^0}}\int_\epsilon^s \frac{\mathrm{d}M^2}{(M^2)^{\alpha_{\mathrm{P}}{}^0}}\int_{-\infty}^0 \mathrm{d}t\, G_{13,2}^{\mathrm{PP,P}}(t)\,\mathrm{e}^{2\alpha_{\mathrm{P}}' t \log(s/M^2)} \tag{10.8.17}$$

The boundary $M^2 = s$ is where $x = 1$, and ϵ marks the lower limit below which the triple-Regge approximation breaks down. Then putting say $G_{13,2}^{\mathrm{PP,P}}(t) = G\,\mathrm{e}^{at}$ for simplicity (see fig. 10.25)

$$\sigma_{12}^{\mathrm{D}}(s) = \frac{G s^{2\alpha_{\mathrm{P}}{}^0-2}}{16\pi^2 (s_0)^{\alpha_{\mathrm{P}}{}^0}}\int_\epsilon^s \frac{\mathrm{d}M^2}{(M^2)^{\alpha_{\mathrm{P}}{}^0}(a+2\alpha'_{\mathrm{P}}\log(s/M^2))} \tag{10.8.18}$$

$$\sim s^{2\alpha_{\mathrm{P}}{}^0-2}$$

if $\alpha_{\mathrm{P}}^0 < 1$. But if $\alpha_{\mathrm{P}}^0 = 1$, using

$$\int \frac{\mathrm{d}x}{x\log x} = \log(\log x), \tag{10.8.19}$$

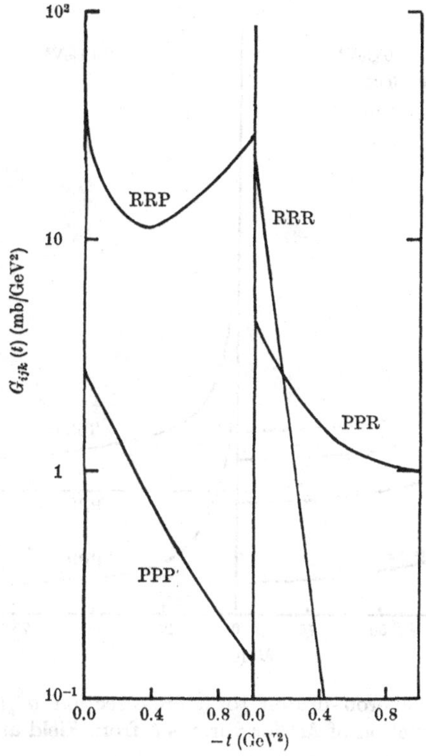

FIG. 10.25 The various triple-Regge couplings, $G^{ij,k}$, found by Field and Fox (1974) in a fit to the pp $\to$ pX data.

we find $$\sigma_{12}^{\mathrm{D}}(s) \propto \frac{1}{2\alpha'_{\mathrm{P}}} \log\left(1+\frac{2\alpha'_{\mathrm{P}}}{a}\log s\right) \sim \log(\log s) \qquad (10.8.20)$$

Though this behaviour is compatible with the Froissart bound (2.4.10) there is evidently an inconsistency because $\alpha_{\mathrm{P}}^0 = 1$ gives

$$\sigma_{12}^{\mathrm{tot}}(s) \to \text{constant} - O((\log s)^{-1})$$

(see (8.6.9)) and clearly we must have $\sigma_{12}^{\mathrm{D}}(s) < \sigma_{12}^{\mathrm{tot}}(s)$ as $s \to \infty$. Indeed no ordinary Regge singularity can give $\sigma^{\mathrm{tot}} \sim \log(\log s)$. On the other hand if $G_{13,2}^{\mathrm{PP,P}}(t)$ vanished at $t = 0$, for example

$$G_{13,2}^{\mathrm{PP,P}}(t) = (-t)\, G\, \mathrm{e}^{at}$$

say, then (10.8.17) would give

$$\sigma^{\mathrm{D}} \propto \int_{\epsilon}^{s} \frac{\mathrm{d}M^2}{(M^2)^{\alpha_{\mathrm{P}}^0}(a+2\alpha'_{\mathrm{P}}\log(s/M^2))^2} \propto \frac{1}{2\alpha'_{\mathrm{P}}a} - \frac{1}{2\alpha'_{\mathrm{P}}(a+2\alpha'_{\mathrm{P}}\log s)}$$
$$\to \text{constant} - O((\log s)^{-1}) \qquad (10.8.21)$$

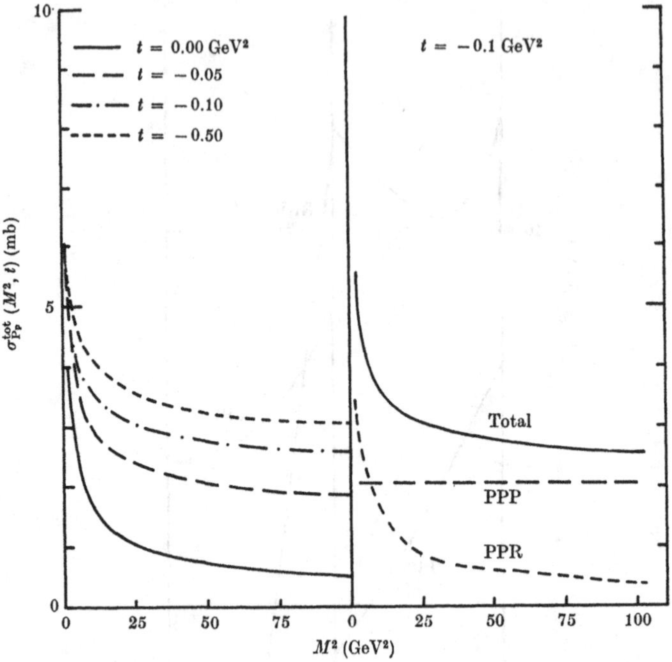

FIG. 10.26 The Pomeron–proton total cross-section $\sigma_{\mathrm{Pp}}^{\mathrm{tot}}(M^2, t)$ defined in (10.8.15) as a function of M^2 for various t, from Field and Fox (1974).

which would be compatible with P dominance. This problem, first noted in the context of the multi-peripheral model (see section 11.4 below) by Finkelstein and Kajantie (1968*a*, *b*), has been re-examined by many authors, for example Arbanel *et al.* (1971), Goddard and White (1972), Arbarbanel and Bronzan (1974*a*). A useful review of these arguments has been given by Brower and Weis (1975). Thus even though $\gamma^{\mathrm{PP,P}}(t)$ is small, the fact that empirically it appears to be non-zero at $t = 0$ raises an important difficulty which we shall examine further in the next chapter.

10.9 Finite-mass sum rules

In combining a Regge exchange model for the fragmentation region with Mueller's theorem in fig. 10.23 we have been led to study the discontinuity in M^2 of the Reggeon–particle scattering amplitude $A(i2 \rightarrow j2)$. From this viewpoint the function of particles 1 and 3 is simply to produce the virtual Reggeons, i, j. This is very analogous to the way in which virtual photon amplitudes are produced in electroproduction (cf. fig. 12.1 below).

The centre-of-mass energy for this Reggeon–particle amplitude is just M, the missing mass in $1+2\to 3+X$, and since to maintain the limit $s/M^2 \to \infty$ it is frequently necessary to consider rather small M^2 data it is useful to be able to obtain information about the Regge singularities, α_k, by using FESR to average over the resonance region of M^2, in analogy with section 7.2, rather than trying to make Regge fits at high M^2. These sum rules are called 'finite-mass sum rules', FMSR (see Hoyer 1974).

We begin by introducing the crossing-symmetric variable (cf. (7.2.3))

$$\nu \equiv p_2 \cdot (p_1 - p_3) \tag{10.9.1}$$

and, since

$$s \equiv (p_1+p_2)^2 = m_1^2+m_2^2+2p_1\cdot p_2, \quad u \equiv (p_2-p_3)^2 = m_2^2+m_3^2-2p_2\cdot p_3 \tag{10.9.2}$$

this can be rewritten, using (10.2.12), as

$$\nu = \tfrac{1}{2}(M^2-t-m_2^2) \to \tfrac{1}{2}M^2 \quad \text{for} \quad M^2 \gg t, m_2^2 \tag{10.9.3}$$

Then from (10.8.6), taking just the leading $1\overline{3}$ trajectory $i=j$,

$$\frac{\mathrm{d}^2\sigma}{\mathrm{d}t\,\mathrm{d}M^2} = \frac{1}{16\pi^2 s^2}|\gamma_{1\overline{3}}^{i}(t)|^2\,|\xi_i(t)|^2\left(\frac{s}{M^2}\right)^{2\alpha_i(t)} \times \mathrm{Disc}_{M^2}\{A(i2\to i2;\,t,M^2,0)\} \tag{10.9.4}$$

and with (10.8.5) for $\mathrm{Disc}_{M^2}\{A(i2\to i2)\}$ we obtain, for an even-signature trajectory $\mathscr{S}_k = +1$ (cf. (7.2.8), (7.2.15)),

$$\int_0^N \nu\,\mathrm{d}\nu\left(\frac{\mathrm{d}^2\sigma(12\to 3X)}{\mathrm{d}t\,\mathrm{d}M^2}+\frac{\mathrm{d}^2\sigma(32\to 1X)}{\mathrm{d}t\,\mathrm{d}M^2}\right) = \sum_k \frac{G_{1\overline{3},2}^{ii,k}(t)}{16\pi^2(s_0)^{\alpha_k(0)}}\, s^{(2\alpha_i(t)-2)}\, 2\int_0^N (M^2)^{\alpha_k(0)-2\alpha_i(t)}\,\tfrac{1}{4}M^2\,\mathrm{d}M^2 \tag{10.9.5}$$

The factor 2 appears on the right-hand side because, as in (7.2.9), we are adding the cuts for positive M^2 and for negative M^2, which describe the processes $12\to 3X$ and $32\to 1X$ respectively, at fixed $t_{22'}=0$. These are the two discontinuities of the even-signature k trajectory (see fig. 10.27). And on performing the integration we obtain for the right-hand side

$$\sum_k \frac{G_{1\overline{3},2}^{ii,k}(t)\, s^{2\alpha_i(t)-2}}{16\pi^2(s_0)^{\alpha_k(0)}}\,\frac{1}{2}\,\frac{N^{\alpha_k(0)-2\alpha_i(t)+2}}{\alpha_k(0)-2\alpha_i(t)+2} \tag{10.9.6}$$

In practice it is not usually possible to go to sufficiently high energies

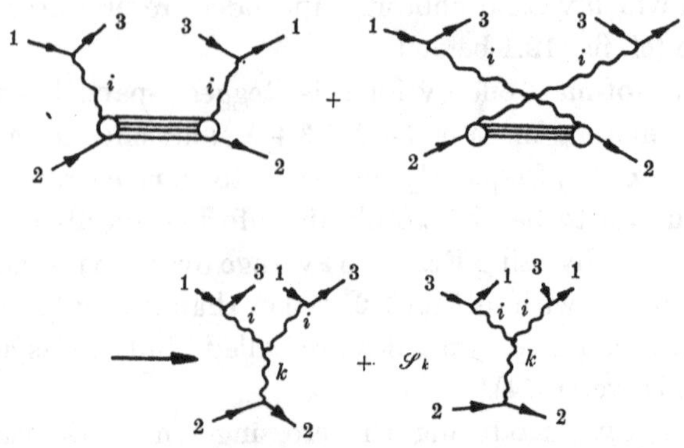

FIG. 10.27 Triple-Regge representations for $1+2\to 3+X$ and $3+2\to 1+X$ used for FMSR evaluations.

for a single trajectory i to contribute, and so it is necessary to replace $\sum_k$ by $\sum_{i,j,k}$ in (10.9.6). Also we can take higher moments (like (7.2.14) and (7.2.16)) and obtain (setting $s_0 \equiv 1$ for convenience)

$$\int_0^N \nu^n \, d\nu \left[\frac{d^2\sigma(12\to 3X)}{dt\,dM^2} + (-1)^{n+1}\frac{d^2\sigma(32\to 1X)}{dt\,dM^2}\right]$$

$$= \sum_{i,j,k} \frac{G^{ij,k}_{13,2}(t)}{32\pi^2} s^{\alpha_i(t)+\alpha_j(t)-2} \frac{N^{\alpha_k(0)-\alpha_i(t)-\alpha_j(t)+n+1}}{\alpha_k(0)-\alpha_i(t)-\alpha_j(t)+n+1} \quad (10.9.7)$$

where $n = 1, 3, 5, \ldots$, for $\mathcal{S}_k = 1$ and $n = 0, 2, 4, \ldots$ for $\mathcal{S}_k = -1$.

These FMSR were introduced by Einhorn *et al.* (1972*a*) and Sanda (1972) and have been widely employed to complement triple-Regge fits. For example Roy and Roberts (1974) and Field and Fox (1974) used them in the fits described in the previous section.

The duality properties of these sum rules are rather interesting. For $i, j = R$ (i.e. ordinary Reggeons, not P) we can expect the usual two-component duality of two-body reactions (section 7.3), i.e. resonances in M^2 will be dual to k = R, while the non-resonant background should be dual to k = P, since all we have done is move out in t along the i, j trajectories away from the physical particles. This seems to be well verified (see Hoyer 1974). But what about the Pomeron–particle amplitude P + 2 → P + 2? On the basis of the duality diagrams, fig. 10.28 (*a*), (*b*), Einhorn *et al.* (1972) argued that (unlike R + 2 → R + 2) the resonances in M^2 build up the P exchange. But on the other hand if the P couples through the f, the resonances should be dual to the R and P is dual to the background as in fig. 10.28 (*c*).

However, this diagram contains a closed loop and so would normally be excluded from consideration. The 'theory' is thus ambiguous, and so unfortunately is the phenomenology at present (see Hoyer 1974).

By taking wrong-moment sum rules (i.e. n even for $\mathscr{S}_k = +1$, and n odd for $\mathscr{S}_k = -1$) we can explore the fixed poles which may be present in the Reggeon–particle scattering amplitudes (cf. (7.2.21)). For example if in an even-signature amplitude we take the zeroth moment we obtain (with $j = i$, and again setting $s_0 = 1$)

$$\int_0^N \mathrm{d}\nu \left(\frac{\mathrm{d}^2\sigma(12\to 3X)}{\mathrm{d}t\,\mathrm{d}M^2} + \frac{\mathrm{d}^2\sigma(32\to 1X)}{\mathrm{d}t\,\mathrm{d}M^2}\right)$$
$$= \sum_{i,k} \frac{1}{16\pi^2 s^2} |\gamma^i_{13}(t)|^2 |\xi_i(t)|^2 s^{2\alpha_i(t)} \gamma^k_{22}(0)$$
$$\times \frac{1}{2}\left[G^{ii}_1(t) + \gamma^{ii,k}(t,0)\frac{N^{\alpha_k(0)-2\alpha_i(t)+1}}{\alpha_k(0)-2\alpha_i(t)+1}\right] \quad (10.9.8)$$

where $G^{ii}_m(t)$ are the residues of the fixed poles in the Reggeon–particle amplitude $i2 \to i2$ at the nonsense points $J - 2\alpha_i(t) = -m$, $m = 1, 3, 5, \ldots$ (since the t-channel helicities of the trajectories are $\alpha_i(t)$). $G^{ii}_1(t)$ is related to the Reggeon–particle fixed-pole coupling $N_1(t, t_1, t_2)$ which occurs in the expressions (8.2.37) and (8.3.8) for a Regge cut in the Gribov calculus by (see (8.2.39))

$$G^{ii}_1(t) = N^{ii}_1(0, t, t)\,(\gamma^i_{22})^{-1} \quad (10.9.9)$$

Thus by comparing right- and wrong-moment sum rules one can in principle evaluate N and substitute it into (8.4.1) and obtain an expression for the Regge cut. This has been attempted by Roberts and Roy (1972) who used inclusive data on $K^+ \xrightarrow{p} K^0$ and $K^- \xrightarrow{p} \bar{K}^0$ to evaluate $\rho \otimes \rho$ and $A_2 \otimes A_2$ cuts in $pp \to pp$, and by Muzinich *et al.* (1972) who have tried to estimate the $P \otimes P$ cut in $pp \to pp$. They find that the cut has a strength of only about 40 per cent of the eikonal/absorption prescription ($N^{PP}_1(t, t_1, t_2) = 1$, see section 8.4). However, the uncertainties in the triple-Reggeon couplings make the errors in these evaluations rather large. Also the procedure is not self-consistent since the cuts have been omitted from the inclusive sum rules, so this approach can only be even approximately successful if cuts $\ll$ poles.

It will be evident from the preceding sections that, despite being restricted to $t_{22'} = 0$, this triple-Regge regime should eventually provide many useful insights into Reggeon dynamics.

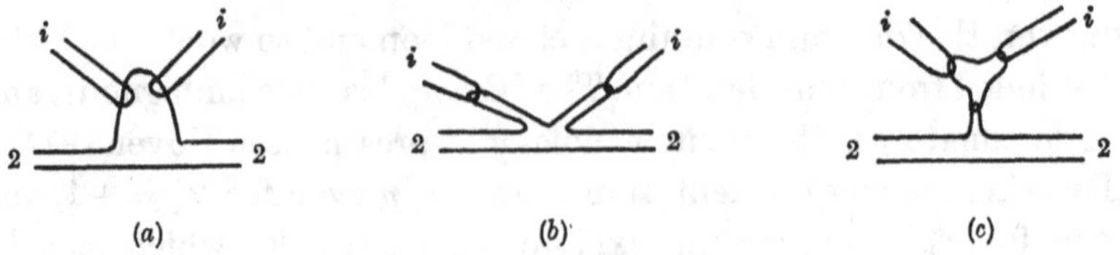

FIG. 10.28 (*a*) Duality diagram for P exchange in $i2 \to i2$. (*b*) A redrawing of (*a*) suggesting that the P-exchange coupling might be dual to the resonances. (*c*) An alternative duality diagram, involving a closed loop, which suggests that P exchange is dual to the background as usual.

10.10 Correlations and the correlation length

The two-particle inclusive distribution for $1+2 \to 3+4+X$ was defined in (10.3.12). The dynamics of particle production can obviously be explored further by observing any correlations there may be between the two observed final-state particles. For example if 3 and 4 were mainly produced through a resonance decay, $1+2 \to r+X$, $r \to 3+4$, then the momenta of these particles would be closely related.

We can define the two-particle correlation function by

$$c_2(\mathbf{p}_3, \mathbf{p}_4, s) \equiv \rho_2(\mathbf{p}_3, \mathbf{p}_4, s) - \rho_1(\mathbf{p}_3, s)\rho_1(\mathbf{p}_4, s) \tag{10.10.1}$$

where the ρ's are defined in (10.3.10) and (10.3.15). If there is no correlation between the production of particles 3 and 4 the probability of producing both must be just the product of the individual production probabilities, i.e.

$$\rho_2(\mathbf{p}_3, \mathbf{p}_4, s) = \rho_1(\mathbf{p}_3, s)\rho_1(\mathbf{p}_4, s) \tag{10.10.2}$$

giving $c_2 = 0$ as required. It is also convenient to introduce

$$C_2(s) \equiv \int c_2(\mathbf{p}_3, \mathbf{p}_4, s)\frac{\mathrm{d}^3\mathbf{p}_3}{16\pi^2 E_3}\frac{\mathrm{d}^3\mathbf{p}_4}{16\pi^2 E_4} = \langle n_3 n_4 - \delta_{34} n_3\rangle - \langle n_3\rangle\langle n_4\rangle \tag{10.10.3}$$

from (10.10.1), (10.3.16) and (10.3.11). If 3 and 4 are identical particles

$$C_2(s) = F_2(s) - F_1^2(s) \tag{10.10.4}$$

We have seen in fig. 10.5 that $F_1 \sim \log s$ approximately, and similarly (fig. 10.29) $C_2(s) \sim (\log s)^2$ approximately (or it could be $\sim$ a small power of s).

Likewise we can define the three-particle correlation by

$$\begin{aligned} c_3(\mathbf{p}_3, \mathbf{p}_4, \mathbf{p}_5, s) = {} & \rho_3(\mathbf{p}_3, \mathbf{p}_4, \mathbf{p}_5, s) - \rho_1(\mathbf{p}_3, s)\, c_2(\mathbf{p}_4, \mathbf{p}_5, s) - \rho_1(\mathbf{p}_4, s) \\ & \times c_2(\mathbf{p}_3, \mathbf{p}_5, s) - \rho_1(\mathbf{p}_5, s)\, c_2(\mathbf{p}_3, \mathbf{p}_4, s) - \rho_1(\mathbf{p}_3, s)\rho_1(\mathbf{p}_4, s)\rho_1(\mathbf{p}_5, s) \end{aligned} \tag{10.10.5}$$

and so on.

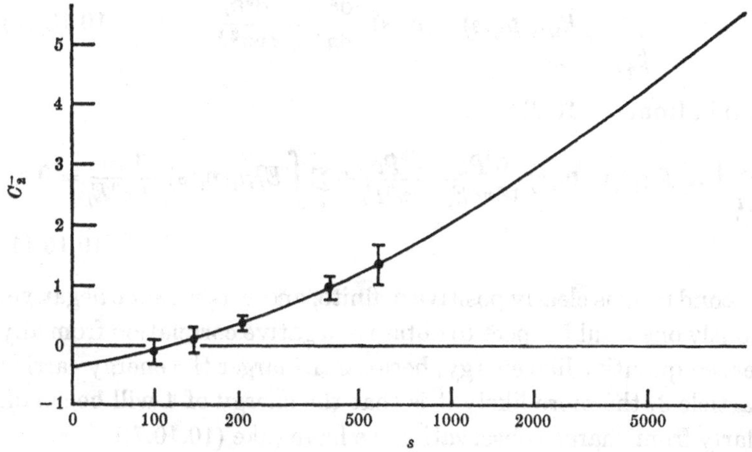

FIG. 10.29 Data on $C_2(s)$ against s for negatively charged particles, from Harari (1974). The curve is a fit with $C_2^- = 0.14(\log s)^2 - 0.65 \log s + 0.06$.

Some correlations have to be present because of kinematics (i.e. conservation of E, p, etc.), or quantum number conservation (of B, Q, S, I, G etc.): see de Tar, Freedman and Veneziano (1971). For example, since in $1+2 \to 3+4 \ldots (m+2)$ we have

$$\sum_{n=3}^{m+2} E_n = \sqrt{s} \tag{10.10.6}$$

i.e. the total centre-of-mass energy of all the outgoing particles must equal that of the initial state, and there is an energy conservation sum rule

$$\sum_l \int E_l \rho_1(p_l, s) \frac{d^3p_l}{16\pi^2 E_l} = \sqrt{s} \tag{10.10.7}$$

since the left-hand side gives the probability of producing a particle of type l with energy E_l, integrated over all possible energies, and summed over all possible types of particles. Also since

$$\left(\sum_{n=3}^{m+2} E_n\right)^2 = s \tag{10.10.8}$$

we have similarly

$$\sum_{\substack{k,l \\ k \neq l}} \int E_k E_l \rho_2(p_k, p_l, s) \frac{d^3p_k}{16\pi^2 E_k} \frac{d^3p_l}{16\pi^2 E_l} + \sum_l \int E_l^2 \rho_1(p_l, s) \frac{d^3p_l}{16\pi^2 E_l} = s \tag{10.10.9}$$

But since from (10.10.1) we can express ρ_2 in terms of c_2 and ρ_1, and

since
$$\sum_{\substack{k,l \\ k \neq l}} E_k E_l \rho_1(\mathbf{p}_k, s)\, \rho_1(\mathbf{p}_l, s) \frac{\mathrm{d}^3\mathbf{p}_k}{16\pi^2 E_k} \frac{\mathrm{d}^3\mathbf{p}_l}{16\pi^2 E_l} = s \qquad (10.10.10)$$

we obtain from (10.10.9)

$$\sum_{\substack{k,l \\ k \neq l}} \int E_k E_l c_2(\mathbf{p}_k, \mathbf{p}_l, s) \frac{\mathrm{d}^3\mathbf{p}_k}{16\pi^2 E_k} \frac{\mathrm{d}^3\mathbf{p}_l}{16\pi^2 E_l} + \sum_l \int E_l^2 \rho_1(\mathbf{p}_l, s) \frac{\mathrm{d}^3\mathbf{p}_l}{16\pi^2 E_l} = 0 \qquad (10.10.11)$$

The second term is clearly positive definite, and so c_2 must be negative. Obviously one would expect to obtain a negative correlation from any conserved quantity like energy, because the larger the energy carried by particle 3, the more likely it is that the energy of 4 will be small. Similarly from charge conservation we have (like (10.10.7))

$$\sum_l \int Q_l \rho_1(\mathbf{p}_l, s) \frac{\mathrm{d}^3\mathbf{p}_l}{16\pi^3 E_l} = \sum_l Q_l \langle n_l \rangle = Q_1 + Q_2 \qquad (10.10.12)$$

using (10.3.11), which gives a negative correlation between the charges of the particles produced in a reaction.

In addition to these kinematic correlations there may be dynamical correlations due to the production mechanism, for example the resonance decay mentioned above. Such correlations seem much less likely if the particles occur at very widely spaced points on the rapidity plot (fig. 10.4), and it is useful to try and determine the distance in rapidity over which one can expect there to be strong correlations. This is called the 'correlation length', Λ, defined such that there will be negligible correlation between particles 3 and 4 if

$$|y_3 - y_4| \gg \Lambda \qquad (10.10.13)$$

Thus the projectile fragmentation region of fig. 10.4 (*b*) is

$$y_{3\,\max} \geqslant y_3 > (y_{3\,\max} - \Lambda) = \tfrac{1}{2}\log(s/\mu_3^2) - \Lambda$$

and the target fragmentation region is

$$y_{3\,\min} \leqslant y_3 < (y_{3\,\min} + \Lambda) = -\tfrac{1}{2}\log(s/\mu_3^2) + \Lambda.$$

Note that since we are taking Λ to be independent of s we are assuming that scaling holds in the central region. But for low s, $\Lambda > \log(s/\mu_3^2)$, so the two fragmentation regions overlap and scaling is not expected.

In the central region the Mueller–Regge diagram for

$$1 + 2 \rightarrow 3 + 4 + X$$

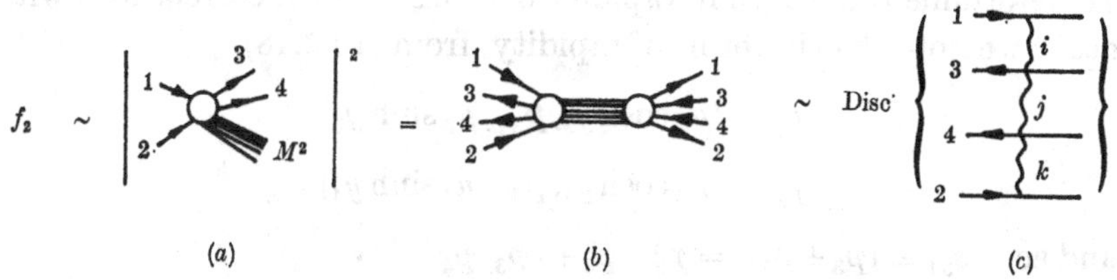

FIG. 10.30 Treble-Regge representation of the two-particle inclusive process $1+2 \to 3+4+X$.

is the treble-Regge diagram fig. 10.30, where $s_{12}, t_{13}, s_{34}, t_{24} \to \infty$ with $(t_{13}s_{34}t_{24})/s_{12}$ etc. fixed. And so (cf. (10.6.4))

$$f_2(p_3, p_4, s) \to \sum_{i,j,k} \gamma_{ijk}(\mu_3^2, \mu_4^2) \left|\frac{t_{13}}{s_0}\right|^{\alpha_i(0)-1} \left|\frac{t_{34}}{s_0}\right|^{\alpha_j(0)-1} \left|\frac{t_{42}}{s_0}\right|^{\alpha_k(0)-1} \tag{10.10.14}$$

Once the energy is high enough for the central region to be really well separated from the fragmentation regions, we need only include the P for i and k, so for $t_{13}, t_{24} \to \infty$, if $\alpha_{\mathrm{P}}(0) = 1$,

$$f_2(p_3, p_4, s) \to \sum \gamma_{\mathrm{P}j\mathrm{P}}(\mu_3^2, \mu_4^2) \left(\frac{s_{34}}{s_0}\right)^{\alpha_j(0)-1} \xrightarrow[s_{34}\to\infty]{} \gamma_{\mathrm{PPP}}(\mu_3^2, \mu_4^2) \tag{10.10.15}$$

which gives the scaling behaviour expected in the central region. How fast the latter limit is approached depends on the spacing of the secondary trajectories, R, in the sum over j.

Using factorization we can write (cf. (10.6.6))

$$\gamma_{\mathrm{P}j\mathrm{P}}(\mu_3^2, \mu_4^2) = \gamma_{11}^{\mathrm{P}} \gamma_{33}^{\mathrm{P}j}(\mu_3^2) \gamma_{44}^{j\mathrm{P}}(\mu_4^2) \gamma_{22}^{\mathrm{P}} \tag{10.10.16}$$

So using (10.5.7) we can write, from (10.3.15) and (10.10.15),

$$\rho_2(p_3, p_4, s) \to \gamma_{33}^{\mathrm{PP}}(\mu_3^2) \gamma_{44}^{\mathrm{PP}}(\mu_4^2) \tag{10.10.17}$$

which is independent of the nature of particles 1 and 2. Then because of (10.6.7) we find

$$\rho_2(p, p_4, s) \to \rho_1(p_3, s) \rho_1(p_4, s) \tag{10.10.18}$$

and so from (10.10.1) $c_2(p_3, p_4, s) \to 0$ and there is no correlation. This is because we have assumed that asymptotically a single factorizable pole dominates, and so each vertex is completely independent.

However, at lower s_{34} we can expect corrections to the Regge behaviour from the lower-lying R trajectories, and these will produce correlations between the particles at non-asymptotic sub-energies.

To determine the length in rapidity over which such correlations will occur we note that in terms of rapidity, from (10.2.18),

$$p_3 = (\mu_3 \cosh y_3, \mathbf{p}_{3\mathrm{T}}, \mu_3 \sinh y_3)$$

$$p_4 = (\mu_4 \cosh y_4, \mathbf{p}_{4\mathrm{T}}, \mu_4 \sinh y_4)$$

and so

$$\begin{aligned} s_{34} &\equiv (p_3+p_4)^2 = p_3^2+p_4^2+2p_3\cdot p_4 \\ &= m_3^2+m_4^2+2\mu_3\mu_4 \cosh y_3 \cosh y_4 - 2\mathbf{p}_{3\mathrm{T}}\cdot \mathbf{p}_{4\mathrm{T}} \\ &\quad - 2\mu_3\mu_4 \sinh y_3 \sinh y_4 \\ &= m_3^2+m_4^2+2\mu_3\mu_4 \cosh(y_3-y_4) - 2\mathbf{p}_{3\mathrm{T}}\cdot \mathbf{p}_{4\mathrm{T}} \\ &\xrightarrow[s_{34}\to\infty]{} 2\mu_3\mu_4 \cosh(y_3-y_4) \to \mu_3\mu_4 \,\mathrm{e}^{|y_3-y_4|} \end{aligned} \tag{10.10.19}$$

Hence (10.10.15) gives

$$f_2(\mathbf{p}_3, \mathbf{p}_4, s) \to \sum_j \gamma_{\mathrm{P}j\mathrm{P}}(\mu_3^2, \mu_4^2) \left(\frac{\mu_3\mu_4}{s_0}\right)^{\alpha_j(0)-1} \mathrm{e}^{(\alpha_j(0)-1)|y_3-y_4|} \tag{10.10.20}$$

The first term with $j = \mathrm{P}$, $\alpha_{\mathrm{P}}(0) = 1$, gives no correlation as we have seen, but the second term with $j = \mathrm{R}$, $\alpha_{\mathrm{R}}(0) \approx 0.5$, gives a contribution

$$\rho_2(\mathbf{p}_3, \mathbf{p}_4, s) \propto \mathrm{e}^{-\frac{1}{2}|y_3-y_4|} \tag{10.10.21}$$

which in (10.10.1) gives

$$c_2(\mathbf{p}_3, \mathbf{p}_4, s) \propto \mathrm{e}^{-\frac{1}{2}|y_3-y_4|} \tag{10.10.22}$$

and so if we define the correlation length Λ as the distance in rapidity within which the correlation has fallen to e^{-1} of its maximum value, then Regge theory predicts that

$$\Lambda = (\alpha_{\mathrm{P}}(0) - \alpha_{\mathrm{R}}(0))^{-1} = 2 \tag{10.10.23}$$

This seems to be quite well verified in many processes. See for example fig. 10.31 which shows how the events peak in a ridge where $y_3 \approx y_4$. This number is quite important as it gives the width in rapidity of the fragmentation regions, and shows that we need $Y \approx 8$ (as at the CERN-ISR) before the central region is well separated from them.

This prediction depends crucially on the fact that each Regge pole contribution must factorize, so that only the non-factorizability of a sum of Regge poles produces correlations. However, Regge cut contributions will in general not factorize, and so for example $\mathrm{P}\otimes\mathrm{P}$ cuts could produce correlations of infinite correlation length. The apparent absence of very strong long-range correlations must mean

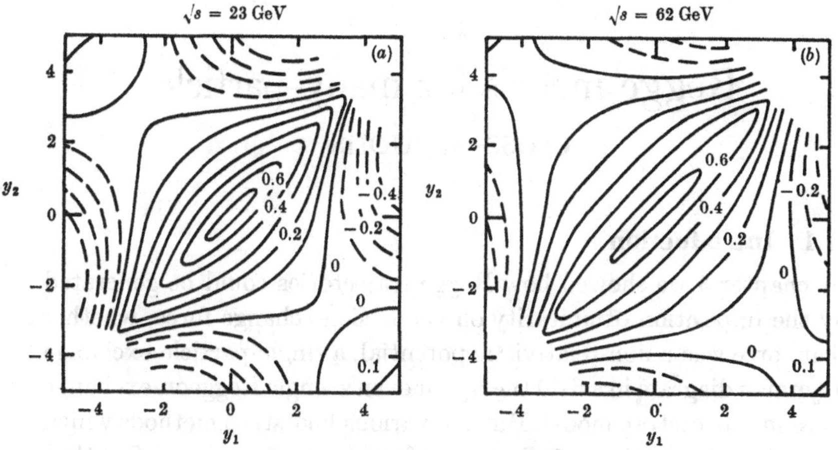

FIG. 10.31 Contours of constant correlation $c_2(y_3, y_4; s)$, in the y_3–y_4 plane, for charged particle pairs (mainly pions) produced in pp collisions at CERN-ISR, from Zalewski (1974).

that the P singularity is at least approximately factorizable, and lends support to the view that it is effectively a pole at available energies. However, we shall see in the next chapter that there are some long-range correlation effects.

11
Regge models for many-particle cross-sections

11.1 Introduction

In chapter 3 we showed how Regge trajectories could be generated by the imposition of unitarity on the basic exchange force, whether that force was a non-relativistic potential, a single-particle-exchange Feynman diagram in a field theory, or even a single Reggeon-exchange force in a bootstrap model. But the various bootstrap methods which we reviewed in section 3.5 all suffered from the very serious defect that they were limited to two-body unitarity in one channel or another. In chapters 9 and 10 we have found that Regge theory can also predict successfully the sort of behaviour to be expected in many-particle scattering amplitudes, so it is now possible to return to some of the most fundamental questions of Regge theory, such as how the Regge singularities are self-consistent under unitarity, and whether the bootstrap idea introduced in section 2.8 can be correct.

For this purpose we need models for many-particle production processes, and in the next two sections we examine two such models. One, the diffraction model, though inadequate by itself, does describe Pomeron-exchange effects and the fragmentation region, while the other, the multi-peripheral model, though applicable only in certain regions of phase space, allows one to approximate the effect of multi-Reggeon exchange. The so-called 'two-component model' which incorporates both these contributions seems to account quite well for the basic structure of many-particle cross-sections, if not all the details.

The next step is to try and convert this success into a self-consistent bootstrap model combining both duality and unitarity. This is a major task which has certainly not yet been completed satisfactorily. But in the final sections of this chapter we review some of the progress which has been made.

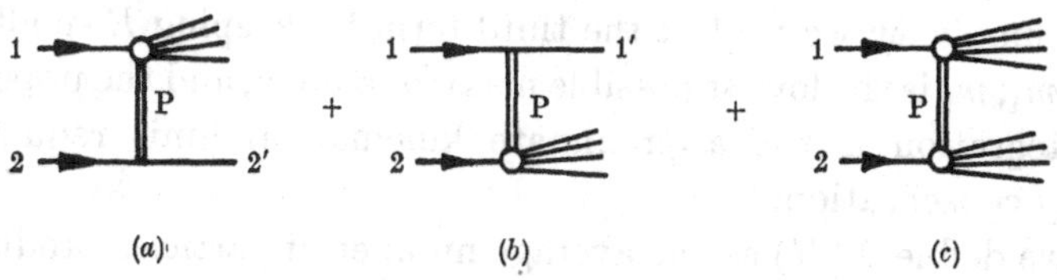

FIG. 11.1 The diffraction model in which the incoming particles are excited by P exchange to high-mass 'novae' which subsequently decay into particles.

11.2 The diffraction model

This model was proposed by various authors under a variety of names such as 'the diffractive excitation model' (Good and Walker 1960, Adair 1968, Hwa *et al.* 1970, 1971, 1972), 'the limiting fragmentation model' (Benecke *et al.* 1969), 'the fireball' model (Hagedorn 1965, 1970) and 'the nova model' (Jacob and Slansky 1972, Jacob, Berger and Slansky 1972), each with a somewhat different physical motivation. Originally it may have been hoped that the model might account for most of the high energy cross-section, but this is now known not to be true. It does, however, provide a significant fraction ($\approx 20\%$) of the events as we shall see. Our presentation will be based mainly on the nova version (see for example Berger 1971*b*).

The model incorporates the three Pomeron-exchange diagrams of fig. 11.1 in which the incoming particles are excited to form 'novae' or 'fireballs' which then decay into the observed final-state particles. This clearly reproduces the leading-particle effect. The three diagrams are supposed to add incoherently. It is assumed that the inelasticity is small so that rather few particles are produced (which is true, since empirically $\langle n\rangle \propto \log s$), that most particles are produced only with small p_T (also true – see fig. 10.17), and that only the energy-independent, scaling, single-P exchange is important (which is in fact wrong).

The cross-section for producing a fireball of mass M from particle i is denoted by $\rho_i(M)$, so that the total inelastic cross-section can be written as the sum of figs. 11.1, in the form

$$\begin{aligned}\sigma_{12}^{\text{in}}(s) &= \int_{M_0}^{\sqrt{s}-m_2} \rho_1(M)\,\mathrm{d}M + \int_{M_0}^{\sqrt{s}-m_1} \rho_2(M)\,\mathrm{d}M \\ &\quad + \iint_{M_0}^{M_1+M_2=\sqrt{s}} \rho_1(M_1)\,\rho_2(M_2)\,R(M_1,M_2)\,\mathrm{d}M_1\,\mathrm{d}M_2 \\ &\approx \sum_{i=1,2} \int^{\sqrt{s}} \rho_i(M)\,\mathrm{d}M \qquad (11.2.1)\end{aligned}$$

if for simplicity we neglect the third term by keeping R small. Here $M_0 \geqslant m_1, m_2$ is the lowest possible mass for a nova, and the upper limit of integration is the approximate kinematical limit required by energy conservation.

If we define $N(M)$ as the average number of particles produced in the decay of a nova of mass M, then the average multiplicity in an event will be

$$\langle n \rangle = \sum_i \frac{\int N(M)\rho_i(M)\,\mathrm{d}M}{\int \rho_i(M)\,\mathrm{d}M} = \frac{2\int N(M)\rho(M)\,\mathrm{d}M}{\sigma_{12}^{\mathrm{in}}} \tag{11.2.2}$$

if the two ρ_i are taken to be identical.

The decay of a nova into, say, pions is described by the function $\mathrm{d}^3D/\mathrm{d}^3\boldsymbol{q}$, giving the probability that a given pion is emitted into the phase-space volume element $\mathrm{d}^3\boldsymbol{q}$ in the nova's rest frame. So the centre-of-mass frame distribution of pions will be (for each nova)

$$\frac{\mathrm{d}^3\sigma}{\mathrm{d}^3\boldsymbol{p}} = \int N(M)\rho(M)\frac{\mathrm{d}^3D}{\mathrm{d}^3\boldsymbol{q}}\left(\frac{\partial^3\boldsymbol{q}}{\partial^3\boldsymbol{p}}\right)\mathrm{d}M \tag{11.2.3}$$

the last factor being the Jacobian for the Lorentz transformation from the nova's rest frame to the centre-of-mass, a transformation which clearly depends on M. So (11.2.3) gives us the pion distribution in terms of three functions, N, ρ and $\mathrm{d}^3D/\mathrm{d}^3\boldsymbol{q}$, which have to be determined.

Since we are not concerned with the p_{T} distribution, which will simply be built into $\mathrm{d}^3D/\mathrm{d}^3\boldsymbol{q}$, and since q_{T} is unchanged by a Lorentz transformation along the z axis, it is convenient to define

$$A(M, y) \equiv \int \frac{\mathrm{d}^3D}{\mathrm{d}^3\boldsymbol{q}}\left(\frac{\partial^3\boldsymbol{q}}{\partial^3\boldsymbol{p}}\right)\mathrm{d}^2\boldsymbol{p}_{\mathrm{T}} \tag{11.2.4}$$

and then neglect any transverse motion of the nova so that $\boldsymbol{q}_{\mathrm{T}} = \boldsymbol{p}_{\mathrm{T}}$, which gives

$$\frac{\mathrm{d}\sigma}{\mathrm{d}y} \approx \int^{\sqrt{s}} N(M)\rho(M)A(M, y)\,\mathrm{d}M \tag{11.2.5}$$

But these approximations are certainly not essential and more exact kinematics can be employed if desired.

It is simplest to assume an isotropic decay of the nova in its rest frame, so one can put

$$\frac{\mathrm{d}^3D}{\mathrm{d}^3\boldsymbol{q}} \propto \mathrm{e}^{-q^2/K^2} = \mathrm{e}^{-(q_{\mathrm{L}}^2+q_{\mathrm{T}}^2)/K^2} \approx \mathrm{e}^{-q_{\mathrm{L}}^2/K^2}\,\mathrm{e}^{-p_{\mathrm{T}}^2/K^2} \tag{11.2.6}$$

where K must be $\approx 0.45\,\mathrm{GeV}/c$ to fit the observed p_{T} distribution (see for example fig. 10.17). Then writing (see (10.2.18))

$$q_{3\mathrm{L}} = \mu_3 \sinh y_0 \tag{11.2.7}$$

where y_0 is the pion's rapidity in the nova's rest frame, and integrating over q_{T}^2, we get

$$\frac{\mathrm{d}D}{\mathrm{d}y_0} \propto \mathrm{e}^{(\mu_3 \sinh y_0/K)^2} \tag{11.2.8}$$

Now in the centre-of-mass system y_0 is boosted to $y = y_0 \pm y_M$, where y_M is the nova's rapidity ($\pm$ for 1, 2 fragmentation), and from (10.2.17) and (10.2.7), neglecting the transverse motion of the nova, we have for fragments of 1,

$$\begin{aligned}\sinh y_M &= \frac{p_{\mathrm{LM}}}{M} \approx \frac{p_{\mathrm{M}}}{M} \approx \frac{[s^2 - 2(M^2 + m_2^2)\,s + (M^2 - m_2^2)^2]^{\frac{1}{2}}}{2(\sqrt{s})\,M} \\ &\underset{M^2 \gg m_2^2}{\approx} \frac{s - M^2}{2(\sqrt{s})\,M} \underset{s \gg M^2}{\approx} \frac{\sqrt{s}}{2M}\end{aligned} \tag{11.2.9}$$

and so, since $\sinh y_M \approx \frac{1}{2}\mathrm{e}^{y_M}$ for $y_M \gg 1$, we get

$$y_M \approx \tfrac{1}{2} \log \frac{s}{M^2} \tag{11.2.10}$$

for heavy novae at very high energies.

The mean value of $q_{x,y,z}$ in (11.2.6) is K, so the typical energy available to a pion in a nova decay must be

$$Q \approx \sqrt{\tfrac{3}{2}}K \approx 0.5\,\mathrm{GeV} \tag{11.2.11}$$

(neglecting the pion mass) which is in agreement with observation, so if only pions are emitted, the average number produced by a nova of mass M will be

$$N(M) = \gamma(M - M_0) \tag{11.2.12}$$

where M_0 is the ground state energy ($= m_{1,2}$ probably), and $\gamma = 1/Q \simeq 2$. However, we want the average multiplicity of pions to increase only slowly with s, and to achieve this given (11.2.12) it is essential that the probability of producing high-mass novae be small. In fact if (11.2.12) is inserted in (11.2.2) it is clear that we must have $\rho(M) \sim 1/M^2$ if the average multiplicity is to increase logarithmically, for then

$$\langle n \rangle \to \int^{\sqrt{s}} \gamma \frac{\mathrm{d}M}{M} \to \frac{\gamma}{2} \log s \tag{11.2.13}$$

So the single empirical constant, K, determines the form of the functions A, N, and ρ.

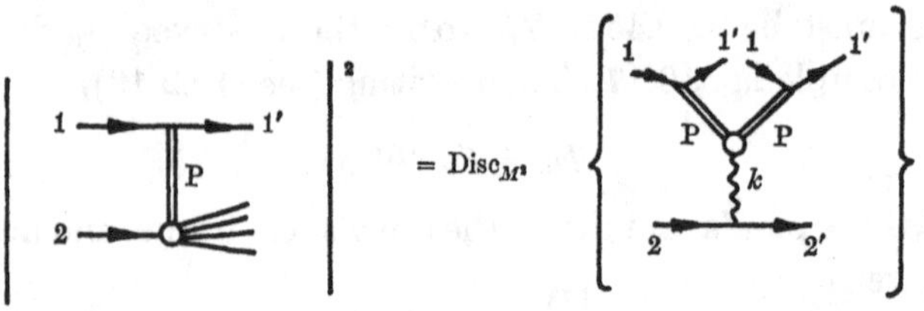

FIG. 11.2 The contribution of fig. 11.1 (b) to the Mueller optical theorem; cf. fig. 10.23.

It is interesting to look at these requirements from the Regge viewpoint since for example fig. 11.1 (b) gives the cross-section for the inclusive process $1+2 \to 1'+X$, with $M_X = M$, in the triple-Regge region $x_1 \approx 1$, so that (see fig. 11.2)

$$\rho_2(M) = \int_{-\infty}^{0} \frac{\mathrm{d}^2\sigma}{\mathrm{d}t\,\mathrm{d}M}\mathrm{d}t = \int_{-\infty}^{0} \frac{2M\mathrm{d}^2\sigma}{\mathrm{d}t\,\mathrm{d}M^2}\mathrm{d}t$$

$$= \sum_k \frac{1}{16\pi^2}\int_{-\infty}^{0} G_{11,2}^{\mathrm{PP},k}(t)\, s^{2\alpha_{\mathrm{P}}(t)-2}\, 2(M^2)^{\alpha_k(0)-2\alpha_{\mathrm{P}}(t)+\frac{1}{2}}\,\mathrm{d}t \tag{11.2.14}$$

from (10.8.6). The dominant region of the t integration will be $t \approx 0$, since $G(t)$ falls exponentially with $-t$, where $\alpha_{\mathrm{P}}(t) \approx 1$. The leading trajectory k should be the Pomeron, but $\alpha_k(0) = 1$ gives too slow a fall of (11.2.14) with M^2. However, we can perhaps neglect this term on the grounds that the triple-Pomeron coupling is small (remembering also that a finite $\gamma^{\mathrm{PP,P}}(t=0)$ is not self-consistent, at least in the pole approximation which we are employing) so that for moderate values of M^2 the dominant contribution will be $k = \mathrm{R}(=\rho, \omega, \mathrm{A}_2, \mathrm{f})$ with $\alpha_{\mathrm{R}}(0) \approx 0.5$, giving

$$\rho_2(M) \sim \frac{1}{M^2} \tag{11.2.15}$$

So from this point of view it looks as though the model may work for intermediate M^2, but not large M^2, though we must also remember that $M^2 \to s$ takes us outside the triple-Regge Region.

Jacob *et al.* (1972) used the parameterization

$$\rho_i(M) = C_i \frac{\mathrm{e}^{-\beta_i/(M-m_i)}}{(M-m_i)^2}, \quad i = 1, 2 \tag{11.2.16}$$

which has the required M^{-2} asymptotic behaviour, with a peak at $M = m_i + \frac{1}{2}\beta_i$; C_1 and β_i are free parameters to be adjusted to fit the data on $\sigma_{12}^{\mathrm{in}}$, the inclusive distributions, etc.

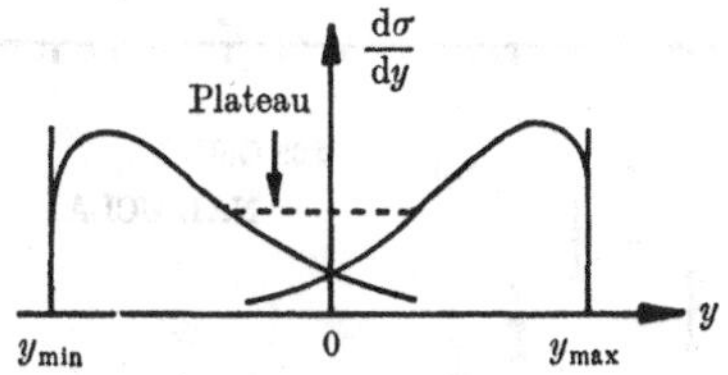

FIG. 11.3 The tails of the two nova distributions produce a central plateau in rapidity.

It is possible to reproduce the inclusive distributions, with their flat central plateau, only because of the M^{-2} tail of $\rho(M)$. From (11.2.8), since $y = y_0 \pm y_M$, the central region $y \approx 0$ requires $y_M \approx 0$, which from (11.2.9) means $M \approx \sqrt{s}$. So this region is occupied by novae which are as heavy as energy conservation permits. Since $\rho(M)\, N(M) \sim M^{-1}$ there is a finite contribution from this region of integration in (11.2.5) and so a central plateau can develop as in fig. 11.3.

Since, from (11.2.2), (11.2.13),

$$\langle n \rangle = \frac{\int \frac{\mathrm{d}\sigma}{\mathrm{d}y}\,\mathrm{d}y}{\sigma_{12}^{\mathrm{in}}} \to \frac{\gamma}{2}\log s \tag{11.2.17}$$

we have

$$\left.\frac{1}{\sigma_{12}^{\mathrm{in}}}\frac{\mathrm{d}\sigma}{\mathrm{d}y}\right|_{\mathrm{plateau}} \approx \frac{\gamma}{2} \approx 1$$

which is compatible with the data to within a factor of 2.

Of course the third term of fig. 11.1 may also be included, and is regarded by some authors (e.g. Hwa) as the most important, and by others as at least equally important at high energies. However, since even with such modifications the model is unable to account for many of the crucial features of many-particle production we shall not pursue these variants here.

The first problem concerns particle correlations. From (10.3.4) and (11.2.2) we obtain with (11.2.12) and (11.2.15)

$$\langle n(n-1)\rangle = 2\int \frac{N(M)\,(N(M)-1)\,\rho(M)\,\mathrm{d}M}{\sigma_{12}^{\mathrm{in}}} \underset{s\to\infty}{\sim} \int^{\sqrt{s}} \mathrm{d}M \to \sqrt{s} \tag{11.2.18}$$

so, even though $F_1 \sim \log s$, $F_2 \sim \sqrt{s}$ and hence from (10.10.4) $C_2 \sim \sqrt{s}$ as well. In fact it is obvious that the model predicts

$$C_n \sim F_n \sim (\sqrt{s})^{n-1}, \quad n > 1 \tag{11.2.19}$$

which is incompatible with the high energy data (e.g. fig. 10.29).

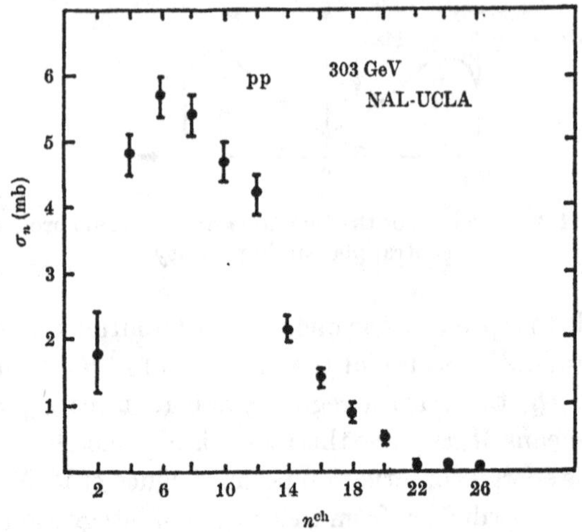

FIG 11.4 Date for σ_n against n at fixed s for charged particles.

Also, since $\rho(M) \sim M^{-2}$ and $N(M) \sim M$, the cross-section for producing n particles, $\sigma_n(s)$, has the behaviour for large n, at fixed s,

$$\sigma_n(s) \underset{n\to\infty}{\longrightarrow} \frac{d\sigma_{12}^{in}}{dn} \propto \frac{d\sigma_{12}^{in}}{dM} \sim \frac{1}{M^2} \sim \frac{1}{n^2} \qquad (11.2.20)$$

from (11.2.1). But experimentally (fig. 11.4) it is falling much faster than this for large n. Part of the problem could be just the failure of this simple version of the model to take into account the phase-space restrictions on producing large numbers of particles, but it has been shown by Le Bellac and Meunier (1973) that even using proper kinematics it is not possible to fit simultaneously the flat $d\sigma/dy$ for $y \approx 0$ and σ_n.

If we include the triple-Pomeron term in (11.2.14) for large M^2, then clearly $N(M) \propto M$ is impossible if we also wish to retain $\langle n\rangle \sim \log s$. If we regard the Pomeron as an ordinary particle then fig. 11.1 (*b*) is just the process $P_v + 2 \to X$ where P_v is the virtual Pomeron, and as M is the total energy for this process we would expect

$$\langle n\rangle \approx C \log (M^2) \qquad (11.2.21)$$

where C is some constant, which seems to be true experimentally (fig. 11.5). Then $dn = 2C\, dM/M$ and so for large n

$$\sigma_n(s) \underset{n\to\infty}{\longrightarrow} \frac{d\sigma_{12}^{in}}{dn} \propto M\frac{d\sigma}{dM} = 2M^2\frac{d\sigma}{dM^2} = \int_{-\infty}^{0} 2M^2 \frac{d^2\sigma}{dt\, dM^2}\, dt \qquad (11.2.22)$$

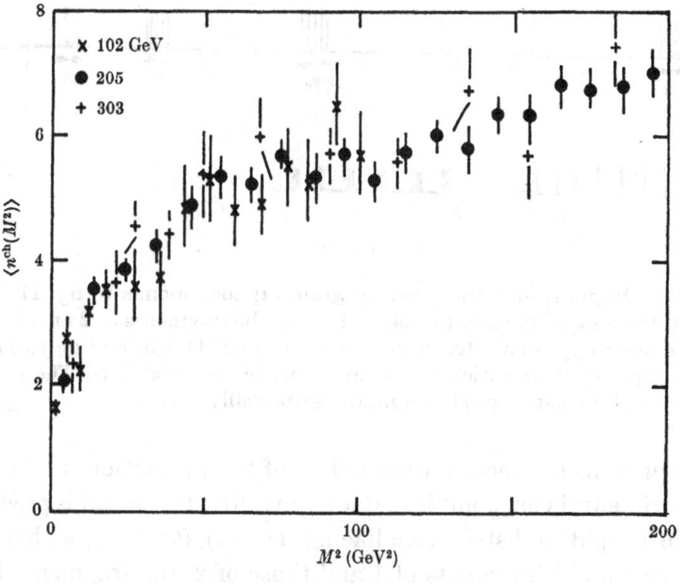

FIG. 11.5 The mean number of charged particles produced in $pp \to pX$ as a function of M^2 at various energies. This is consistent with the form

$$\langle n\rangle = B + C\log(M^2).$$

(From Fox 1973.)

so using (10.8.6) with $G^{PP,P}_{11,2}(t) \propto e^{at}$ and $\alpha_P(t) = 1+\alpha'_P t$ we find

$$\sigma_n \propto \left(a + 2\alpha'_P \log\left(s - \frac{n}{c}\right)\right)^{-1} \qquad (11.2.23)$$

So each $\sigma_n \sim (\log s)^{-1}$ even though

$$\langle n\rangle \propto \int n\sigma_n \, dn \propto \log s$$

Alternatively, with a vanishing triple-Pomeron coupling,

$$G^{PP,P}_{11,2}(t) \propto (-t)\, e^{at}$$

we get
$$\sigma_n \propto \left(a + 2\alpha'_P \log\left(s - \frac{n}{c}\right)\right)^{-2} \qquad (11.2.24)$$

These results are like those of the multi-peripheral model, to be described in the next section, and it is clear that P exchange cannot give a consistent view of σ_n versus n. So, even bearing in mind the fact that the triple-Regge formation is strictly applicable only for $M^2/s \ll 1$, this does appear to help us to understand why the nova model is incorrect.

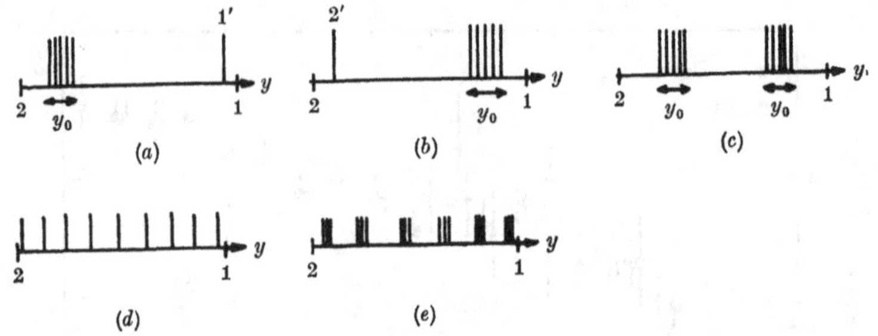

FIG. 11.6 Rapidity distributions: Diagram (*a*) corresponds to fig. 11.1 (*b*) in which 1′ has a rapidity close to that of 1, while the fragments of 2 are clustered within a length y_0. Similarly, (*b*) corresponds to fig. 11.1 (*a*) and (*c*) to 11.1 (*c*). (*d*) The rapidity distribution in the multi-peripheral model. (*e*) The rapidity distribution for clusters produced multi-peripherally.

But perhaps the most serious defect of the diffraction model from an experimental viewpoint is that a given diffractive event is predicted to have a rapidity distribution like fig. 11.6 (*a*), (*b*) or (*c*), with a large gap between the fragments of 1 and those of 2, the fragments being clustered within a range y_0 (see (11.2.8)), even though, when one averages over a large number of events, a flat rapidity distribution may be obtained. In fact only a fraction of the observed events have this structure, many more having the more uniform distribution characteristic of the multi-peripheral model (fig. 11.6 (*d*), (*e*)).

So it is clear that the diffraction model can at best account for only a small part of the high energy cross-section. In section 11.6 we shall combine this diffractive P contribution with the more dominant multi-peripheral amplitude.

11.3 The multi-peripheral model

The basic idea behind the multi-peripheral model is that at high energy the dominant production mechanism should be like fig. 11.7, in which each particle along the chain is produced peripherally, i.e. at small momentum transfer with respect to those adjacent to it. The original version (Bertocchi, Fubini and Tonin 1962, Amati *et al.* 1962*a*, *b*) often referred to as 'the ABFST model' after the initials of the authors, involved elementary pion exchange between successive particles, but we should now think it more appropriate to use Reggeon exchanges instead (Chew *et al.* 1968, 1969, 1970, Halliday 1969, Halliday and Saunders 1969, de Tar 1971), and we might eventually want to include Regge cuts as well.

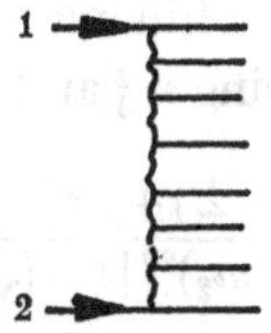

FIG. 11.7 The multi-peripheral model with Reggeon exchanges.

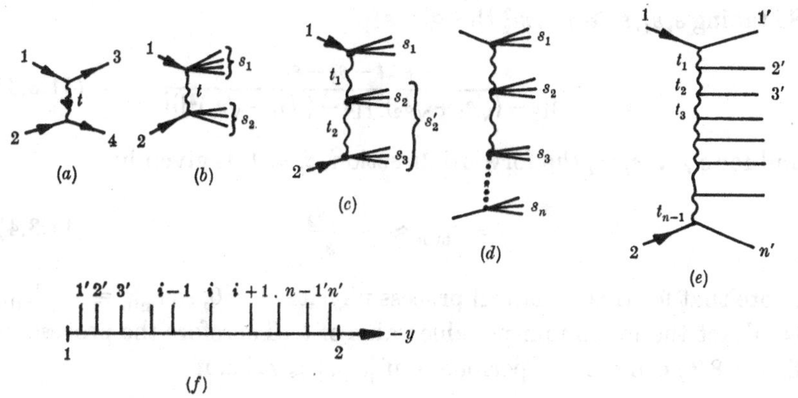

FIG. 11.8 (*a*) Peripheral exchange in $2 \to 2$ scattering. (*b*) Peripheral $2 \to n$ amplitude. (*c*) Doubly peripheral process. (*d*) Multi-peripheral process. (*e*) Multi-peripheral production of single particles, $1+2 \to 1'+2'+\ldots+n'$. (*f*) Strong ordering which occurs when the ordering of the particles in rapidity is the same as the ordering of their couplings, i.e. the same as in (*e*).

A two-body amplitude like fig. 11.8(*a*) can often be represented at fixed s by (cf. (6.8.11))

$$A(s,t) \propto e^{ct} \tag{11.3.1}$$

with $c \approx 2$–$6\,\text{GeV}^{-2}$, indicating the dominance of low-t singularities, i.e. the longer range forces, so that as discussed in section 2.4 the beam can be thought of as interacting strongly with the periphery of the target, and the amplitude is rapidly damped in t. So we can regard an interaction as peripheral, in this sense, if say $|t| \leqslant \tau \equiv 0.5\,\text{GeV}^2$ includes the bulk of the events. (The reader should note that this is a somewhat different use of 'peripheral' from that of section 8.6 where the word meant dominance of impact parameter $b \approx R = 1\,\text{fm}$, producing t dependence of the form $J_n(R\sqrt{-t})$. It is rather unfortunate that both meanings of the word are in current use.)

Similarly the many-particle amplitude, fig. 11.8(*b*), is said to be peripheral if $|t| \leqslant \tau$, and we can expect this to be the dominant t-region for $s \gg s_1, s_2$. However, the minimum possible value of $|t|$, i.e. $|t_{min}|$, is determined by the kinematics and depends on s_1 and s_2. From (1.7.17) we have, replacing m_3^2 and m_4^2 by s_1 and s_2 respectively,

$$z_s = \frac{s^2+s(2t-\Sigma)+(m_1^2-m_2^2)(s_1-s_2)}{\{[s-(m_1+m_2)^2][s-(m_1-m_2)^2][s-(\sqrt{s_1}+\sqrt{s_2})^2][s-(\sqrt{s_1}-\sqrt{s_2})^2]\}^{\frac{1}{2}}} \tag{11.3.2}$$

$$\Sigma \equiv m_1^2+m_2^2+s_1+s_2$$

So taking $s, s_1, s_2 \gg m_1^2, m_2^2$ this gives

$$z_s \approx \frac{s+2t-s_1-s_2}{\{[s-(\sqrt{s_1}+\sqrt{s_2})^2][s-(\sqrt{s_1}-\sqrt{s_2})^2]\}^{\frac{1}{2}}} \tag{11.3.3}$$

and for $s \gg t, s_1, s_2$ the forward direction, $z_s = 1$, is given by

$$t = t_{min} \approx -\frac{s_1 s_2}{s} \tag{11.3.4}$$

(Note that for this s-channel process physical $t < 0$, so $t_{min} \equiv -|t|_{min}$ is in fact the maximum possible value of t.) Therefore the process in fig. 11.8 (*b*) can only be peripheral if $|t_{min}| \leqslant \tau$, i.e. if

$$\frac{s_1 s_2}{s} \leqslant \tau, \tag{11.3.5}$$

which corresponds to the single-Regge limit of section 9.3.

Extending this idea, a process can be doubly peripheral, like fig. 11.8 (*c*) if

$$|t_{1\,min}| = \frac{s_1 s_2'}{s} \leqslant \tau \quad \text{and} \quad |t_{2\,min}| = \frac{s_2 s_3}{s_2'} \leqslant \tau \tag{11.3.6}$$

and so

$$\frac{s_1 s_2'}{s} \cdot \frac{s_2 s_3}{s_2'} = \frac{s_1 s_2 s_3}{s} \leqslant \tau^2 \tag{11.3.7}$$

Note that because of the way we have chosen to analyse the diagram s_2' is the energy appropriate to t_2 exchange, not s, but the final result (11.3.7) treats $s_1 s_2 s_3$ symmetrically. And for n clusters, fig. 11.8 (*d*), we need

$$\frac{s_1 s_2 \ldots s_n}{s} \leqslant \tau^{n-1} \tag{11.3.8}$$

An immediate consequence of this hypothesis is that if we suppose all the clusters to have some average mass, so $\langle s_i \rangle = s_A$, say, $i = 1, \ldots, n$ then (11.3.7) gives

$$\frac{s_A^{\langle n \rangle}}{s} \leqslant \tau^{\langle n \rangle - 1} \tag{11.3.9}$$

where $\langle n \rangle$ is the average number of clusters produced, and so

$$\langle n \rangle \log s_{\mathrm{A}} - \log s \leqslant (\langle n \rangle - 1) \log \tau$$

or

$$\langle n \rangle \leqslant \log \left(\frac{s}{\tau}\right) \log \left(\frac{\tau}{s_{\mathrm{A}}}\right) \tag{11.3.10}$$

So the average number of clusters increases at most logarithmically with s, an experimentally desirable result, particularly if we take the 'clusters' to be single particles, as in fig. 11.8 (e).

In this case we can write a multi-peripheral model for the amplitude $1+2 \rightarrow 1'+ \ldots +n'$ in the form suggested by the multi-Regge model (9.3.10):

$$A^{2\rightarrow n}(p_1, p_2; p'_1, \ldots, p'_n) = \gamma(t_1)\, R(t_1, s_{12})\, G(t_1, t_2, \eta_{12}) \times R(t_2, s_{23})\, G(t_2, t_3, \eta_{12}) \ldots R(t_{n-1}, s_{n-1,n})\, \gamma(t_{n-1}) \tag{11.3.11}$$

where the γ's and G's are the couplings and

$$R(t_i, s_{i,i+1}) \equiv R_i \tag{11.3.12}$$

represents the ith Reggeon exchange. Except at the ends the couplings depend both on the Reggeon masses t_i, t_{i+1} and on the Toller angle variable (9.2.31),

$$\eta_{i,i+1} = \frac{s_{i,i+1,i+2}}{s_{i,i+1}\, s_{i+1,i+2}} \tag{11.3.13}$$

Clearly we have assumed factorization in writing (11.3.11) as well as multiperipherality. The equation is rather complicated because of the signature properties of the Reggeons. The simplest version of the model with an elementary scalar-particle-exchange amplitude would just have all γ's and G's $= g$, the coupling strength, and $R_i = 1/(t - m_i^2)$, corresponding to the Feynman rules of section 1.12.

Equation (11.3.11) may be approximately valid for $|t_i| \leqslant \tau$, $s_{i,i+1} \gg s_0$ for $i = 1, \ldots, (n-1)$, but this is only a small part of the available phase space, and as discussed in section 9.2 many events will probably have low sub-energies, due for example to resonance production. So to apply the model more widely, as we shall do below, it is necessary to make some sort of duality assumption, that this high-sub-energy form of the amplitude also applies, at least in some average sense, for low $s_{i,i+1}$ as well.

If we assume that the model is approximately valid for all phase space, from (1.8.5) we can calculate the cross-section for producing n particles as

$$\sigma_n \equiv \sigma_{12\rightarrow n} \approx \frac{1}{2s} \int \mathrm{d}\Phi_n \, |A^{2\rightarrow n}|^2 \tag{11.3.14}$$

where $d\Phi_n$ is the n-particle phase-space volume element of (1.8.6). If we work in the rest frame of particle 1 we can write (see Halliday and Saunders 1969)

$$\left.\begin{aligned} p_1 &= (m_1, 0, 0, 0) \\ p_2 &= (m_2 \cosh Y, 0, 0, m_2 \sinh Y) \end{aligned}\right\} \tag{11.3.15}$$

where (see (10.2.22))

$$Y \equiv y_2 - y_1 \xrightarrow[s\to\infty]{} \log \frac{s}{m_1 m_2} \tag{11.3.16}$$

and for the final-state particles (see (10.2.18))

$$p_i' = (\mu_i \cosh y_i, \mathbf{p}_{iT}, \mu_i \sinh y_i) \tag{11.3.17}$$

Then from (1.8.6) and (10.3.5)

$$d\Phi_n = \prod_{i=1}^{n} \left(\frac{d^2\mathbf{p}_{iT}\, dy_i}{16\pi^3}\right) (2\pi)^2 \delta^2 \left(\sum_{i=1}^{n} \mathbf{p}_{iT}\right) \tfrac{1}{2} 2\pi \delta \left(\sum_{i=1}^{n} \mu_i e^{y_i} - m_1 - m_2 e^{Y}\right)$$
$$\times\, 2\pi\, \delta \left(\sum_{i=1}^{n} \mu_i e^{-y_i} - m_1 - m_2 e^{-Y}\right) \tag{11.3.18}$$

To simplify we approximate the Reggeon amplitude by

$$\gamma_i R_i \approx g(s_{i,i+1})^{\alpha} \tag{11.3.19}$$

completely ignoring the dependence of the γ's, G's and α's on the t_i, and on the Toller angles (11.3.13), and so (11.3.11) becomes

$$A^{2\to n} \approx g^n \prod_{i=1}^{n-1} (s_{i,i+1})^{\alpha} \tag{11.3.20}$$

Now $$s_{i,i+1} \equiv (p_i' + p_{i+1}')^2 \approx \mu_i \mu_{i+1} e^{y_{i+1} - y_i} \tag{11.3.21}$$

(see (10.2.22)) and if each $s_{i,i+1}$ is large then $y_{i+1} \gg y_i$ for all i. In this region of phase space we have what is called 'strong ordering' in rapidity, i.e. the ordering of the particles in rapidity, fig. 11.8(f), corresponds exactly to the ordering of their couplings in fig. 11.8(e), but clearly this is true only in part of phase space. Then

$$\prod_{i=1}^{n-1} s_{i,i+1} = \mu_1 \mu_2^2 \ldots \mu_{n-1}^2 \mu_n e^{y_n - y_1} \tag{11.3.22}$$

and the maximum contribution in the integral over (11.3.18) comes from $\mathbf{p}_{iT}^2 \approx 0$, so (from (10.2.2)) $\mu_i^2 \approx m_i^2 = m^2$ if we take all the particles

$1', \ldots, n'$ to have the same mass. Hence

$$A^{2\to n} \approx g^n[(m^2)^{n-2}\mu_1\mu_n \mathrm{e}^{y_n-y_1}]^\alpha \tag{11.3.23}$$

From the δ functions (11.3.18) we need

$$\mu_1 \mathrm{e}^{-y_1} \approx m_1 \quad \text{and} \quad \mu_n \mathrm{e}^{y_n} \approx m_2 \mathrm{e}^Y$$

(remember $y_{i+1} \gg y_i$ for all i so all the other terms in the δ functions can be ignored) and hence

$$A^{2\to n} \approx g^n[(m^2)^{n-2} m_1 m_2 \mathrm{e}^Y]^\alpha = g^n[(m^2)^{n-2} s]^\alpha \tag{11.3.24}$$

from (11.3.16). So our approximations have eliminated all the dependence of $A^{2\to n}$ on the sub-energies and momentum transfers.

Then putting

$$z_i = y_{i+1} - y_i \tag{11.3.25}$$

and ignoring p_{iT} in (11.3.18) we get after some manipulation

$$\mathrm{d}\Phi_n \propto \frac{\mathrm{e}^{-Y}}{2m_1 m_2} \prod_{i=1}^{n-1} \mathrm{d}z_i\, \delta\left(Y - \sum_{i=1}^{n-1} z_i\right) \tag{11.3.26}$$

and so

$$\sigma_n \propto g^{2n} s^{2\alpha-2} \int_0^Y \prod_{i=1}^{n-1} \mathrm{d}z_i\, \delta\left(Y - \sum_{i=1}^{n-1} z_i\right) \tag{11.3.27}$$

Now from the Feynman relation (1.12.4)

$$\int_0^1 \mathrm{d}\alpha_1 \ldots \mathrm{d}\alpha_n\, \delta(\Sigma\alpha - 1) = \frac{1}{(n-1)!}$$

and substituting $\alpha_i = z_i/Y$ we get

$$\int_0^Y \mathrm{d}z_i \ldots \mathrm{d}z_{n-1}\, \delta(\Sigma z_i - Y) = \frac{Y^{n-2}}{(n-2)!} \tag{11.3.28}$$

So replacing g by $\bar{g}$,the average of g over the phase-space integration, we obtain

$$\sigma_n = s^{2\alpha-2}\bar{g}^4 \frac{(\bar{g}^2 Y)^{n-2}}{(n-2)!} \tag{11.3.29}$$

and hence

$$\sigma_{12}^{\mathrm{tot}} = \sum_n \sigma_n = \sum_{n=2}^{\infty} s^{2\alpha-2}\bar{g}^4 \frac{(\bar{g}^2 Y)^{n-2}}{(n-2)!} = \bar{g}^4 \mathrm{e}^{Y(2\alpha-2+\bar{g}^2)} \tag{11.3.30}$$

So to get a constant total cross-section we need

$$2\alpha - 2 + \bar{g}^2 = 0, \quad \text{i.e. } \alpha = 1 - \frac{\bar{g}^2}{2} \tag{11.3.31}$$

Hence $\alpha < 1$, and the amplitude cannot be dominated by multiple P exchange. Successive P exchange would give $\sigma_n \sim (\log s)^{n-2}$ and $\sigma_{12}^{\text{tot}} \sim s^{\bar{g}^2}$ in violation of the Froissart bound (Finkelstein and Kajantie 1968*a*, *b*).

If (11.3.31) is substituted back into (11.3.29) we find ($Y = \log s$)

$$\sigma_n = \bar{g}^4 \frac{(\bar{g}^2 Y)^{n-2}}{(n-2)!} \mathrm{e}^{-\bar{g}^2 Y} \tag{11.3.32}$$

so
$$\langle n \rangle = \frac{\sum_n n\sigma_n}{\sum_n \sigma_n} = \bar{g}^2 Y \sim (2-2\alpha) \log s \tag{11.3.33}$$

which gives the required logarithmic increase of the average multiplicity with s (Chew and Pignotti 1968). In fact this result does not really depend in any important way on the details of the model. For if we put say (see Fubini 1963)

$$\sigma_2 = \lambda \bar{\sigma}_2 \tag{11.3.34}$$

where λ is some variable coupling parameter (for example $\lambda =$ the coupling g^2), then by factorization (see fig. 11.9),

$$\sigma_n = \lambda^n \bar{\sigma}_n \tag{11.3.35}$$

so
$$\langle n \rangle = \frac{\sum_n n\lambda^n \bar{\sigma}_n}{\sum_n \lambda^n \bar{\sigma}_n} = \lambda \frac{\mathrm{d}\sigma/\mathrm{d}\lambda}{\sigma}\bigg|_{\lambda=1} \tag{11.3.36}$$

Hence if
$$\sigma(s) = \beta(\lambda)\, s^{\alpha(\lambda)} = \mathrm{e}^{\alpha(\lambda)\log s + \beta(\lambda)} \tag{11.3.37}$$

where α, β are arbitrary functions of λ, then

$$\langle n \rangle = \left(\lambda \frac{\mathrm{d}\alpha}{\mathrm{d}\lambda}\right) \log s + \lambda \frac{\mathrm{d}\beta}{\mathrm{d}\lambda}\bigg|_{\lambda=1} \tag{11.3.38}$$

Thus so long as there is some (unspecified) dynamical relation between the power behaviour of $\sigma(s)$ and the magnitude of some factorizable coupling strength we shall always find

$$\langle n \rangle \sim \log s \tag{11.3.39}$$

independent of the details of the model.

Putting (11.3.33) into (11.3.32) gives

$$\sigma_n = \bar{g}^4 \frac{\langle n \rangle^{n-2} \mathrm{e}^{-\langle n \rangle}}{(n-2)!} \tag{11.3.40}$$

so, at fixed s, σ_n against n has a Poisson distribution whose width

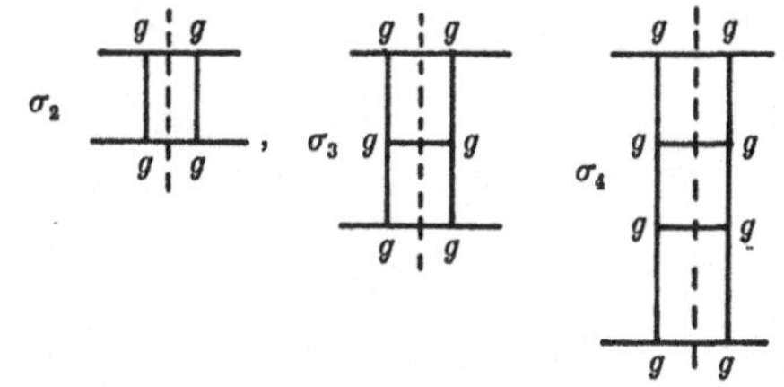

FIG. 11.9 The discontinuities which give the various many-particle cross-sections in the multi-peripheral model. Each successive term contains a factor g^2 relative to previous one.

increases like $\log s$. And the average spacing of the particles in rapidity is

$$\Delta y = \frac{Y}{\langle n \rangle} = \frac{1}{\bar{g}^2} \tag{11.3.41}$$

from (11.3.33).

The probability that the ith particle has rapidity $y = y_i = \sum_{j=1}^{i-1} z_j$ is

$$\frac{d\sigma_{n,i}}{dy} = e^{Y(2\alpha-2)}\bar{g}^{2n} \int_0^Y \prod_{j=2}^{n-1} dz_j\, \delta\left(Y - \sum_{j=1}^{n-1} z_j\right) \delta\left(y - \sum_{j=1}^{i-1} z_j\right)$$

$$= e^{Y(2\alpha-2)}\bar{g}^{2n} \frac{(y)^{i-1}}{(i-1)!} \frac{(Y-y)^{n-i-2}}{(n-i-2)!} \tag{11.3.42}$$

from (11.3.28), the first part coming from the $i-1$ particles with $y < y_i$ and the second from the $n-i$ particles with $y_i < y < Y$, as in fig. 11.8(f). This distribution is shown in fig. 11.10. So the full inclusive distribution is

$$\frac{d\sigma}{dy} = \sum_{i=1}^{n-2} \sum_{n=3}^{\infty} \frac{d\sigma_{n,i}}{dy} = \sum_{n=3}^{\infty} e^{Y(2\alpha-2)}\bar{g}^{2n} \frac{Y^{n-3}}{(n-3)!} \tag{11.3.43}$$

since the binomial expansion gives

$$\frac{Y^{n-3}}{(n-3)!} = \frac{(Y-y+y)^{n-3}}{(n-3)!} = \sum_{i=1}^{n-2} \frac{(Y-y)^{n-i-2}(y)^{i-1}}{(n-i-2)!\,(i-1)!},$$

and so

$$\frac{d\sigma}{dy} = \bar{g}^6 s^{2\alpha-2+\bar{g}^2} = \bar{g}^6 \tag{11.3.44}$$

if (11.3.31) holds to give $\sigma_{12}^{tot}(s) \to \bar{g}^4$. And so we get a flat, uniform scaling distribution of particles in the central region. And combining (11.3.30) with (11.3.41)

$$\frac{1}{\sigma_{12}^{tot}} \frac{d\sigma}{dy} = \bar{g}^2 = \frac{\langle n \rangle}{\log s} \tag{11.3.45}$$

which is the same as the diffraction model result (11.2.17). Of course,

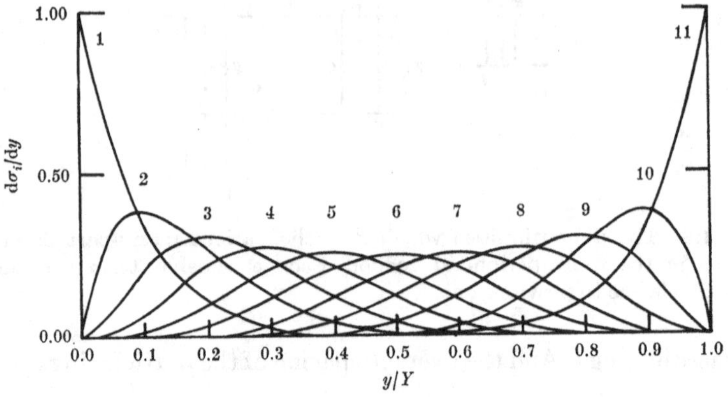

FIG. 11.10 The rapidity distribution of the ith produced particle $d\sigma_i/dy$ (in arbitrary units) in the multi-peripheral model, for 11 produced particles. From de Tar (1971).

since the amount of rapidity available is increasing like $\log s$, any model with a scaling central distribution and $\langle n \rangle \sim \log s$ must obey (11.3.45).

Similarly one can evaluate $d^2\sigma/dy_i\,dy_i$, and it is found, not surprisingly, that there are no correlations between the produced particles in this factorizing model. We shall show this more simply in section 11.5 below.

The most obvious defect of the model is that it does not give any leading-particle effect, i.e. there is no special enhancement of the probability distribution for particles having a similar rapidity to that of the beam or target particles, which the diffraction model produces so naturally. So in section 11.6 we shall attempt to combine the two models. However, first it is useful to examine the internal self-consistency of the multi-peripheral model.

11.4 The multi-peripheral bootstrap

In writing the multi-Regge form for the multi-peripheral amplitude (11.3.11) we can insert arbitrary Regge poles, α_R. And then in 'squaring' the amplitude in (11.3.14), and summing over n we obtain the behaviour (11.3.30) for the total cross-section. Thus in (11.3.31) we obtained the condition on the trajectory for constancy of the total cross-section. But obviously this is not self-consistent because a constant $\sigma^{tot}(s)$ requires P exchange with $\alpha_R(0) = 1$, whereas (11.3.31) demands

$$\alpha_{\mathrm{R}}(0) = 1 - \frac{\bar{g}^2}{2} < 1 \tag{11.4.1}$$

Instead we could demand self-consistency of the input and output Reggeons and require, comparing (6.8.4) and (11.3.30),

$$\sigma_{12}^{\mathrm{tot}}(s) \sim s^{\alpha-1} = \sum_n \sigma_n \sim s^{2\alpha-2-\bar{g}^2} \tag{11.4.2}$$

and so (Chew and Pignotti 1968)

$$\alpha(0) = 1 - \bar{g}^2 < 1 \tag{11.4.3}$$

This is a simple example of a bootstrap calculation. The input Reggeons in the multi-peripheral chain are used in the unitarity equation to build up ladders (see fig. 11.11) which, when summed, give back a Reggeon; and this should, for self-consistency, be identical with the input Reggeons. It is clear from the outset, however, that this can be, at best, only an approximation, because for complete self-consistency we should include cuts in the multi-peripheral chains, and consider diagrams with crossed rungs which give back cuts in the output as well. We shall reconsider this problem in the final section. But here we want to examine a bit more closely the pole-dominance approximation, and so we shall stick to the strong ordering of (11.3.21) *et seq.* with no crossing rungs.

If we adopt the Regge exchange model (11.3.11) for all $2 \to n$ amplitudes, the discontinuity across the two-particle cut (fig. 11.12 (*b*)) is given by (cf. (8.2.11))

$$D_2(s,t) = \frac{1}{16\pi^2 s} \int\int_{-\infty}^{0} \frac{\mathrm{d}t_1\, \mathrm{d}t_1'\, \theta(-\lambda)}{(-\lambda(t,t_1,t_1'))^{\frac{1}{2}}} \gamma^2(t_1)\, R(t_1,s)\, \gamma^{*2}(t_1')\, R^*(t_1',s) \tag{11.4.4}$$

(say) and the complete s-channel discontinuity equation of fig. 11.12 is

$$D(s,t) = \sum_{n=2}^{\infty} D_n(s,t) = D_2(s,t) + \sum_{n=3}^{\infty} \int \mathrm{d}\Phi_n\, \gamma(t_1)\, RGR \ldots \gamma(t_{n-1}) \times \gamma^*(t_1')\, R^* G^* R^* \ldots \gamma^*(t_{n-1}') \tag{11.4.5}$$

Since this infinite sum involves repetition of the same basic two-Reggeon exchange contribution we can rewrite it recursively (cf. (1.13.27), (3.4.20)), as in fig. 11.12 (*f*), in the form (see Chew *et al.* 1969, Goldberger 1969)

$$D(s,t) = D_2(s,t) + D_2 \otimes D \tag{11.4.6}$$

where $\otimes$ implies integration over t_2, t_2' in a similar fashion to (11.4.4). Strictly D_2 and D in this integration may be expected to depend on

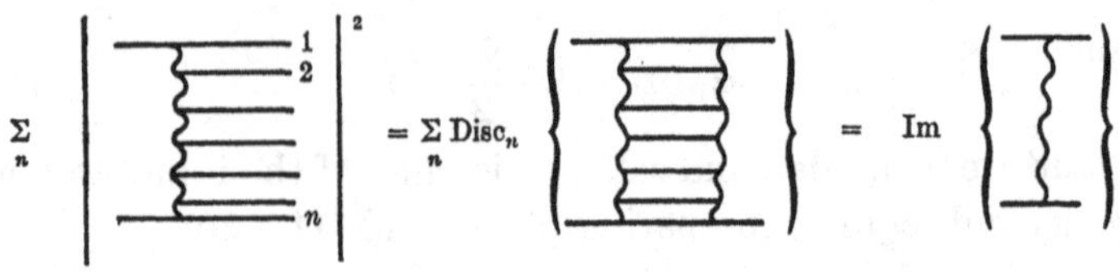

FIG. 11.11 Multi-peripheral bootstrap for a Regge trajectory.

t_2, t'_2 but for simplicity we ignore any such dependence here. This integration is simplified if we project into t-channel partial waves, defining (from (2.5.3))

$$\left.\begin{aligned} A_2(J,t) &= \frac{1}{16\pi^2}\int^{\infty} D_2(s,t)\, Q_J(z_t)\,\mathrm{d}s \\ A(J,t) &= \frac{1}{16\pi^2}\int^{\infty} D(s,t)\, Q_J(z_t)\,\mathrm{d}s \end{aligned}\right\} \tag{11.4.7}$$

and (11.4.6) becomes (cf. (2.2.7))

$$A(J,t) = A_2(J,t) + A_2(J,t)\,A(J,t)$$

or

$$A(J,t) = \frac{A_2(J,t)}{1-A_2(J,t)} \tag{11.4.8}$$

which gives $A(J,t)$ in terms of $A_2(J,t)$ (provided we accept the drastic approximations made en route). Note that we are using t-channel partial waves in the s-channel physical region, so really this is an O(2, 1) not an O(3) projection (see section 6.6).

A rather disturbing feature of fig. 11.12(*b*), and (11.4.4), is that they clearly generate an AFS cut (8.2.17), which we know should be cancelled by higher order discontinuities taken through the Reggeons themselves (see section 8.2). But if we overlook this difficulty, then a fixed-pole input in (11.4.4), i.e.

$$\gamma^2(t)\,R(t,s) \approx \gamma^2(t)\,s^{\alpha_0} \tag{11.4.9}$$

gives

$$D_2(s,t) = \bar{\beta}(t)\,s^{2\alpha_0-1} \tag{11.4.10}$$

where

$$\bar{\beta}(t) \equiv \frac{1}{16\pi^2}\iint_{-\infty}^{0} \frac{\mathrm{d}t_1\,\mathrm{d}t'_1\,\theta(\lambda)}{(-\lambda(t,t_1,t'_1))^{\frac{1}{2}}}\,\gamma^2(t_1)\,\gamma^2(t'_1) \tag{11.4.11}$$

and so from (11.4.7) and (2.7.2) with $\beta(t) \equiv \bar{\beta}/16\pi^2$

$$A_2(J,t) = \frac{\beta(t)}{J-(2\alpha_0-1)} \tag{11.4.12}$$

which in (11.4.8) gives

$$A(J,t) = \frac{\beta(t)}{J-(2\alpha_0-1)-\beta(t)} \tag{11.4.13}$$

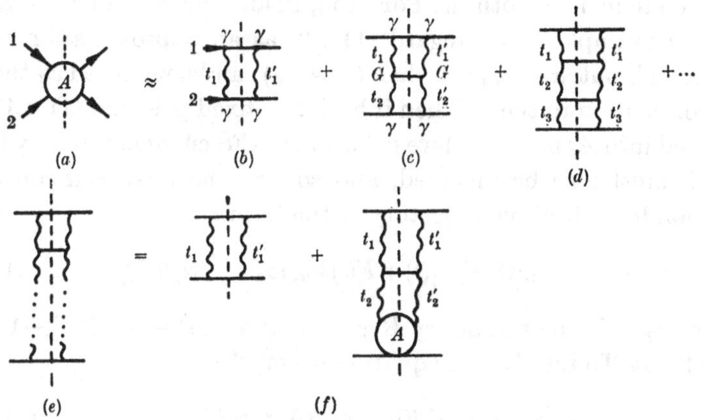

FIG. 11.12 The s-discontinuity of the amplitude (a) is expressed in the multi-Regge approximation (b)–(e). This is rewritten recursively in (f).

i.e. a moving pole at

$$J = \alpha(t) \equiv 2\alpha_0 - 1 + \beta(t) \qquad (11.4.14)$$

Note that if $\alpha_0 = 0$ this becomes $\alpha(t) = -1 + \beta(t)$ in accord with the field-theory result (3.4.19).

So unitarity replaces the input fixed cut (11.4.10) by a moving pole. Self-consistency of input and output at $t = 0$ (the dominant region of (11.4.11) if $\gamma(t)$ falls rapidly with $-t$) demands that

$$\alpha_0 = \alpha(0) = 2\alpha_0 - 1 + \beta(0), \quad \text{i.e.} \quad \alpha_0 = 1 - \beta(0) \qquad (11.4.15)$$

so $\alpha(0) < 1$ in agreement with (11.4.3).

If alternatively we try a moving-pole input

$$\gamma^2(t)\, R(t,s) \approx \gamma^2(t)\, s^{\alpha(t)} \qquad (11.4.16)$$

then from (8.2.17)

$$D_2(s,t) = \bar{\beta}(t)\frac{s^{\alpha_c(t)}}{\log s} \quad \text{where} \quad \alpha_c(t) \equiv 2\alpha\left(\frac{t}{4}\right) - 1 \qquad (11.4.17)$$

giving, through (2.7.4),

$$A_2(J,t) = \beta(t)\log(J - \alpha_c(t)) \qquad (11.4.18)$$

and so

$$A(J,t) = \frac{\beta(t)\log(J-\alpha_c(t))}{1-\beta(t)\log(J-\alpha_c(t))} \qquad (11.4.19)$$

So the output is an AFS cut which has moved from its original position at $J = \alpha_c(t)$, so again self-consistency is not achieved.

The problem is presumably due, at least in part, to the fact that to get even a crudely correct description of the scattering amplitude we need to include both the Pomeron, P, and the secondary Reggeons R. For example, if we regard (11.4.9) as an approximation to the Reggeon input, $\alpha_0 = \alpha_R(0)$, then (11.4.14) can be regarded as the first approximation to the P. Then if both this fixed pole and (11.4.13) are inserted into A_2 in (11.4.8) we get also an AFS cut generated by the P, which must also be included, and so on. The final, self-consistent solution has a leading trajectory of the form

$$\alpha_P(t) = \alpha_R(t) + F(t) \log(\alpha_P(t) - \alpha_c(t)) \tag{11.4.20}$$

where $\alpha_R(t)$ is the secondary Reggeon, and $\alpha_c(t) \equiv 2\alpha_P(t/4) - 1$ is the $P \otimes P$ cut. To satisfy this equation we must have

$$\alpha'_P(0) < \alpha'_R(0), \quad \alpha_R(0) < \alpha_P(0) < 1 \tag{11.4.21}$$

(otherwise $\alpha_c(0) > \alpha_P(0)$). The properties of $\alpha_P(t)$ in (11.4.20) are very different for $t > 0$ and $t < 0$, and it has been called the 'schizophrenic Pomeron' by Chew and Snider (1971).

However, since cross-sections are found still to be rising at high energies this sort of solution of the Pomeron self-consistency problem no longer seems so attractive. Many variants of this approach have been suggested, but quite apart from their computational complexity, which generally necessitates over-simplification of the phase-space integrations, there seem to be two crucial difficulties. One is the generation of AFS cuts, which we know from section 8.2 would not be present if the s-discontinuities of the Reggeons themselves were also incorporated, and the other is the necessity for the strong-ordering assumption, which ensures that only planar diagrams are included. But since low sub-energies generally give the most important contributions to the integrals this is implausible, especially since we know that non-planar diagrams are essential if the correct Regge cut structure is to be obtained as well (see Halliday 1969).

11.5 The generating function

A very useful way of discussing the correlations in models of this sort is the generating function method of Mueller (1971).

In analogy with statistical mechanics the generating function or 'partition function' $Q(z, Y)$ is defined by

$$Q(z, Y) \equiv \sum_{n=0}^{\infty} z^n \sigma_{n+2}(Y) \tag{11.5.1}$$

where $\sigma_{n+2}(Y)$ is the cross-section for producing n particles (so that there are $n+2$ in the final state) at a given $Y \equiv \log(s/m_1 m_2)$ (which gives the length of the rapidity plot), and z is an arbitrary parameter.

Clearly the point $z = 1$ has special significance in that

$$Q(1, Y) = \sum_n \sigma_{n+2}(Y) = \sigma_{12}^{\text{tot}}(Y) \tag{11.5.2}$$

$$\left(\frac{\mathrm{d}Q}{\mathrm{d}z}\right)_{z=1} = \sum_n n\sigma_{n+2}(Y) = \langle n \rangle \sigma_{12}^{\text{tot}}(Y) = F_1 \sigma_{12}^{\text{tot}} \tag{11.5.3}$$

$$\left(\frac{\mathrm{d}^2Q}{\mathrm{d}z^2}\right)_{z=1} = \sum_n n(n-1)\,\sigma_{n+2}(Y) = \langle n(n-1)\rangle\, \sigma_{12}^{\text{tot}}(Y) = F_2 \sigma_{12}^{\text{tot}} \tag{11.5.4}$$

etc., using (10.3.11) and (10.3.16). So the behaviour of Q in the neighbourhood of $z = 1$ gives average multiplicity of produced particles, and we can rewrite (11.5.1) as

$$Q(z, Y) = \sigma_{12}^{\text{tot}}(Y) \sum_{n=0}^{\infty} F_n(Y) \frac{(z-1)^n}{n!} \tag{11.5.5}$$

(we define $F_0 \equiv 1$).

Also by differentiating (11.5.1) with respect to z n times and then setting $z = 0$

$$\sigma_{n+2}(Y) = \frac{1}{n!}\left(\frac{\mathrm{d}^n Q(z, Y)}{\mathrm{d}z^n}\right)_{z=0} \tag{11.5.6}$$

so (11.5.1) can be regarded as a Taylor series for $Q(z, Y)$ about $z = 0$. Hence $z = 0$ is also a special point in that the behaviour of Q in this neighbourhood gives all the multi-particle cross-sections.

Another useful set of relations is obtained by taking

$$\log(Q(z, Y)) = \log\left(\sum_n z^n \sigma_{n+2}(Y)\right) \tag{11.5.7}$$

since

$$\left(\frac{\mathrm{d}(\log Q)}{\mathrm{d}z}\right)_{z=1} = \frac{1}{Q}\left(\frac{\mathrm{d}Q}{\mathrm{d}z}\right)_{z=1} = \left.\frac{\sum_n n z^{n-1}\sigma_{n+2}}{\sum_n z^n \sigma_{n+2}}\right|_{z=1}$$

$$= \frac{\sum_n n\sigma_{n+2}}{\sum \sigma_{n+2}} = \langle n \rangle = F_1 = C_1 \tag{11.5.8}$$

$$\left(\frac{\mathrm{d}^2(\log Q)}{\mathrm{d}z^2}\right)_{z=1} = \left[-\frac{1}{Q^2}\left(\frac{\mathrm{d}Q}{\mathrm{d}z}\right)^2 + \frac{1}{Q}\left(\frac{\mathrm{d}^2Q}{\mathrm{d}z^2}\right)\right]_{z=1}$$

$$= -\frac{\left(\sum_n n\sigma_{n+2}\right)^2}{\left(\sum_n \sigma_{n+2}\right)^2} + \frac{\sum_n n(n-1)\,\sigma_{n+2}}{\sum_n \sigma_{n+2}}$$

$$= -\langle n \rangle^2 + \langle n(n-1)\rangle = C_2(s) \tag{11.5.9}$$

and in general $$\left(\frac{\mathrm{d}^m(\log Q)}{\mathrm{d}z^m}\right)_{z=1} = C_m(s) \qquad (11.5.10)$$

So $Q(z, Y)$ also gives directly all the correlation coefficients, and provides a simple way of deducing the C's from the σ_n's and vice versa.

A trivial example is provided by the multi-peripheral model, from which we expect no correlations because each particle is emitted independently. From (11.3.32)

$$\sigma_{n+2} = \bar{g}^4 \frac{(\bar{g}^2 Y)^n}{n!} \mathrm{e}^{-\bar{g}^2 Y} \qquad (11.5.11)$$

and so (11.5.1) gives

$$Q(z, Y) = \bar{g}^4 \mathrm{e}^{-\bar{g}^2 Y} \sum_n \frac{z^n (\bar{g}^2 Y)^n}{n!} = \bar{g}^4 \mathrm{e}^{\bar{g}^2 Y(z-1)} \qquad (11.5.12)$$

Hence, in agreement with (11.3.33),

$$\langle n \rangle = \left(\frac{\mathrm{d}(\log Q)}{\mathrm{d}z}\right)_{z=1} = \bar{g}^2 Y \qquad (11.5.13)$$

but $$C_2 = \left(\frac{\mathrm{d}^2(\log Q)}{\mathrm{d}z^2}\right)_{z=1} = \left[-\frac{1}{Q^2}\left(\frac{\mathrm{d}Q}{\mathrm{d}z}\right)^2 + \frac{1}{Q}\left(\frac{\mathrm{d}^2 Q}{\mathrm{d}z^2}\right)\right]_{z=1} = 0 \qquad (11.5.14)$$

and similarly all the other C_m are zero because of the factorization built into the model.

More generally, if there are only short-range correlations we can expect all the C's to increase like $\log s$, since for example if $c_2(y_3, y_4, s)$ in (10.10.1) vanishes for $|y_3 - y_4| > \Lambda$ (the correlation length), then the integral in (10.10.3) will be proportional to the length of the rapidity plot. This implies that we can write

$$\log(Q(z, Y)) = P(z)\, Y + S(z) \qquad (11.5.15)$$

where P, S are polynomials in z. The multi-peripheral model has, from (11.5.12), $$P(z) = \bar{g}^2(z-1), \quad S(z) = \log \bar{g}^4 \qquad (11.5.16)$$

This expression (11.5.15) is reminiscent of the statistical mechanics of a gas (see Harari 1974). The grand partition function, Q, is related to the Helmholtz free energy, A, by

$$A = kT \log Q \qquad (11.5.17)$$

This Helmholtz energy can be expressed as the sum of the volume energy PV and the surface energy S, i.e.

$$A = PV + S = kT \log Q \qquad (11.5.18)$$

and $$P = kT\, \partial/\partial V\, (\log Q) \qquad (11.5.19)$$

Now if we regard the rapidity plot (e.g. fig. 11.10) as representing a one-dimensional 'gas' in a container of length $V = Y$, the walls of the container being defined by the rapidities of the incoming particles, then (11.5.18) can be identified with (11.5.15) (if the energies are measured in units such that $kT = 1$). The statistical mechanics result (11.5.18) assumes that there are only short-range correlations between the motions of the gas molecules, due to short-range interactions both between the different molecules and between the molecules and the walls of the container, so that $\log Q \propto V$ as $V \to \infty$.

Of course the applicability of these statistical ideas at present energies is rather doubtful because even at CERN-ISR

$$\log s_{\max} \approx Y_{\max} = 8,$$

and we have seen that the correlation length is $\Lambda \approx 2$ (see (10.10.23)). We would hardly feel justified in employing the methods of statistical mechanics for a gas in a container whose length was only four times the range of the inter-molecular forces. But, as we shall see below, the generating-function method is a useful technique for calculating the correlations, etc., to be expected from various models.

11.6 The two-component model

We have found that though both the diffraction and multi-peripheral models have many features in accord with nature, neither is able to account for all the facts. This is not really surprising because we have seen that duality gives the Pomeron, P, which accounts for diffractive scattering in Regge language, a quite different status from that of the other Reggeons, R. And indeed, two-component duality, in which one adds the P and R contributions, was found to work quite well not only in two-body scattering (chapter 7) but also for the inclusive distributions (chapter 10). It seems likely, therefore, that models in which one adds diffractive and multi-peripheral components may be fairly successful in reproducing many-particle cross-sections (Harari and Rabinovici (1973), Fialkowski and Miettinen (1973); see Harari (1974) for a review). The obvious problems to be overcome are those of multiple counting in absorptive effects, and the inconsistency of multiple Pomeron exchange (see section 8.6).

We assume that the multi-peripheral component of the $2 \to n$ amplitude, R_n, is given by multiple R exchange (fig. 11.11) and so

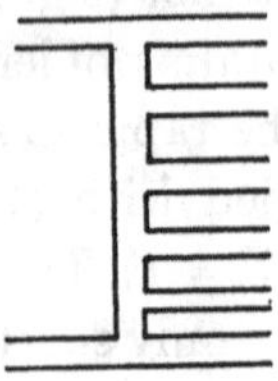

FIG. 11.13 Duality diagram for a multi-peripheral R-exchange amplitude.

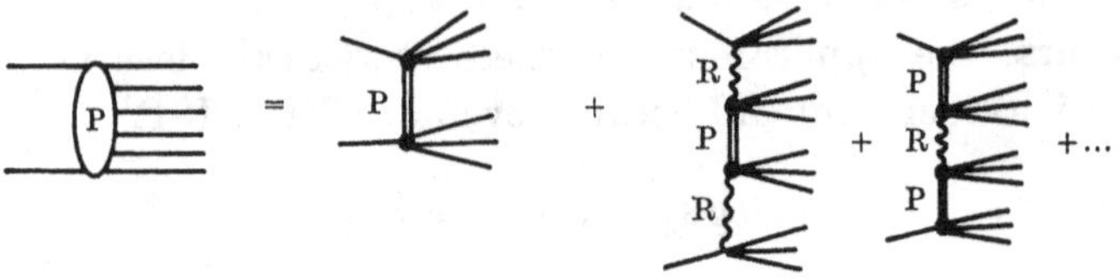

FIG. 11.14 Some of the contributions to the diffractive, P exchange, multiperipheral amplitude. Terms with P and R or many P exchanges are all included in the diffractive component.

from (11.3.29) gives a contribution to the cross-section

$$\sigma_n^{\mathrm{R}} \sim s^{2\alpha_{\mathrm{R}}-2} \tag{11.6.1}$$

(modulo log s factors) where α_{R} is the leading non-Pomeron trajectory, so $\alpha_{\mathrm{R}}(0) \approx 0.5$. The duality diagram for this term is shown in fig. 11.13.

The diffractive component, P_n, will contain many different types of contribution depending on how many P exchanges occur, and where (fig. 11.14), and should give

$$\sigma_n^{\mathrm{P}} \sim \text{constant (modulo log } s) \tag{11.6.2}$$

The two-component hypothesis for the $2 \rightarrow n$ amplitude is that

$$A^{2\rightarrow n} = R_n + P_n \tag{11.6.3}$$

and so the n-body cross-section is symbolically, from (11.3.14),

$$\sigma_n = \frac{1}{s^2}\int \mathrm{d}\Phi_n'(|R_n|^2+|P_n|^2+2\,\mathrm{Re}\,\{R_n P_n^*\})$$
$$\equiv \frac{1}{s^2}(R_n^2+P_n^2+R_n\,.\,P_n) \tag{11.6.4}$$

say, where multiplication implies integration over the n-body phase space, and we have introduced $\mathrm{d}\Phi_n' \equiv \mathrm{d}\Phi_n s/2$; see (11.3.26).

For the elastic $2 \rightarrow 2$ amplitude we have

$$\mathrm{Im}\,\{A^{\mathrm{el}}\} = \mathrm{Im}\,\{P_2+R_2\} \tag{11.6.5}$$

and so from the optical theorem (1.9.6) we obtain the consistency (bootstrap) condition that since

$$\sigma^{\text{tot}} = \frac{1}{s}\,\text{Im}\,\{A^{\text{el}}\} = \sum_n \sigma_n$$

we must have

$$\frac{1}{s}\,\text{Im}\,\{P_2 + R_2\} = \sum_n \frac{1}{s^2}\{R_n^2 + P_n^2 + P_n \,.\, R_n\} \qquad (11.6.6)$$

Now asymptotically $P_2 \sim s$, $R_2 \sim s^{\alpha_R}$ while $R_n^2 \sim s^{2\alpha_R}$, $P_n^2 \sim s^2$ and $R_n \,.\, P_n \sim s^{\alpha_R+1}$ (all modulo $\log s$) but of course we cannot be sure how $\sum_n$ of the right-hand side of (11.6.6) will behave. It seems fairly certain that part of $\text{Im}\,\{P_2\}$ must come from $\sum_n P_n^2$ and part of $\text{Im}\,\{R_2\}$ from $\sum_n R_n^2$, but we have seen how in the multi-peripheral model (11.3.30)

$$\frac{1}{s^2}\sum_n R_n^2 \sim s^{2\alpha_R - 2 + \bar{g}^2} \qquad (11.6.7)$$

so if $\bar{g}^2$ is large enough (i.e. $\bar{g}^2 = 1$ if $\alpha_R = 0.5$) this may also contribute to P_2. In fact it seems likely that this will be a very important contribution because the bulk of the multi-particle cross-section consists of particles with small sub-energies ($s_{i,i+1} < 2\,\text{GeV}^2$) where in $2 \to 2$ scattering R exchange is much bigger than P exchange.

So if we consider processes like $\text{pp} \to \text{pp} + \text{n}(\pi^+\pi^-)$, which will contribute most of the inelastic charged-particle pp events, we can write

$$\sigma_n^{\text{inel}} \approx \frac{1}{s^2}(P_n^2 + R_n^2) \equiv \sigma_n^{\text{P}} + \sigma_n^{\text{R}} \qquad (11.6.8)$$

if we drop the interference term $P_n \,.\, R_n$. This may be justified on the grounds that R contributes mainly to large multiplicities which populate evenly the whole of the rapidity plot (like fig. 11.6 (d)) while P_n gives mainly low multiplicity events in the fragmentation region (fig. 11.6 (a), (b), (c)), so the overlap of the two types of events in the integral (11.6.4) is probably quite small. The relative magnitudes of the two terms will be denoted by p and r respectively, defined by

$$\sum_n \sigma_n^{\text{R}} = r\sigma^{\text{inel}} \quad \text{and} \quad \sum_n \sigma_n^{\text{P}} = p\sigma^{\text{inel}} \qquad (11.6.9)$$

so clearly

$$r + p = 1$$

The multiplicities provided by the two components are defined as

$$\langle n \rangle_{\text{R}} \equiv \frac{\sum_n n\sigma_n^{\text{R}}}{\sum_n \sigma_n^{\text{R}}} = \frac{\sum_n n\sigma_n^{\text{R}}}{r\sigma^{\text{inel}}}, \quad \langle n \rangle_{\text{P}} = \frac{\sum_n n\sigma_n^{\text{P}}}{\sum_n \sigma_n^{\text{P}}} = \frac{\sum_n n\sigma_n^{\text{P}}}{p\sigma^{\text{inel}}} \qquad (11.6.10)$$

and so, from (10.3.8), the average pion multiplicity is

$$\langle n \rangle = p\langle n \rangle_{\mathrm{P}} + r\langle n \rangle_{\mathrm{R}} \tag{11.6.11}$$

i.e. just the weighted average of the multiplicities of the components. Similarly the correlations associated with each term are defined by (see (10.10.3))

$$\left.\begin{aligned} C_{2\mathrm{P}} &= \langle n(n-1)\rangle_{\mathrm{P}} - \langle n \rangle_{\mathrm{P}}^2 = \frac{\sum_n n(n-1)\,\sigma_n^{\mathrm{P}}}{p\sigma^{\mathrm{inel}}} \\ C_{2\mathrm{R}} &= \langle n(n-1)\rangle_{\mathrm{R}} - \langle n \rangle_{\mathrm{R}}^2 = \frac{\sum_n n(n-1)\,\sigma_n^{\mathrm{R}}}{r\sigma^{\mathrm{inel}}} \end{aligned}\right\} \tag{11.6.12}$$

giving
$$C_2 \equiv \frac{\sum_n n(n-1)\,(\sigma_n^{\mathrm{P}}+\sigma_n^{\mathrm{R}})}{\sigma^{\mathrm{inel}}} - \left(\frac{\sum_n n(\sigma_n^{\mathrm{P}}+\sigma_n^{\mathrm{R}})}{\sigma^{\mathrm{inel}}}\right)^2$$

$$= pC_{2\mathrm{P}} + rC_{2\mathrm{R}} + p\langle n \rangle_{\mathrm{P}}^2 + r\langle n \rangle_{\mathrm{R}}^2 - (p\langle n \rangle_{\mathrm{P}} + r\langle n \rangle_{\mathrm{R}})^2$$

$$= pC_{2\mathrm{P}} + rC_{2\mathrm{R}} + rp(\langle n \rangle_{\mathrm{P}} - \langle n \rangle_{\mathrm{R}})^2 \tag{11.6.13}$$

(using $r+p = 1$) which is not the weighted average of (11.6.12). This is a rather important result, because even if $C_{2\mathrm{P}} \sim$ constant, and $C_{2\mathrm{R}} = 0$ (equation (11.5.14)), we shall still get $C_2 \sim \log^2 s$, implying some long-range correlations, provided $\langle n \rangle_{\mathrm{R}} \sim \log s$ as expected from (11.3.33), and this is in much better accord with the data in fig. 10.29. The long-range correlations arise just because we have the sum of two types of exchanges, P and R, so factorization does not hold.

Harari and Rabinovici (1972) (see also Harari (1974)) have fitted the pp data with a model of this sort, assuming that $\sigma_n^{\mathrm{P}} = d_n$ are constants for $n = 0, 1, 2$ (i.e. for 2, 4, 6 prongs) and $\sigma_n^{\mathrm{P}} = 0$ for $n > 3$ (i.e. the diffractive component contributes only to the lowest multiplicities), while $\langle n \rangle_{\mathrm{R}} = c_1 \log(s/s_1)$ and $C_{2\mathrm{R}} = c_2 \log(s/s_2)$, $C_{m\mathrm{R}} = 0$ for $m > 2$. The seven parameters d, d_1, d_2, c_1, c_2, s_1, s_2 enable them to fit $\langle n \rangle$, C_2 and σ_n, $0 \leqslant n \leqslant 6$.

From (11.6.9)
$$p = \frac{d_0 + d_1 + d_2}{\sigma^{\mathrm{inel}}} \tag{11.6.14}$$

They find $p = 0.16$, so the multi-peripheral component dominates, as is rather clear from the fall of the multiplicity cross-sections in fig. 11.15. Also from (11.6.11)

$$\langle n \rangle = rc_1 \log\left(\frac{s}{s_1}\right) + \frac{d_1 + 2d_2}{\sigma^{\mathrm{inel}}} \rightarrow rc_1 \log\left(\frac{s}{s_1}\right) \tag{11.6.15}$$

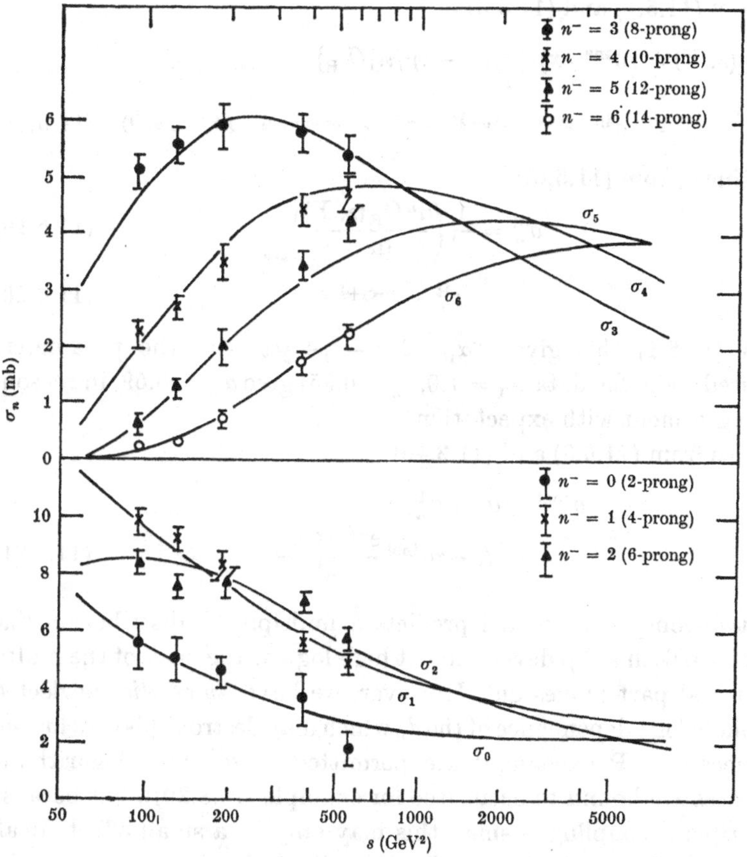

FIG. 11.15 Fit to the energy dependence of the multiplicity cross-sections in $pp \to pp + n(\pi^+\pi^-)$ with the two-component model (from Harari 1974).

and from (11.6.13)

$$C_2 = rpc_1^2\left(\log\left(\frac{s}{s_1}\right)\right)^2 + rc_2 \log\left(\frac{s}{s_2}\right) - 2r\left(\frac{d_1 + 2d_2}{\sigma}\right)\log\left(\frac{s}{s_1}\right)$$

$$+ pC_{2\mathrm{P}} + rp\langle n\rangle_{\mathrm{P}}^2 \to rpc_1^2\left(\log\left(\frac{s}{s_1}\right)\right)^2 \quad (11.6.16)$$

So the two-component model gives

$$\frac{C_2}{\langle n\rangle^2} \to \frac{p}{r} = \text{constant} \quad (11.6.17)$$

which is experimentally quite good, and certainly much better than $C_2/\langle n\rangle^2 \sim (\log s)^{-1}$ from the multi-peripheral type of model, or $\sim (\sqrt{s})(\log s)^{-2}$ from the diffraction model ((11.2.17), (11.2.19)).

From (11.5.1) and (11.5.5)

$$Q_{\mathrm{R}}(z, Y) = r\sigma^{\mathrm{inel}} \exp\{\sum_i [(z-1)^i/i!]\, C_{i\mathrm{R}}\}$$

$$= r\sigma^{\mathrm{inel}}\, \mathrm{e}^{(z-1)\langle n\rangle_{\mathrm{R}} + \frac{1}{2}(z-1)^2 C_{2\mathrm{R}}} \sim s^{-c_1+\frac{1}{2}c_2} \quad \text{at} \quad z = 0 \qquad (11.6.18)$$

and since (from (11.5.6))

$$\sigma_n^{\mathrm{R}} = \frac{1}{n!}\left(\frac{\mathrm{d}^n Q_{\mathrm{R}}(z, Y)}{\mathrm{d}z^n}\right)_{z=0} \qquad (11.6.19)$$

all

$$\sigma_n^{\mathrm{R}} \sim s^{-c_1+\frac{1}{2}c_2} \qquad (11.6.20)$$

With (11.6.1) this gives $2\alpha_{\mathrm{R}} - 2 = -c_1 + \frac{1}{2}c_2$, and the parameters required to fit the data ($c_1 = 1.0$, $c_2 = 0.35$) give $\alpha_{\mathrm{R}} = 0.59$, in reasonable agreement with expectation.

Since from (11.6.8) and (11.3.40)

$$\sigma_n^{\mathrm{inel}} = \sigma_n^{\mathrm{P}} + \sigma_n^{\mathrm{R}}$$

$$= d_n + r\sigma^{\mathrm{inel}} \frac{\mathrm{e}^{-\langle n\rangle} \langle n\rangle^n}{n!} \qquad (11.6.21)$$

the two-component model predicts a multiplicity distribution like fig. 11.16, with a dip developing at high $\log s$ as the peak of the multi-peripheral part moves out. However, we have, *inter alia*, neglected the likely $\log s$ dependence of the d_n which may destroy this conclusion. If successive P exchanges are permitted in σ_n^{P} such logarithmic increases are bound to occur (see for example (10.8.20)), but because the triple-P coupling is small this may only be a small effect. It all depends on how one tries to solve the self-consistency problem of the P_n part of (11.6.6) – with $\alpha_{\mathrm{P}}(0) < 1$ as in (11.4.3), $= 1$ as in (8.6.9), or > 1 as in (8.6.14), which all give different behaviours for $\sigma^{\mathrm{tot}}(s)$.

But if we are willing to push such problems to the back of our minds, this sort of two-component model seems to provide rather a good first approximation of the data.

11.7 The duality bootstrap

The two-component model, which combines the virtues of the diffraction and multi-peripheral models, and two-component duality, seems to be along the right lines. However, it clearly does not make full use of the content of duality as discussed in sections 7.3 and 10.7. Nor can it be regarded as self-consistent in that both multiple-P exchange and the planar nature of the multi-peripheral model are inconsistent

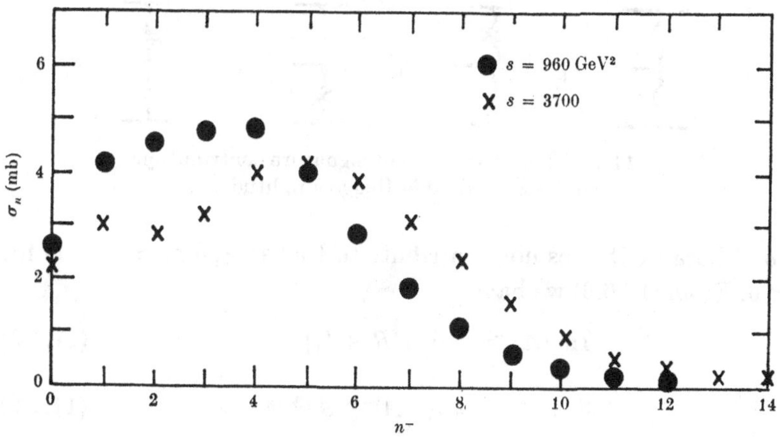

FIG. 11.16 Predictions of the two-component-model fit for the high energy multiplicity distributions (from Harari 1974).

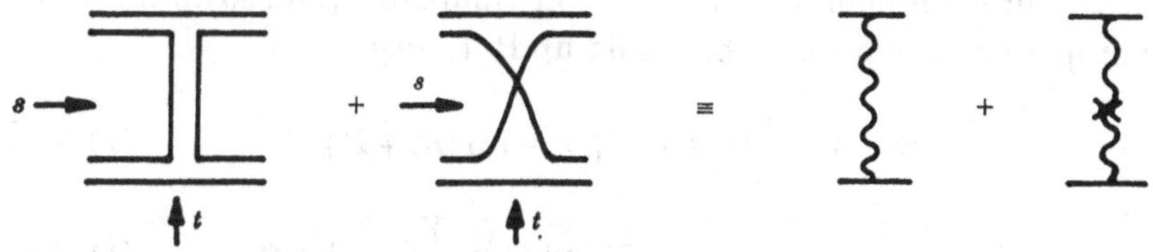

Fig. 11.17 The s–t and t–u planar duality diagrams which provide the two contributions to the signature factor of a t-channel Reggeon.

with unitarity requirements. Recently some progress has been made in overcoming these problems by making better use of duality (Lee 1973, Veneziano 1973, 1974*b*, Chan, Paton and Tsou 1975, Aurenche *et al.* 1975).

In $2 \to 2$ scattering there are just two diagrams for R exchange in the t channel (see figs. 11.17), one s–t planar, the other t–u planar, which give the two discontinuities (s- and u-channel) of a definite-signature Reggeon. Then in $2 \to 3$ we have the four diagrams of figs. 11.18, and so on, there being 2^{n-1} different diagrams for an n-particle final state. Only one of these is s–t planar, and all the other $2^{n-1}-1$ are non-planar, but all the diagrams contribute equally to σ_n, and so, like (11.3.29),

$$\sigma_n = 2^{n-1}\bar{g}^4 \frac{(\bar{g}^2 Y)^{n-2}}{(n-2)!} s^{2\alpha_R - 2} \tag{11.7.1}$$

if we neglect interference between the various terms.

However, the crossed diagram in fig. 11.19(a) does not contribute

FIG. 11.18 The four different signature contributions to the $2 \to 3$ double-Regge amplitude.

to σ^{tot} because it does not contribute to $\text{Im}\{A^{2\to 2}\}$ for $s > 0$, only for $s < 0$. From (11.6.6) we have

$$\text{Im}\{A^{2\to 2}\} = \text{Im}\{R_2 + P_2\} \tag{11.7.2}$$

$$= \sum_{n=2}^{\infty} A^{2\to n} A^{*2\to n} \tag{11.7.3}$$

which is represented by fig. 11.19 (*b*) (where again we have neglected cross terms like fig. 11.19 (*c*), see (11.6.8) *et seq.*). Only the first diagram in each group is planar, and so can contribute to R, and so all the other non-planar ones presumably build up P. Hence

$$\sigma^{\text{tot}}(s) = \frac{1}{s}\text{Im}\{A^{2\to 2}\} = \frac{1}{s}\text{Im}\{R_2 + P_2\} \tag{11.7.4}$$

$$= \sum_n \sigma_n = \sum_n 2^{n-1}\bar{g}^4 \frac{(\bar{g}^2 Y)^{n-2}}{(n-2)!} s^{2\alpha_R - 2} \tag{11.7.5}$$

but for each n only 1 term contributes to R and $(2^{n-1}-1)$ to P, so

$$\frac{1}{s}\text{Im}\{R_2\} \sim s^{\alpha_R - 1} = \sum_n \bar{g}^4 \frac{(\bar{g}^2 Y)^{n-2}}{(n-2)!} s^{2\alpha_R - 2} \sim s^{2\alpha_R - 2 + \bar{g}^2} \tag{11.7.6}$$

$$\frac{1}{s}\text{Im}\{P_2\} \sim s^{\alpha_P - 1} = \sum_n (2^{n-1}-1)\bar{g}^4 \frac{(\bar{g}^2 Y)^{n-2}}{(n-2)!} s^{2\alpha_R - 2} \sim s^{2\alpha_R - 2 + 2\bar{g}^2} \tag{11.7.7}$$

so
$$\alpha_R = 1 - \bar{g}^2, \quad \alpha_P = 2\alpha_R - 1 + 2\bar{g}^2 = 1 \tag{11.7.8}$$

Thus, unlike (3.4.13), (11.3.30) and (11.4.14), the height of the trajectory α_R decreases as the strength of the coupling, $\bar{g}^2$, increases, and, even more remarkably, $\alpha_P = 1$ independent of the coupling strength.

Much more detailed calculations along these lines have been attempted by Chan and co-workers (Aurenche *et al.* 1975). For each $A^{2\to n}$ they use a dual amplitude, but for small sub-energies $s_{i,i+1} < \bar{s}$, say, they approximate the dual amplitude by its resonance contributions in the $s_{i,i+1}$ channel, while for $s_{i,i+1} > \bar{s}$ they use the Regge

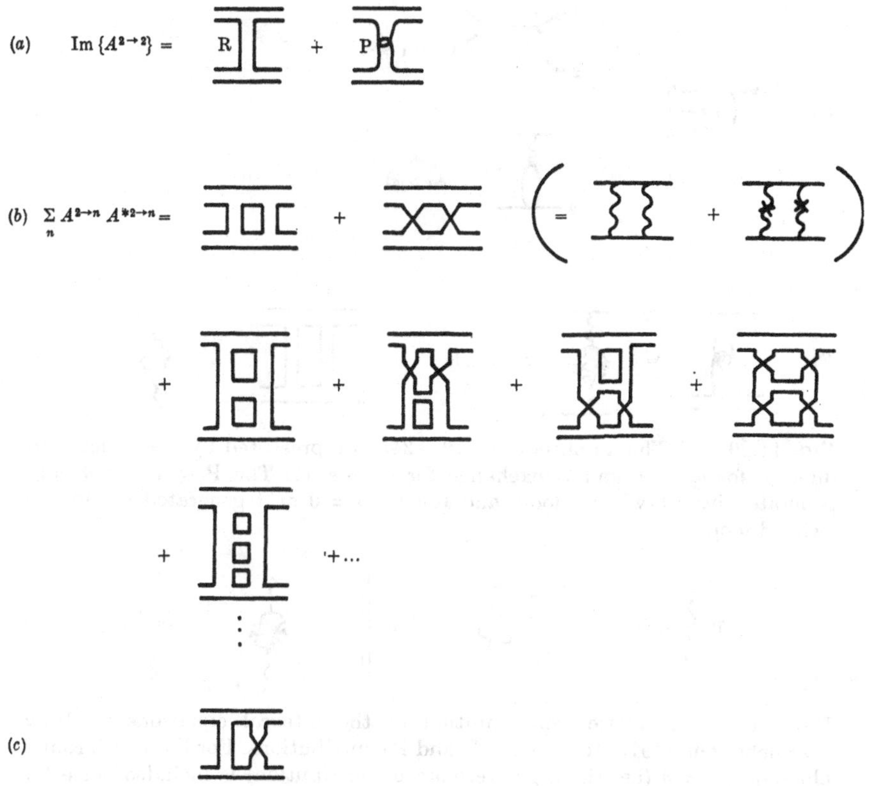

FIG. 11.19 (*a*) R + P contributions to Im $\{A^{2\to 2}\}$. (*b*) Duality diagrams for the multi-Regge contributions to $\sum_n A^{2\to n}A^{*2\to n}$. (*c*) A cross term of the type neglected in (*b*).

exchange approximation (see fig. 11.20(*a*)). Also they include the SU(N), $N = 2$ (or 3), symmetry by including τ (or λ) matrices for each quark, as we described when obtaining (9.4.26), which ensures $I = 0$ for the P, and degenerate $I = 0, 1$ trajectories for R etc. The structure naturally gives

$$\alpha_{\rm P}(0) > \alpha_{\rm R}(0), \quad \alpha'_{\rm P} < \alpha'_{\rm R}$$

They insert the Reggeon into integral equations like (11.4.6) (see fig. 11.21), perform loop integrations similar to (11.4.4), and insist that the output Reggeon be the same as the input. This gives the parameters of the P trajectory, which for $\alpha^0_{\rm R} = 0.5$, $\alpha'_{\rm R} = 1$ input gives $\alpha^0_{\rm P} = 1.12$, $\alpha'_{\rm P} = 0.1$ output, which are not too far from the observed values.

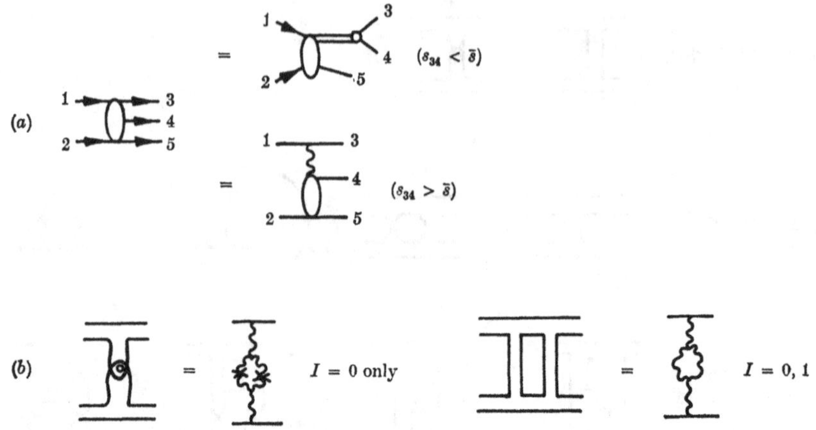

FIG. 11.20 (*a*) The amplitude for $12 \rightarrow 345$ is represented by resonance production for $s_{34} < \bar{s}$, and R exchange for $s_{34} > \bar{s}$. (*b*) The P with $I = 0$ only generated by a 'twisted' loop, and R with $I = 0$ or 1 generated by an untwisted loop.

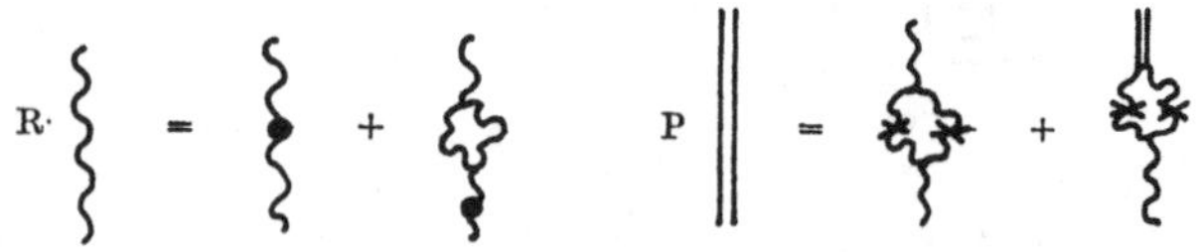

FIG. 11.21 Schematic representation of the integral equations used by Aurenche *et al.* (1975) to generate R and P contributions. For lines with round blobs only $s < \bar{s}$ (i.e. the input resonance contribution) is included since for $s > \bar{s}$ the resonances are equivalent to R exchange (cf. fig. 11.12(*f*)).

But of course the next iteration with the P included in the input will produce cuts with $\alpha_c(0) > 1$, which brings us back to the problem which has featured several times in our discussion, how the P with $\alpha_P(0) \approx 1$ can be made consistent with unitarity (see for example (11.3.31) *et seq.*). If $\alpha_P(0) < 1$ there is, in principle, no problem, because at very high energies the self-consistent solution will look rather like the multi-peripheral bootstrap, with a single dominant P pole exchange and $\alpha_P(0) = 1-\bar{g}^2$ as in (11.4.3). The continuing rise of $\sigma^{tot}(s)$ at CERN-ISR energies has to be regarded as a non-asymptotic effect, and eventually $\sigma^{tot}(s) \rightarrow 0$ as $s \rightarrow \infty$. If $\alpha_P(0) = 1$ then the pole cannot be dominant asymptotically unless the triple-P coupling $\gamma^{PP,P}(t) \rightarrow 0$ as $t \rightarrow 0$ (see (10.8.21)) to forbid multi-Pomeron exchange, which phenomenologically seems untrue. So a self-consistent solution must have dominant cuts, as in the Reggeon field theory mentioned in section 8.3, and $\sigma^{tot}(s) \sim (\log s)^\nu$, $\nu > 0$, and $\alpha_P(t) = 1+\alpha'(t)^\kappa$ (see

Abarbanel *et al.* 1975) so the trajectory is quite unlike fig. 6.6(*b*). Or if $\alpha_P(0) > 1$ we have cut dominance, and all the absorption problems discussed in section 8.6 occur not only for $P \otimes P$ cuts but for $R \otimes P$ as well, so the apparent dominance of poles at available energies becomes a non-asymptotic effect. In fact, as we noted in sections 8.3 and 10.8 these self-consistency problems require a consideration of what happens not just at large $\log s$, but large $\log(\log s)$, which is not achievable even in principle.

At present Regge poles seem to fit the data far better than one has any right to expect, which is pleasant for the phenomenologist. But it means that one can gain rather little insight from experiment as to the nature of the unitarity constraints which must inter-relate poles and cuts, and restrict the Reggeon parameters, and may even uniquely determine them in the full bootstrap sense. The models discussed in this chapter take us only a little way towards such a self-consistent unitarization, and although the incorporation of duality has produced a useful advance towards building up the Pomeron we are still as far as ever from understanding how it can be made consistent. The approach is still a perturbative one, except that the effective expansion parameter is $\gamma^{PP,P}(t)\log s$ (rather than the residue in, say, (11.3.20), see Chew (1973)) so that, since $\gamma^{PP,P}(0)$ is small, there is quite good convergence for small $\log s$, but the expansion will not converge for large $\log s$, and so we do not attain a self-consistent asymptotic behaviour.

These problems make it hard to understand why dual models, which are based on imposing the desired Regge asymptotic behaviour on non-unitarity narrow-resonance amplitudes, are so successful. In particular, what is the significance of the fact that they require, even in the Born approximation, linear trajectories $\alpha_B(t) = \alpha_B^0 + \alpha'_B t$, where $\alpha'_B \approx 1\,\text{GeV}^{-2}$ sets the scale for hadronic interactions? If unitarity is to make only a small change in these trajectory functions ($\alpha(t) \approx \alpha_B(t)$ for all t) they must satisfy the twice subtracted dispersion relation (3.2.12) with $\text{Im}\{\alpha\}$ small. However, the Regge trajectories which are generated by the iteration of a basic exchange force in some sort of ladder, as in potential scattering (section 3.3), field theory (section 3.4), or the Reggeized multi-peripheral model (section 11.3), all obey the singly subtracted dispersion relation (3.3.11), where n is a constant which depends on the asymptotic behaviour of the Born approximation (see (3.3.32), (3.4.19)) but the position of the trajectory (and hence $\alpha'(t)$) depends on the coupling strength g^2 through unitarity, and $\alpha' \to 0$ as $g^2 \to 0$. It has been suggested (see Veneziano 1974) that perhaps one should regard $\sqrt{\alpha'} \approx 2 \times 10^{-14}$ cm as a fundamental length, below which the concept of point-like particles does not make any sense. But if $\alpha_B'^{-1}$ is the fundamental energy scale of hadronic physics it is hard to see how the trajectories can possibly be built up through unitarity as bootstrap models require (Collins *et al.* 1968*a*, Collins 1971).

Even more obscure is the relation between the quark model, which describes the internal symmetry structure of the dual Born approximation so well (for example in duality diagrams), and the dynamics of unitary models. The harmonic oscillator type of potential between quarks, which is needed to generate linear trajectories and reproduce the resonance spectrum (see section 3.3), and which must also prevent quarks from actually being produced in scattering experiments, is not evident in particle scattering at all. The forces between the particles (due to Reggeon exchange) seem quite different from the forces between the quarks, despite the fact that the particles are supposed to be composed of quarks. Various schemes for confining quarks in 'bags' have been proposed, but their significance for Regge dynamics is not yet clear (see Chados *et al.* 1974).

So we are still some way from understanding why Regge theory, and in particular Regge pole dominance, works so well, yet unitarization, which first motivated the introduction of Reggeons rather than fixed-spin elementary particles, seems comparatively unimportant. But at least it has become much clearer what are the relevant questions to ask about hadronic interactions, which gives us reason for anticipating that some of these fundamental questions may be solved before very long.

12

Regge poles, elementary particles and weak interactions

12.1 Introduction

So far in this book we have been solely concerned with hadronic interactions, which are the principal field in which Regge theory has been used. We have ignored electromagnetic effects in assuming that isospin is an exact symmetry of the scattering processes, and have not needed to mention the weak-interaction properties of the particles such as β-decay, etc. But of course any discussion of the electromagnetic or weak interactions of hadrons necessarily involves consideration of their hadronic properties too, because it is the strong interaction which is mainly responsible for the composite structure of the hadrons. Regge theory has played a small but not insignificant role in the development of theories of these weaker interactions, and clearly if there is to be any chance of unifying all the interactions they must be reconciled with Regge theory. In this chapter we shall look rather briefly at the problems which may arise in so doing.

Basically there are two such problems. First, weak interactions (and from now on we shall usually use the word 'weak' to refer to both electromagnetism and the weak interaction) are generally formulated in terms of a Lagrangian field theory for the interaction of a basic set of elementary particles. These are the leptons, l (i.e. electron e, muon μ, and neutrinos ν_e, ν_μ), photon γ, vector boson W, etc., and elementary hadrons (which at least initially do not lie on Regge trajectories but occur as Kronecker $\delta_{\sigma J}$ terms in the J plane). Alternatively the hadrons may be composed of elementary quarks bound together by the exchange of 'gluon' particles. The question then arises as to whether these elementary particles can be 'Reggeized' as a result of the interaction, i.e. whether they can be made to lie on Regge trajectories. This problem is obviously fundamental to attempts to marry field theory to Regge physics, and we examine it in section 12.3.

Secondly, theories of the coupling of the weak interactions to hadrons are generally used only to first order in the weak coupling constant (e^2 or G) and so the constraints of unitarity are inoperative. This means that fixed poles in the J plane, $\sim (J-J_0)^{-1}$, are not neces-

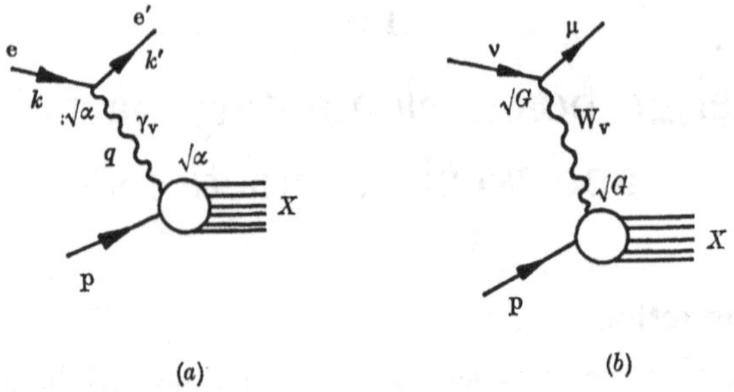

FIG. 12.1 (a) Deep inelastic electron scattering on a proton, $ep \to eX$, in the one-photon exchange approximation. The coupling is $\sqrt{\alpha} \approx 137^{-\frac{1}{2}}$ at each vertex, and the bottom part of the diagram is the amplitude for $\gamma_v p \to X$, where γ_v is the 'virtual' photon of 'mass' q^2. (Real photons have $q^2 = 0$ of course.) (b) Deep inelastic neutrino scattering $\nu p \to \mu X$ in the single virtual vector boson exchange approximation. The Fermi weak-interaction coupling $\sqrt{G}$ appears at each vertex.

sarily forbidden, and some theories such as current algebra actually require them. However, one is then led to wonder what would happen if one tried to work to all orders in the coupling since the results of sections 3.4 and 4.7 suggest that such fixed poles must be Reggeized by unitarity. But if so, what particles lie on the resulting trajectories? These questions will be examined in section 12.4.

But the main significance of Regge theory is that it tells us about the asymptotic behaviour to be expected in scattering amplitudes, and we conclude with a very short review of Regge predictions for weak scattering amplitudes. These include electromagnetic processes like 'deep inelastic' electron scattering, $ep \to eX$ (fig. 12.1 (a)) which, when the known electron–photon coupling, $\sqrt{\alpha}$, and the photon propagator have been extracted, depends just on the cross-section for the absorption of a virtual photon by a proton, i.e. $\gamma_v p \to X$. Or we may have neutrino scattering $\nu p \to lX$ (where $l = e$ or μ depending on whether ν is an electron- or muon-type neutrino) which can be described, at least as a matter of convenience, as $W_v p \to X$, where $W^{\pm}$ is the hypothetical 'intermediate vector boson' which in some theories is regarded as the mediator of the weak interaction (fig. 12.1 (b)). So γ_v and W_v couple to the electromagnetic and weak 'currents' of the hadrons respectively, and in this chapter we are concerned with such hadronic currents.

We shall not, however, attempt a full introduction to weak interaction theories, and the reader who is unfamiliar with these topics will find the books by Bransden, Evans and Major (1973), Gasiorowicz (1966), Bernstein (1968) and Feynman (1972) very useful, in addition to the references appearing later in the text.

12.2 Photo-production and vector dominance

There has been one exception to our exclusion of non-hadronic interactions from consideration thus far. In table 6.5, and at various points where we have discussed Regge phenomenology, we have included photo-production processes like $\gamma p \to \pi^+ n$ among those to be examined. The reason for this is that at high energies photons seem to behave almost exactly like hadrons, except for their weaker coupling. The explanation for this behaviour seems to be that photons couple to hadrons mainly via the vector mesons, as in fig. 12.2(*a*) (see for example Gilman (1972)).

The photon has $Q = B = S = 0$, $(J^P)C_n = (1^-)-$, but not being a hadron it does not have a definite isospin. It is found to behave like a mixture of $I = 0$ and 1, with no strong evidence for $I > 1$ components. The hadrons which share these properties are the vector mesons, the ρ with $I = 1$, and ω and ϕ with $I = 0$ (together with any daughters these may have). Fig. 12.2(*a*) suggests that one should write

$$A_H(\gamma 2 \to 34) = \sum_V \frac{e}{f_V} A_H(V2 \to 34) \tag{12.2.1}$$

where
$$e = \sqrt{(4\pi\alpha)} \equiv \left(\frac{e^2}{\hbar c}\right)^{\frac{1}{2}} \approx \left(\frac{4\pi}{137}\right)^{\frac{1}{2}} \tag{12.2.2}$$

(since in our units $\hbar = c = 1$). In (12.2.1) f_V is the coupling between vector meson and photon, and $V = \rho, \omega, \phi$, plus any other vector mesons one may care to add. The coupling f_V is directly related to the partial decay width $\Gamma(V \to e^+e^-)$ through fig. 12.2(*b*), which gives

$$\Gamma = \frac{4\pi\alpha m_V}{3f_V^2} \tag{12.2.3}$$

so f_V can be determined independently. Also the electromagnetic form factors which describe the photon coupling to a given hadron can be approximated by vector meson exchange, like fig. 12.2(*c*). Thus the pion's electromagnetic form factor can be written

$$F_\pi(q^2) \approx \frac{eg_{\rho\pi\pi}}{f_\rho} \frac{m_\rho^2}{m_\rho^2 - q^2} \tag{12.2.4}$$

exhibiting the pole at $q^2 = m_\rho^2$.

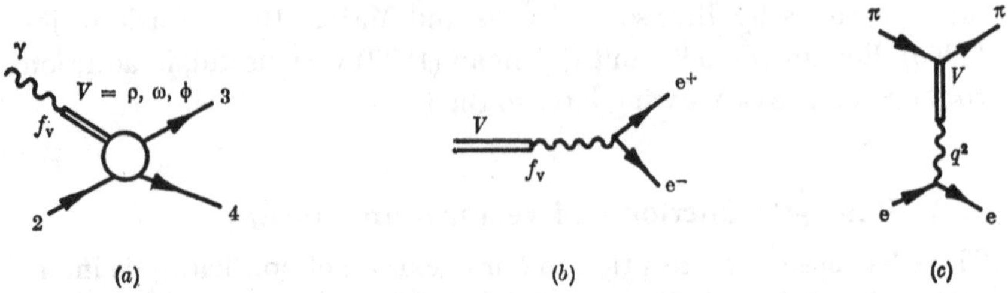

FIG. 12.2 (a) Vector dominance hypothesis in $\gamma 2 \to 34$. The photon couples to the hadrons via the vector mesons $V = \rho, \omega, \phi$. (b) The decay $V \to e^+e^-$. (c) The pion electromagnetic form factor determined in $e\pi \to e\pi$. It is assumed that the pion couples to the virtual photon exchange via V.

The obvious difficulty with (12.2.1) is that the photon, being massless, has helicities $\mu_\gamma = \pm 1$ only, from gauge invariance, whereas the vector mesons have $\mu = 1, 0, -1$, so the relation can only be true for transversely polarized mesons. It is not clear in which Lorentz frame the equality should hold, but it is generally supposed, and seems to be true experimentally, that the relation applies in the s-channel centre-of-mass frame, i.e. the helicity frame.

So we can make Regge hypotheses about photo-production amplitudes simply by treating the photon as a mixture of $I = 0, 1$ vector mesons as in fig. 12.3. This can be tested using for example the relation (Beder 1966, Dar *et al.* 1968)

$$\frac{1}{2}\left[\frac{d\sigma_{\perp,\parallel}}{dt}(\gamma p \to \pi^+ n) + \frac{d\sigma_{\perp,\parallel}}{dt}(\gamma n \to \pi^- p)\right]$$
$$= \frac{e^2}{f_\rho^2}(\rho_{11} \pm \rho_{1-1})\frac{d\sigma}{dt}(\pi^- p \to \rho^0 n) \qquad (12.2.5)$$

where $\perp, \parallel \equiv$ photon polarization perpendicular/parallel to the production plane. It has been assumed that the ω and ϕ contributions can be neglected because of their small couplings, and by taking the sum of π^+ and π^- photo-production the ρ–ω interference term in the square modulus of (12.2.1) is eliminated. The density matrix combination $(\rho_{11} \pm \rho_{1-1})$ for the ρ decay gives the required ρ helicities (see section 4.2). Such relations work rather well in general.

Another interesting consequence of (12.2.1) is that

$$A_H(\gamma 2 \to V2) = \sum_V \frac{e}{f_V} A_H(V2 \to V2) \qquad (12.2.6)$$

so, neglecting the spin dependence, and the possibility of transitions

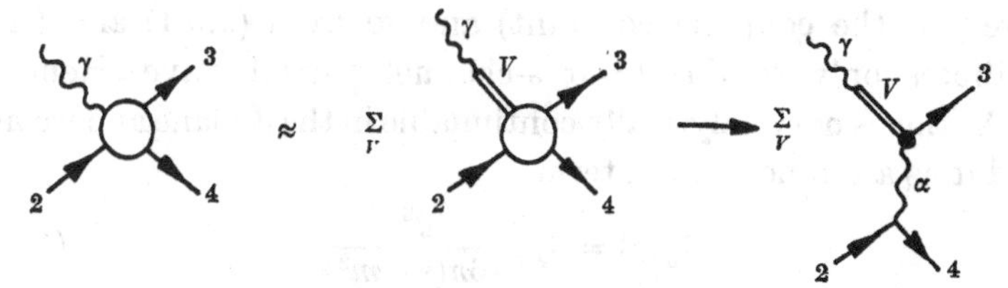

FIG. 12.2 Regge pole approximation to photo-production using vector dominance.

like $V_i 2 \to V_j 2$, $i \neq j$, we have

$$\frac{d\sigma}{dt}(\gamma 2 \to V2) = \frac{e^2}{f_V^2}\frac{d\sigma}{dt}(V2 \to V2) \tag{12.2.7}$$

and for $t = 0$, using the optical theorem (1.9.6),

$$\frac{d\sigma}{dt}(\gamma 2 \to V2)_{t=0} = \frac{e^2}{f_V^2}\frac{1}{16\pi}(\sigma_{V2}^{\text{tot}})^2 \tag{12.2.8}$$

assuming (for simplicity) that at high energies $A(V2 \to V2)$ is pure imaginary due to P exchange. So the differential cross-section for photo-producing vector mesons on protons, say, gives the ρp total cross-section. A further step is to take

$$A_H(\gamma 2 \to \gamma 2) = \sum_V \frac{e}{f_V} A_H(\gamma \text{p} \to V\text{p}) \tag{12.2.9}$$

which, again neglecting the spin dependence and real parts for simplicity, gives

$$\frac{d\sigma}{dt}(\gamma\text{p} \to \gamma\text{p}) = e^2\left\{\sum_V \frac{1}{f_V}\left[\frac{d\sigma}{dt}(\gamma\text{p} \to V\text{p})\right]^{\frac{1}{2}}\right\}^2 \tag{12.2.10}$$

(though this relation does not seem to work so well).

The success of the vector-dominance hypothesis allows us to treat high-energy photo-production processes just like ordinary hadronic processes.

12.3 The Reggeization of elementary particles*

In a Lagrangian field theory the contribution of an elementary particle propagator for a particle of mass m and spin σ takes the form (cf. (2.3.1), and Appendix B of Bjorken and Drell (1965))

$$\frac{(2\sigma+1)g^2 P_\sigma(z_s)}{s-m^2} \tag{12.3.1}$$

* This section may be omitted at first reading.

(where g is the coupling constant) and so from (2.2.1) and (A.20) contributes only to the $J = \sigma$ s-channel partial wave. Hence its contribution is not analytically continuable in the J plane and we must regard it as a Kronecker $\delta_{J\sigma}$ term

$$A_J(s) = \delta_{J\sigma}\frac{g^2}{16\pi(s-m^2)} \tag{12.3.2}$$

We have found that there is no evidence for such terms in hadronic physics, which suggests that Lagrangian field theories are inapplicable to strong interactions.

This conclusion may be too hasty, however, because (12.3.1) is only the Born approximation, the first term in a perturbation expansion of the theory, and it is possible that other terms might appear to cancel the $\delta_{J\sigma}$ and replace it by a moving Regge pole

$$A_J(s) \approx \frac{\beta(s)}{J-\alpha(s)}, \quad \alpha(m^2) = \sigma \tag{12.3.3}$$

instead, in which case the input elementary particle would be 'Reggeized' by unitarization of the field theory. For this to happen the theory must be able to generate a Kronecker δ to cancel the input, and in fact such $\delta_{J\sigma}$ terms may well arise at nonsense points (Gell-Mann and Goldberger 1962, Gell-Mann *et al.* 1962, 1964).

In section 4.8 we found that at right-signature sense–nonsense (sn) points J_0, since $e^J_{\mu\mu'} \sim (J-J_0)^{-\frac{1}{2}}$ we need a SCR to cancel the infinity, which would be incompatible with unitarity, giving $A_J \sim (J-J_0)^{\frac{1}{2}}$. This causes sense–nonsense decoupling as described in section 6.3. However, suppose we consider just the left-hand cut of the partial-wave amplitude, $A^{\mathrm{L}}_{HJ}(s)$ (cf. (3.5.1)), which stems from the crossed t- and u-channel singularities, and may be regarded as the input 'potential' for the N/D method of calculating partial-wave amplitudes (section 3.5). $A^{\mathrm{L}}_{HJ}(s)$ is not restricted by unitarity and so from the Froissart–Gribov projection (4.5.7) we can expect

$$\left.\begin{aligned} \langle s|\,A^{\mathrm{L}}_J\,|s\rangle &\sim \text{const nt} \\ \langle s|\,A^{\mathrm{L}}_J\,|n\rangle &\sim (J-J_0)^{-\frac{1}{2}} \\ \langle n|\,A^{\mathrm{L}}_J\,|n\rangle &\sim (J-J_0)^{-1} \end{aligned}\right\} \tag{12.3.4}$$

for $J \to J_0$, where $|s\rangle$ and $|n\rangle$ are respectively sense and nonsense helicity states for $J = J_0$.

For example in spins $1+\frac{1}{2}\to 1+\frac{1}{2}$ (fig. 12.4(a)) the u-channel spin $= \frac{1}{2}$ exchange Born term (fig. 12.4(c)) $\sim s^{-\frac{1}{2}}$, and gives a fixed

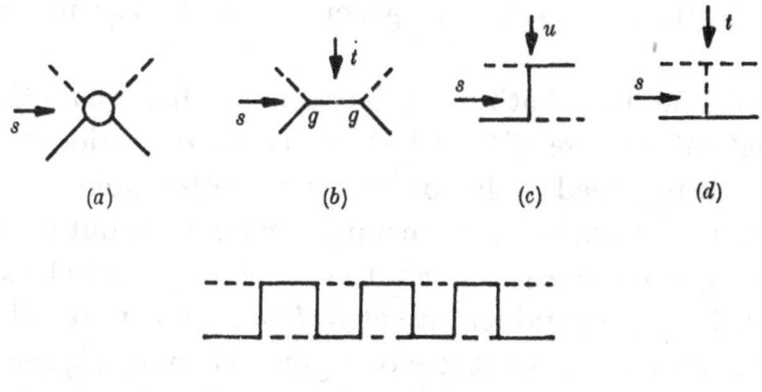

FIG. 12.4 (*a*) The amplitude for spins $1+\frac{1}{2}\to 1+\frac{1}{2}$ in the s channel (— spin $=\frac{1}{2}$, --- spin $=1$ particles). (*b*) The s-channel Born term. (*c*) The u-channel Born term. (*d*) The t-channel Born term. (*e*) Unitarization of the u-channel Born term.

singularity like (12.3.4) at $J=\frac{1}{2}$, which is a sense–nonsense point for the helicity $1+\frac{1}{2}\to 1+\frac{1}{2}$ amplitude. With composite particles one would expect this singularity to be cancelled by other contributions to give the SCR of (4.8.3), but with elementary particles there is no need for this to happen. In fact as $g^2\to 0$ the Born terms must be dominant.

Then if we treat fig. 12.4 (*c*) as the first term of a perturbation expansion in g^2, with higher order terms like fig. 12.4 (*e*), we can write the full solution in the form (Calogero *et al.* 1963*a*)

$$\langle s|\,A_J\,|s\rangle = \langle s|\,A_J^{\text{nn}}\,|s\rangle\,[1+\sum_n \langle s|\,A_J^{\text{L}}\,|n\rangle\langle n|\,A_J\,|s\rangle] \quad (12.3.5)$$

where A_J^{nn} is the amplitude obtained when nonsense intermediate states are excluded from the perturbation series, while $\sum_n$ is over nonsense states only. Now $\langle s|\,A_J^{\text{L}}\,|n\rangle \sim (J-J_0)^{-\frac{1}{2}}$ from (12.3.4) but unitary requires $\langle n|\,A_J\,|s\rangle \sim (J-J_0)^{\frac{1}{2}}$ so the second term in the bracket is finite but non-zero. So for elementary-particle theories the nonsense states give a finite contribution to the analytically continued (in J) ss partial-wave amplitude $\langle s|\,A_J\,|s\rangle$ which makes it different from the physical partial-wave amplitude, which is just $\langle s|\,A_J^{\text{nn}}\,|s\rangle$.

In some circumstances this difference may be exactly equal to the elementary-particle $\delta_{J\sigma}$ term (12.3.2), $\sigma=\frac{1}{2}$, from the s-channel pole, fig. 12.4 (*b*), so that

$$\langle s|\,A_J^{\text{nn}}\,|s\rangle + \delta_{J\sigma}\langle s|\,A_\sigma\,|s\rangle = \langle s|\,A_J\,|s\rangle \quad (12.3.6)$$

Then the physical amplitude is after all equal to the analytically continued amplitude, and the solution will exhibit Regge behaviour.

This will clearly not happen in general, but it may in particular theories.

Though A_J^{nn} and A_J both have the same left-hand cuts they need not be identical because of the CDD ambiguity of section 3.5, and in fact A_J can be regarded as the solution to the N/D equations with the nonsense states excluded from the unitarity relation, but with a CDD pole for $J = J_0 = \sigma$ corresponding to the elementary-particle exchange, fig. 12.4 (*b*). It is a general feature of N/D solutions that poles which arise as bound or resonant states of a given channel appear as CDD poles in channels which are coupled thereto (see for example Squires (1964), Atkinson, Dietz and Morgan (1966), Jones and Hartle (1965)). The number of CDD poles needed is equal to the number of independent helicity amplitudes for which $J = J_0$ is a nonsense value. However, partial-wave amplitudes must also have the correct threshold behaviour (determined by l rather than J, see (4.7.6)) and Mandelstam (1965) pointed out that in some situations there is a unique amplitude containing no more than one CDD pole which satisfies these threshold conditions, in which case (12.3.6) must be satisfied.

This is in fact true of the spins $= 1 + \frac{1}{2}$ example mentioned above, though it is not true for the elementary vector meson in the t channel (fig. 12.4 (*d*)) which does not Reggeize in this way. Abers and Teplitz (1967) (see also Abers, Keller and Teplitz (1970)) have analysed the general spin problem, and find that the cases $(\sigma_1, \sigma_2) = (0, 0)$, $(0, \frac{1}{2})$ and $(\frac{1}{2}, \frac{1}{2})$ do not work, but that higher spins, like for example $(\frac{1}{2}, \sigma)$, $\sigma \geqslant 1$, which have suitable nonsense states, often will obey (12.3.6). But as such high-spin field theories are generally un-renormalizable it is not clear whether these results are useful.

Of course even if (12.3.6) is not automatically satisfied by the CDD solution it may actually be satisfied when the masses and couplings take on particular values so as to make the SCR hold, giving a bootstrap type of solution, but this cannot happen for weak coupling theories.

So only in certain field theories is Reggeization of the input elementary particles likely. However, it has been pointed out by Grisaru, Schnitzer and Tsao (1973) that these Reggeization rules may be applicable in re-normalizable unified gauge theories of strong, electromagnetic and weak interactions (see Iliopoulos (1974) for a review and references) in which the hadrons are viewed as composed initially of elementary spin $= \frac{1}{2}$ quarks bound together by elementary spin $= 1$

vector gluons. In gauge theories both the spin $= \frac{1}{2}$ and spin $= 1$ particles may be Reggeized though their interactions, unlike the case considered above. So the fact that only Reggeons are observed in hadronic physics does not necessarily preclude the existence of an elementary-particle sub-structure.

12.4 Fixed poles*

The diagrams of figs. 12.1 and 12.3 differ from hadronic scattering amplitudes in that though the blobs are assumed to contain the full set of hadronic singularities required by unitarity, the weak coupling constant $e^2/4\pi \equiv \alpha \approx \frac{1}{137}$ or $G \approx 1 \times 10^{-5} m_N^{-2}$ (the Fermi weak interaction coupling constant) appears explicitly only to the first order. The only γ (or W) to appear is an external particle to this blob. So for example the unitarity equation for the amplitude $\gamma + 2 \to 3 + 4$ is (fig. 12.5) usually taken to be

$$\operatorname{Im}\{A(\gamma+2\to 3+4)\} = \sum_n A(\gamma+2\to n)\, A^*(n\to 3+4) \quad (12.4.1)$$

with a sum over hadronic intermediate states only. Were we to include photon intermediate states as well, as in the other terms on the right-hand side of fig. 12.5, to give

$$\operatorname{Im}\{A(\gamma+2\to 3+4)\} = \sum_n A(\gamma+2\to n)\, A^*(n\to 3+4) + \sum_n A(\gamma+2\to \gamma+n)\, A^*(\gamma+n\to 3+4) + \dots \quad (12.4.2)$$

the terms in the second summation would be smaller than those in the first by another factor α (or G for weak interactions), which is why they are generally neglected. So if we specialise to the two-body intermediate state $3+4$ by remaining below the inelastic threshold (fig. 12.6) (e.g. $\gamma p \to \pi^+ n$ below the $\pi\pi n$ threshold), and project into partial waves, we have, instead of (4.7.4),

$$A_{HJ}(s) - (A_{HJ}(s))^* = 2i\rho_H(s)\, A_{HJ}(s)\, (A^{\mathrm{el}}_{HJ^*}(s))^* \quad (12.4.3)$$

where $A^{\mathrm{el}} \equiv A(34 \to 34)$. A similar equation holds in the t channel for $\gamma\bar{3} \to \bar{2}4$.

The fact that $A_{HJ}(s, \gamma 2 \to 34)$ appears only linearly on the right-hand side of (12.4.3) means that the theorems we enunciated in section 4.7 on the impossibility of real-axis fixed poles (except those at wrong-signature nonsense points which are shielded by cuts) do not

* This section may be omitted at first reading.

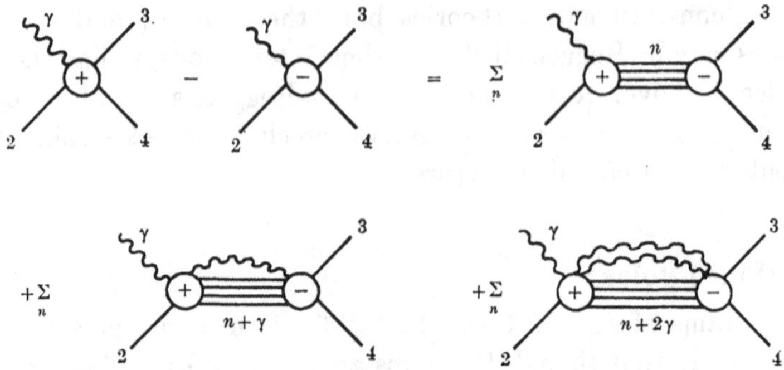

FIG. 12.5 The unitarity equation for $A(\gamma 2 \to 34)$ including higher order terms in the weak coupling.

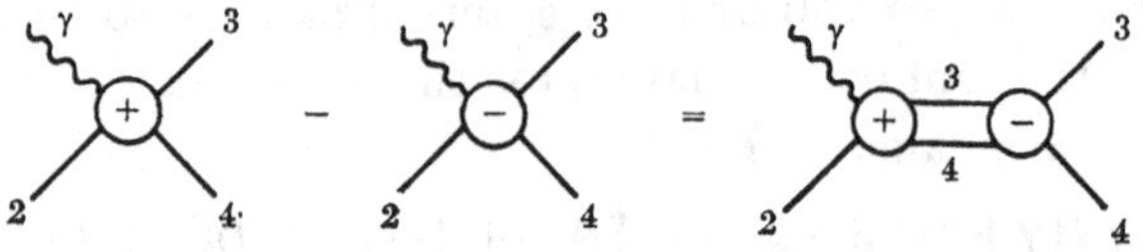

FIG. 12.6 The two-body unitarity equation for $\gamma 2 \to 34$ valid below the inelastic threshold.

apply to these weak amplitudes. So fixed poles might occur at right-signature points, in which case they would contribute to the asymptotic behaviour. In other words the SCR which must hold to prevent such fixed poles in hadronic amplitudes may not be satisfied in weak amplitudes.

It was pointed out by Bronzan *et al.* (1967) and Singh (1967) that current algebra predicts the occurrence of such fixed poles. (For an introduction to current algebra see Renner (1968), or Adler and Dashen (1968).) This is because current algebra theory relates the magnitude of the single-current coupling to that of the two-current amplitude, and in particular for $\gamma_V + 2 \to \gamma_V + 2$ (fig. 12.7) it gives (Fubini 1966, Dashen and Gell-Mann 1966)

$$\frac{1}{\pi}\int_{s_T}^{\infty} \mathrm{d}z_t\, D_{sH}^{I_t=1}(s,t,q^2,q'^2) = F(t) \qquad (12.4.4)$$

where γ_V is an iso-vector photon and 2 is a spinless particle (or we can regard (12.4.4) as a spin-averaged equation). $F(t)$ is the iso-vector form factor of particle 2, and $D_{sH}^{I_t=1}$ is the odd-signature discontinuity in s $(= D_s - D_u)$ of the reduced t-channel helicity amplitude $\hat{A}_{H_t}(s,t,q^2,q'^2)$ for the process $\gamma(q^2) + \gamma'(q'^2) \to 2\bar{2}'$. This is the helicity amplitude with

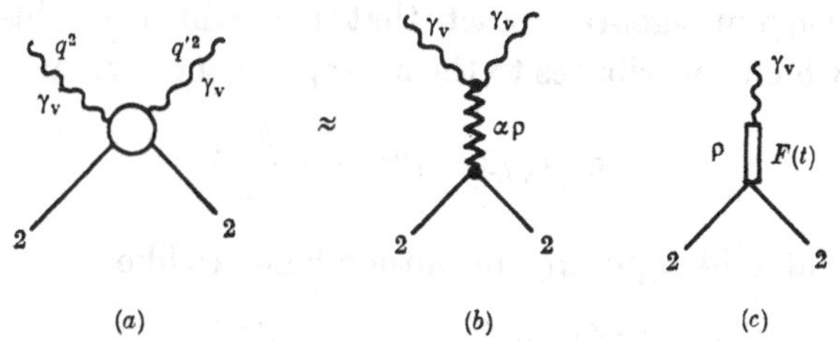

FIG. 12.7 (a) Virtual-photon Compton scattering $\gamma_v+2\to\gamma_v+2$. (b) The ρ-exchange approximation to this amplitude. (c) The ρ pole in the electromagnetic form factor of particle 2.

the half-angle factor

$$\xi_{\lambda\lambda'}(z_t) \equiv \left(\frac{1+z_t}{2}\right)^{\frac{1}{2}|\lambda+\lambda'|}\left(\frac{1-z_t}{2}\right)^{\frac{1}{2}|\lambda-\lambda'|} = [\tfrac{1}{4}(1-z_t^2)]^{\frac{1}{2}\lambda} \quad (12.4.5)$$

extracted (see (4.4.1), with $\lambda_\gamma = \pm 1$, $\lambda_2 = 0$ so $\lambda = 0$ or 2), and with $I = 1$ in the t channel. We expect the dominant $I_t = 1$ exchange to be the ρ trajectory. This is because we are dealing with isovector (charged) photons: for real photons ρ exchange is forbidden by charge conjugation. So for $\lambda_\gamma = -\lambda_{\gamma'} = 1$, i.e. $\lambda = 2$, we expect $D_{sH} \sim s^{\alpha_\rho(t)-2}$ in (12.4.4) giving

$$F(t) \sim \frac{1}{\alpha_\rho(t)-1} \sim \frac{1}{\sigma'(t-m_\rho^2)} \quad \text{for} \quad t\to m_\rho^2, \alpha_\rho\to 1 \quad (12.4.6)$$

so the form factor has the ρ pole as anticipated in fig. 12.7(c). However, the point $J = 1$ is a sense–nonsense point for this amplitude ($\lambda \equiv |\lambda_\gamma - \lambda_{\gamma'}| = 2$, $\lambda' = |\lambda_2 - \lambda_{2'}| = 0$) so we would expect a superconvergence relation to hold. Indeed if 2, 2′ are replaced by protons, and we fix on $t = 0$ where D_s can be replaced by $\sigma_{\gamma p}^{\text{tot}}$ using the optical theorem (1.9.6), then the SCR becomes

$$\int_0^\infty \frac{d\nu}{\nu}(\sigma_P(\nu)-\sigma_A(\nu)) = \frac{2\pi^2 e^2}{m_p^2}\mu'_p \quad (12.4.7)$$

where σ_P and σ_A are the total γp cross-sections with spins parallel and anti-parallel (respectively) and μ'_p is the proton's anomalous magnetic moment (i.e. the form factor at $t = 0$). Equation (12.4.7) is the well-known Drell–Hearn (1966) sum rule, which certainly seems to hold experimentally, and it gives the residue of the ρ pole a nonsense factor $\alpha_\rho(t)-1$ to cancel (12.4.6).

However the left-hand side of (12.4.4) is the residue of a fixed pole at the sn point $J = 1$ so, since the form factor $F(t)$ is certainly nonvanishing, current algebra predicts that there will be a right-signature fixed pole which contributes to the asymptotic behaviour

$$\operatorname{Re}\{\hat{A}(\gamma 2 \to \gamma 2)\} \sim -\frac{F(t)}{z_t} \tag{12.4.8}$$

and so the full odd-signature amplitude behaves like

$$\hat{A}(\gamma \mathrm{p} \to \gamma \mathrm{p}) = \frac{1}{\pi}\int^{\infty} \frac{\mathrm{d}z_t' D_{sH}(s',t)}{z_t' - z_t} = -\frac{1}{z_t}\frac{1}{\pi}\int^{\infty} \mathrm{d}z_t' D_{sH}(s',t)$$
$$+\frac{1}{z_t}\frac{1}{\pi}\int^{\infty} \frac{\mathrm{d}z_t' z_t'}{z_t' - z_t} D_{sH}(s',t) \to -\frac{F(t)}{z_t} + \frac{G_\rho(t)\,(z_t)^{\alpha_\rho(t)-2}}{\sin \pi\alpha_\rho(t)} \tag{12.4.9}$$

So there is a moving Regge pole $\alpha_\rho(t)$, and a fixed pole at $J = 1$, but no singularity at $t = m_\rho^2$, $\alpha_\rho = 1$ because at this point the two terms cancel, with $F(t)$ behaving as in (12.4.6).

Thus current algebra predicts that there will be fixed poles at right-signature nonsense points which do not contribute to D_{sH} and hence do not affect the total cross-section, but do contribute to the asymptotic behaviour of the real part of the amplitude. This is hardly surprising in that fig. 12.7 (*c*) has coupled the ρ to a fixed-spin current.

But the question then arises as to what would happen if we were to work not just to first order in e^2, but included all orders, in the (t-channel) unitarity equation, like (12.4.2). Our experience with weak-coupling field-theory solutions like (3.4.17) suggests that the fixed pole at $J = 1$ would turn into a moving Regge pole which $\to 1$ as $e^2 \to 0$, i.e.

$$\alpha(t) = 1 + e^2 f(t) \tag{12.4.10}$$

where $f(t)$ is some function of t like (3.4.19), of order 1. So we anticipate a trajectory with a slope $\alpha' = O(e^2) = O(\frac{1}{137})$. Such a trajectory is not seen in the asymptotic behaviour, nor has it manifested itself as high-mass particles. If, alternatively, the pole remained fixed it would produce Kronecker δ_{JJ_0} terms in the ss amplitudes as described in the previous section. All this suggests that current algebra may itself be wrong, though of course we do not really know how to deal properly with the zero-mass photon as an intermediate state.

Fixed poles do occur at wrong-signature points, and indeed a $J = 1$ fixed pole may be essential to the asymptotic behaviour of the even-signature Compton scattering amplitude (Abarbanel *et al.* 1967). For $\gamma + 2 \to \gamma + 2$ (with 2 spinless) there are just two independent s-channel helicity amplitudes, $\langle 1,0|\,A^s\,|1,0\rangle$ and $\langle 1,0|\,A^s\,|-1,0\rangle$. The helicity crossing matrix like (4.3.4) (see Ader, Capdeville and Navelet 1968), relating these amplitudes to those for the t-channel process $\gamma\gamma \to 2\bar{2}$, simplifies at $t = 0$ to

$$\left.\begin{aligned} \langle 1,0| A^s |1,0\rangle &= \langle 1,-1| A^t |0,0\rangle, \quad \lambda = 2 \\ \langle 1,0| A^s |-1,0\rangle &= \langle 1,1| A^t |0,0\rangle, \quad \lambda = 0 \end{aligned}\right\} \tag{12.4.11}$$

and, since the helicity-flip amplitude $\langle 1,0| A^s |-1,0\rangle$ must vanish at $t = 0$ by angular-momentum conservation, only $\langle 1,-1| A^t |0,0\rangle$ survives. Now $J_t = 1$ is a sn point for this amplitude. The dominant even-signature exchange for this elastic amplitude will be the Pomeron, so if $\alpha_P(0) = 1$, and the P residue has a nonsense decoupling factor at $\alpha_P(t) = 1$, then the P-exchange contribution to $\langle 1,0| A^s |1,0\rangle$ will vanish at $t = 0$. The optical theorem gives

$$\sigma^{\text{tot}}_{\gamma 2} = 1/s \, \text{Im}\{\langle 1,0| A^s |1,0\rangle\} \tag{12.4.12}$$

so $\sigma^{\text{tot}}_{\gamma 2} \to 0$ as $s \to \infty$ if the P decouples, unlike other total cross-sections.

But this completely contradicts our observation in section 12.2 that the photon behaves like a hadron at high energies, and the observed approximate constancy of $\sigma^{\text{tot}}_{\gamma \text{p}}(s)$ at large s. So either there is a Gribov–Pomeranchuk fixed pole at $J = 1$ which removes the decoupling factor (see table 6.2), in which case $\sigma^{\text{tot}}_{\gamma 2}(s)$ is controlled only by the third double-spectral function, which seems rather odd, or the P residue is singular at $t = 0$. In fact the residue of the $J = 1$ fixed pole in the Froissart–Gribov projection of the $\lambda = 2$ amplitude can be expressed as

$$G_1(t) = \frac{1}{\pi}\int_{m_2^2}^{\infty} \mathrm{d}z'_t D_{sH}(s',t) = \frac{Ke^2}{t} + \int_{s_I}^{\infty} \mathrm{d}z'_t D_{sH}(s,t) \tag{12.4.13}$$

where K is a constant and s_I is the inelastic threshold. The kinematical t^{-1} factor from the Born diagram in fig. 12.8 stems from the kinematics of the massless photon ($q^2_{t13} = t/4$ from (1.7.15)). So if $G_1(t)$ vanishes the residue of the P pole in D_{sH} must behave like t^{-1} so that the right-hand side can vanish. Clearly $t = 0$ is a special point because it coincides with the initial-state threshold of $\gamma\gamma \to 2\bar{2}$, and photon partial-wave amplitudes may well have an unusual threshold behaviour (see Collins and Gault (1972) and below). For a more complete discussion see Landshoff and Polkinghorne (1972).

It has also been suggested that there might be a fixed pole at the sn point $J_t = 0$ in this amplitude (Damashek and Gilman 1970). Since $J = 0$ is a right-signature point this would give a real constant contribution to the Compton scattering amplitude

$$A(\gamma 2 \to \gamma 2, s, t) = \sum_i A^{R_i}(s,t) + G_0(t) \tag{12.4.14}$$

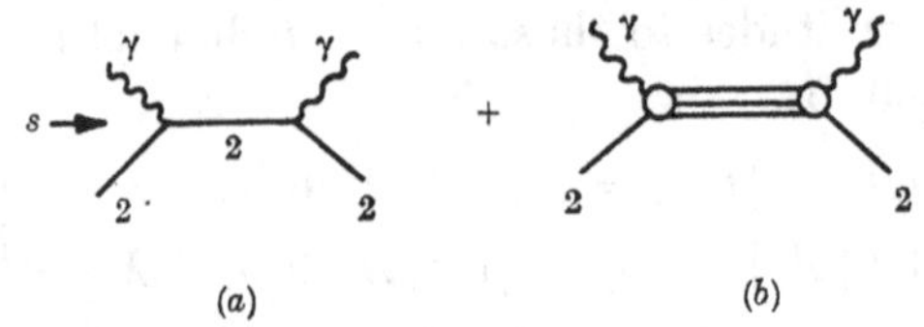

FIG. 12.8 (a) The s-channel Born term in $\gamma 2 \to \gamma 2$. (b) Other s-channel intermediate states which contribute to D_s for $s > s_I$ the inelastic threshold.

where $A^{R_i}(s,t)$ are the usual Reggeon exchange amplitudes, P and A_2 and $G_0(t)$ is the fixed-pole contribution, which is independent of s for all t. Gilman and co-workers have attempted fits of forward $d\sigma/dt(\gamma p \to \gamma p)$ and $\sigma^{tot}_{\gamma p}$ with $\alpha_P(0) = 1$, $\alpha_{A_2}(0) = \frac{1}{2}$, and find that such a real part is needed. In fact they identify

$$G_0(t) = -\frac{e^2}{m_p} \tag{12.4.15}$$

which is the Thompson amplitude for Compton scattering off a proton at zero photon energy, when the proton structure is not penetrated. However, adjusting the values of the trajectory intercepts seems to make this extra term unnecessary (Close and Gunion 1971), so there is no convincing evidence for this fixed pole. See also Brodsky, Close and Gunion (1972) and Landshoff and Polkinghorne (1972).

Fixed poles have also been searched for in photo-production processes like $\gamma p \to \pi^+ n$. In the backward direction one might have elementary nucleon exchange at $J_u = \frac{1}{2}$, giving $d\sigma/du \sim 1/s$ at fixed u, but in fact $d\sigma/du \sim s^{-2.6}$ at $u \approx 0$ corresponding to $\alpha_N(0) \approx -0.3$.

The forward direction is particularly interesting because, as we have discussed in sections 6.8j and 8.7f, this process is controlled by evasive π exchange together with a self-conspiring $\pi \otimes P$ cut. The rapid variation of $d\sigma/dt$ near $t = 0$ demands the presence of the pion pole term (see (8.7.5)). However the right-signature point $\alpha_\pi(t) = 0$ is a nonsense point for all $\gamma\pi \to \bar{p}n$ t-channel amplitudes since $\lambda_\gamma - \lambda_\pi = 1$, so normally one would expect a nonsense factor and no pion pole. At one time it was though that a fixed pole must be present to remove the need for a nonsense factor (as described above for Compton scattering), but since $J = 0$ is a right-signature point such a fixed pole should be seen in the asymptotic behaviour, which it is not. So again it would seem that the photon coupling is unusual. Now $t = m_\pi^2$ is one of the thresholds of $\gamma\pi \to \bar{p}n$, so as with Compton scattering it looks as if the blame can be placed on an unusual threshold behaviour (see Collins and Gault (1972) for references).

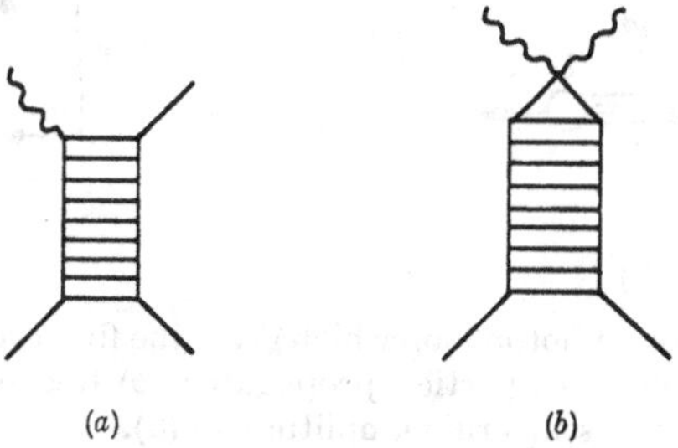

FIG. 12.9 (*a*) Field-theory model for the coupling of a photon to a composite Reggeon exchange. (*b*) The two-photon coupling to a ladder which gives rise to a $J = 1$ fixed pole (to first order in e^2).

To summarize then, we have found no very strong evidence for unusual fixed poles in weak amplitudes (despite current algebra) and some evidence against them. Theoretically (Rubinstein, Veneziano and Virasoro 1968, Dosch 1968, Landshoff and Polkinghorne 1972) there is reason to suppose that when currents couple to composite particles, for example particles built from ladders like figs. 12.9(*a*), (*b*), the only fixed poles occur at the nonsense points $J = \sigma_\gamma - n$ and $J = \sigma_{\gamma1} + \sigma_{\gamma2} - n$, $n = 1, 2, \ldots$ respectively. The latter seem to be closely related to the scaling behaviour seen in deep inelastic electron scattering (section 12.5). But all the arguments in favour of fixed poles arise from working only to first order in e^2, and could be wrong. Hence one can feel fairly secure in treating the photon like a hadron as far as the leading Regge behaviour is concerned.

One interesting consequence of this concerns the electromagnetic mass differences of isotopic multiplets. Cottingham (1963) showed how the first-order electromagnetic contributions to the self energy (= mass) of a particle, given by the photon emission and re-absorption diagram fig. 12.10, can be directly related to the spin-averaged forward Compton scattering amplitude on the given particle by a photon of mass q^2

$$\delta M = \frac{e^2}{2\pi}\int \frac{d^4 q}{(2\pi)^4} \sum_\mu \frac{A_{\mu\mu}(\nu, q^2)}{q^2 - i\epsilon} \tag{12.4.16}$$

where $\nu \equiv p.q$. The $A_{\mu\mu}$ can be expressed in terms of a dispersion relation in ν as

$$\hat{A}_{\mu\mu}(\nu, q^2) = \frac{2}{\pi}\int_0^\infty \frac{\nu' d\nu' D_{sH}(\nu', q^2)}{\nu'^2 - \nu^2} \tag{12.4.17}$$

If the non-flip helicity amplitudes are Regge pole dominated at high energy we expect their even signature $(s+u)$ discontinuities to behave like

$$D_{sH}(\nu, q^2) \to \gamma_i(q^2)\, \nu^{\alpha_i(0)} \tag{12.4.18}$$

FIG. 12.10 (*a*) The one-photon loop which gives the first order electromagnetic mass re-normalization of the particle propagator. (*b*) Reggeon exchange model for the virtual Compton-scattering amplitude in (*a*).

For example the difference between the neutron and proton electromagnetic masses depends on the dominant even-signature $\Delta I = 1$ exchange, i.e. the A_2 trajectory with $\alpha(0) \approx 0.5$ (Harari 1966), so clearly (12.4.17) will diverge, and hence these equations do not permit one to calculate this electromagnetic mass difference unless one can determine the subtraction constant. However, the $\Delta I = 2$ mass difference $(m_{\pi^\pm} - m_{\pi^0})$ is dominated by $I = 2$ exchange and since no such trajectory is known the dominant (Regge cut?) exchange may well have $\alpha(0) < 0$ so the integral should converge. This may help to account for the fact that $m_{\pi^\pm} > m_{\pi^0}$ as one would expect (i.e. electromagnetic effects add to the mass of the pions) but $m_n > m_p$ which contradicts this expectation. This criterion based on the intercept of the exchanged Reggeon seems to work for the signs of other mass differences as well.

This is just one example of the way in which the known Regge asymptotic behaviour is helpful for understanding the weaker interactions, particularly their dispersion sum rules.

12.5 Deep inelastic scattering

Some of the most interesting results on the structure of hadrons have come from deep inelastic scattering experiments on nucleons, $ep \to eX$ and to a lesser extent $\nu p \to \mu X$. These are treated in the one-photon or one-W exchange approximation as in figs. 12.1. (For reviews of these processes see for example Gilman (1972) and Llewellyn-Smith (1972), respectively, and Landshoff and Polkinghorne (1972).)

With the four-momenta indicated in the figure we have

$$k^2 = k'^2 = m_e^2 \approx 0 \quad \text{and} \quad q = k - k' \tag{12.5.1}$$

In the laboratory frame (proton at rest) we can write

$$p = (m_p, 0, 0, 0), \quad k = (E, \mathbf{k}), \quad k' = (E', \mathbf{k}'), \quad q = (E - E', \mathbf{k} - \mathbf{k}'),$$
$$k^2 = E^2 - \mathbf{k}^2 \approx 0, \quad k'^2 = E'^2 - \mathbf{k}'^2 \approx 0 \tag{12.5.2}$$

Hence $$\nu \equiv p.q = (E-E')\,m_{\mathrm{p}} \tag{12.5.3}$$

gives the energy of the virtual photon, while its mass

$$q^2 = (E-E')^2 - (\boldsymbol{k}-\boldsymbol{k}')^2 = E^2 + E'^2 - 2EE' - \boldsymbol{k}^2 - \boldsymbol{k}'^2 + 2|\boldsymbol{k}|\,|\boldsymbol{k}'|\cos\theta$$
$$\approx -4EE'\sin^2\frac{\theta}{2} \tag{12.5.4}$$

(using (12.5.2)) depends on θ, the scattering angle between the directions of motion of the initial- and final-state leptons. (The reader will note that many authors (e.g. Gilman) define ν without the factor m_{p} in (12.5.3) and take the opposite sign for q^2 in (12.5.4).)

For the scattering process in the bottom of the figure, $\gamma_{\mathrm{v}} + \mathrm{p} \rightarrow X$, the effective centre-of-mass energy squared is

$$s \equiv (p+q)^2 = p^2 + q^2 + 2p.q = m_{\mathrm{p}}^2 + q^2 + 2\nu \equiv M_X^2 \tag{12.5.5}$$

using (12.5.3). Averaging over the spins of the electron and proton, the differential cross-section for $\mathrm{ep} \rightarrow \mathrm{e}X$ is found to be (Drell and Walecka 1964)

$$\frac{\mathrm{d}^2\sigma}{\mathrm{d}\Omega\,\mathrm{d}E'} = \frac{4e^4E'^2}{q^4}\left(2W_1(\nu, q^2)\sin^2\frac{\theta}{2} + W_2(\nu, q^2)\cos^2\frac{\theta}{2}\right) \tag{12.5.6}$$

where $\mathrm{d}\Omega$ is the element of solid angle within which the final-state electron of energy E' is detected, and $W_{1,2}$ are the conventionally defined deep inelastic structure functions of the nucleon. They are directly related to the total cross-sections for transversely and longitudinally polarized virtual photons scattering on a proton (σ_{T} and σ_{L} respectively) by

$$\left.\begin{aligned} W_1(\nu, q^2) &= \frac{K_1}{4\pi e^2}\sigma_{\mathrm{T}}(\nu, q^2) \\ W_2(\nu, q^2) &= \frac{K_2}{4\pi e^2}(\sigma_{\mathrm{T}}(\nu, q^2) + \sigma_{\mathrm{L}}(\nu, q^2)) \end{aligned}\right\} \tag{12.5.7}$$

where $$K_1 \equiv \frac{1}{m_{\mathrm{p}}}\left(\nu + \frac{q^2}{2}\right), \quad K_2 \equiv m_{\mathrm{p}}\left(\nu + \frac{q^2}{2}\right)\left(\frac{q^2}{m_{\mathrm{p}}^2 q^2 - \nu^2}\right) \tag{12.5.8}$$

As $q^2 \rightarrow 0$, $\sigma_{\mathrm{L}} \rightarrow 0$ and σ_{T} is the real $\gamma\mathrm{p}$ total cross-section.

Elastic ep scattering (fig. 12.11) clearly requires $M_X^2 = m_{\mathrm{p}}^2$ and so from (12.5.5)

$$q^2 = -2\nu \tag{12.5.9}$$

at which values W_1 and W_2 are related to the proton's electromagnetic

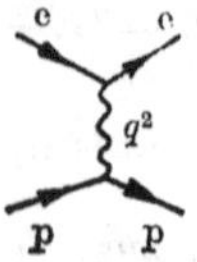

FIG. 12.11 One-photon exchange diagram for ep → ep.

form factors, G_E and G_M, by

$$\left.\begin{aligned} W_1(\nu, q^2) &= -\frac{q^2 G_M^2(q^2)}{4m_p^2}\,\delta\left(\nu+\frac{q^2}{2}\right) \\ W_2(\nu, q^2) &= \left[G_E^2(q^2) - \frac{q^2}{4m_p^2}\,G_M^2(q^2)\right]\left(1-\frac{q^2}{4m_p^2}\right)^{-1}\delta\left(\nu+\frac{q^2}{2}\right) \end{aligned}\right\} \quad (12.5.10)$$

The most remarkable result to come from experiments on deep inelastic scattering is the scaling of W_1 and νW_2 as ν, $|q^2| \to \infty$ (q^2 is negative in the physical region)

$$\left.\begin{aligned} W_1(\nu, q^2) &\to F_1\left(\frac{2\nu}{|q^2|}\right) \\ \nu W_2(\nu, q^2) &\to F_2\left(\frac{2\nu}{|q^2|}\right) \end{aligned}\right\} \quad (12.5.11)$$

where F_1, F_2 are functions which depend only on the dimensionless ratio

$$\omega \equiv \frac{2\nu}{|q^2|} \equiv \frac{1}{x} \quad (12.5.12)$$

and not on the values of ν and $|q^2|$ individually. That is to say, if both ν and $|q^2|$ are varied, keeping their ratio fixed, the values of W_1 and νW_2 are unchanged (see section 10.5 for the concept of scaling).

The most simple explanation of this scaling effect is provided by the parton model (see Feynman 1972) in which the nucleon is imagined to be composed of a number of structureless, point-like, charged particles (partons), i, each carrying a fraction x_i of the total proton momentum

$$p_i = x_i p, \quad 0 \leqslant x_i \leqslant 1 \quad (12.5.13)$$

If the parton is structureless its mass, and its charge, Q_i, will be unchanged by the scattering (fig. 12.12) and so it will give a contribution to the W's like (12.5.10) but without any q^2 dependence of the form factor (so $G_E = Q_i$). So for example

$$W_2^i(\nu, q^2) = Q_i^2\,\delta\left(\nu+\frac{q^2}{2x_i}\right) = \frac{Q_i^2 x_i}{\nu}\,\delta\left(x_i-\frac{|q^2|}{2\nu}\right) \quad (12.5.14)$$

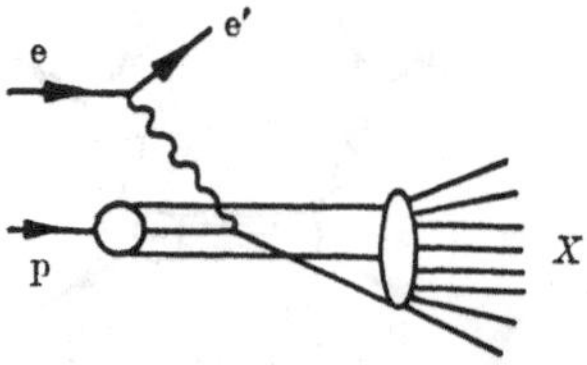

FIG. 12.12 Parton description of ep → eX. The proton is composed of structureless partons (quarks) and the photon is absorbed by one of these partons. In the right-hand blob the quarks recombine to form ordinary hadrons, X.

Then if $f(x)$ is the probability that the parton has a fraction x of the proton's momentum we get

$$\nu W_2 = \sum_i Q_i^2 \int_0^1 dx_i\, f(x_i)\, x_i\, \delta\left(x_i - \frac{|q^2|}{2\nu}\right)$$
$$= \sum_i Q_i^2 x f(x)|_{x=|q^2|/2\nu} \equiv F_2\left(x = \frac{|q^2|}{2\nu}\right) \qquad (12.5.15)$$

This result depends crucially on the partons being structureless since otherwise form factors, functions of q^2, would appear in (12.5.14) as they do in (12.5.10), and would destroy the scaling.

The parton model has enjoyed considerable success, not only because it 'explains' scaling but because if the partons are taken to be spin $= \frac{1}{2}$ quarks many features both of the spin dependence and the internal symmetry properties of the cross-sections are accounted for. The main problem is of course that the quarks are not observed, and it is not clear what mechanism can be responsible for the right-hand blob of fig. 12.12 in which all the quarks, including the scattered one, recombine to form conventional hadrons.

However, our main interest is in the Regge properties of these results. Since the W's are, apart from the kinematical factors in (12.5.7), γp total cross-sections, we can hope to describe these cross-sections by making Regge models of the γ_Vp elastic scattering amplitude, as in fig. 12.13 (cf. fig. 10.23). The Regge limit is $\nu \to \infty$, q^2 fixed (so $x \to 0$), and we expect

$$\sigma_T, \sigma_L \sim \sum_k \nu^{\alpha_k(0)-1} \qquad (12.5.16)$$

so that

$$\left.\begin{aligned} W_1(\nu, q^2) &\sim \nu\sigma_T \to \sum_k \beta_1^k(q^2)\left(\frac{\nu}{s_0}\right)^{\alpha_k(0)} \\ \nu W_2(\nu, q^2) &\sim |q^2|\,(\sigma_T + \sigma_S) \to \sum_k |q^2|\,\beta_2^k(q^2)\left(\frac{\nu}{s_0}\right)^{\alpha_k(0)-1} \end{aligned}\right\} \qquad (12.5.17)$$

So, if the leading singularity is the Pomeron with $\alpha_P(0) = 1$, both

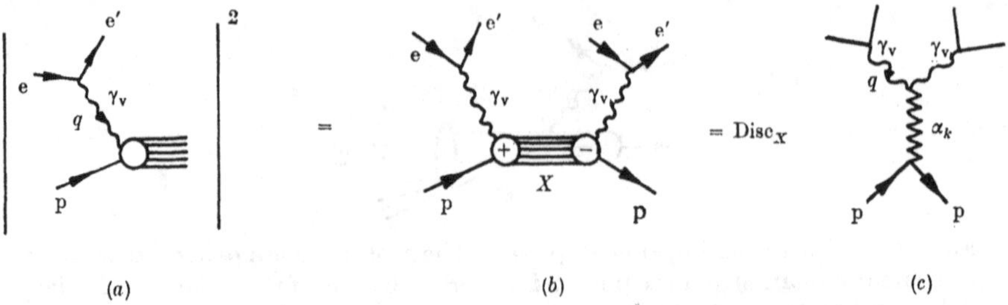

FIG. 12.13 Reggeon exchange description of deep-inelastic ep scattering in terms of $\gamma_v p \to \gamma_v p$. The leading trajectories which can be exchanged in the elastic amplitude are k = P, f and A_2.

νW_2 and $xW_1 \to$ constant as $x \equiv q^2/2\nu \to 0$. The region where scaling occurs ($|q^2|, \nu \to \infty$, x fixed) may overlap the Regge asymptotic region ($\nu \to \infty$, q^2 fixed). This clearly depends on the behaviour of the couplings $\beta^{\mathrm{P}}_{1,2}(q^2)$ as $|q^2| \to \infty$, but if they are to overlap we need

$$\beta^{\mathrm{P}}_1(q^2) \xrightarrow[|q^2|\to\infty]{} \frac{g_1}{(|q^2|)^{\alpha_{\mathrm{P}}(0)}}, \quad \beta^{\mathrm{P}}_2(q^2) \to \frac{g_2}{(|q^2|)^{\alpha_{\mathrm{P}}(0)}} \tag{12.5.18}$$

where g_1, g_2 are constants, so that W_1, $\nu W_2 \to$ scaling function of x only, i.e.

$$\left.\begin{aligned} W_1(\nu, q^2) &\xrightarrow[x\to 0]{} \frac{g_1}{(2s_0 x)^{\alpha_{\mathrm{P}}(0)}} \\ \nu W_2(\nu, q^2) &\xrightarrow[x\to 0]{} \frac{g_2}{(2s_0 x)^{\alpha_{\mathrm{P}}(0)-1}} \end{aligned}\right\} \tag{12.5.19}$$

This accords with the behaviour found in field-theory models with a point-like electromagnetic coupling (Abarbanel, Goldberger and Trieman 1969).

It should be noted that though fig. 12.13 (c) looks like the triple-Regge diagram fig. 10.23 (c) it is really quite different. Fig. 10.23 involves trajectories $\alpha_i(t)$, $\alpha_j(t)$ so that the angular momentum changes as a function of t, while fig. 12.13 (c) involves the photons $\gamma(q^2)$ which remain at $J = 1$ even though q^2 is varied. So rather different information is obtained from electro-production. However, if we adopt a multi-peripheral type of model (fig. 12.14) it is only at the top end that the couplings will be affected by q^2 (because there are only short-range correlations) and many features should be hadron-like, such as multiplicities etc. It is the photon's fragmentation region which should exhibit most of the differences (see Cahn (1974) for a more detailed discussion).

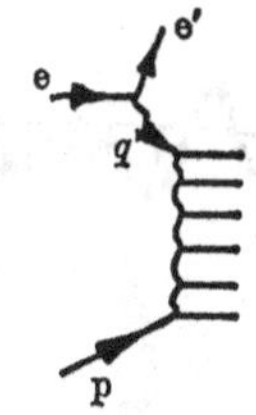

FIG. 12.14 Multi-peripheral model for ep → e'X.

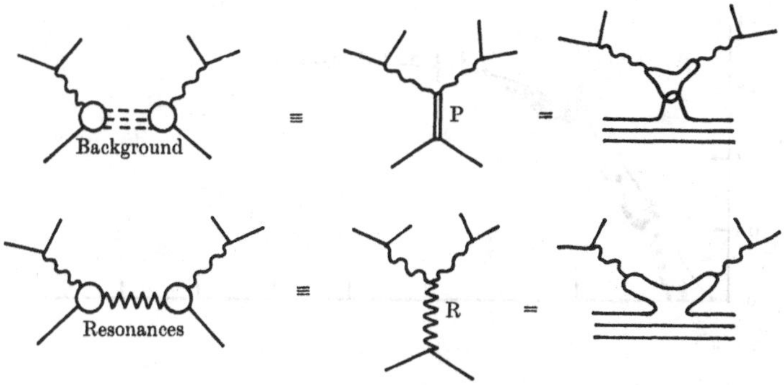

FIG. 12.15 Duality diagrams for ep → eX.

Regge theory also predicts that the principal non-constant corrections to (12.5.19) will stem from $k = \mathrm{R}$ (f and A_2) with $\alpha_{\mathrm{R}}(0) \approx 0.5$. Two-component duality suggests that the duality properties will be as in fig. 12.15, with the P dual to background, b, and R dual to the resonances, r (Harari 1970). Since the Regge region is found to overlap the scaling region this implies strong constraints on the resonance-production cross-sections as a function of q^2, and hence on the transition form factors for $\gamma_{\mathrm{v}}(q^2) + \mathrm{p} \to \mathrm{r}$. The way in which the resonance contributions are smoothed to the scaling behaviour as $|q^2|$ (and hence ν) is increased for fixed ω is shown in fig. 12.16. This is just like the smoothing in the Veneziano model with $\mathrm{Im}\{\alpha\} \neq 0$ (see fig. 7.6). In fact the Veneziano model has enjoyed some success in fitting the ν, q^2 dependence of the R term (Landshoff and Polkinghorne 1970, 1971). There is, however, a rather fundamental problem that the Veneziano model is constructed with factorized Reggeon couplings, whereas here we need a fixed-spin $J = 1$ coupling, so even if we project out $J = 1$ on one leg of the Veneziano model problems concerned with the difference between elementary and composite particle couplings arise (see Drummond 1972).

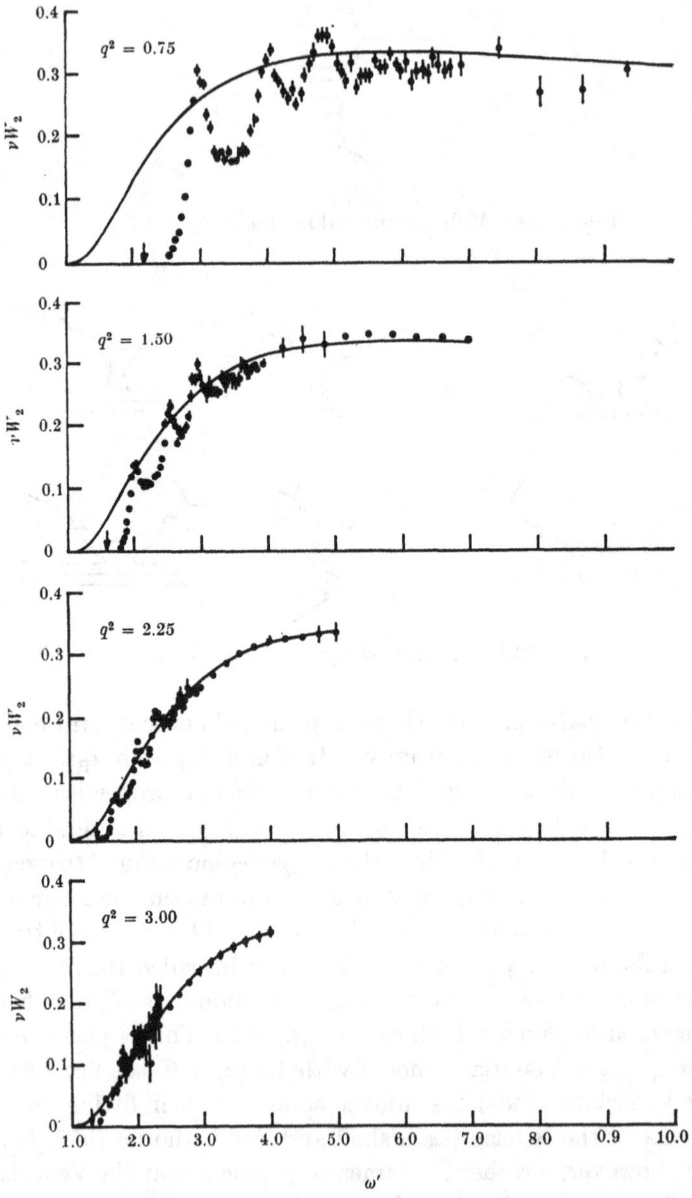

Fig. 12.16 Plot of $\nu W_2(\nu, q^2)$, at various values of $|q^2|$ (GeV units), versus $\omega' \equiv \omega + m_N/q$. The solid line is the scaling curve found for higher values of $|q^2|$ and the resonance oscillations converge to this line as $|q^2|$ increases. (From Gilman 1972.)

In deep inelastic neutrino scattering, $\nu p \to \mu X$, there is an extra structure function because parity is not a conserved quantity, and the differential cross-section reads, instead of (12.5.6),

$$\frac{d^2\sigma}{d\Omega\, dE'} = \frac{G^2E'^2}{2\pi^2 m_p^2}\left(2W_1(\nu, q^2)\sin^2\frac{\theta}{2} + W_2(\nu, q^2)\cos^2\frac{\theta}{2} \right.$$
$$\left. \mp \frac{E+E'}{m_p} W_3(\nu, q^2)\sin^2\frac{\theta}{2}\right) \quad (12.5.20)$$

with $\mp$ for $\nu, \bar{\nu}$ scattering, respectively. The extra function W_3 is odd under C_n and so vanishes as $\nu \to \infty$ since P exchange is not possible. Again the quark–parton model is rather successful, and Regge theory has been used to predict the high ν behaviour (see Llewellyn-Smith 1972). The most interesting results for Regge theory may come at higher energies (if these can be achieved) because the phenomenological Fermi theory, which can be used only for the first order in G, will violate unitarity for $E_{\nu\,\mathrm{lab}} > 10^5$ GeV, and so there must be unitarity corrections, and these will presumably heed the restrictions on fixed poles discussed in section 12.4.

In conclusion, Regge theory has so far had only a modest though honourable role to play in weak interactions, mainly because it is still possible to work only to first order in the weak coupling (e^2 or G). But when the time comes to construct a proper unitary theory of the weak interactions of hadrons, or even perhaps a unified theory of all the interactions, Reggeization will be a crucial ingredient.

Appendix A
The Legendre functions

The representation functions of orbital angular momentum are (Schiff 1968, Edmonds 1960, Rose 1967) the spherical harmonics

$$Y_{lm}(\theta, \phi) = (-1)^m \left[\frac{(2l+1)(l-m)!}{4\pi(l+m)!}\right]^{\frac{1}{2}} P_l^m(z)\, e^{im\phi} \tag{A.1}$$

where

$$z \equiv \cos\theta \tag{A.2}$$

and where the $P_l^m(z)$ are the associated Legendre functions. Their properties are discussed in great detail in Erdelyi *et al.* (1953, vol. 1), which we shall refer to below as E followed by the appropriate page number.

Scattering problems for spinless particles are symmetrical about the beam direction, which is conventionally taken to be the z axis. This eliminates the ϕ dependence, so we are only concerned with

$$Y_{l0}(\theta, \phi) = \left(\frac{2l+1}{4\pi}\right)^{\frac{1}{2}} P_l(z) \tag{A.3}$$

These Legendre functions are eigenfunctions of the operator for the square of the angular momentum, L^2, i.e.

$$L^2 P_l(z) = l(l+1) P_l(z), \quad l = 0, 1, 2, \ldots \tag{A.4}$$

which in the co-ordinate representation becomes

$$\frac{d}{dz}\left[(1-z^2)\frac{dP_l}{dz}\right] + l(l+1) P_l(z) = 0 \tag{A.5}$$

which is Legendre's equation (E, p. 121). For integer l these Legendre functions are polynomials in z, regular in the finite z plane, the first few being

$$P_0(z) = 1, \quad P_1(z) = z, \quad P_2(z) = \tfrac{1}{2}(3z^2 - 1), \quad P_3(z) = \tfrac{1}{2}(5z^3 - 3z) \tag{A.6}$$

However, (A.5) also has solutions for $l \neq$ integer which (E, p. 148) may be expressed in terms of the hypergeometric function

$$P_l(z) = F(-l, l+1; 1; (1-z)/2) \tag{A.7}$$

which is singular at $z = -1$ and ∞. These are called Legendre functions of the first kind.

There are also solutions of ($A.5$) singular at $z = \pm 1$ and ∞ called Legendre functions of the second kind (E, p. 122)

$$Q_l(z) = \pi^{\frac{1}{2}} \frac{\Gamma(l+1)}{\Gamma(l+\frac{3}{2})} (2z)^{-l-1} F(\tfrac{1}{2}l+1, \tfrac{1}{2}l+\tfrac{1}{2}; l+\tfrac{3}{2}; z^{-2}) \qquad (A.8)$$

For integer l the first few are (E, p. 152)

$$\left.\begin{aligned} Q_0(z) &= \tfrac{1}{2}\log\left(\frac{z+1}{z-1}\right), \; Q_1(z) = \tfrac{1}{2}z\log\left(\frac{z+1}{z-1}\right) - 1, \\ Q_2(z) &= \tfrac{1}{2}P_2(z)\log\left(\frac{z+1}{z-1}\right) - \tfrac{3}{2}z. \end{aligned}\right\} \qquad (A.9)$$

These functions satisfy *inter alia* the following relations which we need in this book.

The reflection relation (E, p. 140) gives

$$P_l(-z) = \mathrm{e}^{-i\pi l}P_l(z) - \frac{2}{\pi}\sin \pi l Q_l(z) \qquad (A.10)$$

$$= (-1)^l P_l(z), \quad l = \text{integer} \qquad (A.11)$$

The equation ($A.5$) is invariant under the substitution $l \to -l-1$, so (E, p. 140)

$$P_l(z) = P_{-l-1}(z) \qquad (A.12)$$

Also (E, p. 143) for real l

$$\begin{aligned} \operatorname{Im}\{P_l(z)\} &= -P_l(-z)\sin \pi l \quad & z < -1 \\ &= 0 & z \geqslant 1 \end{aligned} \qquad (A.13)$$

The two types of solution are related by the Neumann relation (E, p. 154) for integer l

$$Q_l(z) = -\frac{1}{2}\int_{-1}^{1} \frac{\mathrm{d}z'}{z'-z} P_l(z'), \quad l = 0, 1, 2, \ldots \qquad (A.14)$$

a 'dispersion relation' for $Q_l(z)$, from which it is obvious that (E, p. 143)

$$\begin{aligned} \operatorname{Im}\{Q\,(z)\} &= 0, \quad |z| > 1, \quad l = 0, 1, 2, \ldots \\ &= -\frac{\pi}{2}P_l(z), \quad -1 < z < 1, \end{aligned} \qquad (A.15)$$

For $l \neq$ integer

$$\begin{aligned} \operatorname{Im}\{Q_l(z)\} &= \sin \pi l Q_l(-z), \quad & -\infty < z < -1 \\ &= -\frac{\pi}{2}P_l(z), & -1 < z < 1 \end{aligned} \qquad (A.16)$$

The reflection relation for the second-type functions is (E, p. 140)

$$Q_l(-z) = -e^{-i\pi l} Q_l(z)$$
$$= (-1)^{l+1} Q_l(z), \quad l = \text{integer} \tag{A.17}$$

Other useful results are (E, p. 140)

$$\frac{P_l(z)}{\sin \pi l} - \frac{1}{\pi} \frac{Q_l(z)}{\cos \pi l} = -\frac{1}{\pi} \frac{Q_{-l-1}(z)}{\cos \pi l} \tag{A.18}$$

and
$$Q_l(z) = Q_{-l-1}(z), \quad l = \text{half-odd-integer} \tag{A.19}$$

The orthogonality relation for Legendre polynomials is (E, p. 170)

$$\int_{-1}^{1} P_{l'}(z)\, P_l(z)\, dz = \frac{2}{2l+1} \delta_{ll'}, \quad l, l' \text{ integers} \tag{A.20}$$

and some other integral relations are (E, p. 170)

$$\int_{-1}^{1} P_\alpha(-z)\, P_l(z)\, dz = \frac{1}{\pi} \frac{2 \sin \pi\alpha}{(\alpha - l)(\alpha + l + 1)}, \quad l \text{ integer, } \alpha \text{ anything} \tag{A.21}$$

$$\int_{1}^{\infty} P_\alpha(z)\, Q_l(z)\, dz = \frac{1}{(l-\alpha)(l+\alpha+1)}, \quad l, \alpha \text{ anything} \tag{A.22}$$

$$P_\alpha(-z) = -\frac{\sin \pi\alpha}{\pi} \int_1^\infty \frac{dz' P_\alpha(z')}{z' - z} \tag{A.23}$$

The asymptotic behaviour as $z \to \infty$ for fixed l may be obtained by rewriting (A.7) as (E, p. 127)

$$P_l(z) = \pi^{-\frac{1}{2}} \frac{\Gamma(l+\frac{1}{2})}{\Gamma(l+1)} (2z)^l F(-\tfrac{1}{2}l, -\tfrac{1}{2}l + \tfrac{1}{2}; -l + \tfrac{1}{2}; z^{-2})$$
$$+ \pi^{-\frac{1}{2}} \frac{\Gamma(-l-\frac{1}{2})}{\Gamma(-l)} (2z)^{-l-1} F(\tfrac{1}{2}l + \tfrac{1}{2}, \tfrac{1}{2}l + 1; l + \tfrac{3}{2}; z^{-2}) \tag{A.24}$$

Then since $F \to 1$ as $z \to \infty$ we have (E, p. 164)

$$P_l(z) \xrightarrow[z \to \infty]{} \pi^{-\frac{1}{2}} \frac{\Gamma(l+\frac{1}{2})}{\Gamma(l+1)} (2z)^l, \quad \text{Re}\{l\} \geqslant -\tfrac{1}{2} \tag{A.25}$$

$$\xrightarrow[z \to \infty]{} \pi^{-\frac{1}{2}} \frac{\Gamma(-l-\frac{1}{2})}{\Gamma(-l)} (2z)^{-l-1}, \quad \text{Re}\{l\} \leqslant -\tfrac{1}{2} \tag{A.26}$$

Similarly, from (A.8),

$$Q_l(z) \xrightarrow[z \to \infty]{} \pi^{\frac{1}{2}} \frac{\Gamma(l+1)}{\Gamma(l+\frac{3}{2})} (2z)^{-l-1} \tag{A.27}$$

The asymptotic behaviour as $l \to \infty$ for fixed z is rather more difficult (E, pp. 142, 162; Newton 1964):

$$P_l(z) \underset{l\to\infty}{\longrightarrow} (2\pi l)^{-\frac{1}{2}} (z^2-1)^{-\frac{1}{4}} \mathrm{e}^{\xi}, \quad \mathrm{Re}\{l\} \geqslant 0 \qquad (A.28)$$

where $\xi \equiv 2(\mathrm{Re}\{l\}+1)\log\left[\left(\frac{z+1}{2}\right)^{\frac{1}{2}}+\left(\frac{z-1}{2}\right)^{\frac{1}{2}}\right], \quad z>1$

$$\equiv 2|\mathrm{Im}\{l\}|\tan^{-1}\left(\frac{1-z}{1+z}\right)^{\frac{1}{2}} \quad z^2<1$$

so

$$|P_l(z)| \underset{l\to\infty}{<} l^{-\frac{1}{2}} \mathrm{e}^{|\mathrm{Im}\{l\}\,\mathrm{Re}\{\theta\}+\mathrm{Re}\{l\}\,\mathrm{Im}\{\theta\}|} f(z) \qquad (A.29)$$

$$\left|\frac{P_l(z)}{\sin \pi l}\right| \underset{l\to\infty}{<} l^{-\frac{1}{2}} \mathrm{e}^{|\mathrm{Im}\{l\}\,\mathrm{Re}\{\theta\}+\mathrm{Re}\{l\}\,\mathrm{Im}\{\theta\}|-\pi|\mathrm{Im}\{l\}|} f(z) \qquad (A.30)$$

Also

$$Q_l(z) \underset{|l|\to\infty}{\longrightarrow} l^{-\frac{1}{2}} \mathrm{e}^{-(l+\frac{1}{2})\zeta(z)} \qquad (A.31)$$

where $\zeta(z) \equiv \log[z+(z^2-1)^{\frac{1}{2}}]$.

From (A.7) we see that $P_l(z)$ is an entire function of l, while from (A.8) it is clear that $Q_l(z)$ has poles in l at negative integer values due to the Γ-function in the numerator, and from (A.18)

$$Q_l(z) \approx \pi \frac{\cos \pi l}{\sin \pi l} P_{-l-1}(z), \quad l = -1, -2, -3, \ldots \qquad (A.32)$$

Appendix *B*
The rotation functions

A state of angular momentum J, and z component of angular momentum m, is transformed under a rotation by the Euler angles α, β, γ according to (see Edmonds (1960) p. 54)

$$D(\alpha,\beta,\gamma)\,|Jm\rangle = \sum_{m'=-J}^{J} |Jm'\rangle\langle Jm'|\,D(\alpha,\beta,\gamma)\,|Jm\rangle \tag{B.1}$$

where the rotation operator is

$$D(\alpha,\beta,\gamma) \equiv e^{i\alpha J_z}\,e^{i\beta J_y}\,e^{i\gamma J_z} \tag{B.2}$$

i.e. a rotation by angle γ about the z axis, followed by a rotation by β about the y axis, followed by a further rotation by α about the z axis.

Since the eigenvalue of J_z is m, the matrix elements of $D(\alpha,\beta,\gamma)$ can be written

$$\langle Jm'|\,D(\alpha,\beta,\gamma)\,|Jm\rangle \equiv \mathscr{D}^J_{m'm}(\alpha,\beta,\gamma) \tag{B.3}$$

$$= e^{im'\alpha}\,d^J_{m'm}(\beta)\,e^{im\gamma} \tag{B.4}$$

where the rotation matrices are defined by

$$d^J_{m'm}(\beta) \equiv \langle Jm'|\,e^{i\beta J_y}\,|Jm\rangle \tag{B.5}$$

These matrix elements can readily be evaluated for $J = \frac{1}{2}$ by substituting the Pauli matrix for J_y and expanding the exponential (see (B.19) below), and then higher J values can be derived using the Clebsch–Gordan series (see for example Wigner (1959) p. 167). It is found that (Edmonds (1960) p. 57)

$$d^J_{m'm}(\beta) = \left[\frac{(J+m')!\,(J-m')!}{(J+m)!\,(J-m)!}\right]^{\frac{1}{2}} \sum_\sigma \binom{J+m}{J-m'-\sigma}\binom{J-m}{\sigma}(-1)^{J-m'-\sigma}$$
$$\times\left(\cos\frac{\beta}{2}\right)^{2\sigma+m'+m}\left(\sin\frac{\beta}{2}\right)^{2J-2\sigma-m'-m} \tag{B.6}$$

If the scattering plane is taken to be the x–z plane, then the angle β here corresponds to the scattering angle θ between the directions of motion in the initial and final states, and it is more convenient to write the rotation matrices as functions of $z \equiv \cos\theta$ rather than θ. Also for two-particle helicity states m' and m correspond to the helicity differences λ and λ' defined in (4.4.15). So we shall usually replace (*B*.6) by $d^J_{\lambda\lambda'}(z)$ from now on.

The functions defined by (*B*.6) satisfy the symmetry relations

$$\left.\begin{aligned} d^J_{\lambda\lambda'}(z) &= (-1)^{\lambda-\lambda'} d^J_{-\lambda-\lambda'}(z) = (-1)^{\lambda-\lambda'} d^J_{\lambda'\lambda}(z) \\ d^J_{\lambda\lambda'}(\pi-\theta) &= (-1)^{J-\lambda} d^J_{-\lambda\lambda'}(-\theta) = (-1)^{J-\lambda} d^J_{\lambda'-\lambda}(\theta) \end{aligned}\right\} \quad (B.7)$$

The expression (*B*.6) can be rewritten in terms of Jacobi polynomials $P^{(a,b)}_c(z)$ as

$$d^J_{\lambda\lambda'}(z) = \left[\frac{(J+\lambda)!\,(J-\lambda)!}{(J+\lambda')!\,(J-\lambda')!}\right]^{\frac{1}{2}} \left(\frac{1-z}{2}\right)^{\frac{1}{2}(\lambda-\lambda')} \left(\frac{1+z}{2}\right)^{\frac{1}{2}(\lambda+\lambda')} P^{(\lambda-\lambda',\,\lambda+\lambda')}_{J-\lambda}(z) \quad (B.8)$$

but this is only valid for non-negative values of $\lambda-\lambda'$ and $\lambda+\lambda'$. Other values can be obtained from (*B*.8) using the symmetry relations (*B*.7), which may be incorporated by writing

$$d^J_{\lambda\lambda'}(z) = (-1)^{\Lambda} \left[\frac{(J+M)!\,(J-M)!}{(J+N)!\,(J-N)!}\right]^{\frac{1}{2}} \xi_{\lambda\lambda'}\, P^{(|\lambda-\lambda'|,\,|\lambda+\lambda'|)}_{J-M} \quad (B.9)$$

where

$$M \equiv \max\{|\lambda|, |\lambda'|\}, \quad N \equiv \min\{|\lambda|, |\lambda'|\}, \quad \Lambda \equiv \tfrac{1}{2}(\lambda-\lambda'-|\lambda-\lambda'|) \quad (B.10)$$

and the 'half-angle factor' is defined by

$$\xi_{\lambda\lambda'}(z) \equiv \left(\frac{1-z}{2}\right)^{\frac{1}{2}|\lambda-\lambda'|} \left(\frac{1+z}{2}\right)^{\frac{1}{2}|\lambda+\lambda'|} \quad (B.11)$$

Equation (*B*.9) is a very convenient representation because for integer $J-M$ the Jacobi function is an entire function of z, so the only possible singularities of $d^J_{\lambda\lambda'}(z)$ in z stem from the behaviour of the half-angle factor (*B*.11) at $z = \pm 1$.

However, we shall also wish to continue in J, and for this purpose it is more useful to re-express (*B*.9) in terms of the hypergeometric function (see Andrews and Gunson 1964)

$$d^J_{\lambda\lambda'}(z) = (-1)^{\Lambda} \left[\frac{(J+M)!\,(J-M+|\lambda-\lambda'|)!}{(J-M)!\,(J+M-|\lambda-\lambda'|)!}\right]^{\frac{1}{2}} \frac{1}{|\lambda-\lambda'|!}$$

$$\times\, \xi_{\lambda\lambda'}(z)\, F(-J+M, J+M+1, |\lambda-\lambda'|+1; (1-z)/2) \quad (B.12)$$

The hypergeometric function is an entire function of J, so the only singularities stem from the square bracket, when the factorial functions have poles for negative integer values of their arguments.

Also from the asymptotic form of the hypergeometric function we find

$$d^J_{\lambda\lambda'}(z) \underset{z\to\infty}{\longrightarrow} (-1)^A \times \frac{(2J)!}{[(J+M)!\,(J-M+|\lambda-\lambda'|)!\,(J-M)!\,(J+M-|\lambda-\lambda'|)!]^{\frac{1}{2}}} \times \xi_{\lambda\lambda'}(z)\left(\frac{z}{2}\right)^{J-M}(1+O(z^{-2}))+O(z^{-J-1}) \quad (B.13)$$

so, since $\xi_{\lambda\lambda'}(z) \sim z^M$ from $(B.11)$

$$d^J_{\lambda\lambda'}(z) \sim z^J, \quad \text{for} \quad J > -\tfrac{1}{2}$$

for $J-v \neq$ integer $< M$ where $d^J_{\lambda\lambda'}$ vanishes ($v \equiv 0/\frac{1}{2}$ for physical $J =$ integer/half-odd-integer, i.e. for even/odd fermion number).

These functions also satisfy the orthogonality relations

$$\int_{-1}^{1} d^J_{\lambda\lambda'}(z)\, d^{J'}_{\lambda\lambda'}(z)\,\mathrm{d}z = \delta_{JJ'}\frac{2}{2J+1} \quad (B.14)$$

$$\tfrac{1}{2}\sum_J (2J+1)\, d^J_{\lambda\lambda'}(z)\, d^J_{\lambda\lambda'}(z') = \delta(z-z') \quad (B.15)$$

$$\sum_\lambda d^J_{\lambda\lambda'}(z)\, d^J_{\lambda\lambda''}(z) = \delta_{\lambda'\lambda''} \quad (B.16)$$

Some useful special values are

$$d^J_{m0}(z) = \left[\frac{(J-m)!}{(J+m)!}\right]^{\frac{1}{2}} P^m_J(z) \quad (B.17)$$

$$d^J_{00}(z) = P_J(z) \quad (B.18)$$

for integer J, and

$$d^{\frac{1}{2}}_{\frac{1}{2}\frac{1}{2}}(z) = \frac{1+z}{2} = \cos\tfrac{1}{2}\theta, \quad d^{\frac{1}{2}}_{\frac{1}{2}-\frac{1}{2}} = \frac{1-z}{2} = \sin\tfrac{1}{2}\theta \quad (B.19)$$

We shall also make use of the second-type rotation functions $e^J_{\lambda\lambda'}(z)$, analogous to the second-type Legendre functions $Q_l(z)$ introduced in Appendix A (see Andrews and Gunson 1964). They are defined in terms of the second-type Jacobi functions $Q^{(a,b)}_c(z)$ by

$$e^J_{\lambda\lambda'}(z) = (-1)^{A+\lambda-\lambda'}\left[\frac{(J+M)!\,(J-M)!}{(J+N)!\,(J-N)!}\right]^{\frac{1}{2}} \xi_{\lambda\lambda'}(z)\, Q^{(|\lambda-\lambda'|,\,|\lambda+\lambda'|)}_{J-M}(z) \quad (B.20)$$

For integer $J-M \geqslant 0$ they are related to the $d^J_{\lambda\lambda'}$ by the generalized Neumann relation

$$\xi_{\lambda\lambda'}(z)\, e^J_{\lambda\lambda'}(z) = \frac{1}{2}\int_{-1}^{1}\frac{\mathrm{d}z'}{z-z'}\, d^J_{\lambda\lambda'}(z')\,\xi_{\lambda\lambda'}(z') \quad (B.21)$$

and they satisfy the symmetry properties

$$e^J_{\lambda\lambda'}(z) = (-1)^{\lambda-\lambda'} e^J_{-\lambda-\lambda'}(z) = (-1)^{\lambda-\lambda'} e^J_{\lambda'\lambda}(z) \qquad (B.22)$$

$$e^J_{\lambda\lambda'}(-z) = (-1)^{J-\lambda+1} e^J_{\lambda-\lambda'}(z) \qquad (B.23)$$

Equation ($B.20$) can be re-expressed in terms of the hypergeometric function as

$$e^J_{\lambda\lambda'}(z) = (-1)^{A}\frac{1}{(2J+1)!}[(J+M)!\,(J-M)!\,(J+N)!\,(J-N)!]^{\frac{1}{2}}\xi^{-1}_{\lambda\lambda'}(z)$$
$$\times\frac{1}{2}\left(\frac{z-1}{2}\right)^{-J-1+M} F\left(J-M+1, J-M+|\lambda-\lambda'|+1, 2J+2, \frac{2}{1-z}\right) \qquad (B.24)$$

which gives the J-plane singularities directly, and since $F\to 1$ as $z\to\infty$ the asymptotic behaviour is

$$e^J_{\lambda\lambda'}(z)\underset{z\to\infty}{\longrightarrow}(-1)^{\frac{1}{2}(\lambda-\lambda')}\frac{1}{(2J+1)!}$$
$$\times[(J+M)!\,(J-M)!\,(J+N)!\,(J-N)!]^{\frac{1}{2}}\frac{1}{2}\left(\frac{z}{2}\right)^{-J-1} \qquad (B.25)$$

and, cf. ($A.31$),

$$e^J_{\lambda\lambda'}(z)\underset{J\to\infty}{\longrightarrow}\left(\frac{\pi}{2}\right)^{\frac{1}{2}}\frac{e^{\pm i\pi(\lambda-\lambda')}}{J^{\frac{1}{2}}}\frac{1}{(z^2-1)^{\frac{1}{4}}}e^{-(J+\frac{1}{2})\zeta(z)}, \quad \arg J<\pi \qquad (B.26)$$

where $\zeta(z)\equiv\log[z+(z^2-1)^{\frac{1}{2}}]$, and we use $\pm$ for $\mathrm{Im}\{z\}\gtrless 0$.

For half-odd-integer values of $J-v$ they obey the symmetry relation

$$e^J_{\lambda\lambda'}(z) = (-1)^{\lambda-\lambda'} e^{-J-1}_{\lambda\lambda'}(z) \qquad (B.27)$$

Also, analogous to ($A.18$), there is the relation

$$\frac{d^J_{\lambda\lambda'}(z)}{\sin\pi(J-\lambda)} = \frac{e^J_{\lambda\lambda'}(z)}{\pi\cos\pi(J-\lambda)} - \frac{e^{-J-1}_{-\lambda-\lambda'}(z)}{\pi\cos\pi(J-\lambda)} \qquad (B.28)$$

and we find
$$e^J_{\lambda\lambda'}(z)\approx\frac{d^{J_0}_{\lambda\lambda'}(z)}{J-J_0} \qquad (B.29)$$

for $J\to$ integer (J_0-v) when $J_0<-M$, and similar poles or $(J-J_0)^{-\frac{1}{2}}$ factors for integer (J_0-v), $-M\leqslant J_0<M$, from ($B.24$) (Andrews and Gunson 1964).

References

H. D. I. Abarbanel and J. B. Bronzan, *Phys. Rev.* **D9**, 2397 (1974*a*)
H. D. I. Abarbanel and J. B. Bronzan, *Phys. Rev.* **D9** 3304 (1974*b*)
H. D. I. Abarbanel, J. B., Bronzan, R. L. Sugar and A. R. White, *Physics Reports*, **21C**, 120 (1975)
H. D. I. Abarbanel, G. F. Chew, M. L. Goldberger and L. M. Saunders, *Phys. Rev. Letters*, **26**, 937 (1971)
H. D. I. Abarbanel, M. L. Goldberger and S. B. Trieman, *Phys. Rev. Letters*, **22**, 500 (1969)
H. D. I. Abarbanel and C. Itzykson, *Phys. Rev. Letters*, **23**, 53 (1969)
H. D. I. Abarbanel, F. E. Low, I. Muzinich, S. Nussinov and J. Schwarz, *Phys. Rev.* **160**, 1329 (1967)
W. J. Abbe, P. Kaus, P. Nath and Y. N. Srivastava, *Phys. Rev.* **154**, 1515 (1967)
E. S. Abers, H. Burkhardt, V. L. Teplitz and C. Wilkin, *Nuovo Cimento*, **42**, 365 (1966)
E. S. Abers, D. Keller and V. L. Teplitz, *Phys. Rev.* **D2**, 1757 (1970)
E. S. Abers and V. L. Teplitz, *Phys. Rev.* **158**, 1365 (1967)
R. K. Adair, *Phys. Rev.* **172**, 1370 (1968)
M. Ademollo, H. R. Rubinstein, G. Veneziano and M. A. Virasoro, *Phys. Rev.* **176**, 1904 (1968)
J. P. Ader, M. Capdeville and H. Navelet, *Nuovo Cimento*, **56A**, 315 (1968)
S. L. Adler and R. F. Dashen, *Current algebras and applications to particle physics* (Benjamin, 1968)
A. Ahmadzadeh, P. G. Burke and C. Tate, *Phys. Rev.* **131**, 1315 (1963)
V. de Alfaro and T. Regge, *Potential scattering* (North-Holland, 1965)
D. Amati, A. Stanghellini and S. Fubini, *Nuovo Cimento*, **26**, 6 (1962*a*)
D. Amati, S. Fubini and A. Stanghellini, *Phys. Letters*, **1**, 29 (1962*b*)
I. Ambats *et al.*, *Nucl. Phys.* **B77**, 269 (1974)
M. Andrews and J. Gunson, *J. Math. Phys.* **5**, 1391 (1964)
M. Armad, Fayyazuddin and Riazuddin, *Phys. Rev. Letters*, **23**, 504 (1969)
R. C. Arnold, *Phys. Rev.* **153**, 1523 (1967)
D. Atkinson, K. Dietz and D. Morgan, *Ann. Phys.* **37**, 77 (1966)
P. Aurenche *et al.*, Building a general framework for hadron collisions, *Proceedings of the 6th international colloquium on multi-particle reactions* (Oxford U.P., 1975)

L. A. P. Balazs, *Phys. Rev.* **128**, 1939 (1962)
L. A. P. Balazs, *Phys. Rev.* **129**, 872 (1963)
N. F. Bali, G. F. Chew and A. Pignotti, *Phys. Rev.* **163**, 1572 (1967)
N. F. Bali, D. D. Coon and J. W. Dash, *Phys. Rev. Letters*, **23**, 900 (1969)
J. S. Ball, W. R. Frazer and M. Jacob, *Phys. Rev. Letters*, **20**, 518 (1968)
K. Bardakci and H. Ruegg, *Phys. Letters*, **28B**, 342 (1968)
V. Barger, in J. R. Smith (ed.), *Proceedings of the 17th international conference on higher energy physics, London* (Rutherford Laboratory, 1974)

V. Barger and D. Cline, *Phys. Rev. Letters*, **16**, 913 (1966)
V. Barger and D. Cline, *Phys. Rev.* **155**, 1792 (1967)
V. Barger and D. Cline, *Phys, Rev. Letters*, **24**, 1313 (1970)
V. Barger and R. J. N. Phillips, *Phys. Rev.* **167**, 2210 (1969)
V. Barger and R. J. N. Phillips, *Phys. Letters*, **53B**, 195 (1974)
V. Bargmann, *Ann. Math.* **48**, 568 (1947)
M. B. Bari and M. S. K. Razmi, *Phys. Rev.* **D2**, 2054 (1970)
A. O. Barut, *The theory of the scattering matrix* (MacMillan, 1967)
A. O. Barut and D. E. Zwanziger, *Phys. Rev.* **127**, 974 (1962)
D. S. Beder, *Phys. Rev.* **149**, 1203 (1966)
J. Benecke, T. T. Chou, C. N. Yang and E. Yen, *Phys. Rev.* **188**, 2159 (1969)
E. L. Berger, *Phenomenology in particle physics* (California Institute of Technology, 1971*a*)
E. L. Berger, in *Colloquium on multi-particle dynamics, Helsinki* (1971*b*)
E. L. Berger and G. C. Fox, *Phys. Rev.* **188**, 2120 (1969)
J. Bernstein, *Elementary particles and their currents* (Freeman, 1968)
L. Bertocchi, S. Fubini and M. Tonin, *Nuovo Cimento*, **25**, 626 (1962)
H. Bethe and E. E. Salpeter, *Phys. Rev.* **84**, 1232 (1951)
K. M. Bitar and G. L. Tindle, *Phys. Rev.* **175**, 1835 (1968)
J. D. Bjorken and S. Drell, *Relativistic quantum fields* (McGraw-Hill, 1965)
R. Blankenbecler, and H. M. Fried, *Phys. Rev.* **D8**, 678 (1973)
R. Blankenbecler, J. R. Fulco and R. L. Sugar, *Phys. Rev.* **D9**, 736 (1974)
R. Blankenbecler, M. L. Goldberger, N. N. Khuri and S. B. Trieman, *Ann. Phys.* **10**, 62 (1960)
J. M. Blatt and V. F. Weisskopf, *Theoretical nuclear physics* (Wiley, 1952)
M. J. Bloxam, D. I. Olive and J. C. Polkinghorne, *J. Math. Phys.* **10**, 494, 545, 553 (1969)
P. Bonamy *et al.*, *Nucl. Phys.*, **B52**, 392 (1973)
M. Borghini *et al.*, *Phys. Letters*, **31B**, 405 (1970)
M. Borghini *et al.*, *Phys. Letters*, **36B**, 493 (1971)
J. F. Boyce, *J. Math. Phys.* **8**, 675 (1967)
B. H. Bransden, D. Evans and J. V. Major, *The fundamental particles* (Van Nostrand-Reinhold, 1973)
W. E. Britten and A. O. Barut (eds.), *Lectures in theoretical physics*, vol. 7*A* (University of Colorado Press, 1964)
S. J. Brodsky, F. E. Close and J. F. Gunion, *Phys. Rev.* **D5**, 1384 (1972)
J. B. Bronzan, *Phys. Rev.* **D10**, 746 (1974)
J. B. Bronzan, I. S. Gerstein, B. W. Lee and F. E. Low, *Phys. Rev.* **157**, 1448 (1967)
J. B. Bronzan and C. E. Jones, *Phys. Rev.* **160**, 1494 (1967)
R. C. Brower, R. N. Cahn and J. Ellis, *Phys. Rev.* **D7**, 2080 (1973*a*)
R. C. Brower, M. Einhorn, M. Green, A. Patroscioiu and J. H. Weis, *Phys. Rev.* **D8**, 2524 (1973*b*)
R. C. Brower, C. E. de Tar and J. H. Weis, *Physics Report*, **14C**, 257 (1974)
R. C. Brower and J. H. Weis, *Rev. Mod. Phys.* **47**, 605 (1975)

K. E. Cahill and H. P. Stapp, *Phys. Rev.* **D6**, 1007 (1972)
K. E. Cahill and H. P. Stapp, *Phys. Rev.* **D8**, 2714 (1973)
R. Cahn, in J. R. Smith (ed.), *Proceedings of the 17th international conference on high-energy physics, London* (1974)
F. Calogero, J. M. Charap and E. J. Squires, in *Proceedings of the Sienna conference* (CERN, 1963)

F. Calogero, J. M. Charap and E. J. Squires, *Ann. Phys.* **25**, 325 (1963)
A. Capella, *Phys. Rev.* **D6**, 2047 (1973)
A. Capella, J. Tran Thanh Van and A. P. Contogouris, *Nucl. Phys.* **B12**, 167 (1969)
J. L. Cardy, *Nucl. Phys.* **B28**, 455, 477 (1971)
J. L. Cardy, *Nucl. Phys.* **B75**, 413 (1974*a*)
J. L. Cardy, *Nucl. Phys.* **B79**, 319 (1974*b*)
J. L. Cardy and A. R. White, *Phys. Letters*, **47B**, 445 (1973)
J. L. Cardy and A. R. White, *Nucl. Phys.* **B80**, 12 (1974)
R. Carlitz and M. Kisslinger, *Phys. Rev. Letters*, **24**, 186 (1970)
P. Carruthers, *Introduction to unitary symmetry* (Wiley, 1966)
L. Castillejo, R. H. Dalitz and F. J. Dyson, *Phys. Rev.* **101**, 453 (1956)
A. Chados, R. L. Jaffe, K. Johnson, C. B. Thorn and V. Weisskopf, *Phys. Rev.* **D9**, 3471 (1974)
H.-M. Chan, *Phys. Letters*, **28B**, 425 (1968)
H.-M. Chan, P. Hoyer, H. I. Miettinen and D. P. Roy, *Phys. Letters*, **40B**, 406 (1972*c*)
H.-M. Chan, K. Kajantie and G. Ranft, *Nuovo Cimento*, **49A**, 157 (1967)
H.-M. Chan, J. Loskiewicz and W. W. M. Allison, *Nuovo Cimento*, **57A**, 93 (1968)
H.-M. Chan, H. I. Meittinen and W. S. Lam, *Phys. Letters*, **40B**, 112 (1972*b*(
H.-M. Chan and J. Paton, *Nucl. Phys.* **B10**, 519 (1969)
H.-M. Chan, J. E. Paton and S. T. Tsou, *Nucl. Phys.* **B68**, 479 (1975)
H.-M. Chan, R. O. Raitio, G. H. Thomas and N. A. Tournqvist, *Nucl. Phys.* **B19**, 173 (1970)
H.-M. Chan and S. T. Tsou, *Phys. Letters*, **28B**, 485 (1969)
S. J. Chang and T. M. Yan, *Phys. Rev. Letters*, **25**, 1586 (1970)
J. M. Charap and E. J. Squires, *Phys. Rev.* **127**, 1387 (1962)
H. Cheng and D. Sharp, *Phys. Rev.* **132**, 1854 (1963); *Ann. Phys.* **22**, 481 (1963)
H. Cheng and T. T. Wu, *Phys. Rev.* **186**, 1611 (1969)
H. Cheng and T. T. Wu, *Phys. Rev. Letters*, **24**, 1456 (1970)
G. F. Chew, *S-matrix theory of strong interactions* (Benjamin, 1962)
G. F. Chew, in B. de Witt and M. Jacob (eds.), *High energy physics, les Houches Lectures* (Gordon and Breach, 1965)
G. F. Chew, *Phys. Rev. Letters*, **16**, 60 (1966)
G. F. Chew, *Proceedings of the 5th international conference on high-energy collisions, Stoney Brook* (1973)
G. F. Chew and S. C. Frautschi, *Phys. Rev. Letters*, **7**, 394 (1961)
G. F. Chew and S. C. Frautschi, *Phys. Rev. Letters*, **8**, 41 (1962)
G. F. Chew, M. L. Goldberger, F. Low and Y. Nambu, *Phys. Rev.* **106**, 1337 (1957)
G. F. Chew, F. E. Low and M. L. Goldberger, *Phys. Rev. Letters*, **22**, 208 (1969)
G. F. Chew and S. Mandelstam, *Phys. Rev.* **119**, 476 (1960)
G. F. Chew and A. Pignotti, *Phys. Rev.* **176**, 2112 (1968)
G. F. Chew, T. Rogers and D. R. Snider, *Phys. Rev.* **D2**, 765 (1970)
G. F. Chew and D. R. Snider, *Phys. Rev.* **D3**, 420 (1971)
C. B. Chiu and J. Finkelstein, *Phys. Letters*, **27B**, 510 (1968)
C. B. Chiu and A. Kotanski, *Nucl. Phys.* **B7**, 615; **B8**, 553 (1968)
F. E. Close and J. F. Gunion, *Phys. Rev.* **D4**, 742 (1971)
G. Cohen-Tannoudji, A. Morel and H. Navelet, *Ann. Phys.* **46**, 239 (1968)
G. Cohen-Tannoudji, Ph. Salin and A. Morel, *Nuovo Cimento*, **55A**, 412 (1968)
G. Cohen-Tannoudji, F. Henyey, G. L. Kane and W. J. Zakrzewski, *Phys. Rev. Letters*, **26**, 112 (1971)

S. Coleman and R. E. Norton, *Nuovo Cimento*, **38**, 438 (1965)
P. D. B. Collins, *Phys. Rev.* **142**, 1163 (1966)
P. D. B. Collins, *Physics Reports*, **1C**, 105 (1971)
P. D. B. Collins and A. Fitton, *Nucl. Phys.* **B91**, 332 (1975)
P. D. B. Collins and F. D. Gault, *Nuovo Cimento*, **10A**, 189 (1972)
P. D. B. Collins and F. D. Gault, *A bibliography of Regge phenomenology*, University of Durham report (1975)
P. D. B. Collins, F. D. Gault and A. Martin, *Nucl. Phys.* **B80**, 136; **B83**, 241 (1974)
P. D. B. Collins and R. C. Johnson, *Phys. Rev.* **169**, 1222 (1968)
P. D. B. Collins and R. C. Johnson, *Phys. Rev.* **182**, 1755 (1969)
P. D. B. Collins, R. C. Johnson and E. J. Squires, *Phys. Letters*, **26B**, 223 (1968*a*)
P. D. B. Collins, R. C. Johnson and E. J. Squires, *Phys. Letters*, **27B**, 23 (1968*b*)
P. D. B. Collins and K. L. Mir, *Nucl. Phys.* **B19**, 509 (1970)
P. D. B. Collins, G. C. Ross and E. J. Squires, *Nucl. Phys.* **B10**, 475 (1969)
P. D. B. Collins and E. J. Squires, *Regge poles in particle physics* (Springer, 1968)
P. D. B. Collins and R. A. Swetman, *Lettere al Nuovo Cimento*, **5**, 793 (1972)
W. N. Cottingham, *Ann. Phys.* **25**, 424 (1963)
R. E. Cutkosky, *J. Math. Phys.* **1**, 429 (1960)
R. E. Cutkosky, *Rev. Mod. Phys.* **33**, 446 (1961)
R. E. Cutkosky and B. B. Deo, *Phys. Rev. Letters*, **19**, 1256 (1967)

R. H. Dalitz, *Phil. Mag.* **44**, 1068 (1953)
R. H. Dalitz, in B. de Witt and M. Jacob (eds.), *High energy physics, les Houches Lectures* (Gordon and Breach 1965)
M. Damashek and F. J. Gilman, *Phys. Rev.* **D1**, 1319 (1970)
A. Dar, V. F. Weisskopf, C. A. Levinson and H. J. Lipkin, *Phys. Rev. Letters*, **20**, 1261 (1968)
R. F. Dashen and M. Gell-Mann, *Phys. Rev. Letters*, **17**, 340 (1966)
R. Dolen, D. Horn and C. Schmid, *Phys. Rev.* **166**, 1768 (1968)
G. Domokos and G. L. Tindle, *Phys. Rev.* **165**, 1906 (1968)
A. Donnachie and R. G. Kirsopp, *Nucl. Phys.* **B10**, 433 (1969)
H. Dosch, *Phys. Rev.* **173**, 1595 (1968)
W. Drechsler, *Nuovo Cimento*, **53A**, 115 (1968)
S. D. Drell and A. C. Hearn, *Phys. Rev. Letters*, **16**, 908 (1966)
S. D. Drell and J. D. Walecka, *Ann. Phys.* **28**, 18 (1964)
I. T. Drummond, in *Proceedings of the 16th international conference on high-energy physics, Batavia* (Batavia Laboratory, 1972)
I. T. Drummond, P. V. Landshoff and W. J. Zakrzewksi, *Nucl. Phys.* **B11**, 383 (1969*a*)
I. T. Drummond, P. V. Landshoff and W. J. Zakrzewski, *Phys. Letters*, **28B**, 676 (1969*b*)
L. Durand, *Phys. Rev. Letters*, **18**, 58 (1967)
L. Durand and Y. T. Chiu, *Phys. Rev.* **139**, *B*646 (1965)

R. J. Eden, *High energy collisions of elementary particles* (Cambridge U.P., 1967)
R. J. Eden, *Rev. Mod. Phys.* **43**, 15 (1971)
R. J. Eden, P. V. Landshoff, D. I. Olive and J. C. Polkinghorne, *The analytic S-matrix* (Cambridge U.P., 1966)

A. R. Edmonds, *Angular momentum in quantum mechanics* (Princeton U.P., 1960)
M. B. Einhorn, J. Ellis and J. Finkelstein, *Phys. Rev.* **D5**, 2063 (1972*a*)
M. B. Einhorn, M. Green and M. A. Virasoro, *Phys. Rev.* **D6**, 1675 (1972*b*)
A. Erdelyi, W. Magnus, F. Oberhettinger and F. G. Tricomi (eds.), *Higher transcendental functions*, 3 vols. (McGraw-Hill, 1953)

T. Ferbel, *Phys. Rev. Letters*, **29**, 448 (1972)
M. Ferro-Luzzi *et al.*, *Phys. Letters*, **34B**, 534 (1971)
R. P. Feynman, *Phys. Rev. Letters*, **23**, 1415 (1969)
R. P. Feynman, *Photon-hadron interactions* (Benjamin, 1972)
K. Fialkowski and H. I. Miettinen, *Phys. Letters*, **43B**, 61 (1973)
R. D. Field and G. C. Fox, *Nucl. Phys.* **B80**, 367 (1974)
J. Finkelstein and K. Kajantie, *Nuovo Cimento*, **56A**, 659 (1968*a*)
J. Finkelstein and K. Kajantie, *Phys. Letters*, **26B**, 305 (1968*b*)
G. C. Fox, in C. Quigg (ed.), *Proceedings of Stony Brook conference on high energy collisions* (American Institute of Physics, 1973)
W. M. Frank, D. J. Land and R. M. Spector, *Rev. Mod. Phys.* **43**, 36 (1971)
S. C. Frautschi, P. Kaus and F. Zachariasen, *Phys. Rev.* **133**, *B*1607 (1964)
W. R. Frazer *et al.*, *Rev. Mod. Phys.* **44**, 284 (1972)
D. Z. Freeman and J. M. Wang, *Phys. Rev.* **153**, 1596 (1967)
P. G. O. Freund, *Phys. Rev. Letters*, **20**, 235 (1968*a*)
P. G. O. Freund, *Phys. Rev. Letters*, **21**, 1375 (1968*b*)
M. Froissart, *La Jolla conference on weak and strong interactions*, unpublished (1961*a*)
M. Froissart, *Phys. Rev.* **123**, 1053 (1961*b*)
S. Fubini, in R. G. Moorhouse (ed.), *Strong interactions and high energy physics* (Oliver and Boyd, 1963)
S. Fubini, *Nuovo Cimento*, **43**, 475 (1966)
S. Fubini, D. Gordon and G. Veneziano, *Phys. Letters*, **29B**, 679 (1969)
S. Fubini and G. Veneziano, *Nuovo Cimento*, **64A**, 811 (1969)
S. Fubini and G. Veneziano, *Nuovo Cimento*, **67A**, 29 (1970)
S. Fubini and G. Veneziano, *Ann. Phys.* **63**, 12 (1971)

M. K. Gaillard, B. W. Lee and J. L. Rosner, *Rev. Mod. Phys*, **47**, 277 (1975)
S. Gasiorowicz, *Elementary particle physics* (Wiley, 1966)
M. Gell-Mann, in *CERN conference on high-energy physics* (CERN, 1962)
M. Gell-Mann, *Phys. Letters*, **8**, 214 (1964)
M. Gell-Mann and M. L. Goldberger, *Phys. Rev. Letters*, **9**, 275 (1962)
M. Gell-Mann, F. E. Low, E. Marx, V. Singh and F. Zachariasen, *Phys. Rev.* **133**, *B*145, *B*161 (1964)
F. J. Gilman, *Physics Reports*, **4C**, 95 (1972)
R. J. Glauber, in W. E. Britten and L. G. Dunham (eds.), *Lectures in theoretical physics* (Interscience, 1959)
P. Goddard, J. Goldstone, C. Rebbi and C. B. Thorn, *Nucl. Phys.* **B56**, 109 (1973)
P. Goddard and A. R. White, *Nuovo Cimento*, **1A**, 645 (1971)
P. Goddard and A. R. White, *Phys. Letters*, **38B**, 93 (1972)
M. L. Goldberger, *Multiperipheral dynamics*, Princeton University report (1969)
M. L. Goldberger and K. M. Watson, *Collision theory* (Wiley, 1964)
G. R. Goldstein and J. F. Owens, *Nucl. Phys.* **B103**, 145 (1976)

M. L. Good and W. D. Walker, *Phys. Rev.* **120**, 1857 (1960)
K. Gottfried and J. D. Jackson, *Nuovo Cimento*, **33**, 309 (1964)
M. Gourdin, *Unitary symmetries* (North-Holland, 1967)
V. N. Gribov, *JETP* **41**, 667 (1961) (Tr. *Sov. Phys. JETP* **14**, 478 (1962))
V. N. Gribov, *Sov. Jour. Nucl. Phys.* **5**, 138 (1967)
V. N. Gribov, *Sov. Phys. JETP*, **26**, 414 (1968)
V. N. Gribov and A. A. Migdal, *Sov. Jour. Nucl. Phys.* **8**, 583 (1968)
V. N. Gribov and A. A. Migdal, *Sov. Jour. Nucl. Phys.* **6**, 703 (1969)
V. N. Gribov and I. Ya. Pomeranchuk, in *CERN conference on high-energy physics* (CERN, 1962)
V. N. Gribov and I. Ya. Pomeranchuk, *Phys. Rev. Letters*, **9**, 238 (1962)
V. N. Gribov, I. Ya. Pomeranchuk and K. A. Ter-Martirosyan. *Phys. Rev.* **139**, *B*184 (1965)
M. T. Grisaru, H. J. Schnitzer and H. S. Tsao, *Phys. Rev.* **D8**, 4498 (1973)
D. Gross, *Phys. Rev. Letters*, **19**, 1303 (1967)
D. Gross, *Nucl. Phys.* **B13**, 467 (1969)

R. Hagedorn, *Suppl. Nuovo Cimento*, **3**, 147 (1965); **6**, 169 (1968); **6**, 311 (1968).
I. G. Halliday, *Nuovo Cimento*, **60**, 177 (1969)
I. G. Halliday and C. T. Sachrajda, *Phys. Rev.* **8**, 3598 (1973)
I. G. Halliday and L. M. Saunders, *Nuovo Cimento*, **60A**, 115 (1969)
F. Halzen, A. Kumar, A. D. Martin and C. Michael, *Phys. Letters*, **32B**, 111 (1970)
F. Halzen and C. Michael, *Phys. Letters*, **36B**, 367 (1971)
D. Hankins, P. Kaus and C. J. Pearson, *Phys. Rev.* **137**, *B*1034 (1965)
Y. Hara, *Phys. Rev.* **136**, 507 (1964)
H. Harari, *Phys. Rev. Letters*, **17**, 1303 (1966)
H. Harari, *Phys. Rev. Letters*, **20**, 1395 (1968)
H. Harari, *Phys. Rev. Letters*, **22**, 562 (1969)
H. Harari, *Phys. Rev. Letters*, **24**, 286 (1970)
H. Harari, *Ann. Phys.* **63**, 432 (1971)
H. Harari, in R. L. Crawford and R. Jennings (eds.), *Phenomenology of particles at high energies* (Academic Press, 1974)
H. Harari and E. Rabinovici, *Phys. Letters*, **43B**, 49 (1973)
H. Harari and Y. Zarmi, *Phys. Rev.* **187**, 2230 (1969)
D. R. Harrington, *Phys. Rev.* **D1**, 260 (1970)
B. J. Hartley and G. L. Kane, *Nucl. Phys.* **B57**, 157 (1973)
F. Henyey, G. L. Kane, J. Pumplin and M. H. Ross, *Phys. Rev.* **182**, 1579 (1969)
A. Herglotz, *Ber. Verk. Sächs Akad. Wiss. Leipzig. Math. Naturw. Kl.* **63** (1911)
D. Hill *et al.*, *Phys. Rev. Letters*, **30**, 239 (1973)
J. F. L. Hopkinson and E. Plahte, *Phys. Letters*, **28B**, 489 (1968)
D. Horn, *Physics Reports* **4C**, 1 (1972)
P. Hoyer, R. G. Roberts, and D. P. Roy, *Nucl. Phys.* **B56**, 173 (1973).
P. Hoyer, in J. R. Smith (ed.), *Proceedings of 17th international conference on high-energy physics, London* (Rutherford Laboratory, 1974)
R. Hwa, *Phys. Rev.* **162**, 1706 (1967)
R. Hwa, *Phys. Rev.* **D1**, 1790 (1970)
R. Hwa and C. B. Chiu, *Phys. Rev.* **D4**, 224 (1971)
R. Hwa, and W. S. Lam, *Phys. Rev.* **D5**, 766 (1972)

K. Igi and S. Matsuda, *Phys. Rev. Letters*, **19**, 928 (1967)
J. Iliopoulos, in J. R. Smith (ed.), *Proceedings of the 17th international conference on high-energy physics, London* (1974)

T. Inami, *Nucl. Phys.* **B77**, 337 (1974)
E. Inonu and E. P. Wigner, *Nuovo Cimento*, **9**, 707 (1952)
A. C. Irving, A. D. Martin and V. Barger, *Nuovo Cimento*, **16A**, 573 (1973)
A. C. Irving, A. D. Martin and C. Michael, *Nucl. Phys.* **B32**, 1 (1971)

J. D. Jackson, in C. de Witt and M. Jacob (eds.), *High-energy physics, les Houches Lectures* (Gordon and Breach, 1965)
J. D. Jackson and G. E. Hite, *Phys. Rev.* **169**, 1248 (1968)
M. Jacob in *Proceedings of the 16th international conference on high-energy physics, Batavia* (Batavia Laboratory, 1972)
M. Jacob, E. Berger and R. L. Slansky, *Phys. Rev.* **D6**, 2580 (1972)
M. Jacob and R. L. Slansky, *Phys, Rev.* **D5**, 1847 (1972)
M. Jacob and G. C. Wick, *Ann. Phys.* **7**, 404 (1959)
C. J. Jochain and C. Quigg, *Rev. Mod. Phys.* **46**, 279 (1974)
C. E. Jones and J. B. Hartle, *Phys. Rev.* **140**, *B*90 (1965)
C. E. Jones, F. E. Low and J. E. Young, *Phys. Rev.* **D4**, 2358 (1971)
C. E. Jones and V. L. Teplitz, *Phys. Rev.* **159**, 1271 (1967)
H. F. Jones and M. Scadron, *Nucl. Phys.* **B4**, 267 (1967)
H. Joos, in Britten and Barut (eds.), *Lectures in theoretical physics*, vol. 8*A* (University of Colorado, 1964)

G. L. Kane, F. Henyey, D. R. Richards, M. Ross and G. Williamson, *Phys. Rev. Letters*, **25**, 1519 (1970)
A. Kernan and H. K. Sheppard, *Phys. Rev. Letters*, **23**, 1314 (1969)
T. W. B. Kibble, *Phys. Rev.* **131**, 2282 (1963)
Z. Koba and H. B. Nielsen, *Nucl. Phys.* **B10**, 633 (1969)
J. J. J. Kokkedee, *The quark model* (Benjamin, 1969)
H. A. Kramers, *Atti. Congr. Intern. Fisici Como*, **2**, 545 (1927)
R. Kronig, *J. Opt. Soc. Amer.* **12**, 547 (1926)
P. K. Kuo and P. Suranyi, *Phys. Rev.* **D1**, 3416, 3424 (1970)

K. W. Lai and J. Louie, *Nucl. Phys.* **B19**, 205 (1970)
L. D. Landau, *Nucl. Phys.* **13**, 181 (1959)
P. V. Landshoff and J. C. Polkinghorne, *Phys. Rev.* **181**, 1989 (1969)
P. V. Landshoff and J. C. Polkinghorne, *Nucl. Phys.* **B19**, 432 (1970)
P. V. Landshoff and J. C. Polkinghorne, *Physics Reports*, **5C** (1972)
P. V. Landshoff and J. C. Polkinghorne, *Phys. Rev.* **D5**, 2056 (1972)
P. V. Landshoff and W. J. Zakrzewski, *Nucl. Phys.* **B12**, 216 (1969)
E. Leader, *Phys. Rev.* **166**, 1599 (1968)
M. Le Bellac, *Phys. Letters*, **25B**, 524 (1967)
M. Le Bellac, *Nuovo Cimento*, **55A**, 318 (1968)
M. Le Bellac and J. L. Meunier, *Phys. Letters*, **43B**, 127 (1973)
H. Lee, *Phys. Rev. Letters*, **30**, 719 (1973)
J. Lee-Franzini, *Proceedings of 5th international conference on high-energy collisions, Stoney Brook* (1973)
H. Lehmann, *Nuovo Cimento*, **10**, 579 (1958)
D. W. G. S. Leith, in J. Tran Thanh Van (ed.), *The Pomeron* (CNRS, 1973)
N. Levinson, *Kgl. Dansk Videnscк. Selsk., Mat.-Fys. Medd.* **25**, no. 9 (1949)
M. Levy and J. Sucher, *Phys. Rev.* **186**, 1653 (1969)
J. M. Levy-Leblond, *Nuovo Cimento*, **45**, 772 (1966)
Y. C. Liu and S. Okubo, *Phys. Rev. Letters*, **19**, 190 (1967)
C. H. Llewellyn-Smith, *Physics Reports*, **3C**, 261 (1972)

F. K. Loebinger, in J. R. Smith (ed.), *Proceedings of 17th international conference on high-energy physics, London* (1974)
C. Lovelace, *Phys. Letters*, **28B**, 264 (1968)
C. Lovelace and D. Masson, *Nuovo Cimento*, **26**, 472 (1962)

S. MacDowell, *Phys. Rev.* **116**, 774 (1959)
W. Magnus and F. Oberhettinger, *Functions of mathematical physics* (Chelsea, 1949)
S. Mandelstam, *Phys. Rev.* **112**, 1344 (1958)
S. Mandelstam, *Phys. Rev.* **115**, 1741, 1752 (1959)
S. Mandelstam, *Ann. Phys.* **19**, 254 (1962)
S. Mandelstam, *Nuovo Cimento*, **30**, 1127, 1148 (1963)
S. Mandelstam, *Phys. Rev.* **137**, *B*949 (1965)
S. Mandelstam, *Nucl. Phys.* **B64**, 205 (1973)
S. Mandelstam, *Physics Reports*, **13C**, 259 (1974)
J. Mandula, C. Rebbi, R. Slansky, J. Weyers and G. Zweig. *Phys. Rev. Letters*, **22**, 1147 (1969)
J. Mandula, J. Weyers and G. Zweig, *Phys. Rev. Letters*, **23**, 266 (1970)
A. Martin, in G. Moorhouse (ed.), *Strong interactions and high energy physics* (Oliver and Boyd, 1963)
A. Martin, *Nuovo Cimento*, **42**, 930; **44**, 1219 (1966)
A. Martin and P. R. Stevens, *Phys. Rev.* **D5**, 147 (1972)
A. D. Martin and T. D. Spearman, *Elementary particle theory* (North-Holland, 1970)
C. Michael, *Nucl. Phys.* **B13**, 644 (1969*a*)
C. Michael, *Phys. Letters*, **29B**, 230 (1969*b*)
H. I. Miettinen, *Phys. Letters*, **38B**, 431 (1972)
H. I. Miettinen, Helsinki Thesis (1973)
A. A. Migdal, A. M. Polyakov and K. A. Ter-Martirosyan, *Phys. Letters*, **48B**, 239 (1974)
D. R. O. Morrison, *Phys. Letters*, **25B**, 238 (1967)
D. R. O. Morrison, in J. R. Smith (ed.), *Proceedings of 4th international conference on high-energy collisions* (Rutherford Laboratory, 1972)
P. M. Morse and H. Feshbach, *Methods of theoretical physics* (McGraw-Hill, 1953)
A. H. Mueller, *Phys. Rev.* **D2**, 2963 (1970)
A. H. Mueller, *Phys. Rev.* **D4**, 150 (1971)
I. J. Muzinich, F. E. Paige, T. L. Trueman and L.-L. Wang, *Phys. Rev.* **D6**, 1048 (1972)

A. Neveu and J. H. Schwarz, *Phys. Rev.* **D4**, 1109 (1971)
R. G. Newton, *The complex J-plane* (Benjamin, 1964)
R. G. Newton, *Scattering theory of waves and particles* (McGraw-Hill, 1966)

R. Odorico, *Phys. Letters*, **34B**, 65 (1971)
R. Odorico, *Nucl. Phys.* **B37**, 509 (1972)
P. J. O'Donovan, *Phys. Rev.* **185**, 1902 (1969)
R. Oehme and G. Tiktopoulos, *Phys. Letters*, **2**, 86 (1962)
R. Oehme, *Nucl. Phys.* **B16**, 161 (1970)
D. I. Olive and J. C. Polkinghorne, *Phys. Rev.* **171**, 1475 (1968)
D. Olive, in J. R. Smith (ed.), *Proceedings of the 17th international conference on high-energy physics, London* (1974)

Particle Data Group, *Physics Letters*, **50B**, 1 (1974)
Particle Data Group, *Rev. Mod. Phys.* **47**, 535 (1975)
B. Peterson and N. A. Tornqvist, *Nucl. Phys.* **B13**, 629 (1969)
R. J. N. Phillips, *Phys. Letters*, **24B**, 342 (1967)
R. J. N. Phillips, G. A. Ringland and R. P. Worden, *Phys. Letters*, **40B**, 239 (1972)
H. Pilkuhn, *The interactions of hadrons* (Wiley, 1967)
E. Plahte and R. G. Roberts, *Nuovo Cimento*, **60A**, 33 (1969)
J. H. Poincaré, *Acta. Math.* **4**, 213 (1884)
J. C. Polkinghorne, *J. Math. Phys.* **5**, 431 (1964)
J. C. Polkinghorne, *Nuovo Cimento*, **7A**, 555 (1972)
I. Ya. Pomeranchuk, *Sov. Phys. JETP* **7**, 499 (1958)

C. Quigg, *Nucl. Phys.* **B34**, 77 (1971)

T. Regge, *Nuovo Cimento*, **14**, 951 (1959)
T. Regge, *Nuovo Cimento*, **18**, 947 (1960)
B. Renner, *Current algebras and their applications* (Pergamon, 1968)
M. Rimpault and Ph. Salin, *Nucl. Phys.* **B22**, 235 (1970)
G. A. Ringland and R. J. N. Phillips, *Nucl. Phys.* **B13**, 274 (1969)
R. G. Roberts and D. P. Roy, *Phys. Letters*, **40B**, 555 (1972)
M. E. Rose, *Elementary theory of angular momentum* (Wiley, 1957)
J. L. Rosner, *Phys. Rev. Letters*, **21**, 950 (1968)
J. L. Rosner, *Phys. Rev. Letters*, **22**, 689 (1969)
H. J. Rothe, *Phys. Rev.* **159**, 1471 (1967)
D. P. Roy and R. G. Roberts, *Nucl. Phys.* **B77**, 240 (1974)
H. R. Rubinstein, G. Veneziano and M. A. Virasoro, *Phys. Rev.* **167**, 1441 (1968)
A. de Rujula, H. Georgi, S. L. Glashow and H. R. Quinn, *Rev. Mod. Phys.* **46**, 391 (1974)

A. I. Sanda, *Phys. Rev.* **D6**, 280 (1972)
R. Savit and J. Bartels, *Phys. Rev.* **D11**, 2300 (1975)
M. D. Scadron and H. F. Jones, *Phys. Rev.* **173**, 1734 (1968)
L. I. Schiff, *Quantum mechanics* (McGraw-Hill, 1968)
C. Schmid, *Phys. Rev. Letters*, **20**, 689 (1968)
J. H. Schwarz, *Phys. Rev.* **162**, 1671 (1967)
J. H. Schwarz, *Physics Reports*, **8C**, 269 (1973)
J. Scherk, *Rev. Mod. Phys.* **47**, 123 (1975)
H. Sciarrino and M. Toller, *J. Math. Phys.* **8**, 1252 (1967)
V. Singh, *Phys. Rev.* **127**, 632 (1962)
V. Singh, *Phys. Rev. Letters*, **18**, 36 (1967)
D. Sivers and J. Yellin, *Rev. Mod. Phys.* **43**, 125 (1971)
A. Sommerfeld, *Partial differential equations in physics*, p. 282 (Academic Press, 1949)
E. J. Squires, *Complex angular momentum and particle physics* (Benjamin, 1963)
E. J. Squires, *Nuovo Cimento*, **34**, 1751 (1964)
J. K. Storrow, *Nucl. Phys.* **B47**, 174 (1972)
J. K. Storrow, *Nucl. Phys.* **B96**, 77 (1975)
J. J. de Swart, *Nuovo Cimento*, **31**, 420 (1964)
A. R. Swift, *Nucl. Phys.* **B84**, 397 (1975)
A. R. Swift and R. W. Tucker, *Phys. Rev.* **D1**, 2894; **D2**, 2486 (1970)

A. R. Swift and R. W. Tucker, *Phys. Rev.* **D4**, 1707 (1971)

C. E. de Tar, *Phys. Rev.* **D3**, 128 (1971)
C. E. de Tar, D. Z. Freedman and G. Veneziano, *Phys. Rev.* **D4**, 906 (1971)
C. E. de Tar and J. H. Weis, *Phys. Rev.* **D4**, 3141 (1971)
J. C. Taylor, *Nucl. Phys.* **B3**, 504 (1967)
K. A. Ter-Martirosyan, *Nucl. Phys.* **68**, 591 (1965)
G. Tiktopoulos and S. B. Trieman, *Phys. Rev.* **D2**, 805 (1970)
G. Tiktopoulos and S. B. Trieman, *Phys. Rev.* **D3**, 1037 (1971)
E. C. Titchmarsh, *The theory of Fourier integrals* (Oxford U.P., 1937)
E. C. Titchmarsh, *The theory of functions*, 2nd edn. (Oxford U.P., 1939)
M. Toller, *Nuovo Cimento*, **37**, 631 (1965)
M. Toller, *Nuovo Cimento*, **53A**, 671 (1967)
M. Toller, *Nuovo Cimento*, **54A**, 295 (1968)
M. Toller, *Riv. Nuovo Cimento*, **1**, 403 (1969)
T. L. Trueman, *Phys. Rev.* **173**, 1684 (1968) (Erratum *ibid.* **181**, 2154 (1969))
T. L. Trueman and G. C. Wick, *Ann. Phys.* (N.Y.) **26**, 322 (1964)
S. H. Tye and G. Veneziano, *Nuovo Cimento*, **14A**, 711 (1973)

G. Veneziano, *Nuovo Cimento*, **57A**, 190 (1968)
G. Veneziano, *Phys. Letters*, **38B**, 30 (1972)
G. Veneziano, *Phys. Letters*, **43B**, 413 (1973)
G. Veneziano, *Physics Reports*, **9C**, 201 (1974*a*)
G. Veneziano, *Nucl. Phys.* **B74**, 355 (1974*b*)
M. A. Virasoro, *Phys. Rev. Letters*, **22**, 37 (1969)

J. M. Wang and L. L. C. Wang, *Phys. Rev.* **D1**, 663 (1970)
L. L. C. Wang, *Phys. Rev.* **142**, 1187 (1966)
A. E. A. Warburton, *Phys. Rev.* **137**, *B*993 (1964)
G. N. Watson, *Proc. Roy. Soc.* **95**, 83 (1918)
B. R. Webber, *Phys. Rev.* **D3** (1971)
H. A. Weidenmuller, *Phys. Letters*, **24B**, 441 (1967)
S. Weinberg, *Phys. Rev. Letters*, **17**, 616 (1966)
S. Weinberg, *Rev. Mod. Phys.* **46**, 255 (1974)
J. H. Weis, *Phys. Letters*, **43B**, 487 (1973)
J. H. Weis, *Nucl. Phys.* **B71**, 342 (1974)
A. R. White, *Nucl. Phys.* **B39**, 432, 461 (1971)
A. R. White, *Nucl. Phys.* **B50**, 93, 130 (1972)
A. R. White, *Nucl. Phys.* **B50**, 130 (1973*a*)
A. R. White, *Nucl. Phys.* **B67**, 189 (1973*b*)
A. R. White, *Phys. Rev.* **D10**, 1236 (1974)
J. N. J. White, *Lett. al Nuovo Cimento*, **1**, 20 (1971)
E. P. Wigner, *Ann. Math.* **40**, 159 (1939)
E. P. Wigner, *Group theory* (Academic Press, 1959)
P. W. Williams, *Phys. Rev.* **D1**, 1312 (1970)
R. Worden, *Nucl. Phys.* **B58**, 205 (1973)

F. Zachariasen, *Phys. Rev. Letters*, **7**, 112; 268(E) (1961)
F. Zachariasen, *Schladming Lectures*, CERN report Th. 1290 (1971)
K. Zalewski, in J. R. Smith (ed.), *Proceedings of 17th international conference on high-energy physics, London* (1974)
G. Zweig, CERN reports Th. 401, 412 (1964) unpublished

Index

Printed in the United States
by Baker & Taylor Publisher Services

Printed in the United States
by Baker & Taylor Publisher Services